ACS SYMPOSIUM SERIES **682**

Molecular Modeling of Nucleic Acids

Neocles B. Leontis, EDITOR
Bowling Green State University

John SantaLucia, Jr., EDITOR
Wayne State University

Developed from a symposium sponsored by the Division of Computers in Chemistry, at the 213th National Meeting of the American Chemical Society, San Francisco, CA, April 13–17, 1997

American Chemical Society, Washington, DC

Library of Congress Cataloging-in-Publication Data

Molecular modeling of nucleic acids / Neocles B. Leontis, John SantaLucia, Jr.

p. cm.—(ACS symposium series, ISSN 0097–6156; 682)

"Developed from a symposium sponsored by the Division of Computers in Chemistry, at the 213th National Meeting of the American Chemical Society, San Francisco, CA, April, 13–17, 1997."

Includes bibliographical references and indexes.

ISBN 0–8412–3541–4

1. Nucleic acids—Structure—Congresses. 2. Nucleic acids—Structure—Computer simulation—Congresses.

I. Leontis, Neocles B. II. SantaLucia, John, 1964– . III. American Chemical Society. Division of Computers in Chemistry. IV. American Chemical Society. Meeting (213th: 1997: San Francisco, Calif.) V. Series.

QP620.M64 1998
572.8′33—dc21 97–42151
CIP

This book is printed on acid-free, recycled paper.

PRINTED IN THE UNITED STATES OF AMERICA

Advisory Board

ACS Symposium Series

Foreword

THE ACS SYMPOSIUM SERIES was first published in 1974 to provide a mechanism for publishing symposia quickly in book form. The purpose of the series is to publish timely, comprehensive books developed from ACS-sponsored symposia based on current scientific research. Occasionally, books are developed from symposia sponsored by other organizations when the topic is of keen interest to the chemistry audience.

Before agreeing to publish a book, the proposed table of contents is reviewed for appropriate and comprehensive coverage and for interest to the audience. Some papers may be excluded in order to better focus the book; others may be added to provide comprehensiveness. When appropriate, overview or introductory chapters are added. Drafts of chapters are peer-reviewed prior to final acceptance or rejection, and manuscripts are prepared in camera-ready format.

As a rule, only original research papers and original review papers are included in the volumes. Verbatim reproductions of previously published papers are not accepted.

ACS BOOKS DEPARTMENT

Contents

SECONDARY STRUCTURE PREDICTION

MOLECULAR DYNAMICS SIMULATION

Preface

NUCLEIC ACIDS were originally conceived purely as carriers of genetic information in the form of the genetic code. DNA was the repository of genetic information, and RNA served as a temporary copy to be decoded in the synthesis of proteins. The discovery of transfer RNA, the "adapter" molecules that assist in the decoding of genetic messages, broadened awareness of the role of RNA. In the past few years, we have come to appreciate the functional versatility of nucleic acids and their participation in a wide range of vital cellular processes.

As new functions for nucleic acids have been identified and characterized, large numbers of sequences have been determined—so-called primary structural information. The determination of three-dimensional structures, however, has not kept up with the accumulation of primary sequence data. Thus, there is intense interest in developing reliable methods of predicting the three-dimensional structures of polynucleotides based primarily on sequence information, supplemented by readily executed experiments. All efforts directed at elucidating the three-dimensional structure of a nucleic acid molecule on the basis of readily determined sequence data may be broadly defined as "molecular modeling". An intermediate step between primary structure and three-dimensional structure is the determination of secondary structure—the pattern of hydrogen-bonded base–base interactions (base pairing) in a molecule. A hierarchical view of nucleic acid structure views primary structure as determining secondary structure. Tertiary structure emerges as secondary structure elements interact with each other.

This book was developed from a symposium presented at the 213th National Meeting of the American Chemical Society, titled "Molecular Modeling and Structure Determination of Nucleic Acids", sponsored by the ACS Division of Computers in Chemistry, in San Francisco, California, April 13–17, 1997. Our aim in organizing the symposium was to bring together scientists who are employing a variety of theoretical and experimental approaches to understand the structure and dynamics of nucleic acids, DNA, and RNA, with the goal of better understanding biological function. This volume contains contributions that represent the breadth of approaches presented at the symposium.

As discussed in the overview, the synergistic interplay of theoretical molecular modeling approaches and experimental structure determination methods was decisive in the success of Watson and Crick in defining the double helix. As evidenced by the work presented in the symposium, this synergism continues unabated and may be identified as a common underlying theme of this volume.

Other themes that emerged during the symposium included the urgency of dealing with the problem of conformational flexibility and heterogeneity in nucleic acids, particularly for NMR structure determination; the value of treating electrostatic interactions as accurately as possible, and the recent success of the particle mesh Ewald (PME) method in this regard; the need to consider kinetic factors in modeling the final folded conformations of large structures, in addition to purely energetic factors; and, as already mentioned, the value of a hierarchical approach to three-dimensional structure.

It is our hope that this volume will introduce the reader to the wide range of approaches used in modeling nucleic acid structures, the insights into biological function gained by structural and dynamical studies, and the strong interplay between theoretical and experimental methods.

Acknowledgments

We acknowledge the financial support for the symposium provided by the following organizations: the American Chemical Society Petroleum Research Fund (Grant #32048–SE), Glaxo Wellcome, Isis Pharmaceuticals, Molecular Simulations Inc., and Parke-Davis. We thank all the participants, and, in particular, Stephen Harvey, for suggestions on ways to broaden the scope of the symposium. We thank Vassiliki Leontis and Holly SantaLucia for their support, patience, and encouragement. J. SantaLucia, Jr., acknowledges the support of the Department of Chemistry of Wayne State University.

NEOCLES B. LEONTIS
Department of Chemistry
Bowling Green State University
Bowling Green, OH 43403

JOHN SANTALUCIA, JR.
Department of Chemistry
Wayne State University
Detroit, MI 48202

September 2, 1997

Chapter 1

Overview

Neocles B. Leontis[1] and John SantaLucia, Jr.[2]

[1]Chemistry Department, Bowling Green State University, Bowling Green, OH 43402
[2]Department of Chemistry, Wayne State University, Detroit, MI 48202

Molecular modeling of nucleic acids began with James Watson and Francis Crick (*1*). Watson and Crick integrated the experimental findings of many other scientists with their own stereochemical insights to juxtapose the building blocks of DNA, the bases, in a novel way. The now familiar double helix was the result. Although they used no computers for this, theirs was molecular modeling of the highest order! We begin this chapter with an overview of the experimental and theoretical developments which played a role in Watson and Crick's discovery. We continue by highlighting other milestones to our present understanding of nucleic acid structure, dynamics, and function.

Side-by-side with Watson and Crick's first paper on the double helix, there appeared reports of the x-ray fiber diffraction studies of Wilkins and coworkers (*2*) and of Rosalind Franklin and R. G. Gosling (*3*). Without a knowledge of the general nature of these data, it is unlikely that Watson and Crick could have formulated their double helical model. In a more complete paper, Watson and Crick presented their stereochemical reasoning -- their molecular modeling approach (*4*). In support of their model, they cited hydrodynamic data (sedimentation, diffusion, and light-scattering measurements) suggesting that DNA molecules exist as thin rigid fibers 20Å in diameter (*5*), inferences implicit in the fiber diffraction work. These inferences were directly confirmed soon after by electron microscopy (*6*). They took cognizance of the fact that the same x-ray fiber diffraction patterns were observed in DNA from all sources, ranging from viruses to humans, despite large variations in base composition. This gave even greater significance to the careful chemical analyses of Chargaff which showed that the molar ratios of adenine to thymine and of guanine to cytosine are always found to be near unity in DNA from different sources (*7*). Watson and Crick concluded that the three-dimensional structure had to be independent of the base composition and therefore of the sequence. Careful calculations of density led to the realization that DNA helices consist of two strands. The dyad symmetry observed in the diffraction pattern led them to conclude that the chains run in opposite directions.

Important information regarding possible orientations of the bases to their attached sugar rings was provided by the high resolution x-ray structure of cytosine published by Furberg (*8*). Watson and Crick acknowledged their debt to J. Donohue for "constant advice and criticism, especially on interatomic distances." Donohue had published a critical review of hydrogen-bonding in organic crystals as revealed by x-ray crystallography (*9*). Crucial for the base-pairing hypothesis was knowledge of the correct tautomeric form of each of the nucleotide bases. These were derived by comparing electron densities calculated for alternative tautomeric forms of the bases to electron densities obtained from careful x-ray crystallographic analysis, as for example, for adenine (*10*). Watson and Crick were also guided by acid-base titration experiments carried out by Gulland which indicated that in native DNA, the polynucleotide chains are held together by hydrogen bonds involving the bases themselves (*11*). The acidic and basic sites which are accessible to titration in the isolated nucleotides or in denatured DNA are protected from reaction with acid or base in the native structure. Very high or low pH is required to disrupt base-pairing; the two strands separate in a highly cooperative but irreversible manner.

How accurate was the Watson-Crick model compared to models refined against x-ray data? In fact, the model of Watson and Crick did not agree quantitatively with x-ray fiber diffraction data of B-form DNA (*12*). In particular, the diameter of the Watson-Crick duplex was too large and the base-pairs did not pass through the helix axis, as indicated by the experimental data (*2,3*). This is not surprising, as Watson and Crick did not have access to the actual data, but were only aware of the general results. It also illustrates an important point regarding molecular modeling of nucleic acids: it need not be precise to achieve its goal of providing biological insight.

The first high-resolution structure of two hydrogen-bonded DNA bases (1-methylthymine and 9-methyladenine) was obtained by Hoogsteen in 1959 (*13*). In this study the glycosidic bonds connecting the adenine and thymine bases to the deoxyribose sugars were replaced by methyl groups. The structure revealed a surprise: although the thymine hydrogen-bonded as predicted by Watson and Crick, the adenine base was flipped over so that the N7 (instead of the N1) of the adenine base hydrogen-bonded to the thymine N3-H. This arrangement is referred to as Hoogsteen base-pairing and is encountered in certain RNA structures and in triple helical DNA.

The three-dimensional models of double-helical DNA and RNA were incrementally improved as better fiber diffraction data and improved computational methods became available. In the "linked-atom" methods, the nucleotide building blocks of a polynucleotide were modeled using standard bond lengths and angles measured in precise x-ray crystallography of the bases, nucleosides, and nucleotides. Adjustments were made in torsion angles of the polynucleotide until the best fit with the diffraction data was obtained (*14*). This was supplemented with empirical energy functions and energy minimization procedures to relieve bad contacts obtained from hand-built models (*15*). Successive cycles of data collection and refinement led to models which are still accepted today as standard, average A- or B-form helices to which structures obtained for specific sequences by single-crystal x-ray diffraction or by NMR solution methods can be compared (*14,16*). In fact, it was not until 20 years after the Watson-Crick model that base pairing in a short, double-helical segment (consisting of a self-complementary RNA dinucleotide) was viewed at high-resolution by single-crystal x-ray diffraction analysis (*17*). The first high-resolution DNA oligonucleotide structures were obtained a few years later when techniques for synthesis of adequate amounts of pure oligonucleotides with arbitrary sequence were perfected. The very first structure solved also contained an unforeseen surprise -- a left-handed helical conformation, called Z-DNA (*18*). Structures of the expected B-DNA conformation soon followed (*19*). Nearly two decades of crystallographic work on oligonucleotides have revealed that the local structure of DNA is sequence and environment dependent: the local structure at individual base pairs or base-pair steps can deviate significantly from the average structural parameters derived by analysis of fiber diffraction data. Although no simple rules relating local geometry to sequence have emerged, it has become apparent that base-stacking interactions provide the primary stabilizing force (*20*). Sequence-dependent variations observed by x-ray crystallography can arise from effects due to the base sequence itself as well as from effects due to intermolecular contacts in the crystal (crystal-packing forces). Careful analyses of the x-ray structures of the same duplex determined in different crystal environments and of related sequences in the same environment are making it possible to unravel the relative influence of base sequence and crystal packing forces on local structure (*21*).

The biological significance of these high-resolution studies lies in the fact that a wide range of proteins and drugs recognize and bind to specific DNA sequences (for recent reviews see (*22*)). Specific recognition is thought to depend in part on sequence itself (owing to different distributions of hydrogen-bonding donors and acceptors presented in the major or minor grooves of the double helix by different base sequences (*23*)) and also on local helical variations (which of course also depend on sequence). A better understanding of the way sequence affects local structure is therefore necessary to fully understand recognition in DNA (*24*). Crystallographic studies of DNA-protein complexes have further shown that local DNA structure can be severely distorted upon binding of proteins or other ligands. The sequence-dependence of DNA deformability must therefore also be understood for a complete understanding of recognition. Structural changes in the double helix can be expressed as variations in a set of parameters which describe the spatial relationships between paired bases, neighboring base-pairs, and between the local helical axis and the individual bases or base pairs. For example, the twist (w) is defined as the rotation about the helical axis of one base pair relative to the next. Standard names and symbols for helical parameters were agreed upon in 1989 (*25*). Algorithms and computer programs to calculate these parameters based on atomic coordinates are available (*26,27*).

The mean values of local helical parameters obtained from crystallography generally conform to expectations from fiber diffraction studies. What has proved surprising and unexpected is the breadth of the variation for many of the helical parameters (*24*). For example, the helical twist in eight B-DNA dodecamer structures and four decamers gave a mean value of 36.1°, but the values range from 24° to 51° with a standard deviation of 5.9°. Large variations have also been seen in the rise parameter (Dz), mean = 3.36±.46Å, range = 2.5 to 4.4Å, and in the roll angle, mean = 0.6 ±6.0°, range = -18° to +16°. It has been found that twist, rise, cup, and roll are closely correlated, and can be used to categorize base-pair steps into families. Base pair parameters (propeller, buckle, inclination), on the other hand, appear to be mutually uncorrelated. These families have been observed (*24*):

1) High twist profile: High twist, low rise, positive cup, and negative roll. GC, GA, TA steps.
2) Low twist profile: Low twist, high rise, negative cup, and positive roll. All RR except GA.
3) Intermediate twist profile: All RY except GC.
4) Variable twist profile: All YR except TA.

The variability in local helical parameters for specific base-pair steps indicates that DNA is inherently locally polymorphous, many sequences are capable of more than one state of the local helical variables (*21*). The width of the minor groove and the patterns of hydration are other sources of local variation in B-DNA crystal structures. The minor groove is widest whenever phosphates are in B_{II} conformations (ε=*g*-, ζ=*t*) rather than the more common B_I (ε=*t*, ζ=*g*-) conformation. B_{II} conformations are only observed in YR and RR base steps. As regards deformability, x-ray crystallography has revealed that A-tract DNA (sequences containing runs of A-T basepairs) are inherently straight and unbent, whereas junctions between GC and AT regions constitute flexible hinges which can bend, and do so by compression of the major groove (by variation in the roll parameter). Bending at these junctions is not however inherent -- it occurs in response to external forces such as contacts within the crystal or the influence of proteins upon binding.

Computer molecular modeling of duplex DNA will play an important role in sorting out the relative role of crystal packing forces and intrinsic sequence-dependent variations in local helical structure (*28*) and in exploring the deformability of DNA and its sequence dependence.

Conformational Analysis of Nucleic Acids.

Conformational analysis is much more difficult for polynucleotides than for polypeptides, owing to the existence of six single-bond torsion angles per nucleotide along the backbone, compared to only two variable backbone torsion angles per amino acid. Efforts have been made to put limits on the range of possible conformers in DNA and RNA (*29*) in a manner similar to that done for proteins by Pauling (*30*) and Ramachandran (*31*). A seventh important variable in polynucleotides is the glycosidic torsion angle, χ, which determines the relative orientation of the base to its sugar. Donohue and Trueblood recognized that this angle is restricted to two ranges, *syn* and *anti* (*32*). Many early theoretical analyses were concerned with characterizing the relation between the glycosidic angle and the conformations of the sugar ring and phosphodiester backbone (*33*). The backbone torsion angles in polynucleotides are identified as α to ζ according to the IUB-IUPAC recommended nomenclature that is now universally used (*34*). Sundaralingam (1969) analyzed the backbone torsion angles using the atomic coordinates of all the high-resolution, single-crystal x-ray structures of DNA and RNA building blocks known at the time -- nucleosides, nucleotides, phosphodiester model compounds, and the cyclic nucleotides cyclic-UMP and cyclic-AMP (*35*). He also measured torsion angles in models of polynucleotides which had been constructed based on x-ray fiber diffraction data. The important conclusions were 1) that the conformational ranges of the backbone torsion angles are considerably restricted (he identified seven distinct sugar-phosphate chain conformations as possible for right-handed helices) and 2) that the preferred conformation of the nucleotide unit in polynucleotides is the same as that found in monomer single crystals. The comprehensive book by Saenger contains summaries of the conformational analyses of nucleic acids (*36*).

The conformation of the sugar ring itself can be described in terms of puckering, because no more than four of the atoms of the five-membered ring can lie in the same plane without bond angle strain. The puckered atom is the one that is above or below the average plane determined by the coordinates of the other atoms in the ring. For example, in the *C3'-endo* conformation the C3' atom is out of plane and on the same side of the sugar ring as the glycosidic attachment to the base. Sugar pucker is measured by the pseudo-rotation angle P (or Φ) and equivalently by the main-chain torsion angle d (C5'-C4'-C3'-O3'). The values of these two parameters are highly correlated in crystal structures. The concept of pseudo rotation, first developed in 1947 to describe cyclopentane conformation mobility (*37*), was applied to analyze nucleic acid sugar ring conformations by Altona and Sundaralingam (*38*). Two major conformations have been identified from x-ray crystallography, NMR solution studies, and theoretical studies: *C3'-endo* (designated the "Northern" conformation on the pseudo rotation wheel) and *C2'-endo* (designated "Southern"). The energy barrier separating these two most stable conformations is low and the potential minima are broad. Therefore, each conformation represents a family of allowed neighboring conformations (for example *C2'-endo/C1'-exo*) and interconversion between the two conformational families can be rapid. The existence of two low-energy conformations for each sugar ring means that real nucleic acid structures are actually ensembles of related structures in dynamic equilibrium. The contributions from A. Lane and from Ulyanov, et al. in this volume explicitly address the difficulties this introduces for structure determination in solution by NMR.

Modeling of RNA

In the 1950s, it became apparent that most RNA molecules, unlike DNA, are single-stranded (*39*). Hydrodynamic studies indicated that RNA molecules are usually globular rather than extended like DNA. UV-monitored thermal denaturation studies showed that RNA molecules have considerable secondary structure as inferred from the hyperchromicity -- between 40% and 60% of the bases are stacked and paired (*40*). Further evidence that the bases are hydrogen bonded in RNA came from the much slower reactivity with formaldehyde of the amino groups of the bases C, G, and A observed at low temperature as compared to that observed at high temperature. That the paired bases are actually organized into helical domains was indicated by the decrease in optical rotation of the RNA solutions as temperature was increased. This exactly paralleled the changes observed by UV absorbance. Moreover, the direction of optical rotation for RNA was similar to DNA, indicating that RNA also forms right-handed helices. The broad thermal transitions observed by UV spectrophotometry indicated that RNA secondary structure consists of shorter and more heterogeneous helices than DNA, which usually forms one long continuous double helix and therefore melts cooperatively and is more stable. All these data led Fresco, Alberts, and Doty to propose in 1960 a model for RNA secondary structure which has largely stood the test of time (*41*). In their model, the RNA polymer strand folds back on itself locally to form short double-helical base-paired regions connected by short single-stranded loops called hairpin loops because of their U shape. Studies of the stabilities of oligonucleotides and synthetic polymers led them to conclude that the helices had to be at least four base-pairs long; unpaired nucleotides could be accommodated in slightly longer helices. They examined different ways of folding random sequences of up to 90 nucleotides and found that a stable structure is more likely to form by folding to make several shorter helices than one long continuous helix. Their model reproduced the average helical content of authentic RNA samples, as determined by UV melting. Further analysis of the statistical properties of RNA sequences, pioneered by Doty and co-workers, has been pursued by Schuster and co-workers (*42*).

The model of Fresco et al. for RNA structure could be tested once sequences of biological RNA molecules became available. The complete primary sequence of a transfer RNA (tRNA) was obtained by Holley and co-workers in 1965 (*43*). tRNAs consist of single chains of approximately 75 to 90 nucleotides, and thus fall within the range modeled by Fresco et al. They serve as the adapter molecules to which amino acids are specifically attached for decoding the message transcribed from DNA into messenger RNA (mRNA) during protein synthesis in cells. Holley and co-workers considered three models for the secondary structure of their tRNA sequence. The model which proved correct consisted of four short helices. One helix was formed by the two ends of the molecule and the other three by hairpin loops, resulting in the now familiar "clover-leaf" secondary structure model of tRNA. Each double helix was short (4-7 basepairs) and the helical regions were connected by short stretches of unpaired bases and by the hairpin loops, as predicted by Fresco et al. Conclusive evidence for the clover-leaf model was obtained when the primary sequences of other tRNA molecules became available. Nearly all could be folded into the same secondary structure. Zachau and co-workers provided further evidence favoring the clover-leaf model by subjecting tRNA molecules to attack by enzymes which specifically hydrolyze the phosphodiester backbone of single-stranded regions of RNA (*44*). Only the segments of the tRNA corresponding to single-stranded regions in the clover-leaf model were cleaved by the enzymes.

The tRNA story represents the emergence of a general strategy called comparative sequence analysis which has proven enormously successful in deducing secondary structures from primary sequence (*45*). The method rests on the assumption that functionally equivalent molecules will have similar secondary and tertiary structures, even though the primary sequences may vary considerably due to evolutionary drift. Sequences of functionally equivalent molecules from many organisms are obtained and compared. The sequences are aligned in such a way that compensatory changes occur within fragments forming helices (*45*). The comparative approach has played the key role in elucidating the secondary structures of ribosomal RNAs, group I and group II introns, RNase P, and small nuclear RNAs as well as tRNA (*46*). Early comparative sequence analyses of tRNA, 5S rRNA and 16S rRNA were performed manually, but now computer algorithms exist that use both sequence (*47*) and thermodynamic criteria (*48,49*) to assist in a process, that still is not completely automated (*45*).

X-ray fiber diffraction analysis of RNA samples had meanwhile revealed that RNA adopts right-handed helical structures that resemble those of the low-humidity A-form of DNA (*50,51*). Before the x-ray structure of a tRNA molecule (phenylalanine tRNA from yeast) appeared in 1974 (*52,53*), many efforts were recorded to model the 3D structure of tRNA, using as a starting point the clover leaf secondary structure (*54,55*). The structure proposed by Levitt in 1969 is noteworthy since it was the only topologically correct model proposed (*56*). Levitt's success can be attributed to integrating all available physical, chemical, and stereochemical information, and to taking care to maximize base-stacking interactions. The phylogenetic data at the time included 14 tRNA sequences. By carefully comparing these sequences, folded into the cloverleaf secondary structure, he was able to correctly identify a base triple involving positions 9, 12, and 24, and a tertiary Watson-Crick basepair between a conserved purine at position 15 and a conserved pyrimidine at position 48. When the purine was G, the pyrimidine was always C, when the purine was A, the pyrimidine, U. The modeling also utilized the radius of gyration, to establish overall dimensions of the molecule, hydrogen exchange and ORD to establish the number of hydrogen-bonding bases, photocrosslinking to identify a tertiary contact involving U8 and C13 (from which it was deduced that U8 pairs with A14), and chemical modification to define exposed vs. protected residues. Levitt employed a molecular mechanics force field (*57*) similar in functional form to "modern" force fields such as AMBER to enforce proper stereochemistry. No electrostatic terms were included, however. Levitt used hand-held CPK models to minimize solvent accessible surface. In this age of computer graphic workstations, the power of manipulating hand-held models should not be underestimated! Levitt correctly predicted that the terminal amino-acyl helical arm was stacked on the TΨC arm, and the dihydrouracil (D)-arm on the anti-codon arm. He incorporated the prescient 3D model for the anti-codon loop proposed by Fuller and Hodgson on the basis of stacking arguments (*58*). In the model, all bases except 8 pyrimidines are stacked. In the $tRNA^{phe}$ crystal structure, all but five bases, two of which are dihydrouridines, are stacked, even though only 55% of the bases are in double helical stems (*59*).

The techniques of phylogenetic comparison, chemical and enzymatic probing, and thermodynamic prediction have been refined and applied successfully to determine the secondary structures of ever larger RNA molecules. The challenge now is to arrange the many short, irregular, double-helical elements, connected by short single-stranded segments, into a coherent three-dimensional structure. Databases of known 3D structural elements have been assembled, based on crystal studies of tRNA and oligonucleotides. Programs for combinatorially linking nucleotides using the most frequently occurring backbone conformations while simultaneously satisfying the constraints imposed by the secondary structure and various experimental data (such as chemical probing and site-specific mutagenesis) have become available (see for example (*60*)). Phylogenetic methods have been applied to identify correlated, recurring elements of sequence that can function as tertiary contacts. A set of recurring structural elements that take part in tertiary contacts has begun to emerge, making it possible to systematically model 3D structures (*61*). A framework for modeling simultaneously the structure and folding of large RNAs hierarchically has been formulated (*62*).

An example is the group I intron, the first RNA shown to have enzymatic activity (*63*). Michel and Westhof identified several tertiary contacts in the analysis of the structure of the group I intron (*64*). Some involve hairpin loops having GNRA sequences (N = any nucleotide, R = A or G) and the minor grooves of irregular helices. The atomic details of the predicted tertiary contacts were revealed in the recent x-ray crystal structure of the P4-P6 domain of the Tetrahymena thermophila Group I intron (*65*). This is the largest RNA molecule solved by high-resolution x-ray crystallography to date. It contains a wealth of new atomic resolution structural information which provides new structural motifs to employ in modeling new RNA structures.

The Nucleic Acid Database

The Nucleic Acid Database (NDB), accessible by internet (http://ndbserver.rutgers.edu), is a relational database which includes all RNA and DNA structures determined by x-ray crystallography (*66,67*). While the majority of structures are double helical (A-, B-, or Z-form), included also are tRNAs, structures with bound drugs, structures containing chemical modifications or unusual features such as bulges, non-standard base pairs, frayed ends, 3- and 4-strand helices, and ribozymes. Besides primary experimental information (atomic coordinates, crystal data, crystallization conditions, data collection, and refinement methods), the database contains derivative information calculated from the atomic coordinates which is extremely valuable for computer modeling. This includes chemical bond lengths and angles, torsion angles, virtual bond lengths and angles (involving the phosphorus atoms in the backbone), and base morphology parameters calculated according to various algorithms (*27,68,69*). The database allows one to carry out very specific searches and to generate reports on the structures one selects. NMR determined structures, not yet included in NDB, may be found in the Brookhaven Protein Data Bank (http://www.pdb.bnl.gov).

Quantum Mechanical Treatments

The most fundamental level of modeling of any chemical system employs quantum mechanics. Quantum mechanical (QM) treatments are required to understand many important chemical and biological properties of nucleic acids. Moreover, empirical force-field methods, employed to study the conformations of polynucleotides, rely on quantum calculations to obtain crucial parameters that are difficult to measure experimentally, such as atom-centered charges for calculating electrostatic interactions. The obtain a description of a chemical system using QM one solves the time-independent Schrödinger equation with or without the use of empirical parameters.

"*Ab initio*" refers to QM calculational methods that use no empirical procedures or parameters (*70*). Nonetheless, the difficulty of solving the Schrödinger equation necessitates the use of certain approximations, including separation of nuclear and electronic motions (the Born-Oppenheimer approximation), neglect of relativistic effects, and the use of Molecular Orbitals which are expressed as Linear Combinations of Atomic Orbitals (LCAO-MO method). Empirical force-field methods (upon which most conformational analyses, energy minimization, molecular dynamics, and x-ray and NMR structure refinements are based) differ fundamentally from *ab initio* and semi-empirical QM methods in that they are not concerned with solving the Schrödinger equation. Molecules are treated as classical systems composed of atoms held together by bonds modeled as harmonic oscillators. The total energy is calculated as the sum of bond stretching, bond bending, bond torsion rotations, and attraction and repulsion between nonbonded atoms (*71,72*). Since electrons are not treated explicitly, these empirical methods cannot deal with phenomena involving changes in electronic states such as chemical reactivity and the absorption of light.

The Schrödinger equation can only be solved approximately, even for the individual building blocks of a DNA or RNA molecule -- the bases, sugars and phosphates. Early efforts necessarily employed semi-empirical methods or minimal *ab initio* basis sets due to computational limitations. Pioneering work on the application of quantum mechanics for elucidating the chemical and physical properties of nucleic acids was carried out by the Pullmans (*73*). These early studies laid out most of the basic issues which still concern researchers, including: 1) to quantitatively determine base-pairing and base-stacking interaction energies, 2) to predict dipole moments, 3) to predict the relative stabilities of the different tautomeric forms of the bases, 4) to calculate the electronic energy levels, charge distributions, ionization potentials and electron affinities of the bases and to relate these to reactivities toward carcinogenic and mutagenic compounds, 5) to predict the ability of DNA to transport charges, 6) to describe the absorption and emission spectra of the bases and the changes occurring when the bases are stacked, particularly the hypochromism observed in the first UV absorption band (around 260 nm), and 7) to account for the photochemical reactivities of the bases. The results of early investigations were reviewed by Ladik in 1973 (*74*).

A renaissance of quantum mechanical calculations has occurred in recent years owing to the advent of increasingly powerful computers which have made it possible to solve the Schrödinger equation using high-level *ab initio* methods that include electron correlation; this has been shown to be essential for calculating interactions between the DNA bases (*75*). The application of *ab initio* methods to studying nucleic acids was recently reviewed in an article directed to the non-specialist (*76*). One of the most significant results of these studies is that the amino groups of the DNA bases are significantly non-planer (*77*). Interestingly, none of the currently used empirical force fields make use of this finding.

A comprehensive *ab initio* study of base stacking recently appeared in which the stacking interactions of all 10 stacked dimers of the standard bases were calculated as a function of their relative orientations (twist, displacement, and vertical separation) (*78*). The dimers were studied at the second-order Møller-Plesset (MP2) level of theory to treat electron correlation with a medium-sized basis set. This treatment appears to be sufficient to reveal the nature of base-stacking interactions. These calculations indicate that the G-G dimer is most stable while the U-U dimer is least stable; the stability of stacked pairs originates in the electron correlation energy, whereas the most favorable mutual orientation is determined primarily by the Hartree-Fock (HF) energy. Hydrogen-bonding interactions, on the other hand, are dominated by the HF energy. Individually, the HF and electron correlation contributions to the base-stacking energies (intra- and inter-strand) of base pair steps show large sequence-dependent variation, but the overall base-pair stacking energy variations are smaller, ranging from -10 to -15 kcal/mol. A significant finding of these calculations is that the standard coulombic term used in empirical force fields like AMBER (*71*), with point charges localized on the atomic centers, sufficiently describes the electrostatic part of stacking interactions (*78*).

Empirical Approaches to Modeling Nucleic Acid Structure and Dynamics

Unlike QM approaches, the use of empirical energy functions makes it possible to model the structures and simulate the motions of polynucleotides containing thousands of atoms, including solvent molecules and counterions. Several empirical force fields suitable for modeling and simulating nucleic acids are available (see for example (*71,72*)). Empirical force fields are derived by fitting parameters to experimental data and to *ab initio* quantum mechanical calculations. It is important to balance the intramolecular with the intermolecular portions of the potential energy function. Calculation of electrostatic interactions is both crucial and difficult. This is typically done by assigning atom-centered charges and calculating all possible pairwise Coulombic interactions within a given cutoff radius. The difficulties arise from the fact that molecular electron densities can only be calculated approximately and that the way the electron density is partitioned between different atoms in calculating atomic-centered charges is, to some extent, arbitrary and conformation-dependent. Recently, atomic point charges were derived from very high resolution, low temperature, single-crystal X-ray diffraction data of a variety of nucleosides and nucleotides (*79*). These provide nucleic acid modelers with a choice of sets of atomic point charges to employ in potential energy calculations, as well as a valuable point of reference for comparison to *ab initio* fitted charges. A further difficulty stems from the long-range nature of the electrostatic force. To limit the number of pairwise interactions calculated at each step, it has been standard practice to apply a cutoff radius to the Coulombic and van der Waals terms in the potential energy. However, it has been shown that this truncation produces artifacts even for cutoffs as long as 16Å (*80*). An alternative approach, which has met with considerable success, as demonstrated by several contributions in this volume, is based on Ewald summation methods.

As mentioned above, empirical force fields have been employed in conjunction with experimentally determined constraints on inter-atomic distances and torsion angles to refine structural models (see below). The simplest approach is energy minimization, which leads inexorably to the nearest minimum on the multi-dimensional potential energy surface. Techniques have been developed to sample a larger range of conformational space using molecular dynamics or Monte Carlo methods. Molecular dynamics (MD) methods also provide insight into the dynamic behavior and range of conformational flexibility of macromolecules (*81*). MD simulation involves the numerical integration of Newton's equations of motion. Individual atoms or groups of atoms constitute the elements of a classical mechanical system. The gradient of the empirical energy function is calculated to determine the net force on each element of the system at a particular point in time. The forces are integrated to calculate instantaneous velocities from which new positions are calculated. The book by McCammon and Harvey introduces the reader to the theory of MD and its application to proteins and nucleic acids (*82*). The improved quality of available empirical force fields and the ability to calculate electrostatic interactions more accurately using Ewald methods (*83*), is raising the hope of realistically modeling nucleic acid conformations with fewer or even no additional experimental constraints. The contribution from the Kollman group in this volume surveys recent results obtained using these methods. Other chapters in this volume illustrate the use of MD simulation to model how nucleic acid conformation is affected by sequence, changes in ionic environment, chemical modification of the backbone, and photochemical damage.

Thermodynamic Studies

Equilibrium thermodynamics play an important role in RNA folding (*84*) and in DNA metabolism (*85*). Early studies of tRNA thermal denaturation indicated that unfolding of RNA occurs in a step-wise fashion (*86*) with tertiary structure unfolding first, followed by more stable secondary elements at successively higher temperatures. For small RNAs, this process is fully reversible, indicating equilibrium folding and unfolding. This hierarchy of structure suggests that, to a first approximation, tertiary interactions can be neglected when modeling secondary structure. Early observations along these lines led Tinoco and co-workers to postulate that RNA secondary structure could be predicted by using a computer algorithm to calculate folding energies for a given sequence folded into different structures (*87*). Those structures with the lowest free energies were predicted to predominate at equilibrium. A prerequisite to such computations is a data base of empirical folding energies for different structural motifs that occur in RNA, namely base-pairs and loop motifs such as bulges, internal loops, hairpin loops and multi-branched loops (*88*). The dependence of helix stability on sequence is generally modeled using a nearest-neighbor approximation. An essentially complete set of folding rules for RNA has been compiled by Turner's group based on thermal denaturation studies of model oligonucleotides (*89*). A similar set of parameters for folding single-stranded DNA has recently become available (J. SantaLucia, unpublished results). The contribution from Turner's group in this volume summarizes the progress made to date using this approach for predicting RNA secondary structure from sequence. These predictions are typically 70% to 80% correct for RNAs less than 400 nucleotides; they serve as useful starting points for the design of biochemical experiments for secondary structure determination as well as an important first step for the prediction of tertiary structure. The contribution by Gultyaev in this volume underscores the importance of kinetics in the folding of large RNAs which have eluded accurate prediction by equilibrium methods. It is noteworthy that Gultyaev's approach utilizes Turner's thermodynamic database in a genetic algorithm that accounts for kinetically trapped intermediates in the folding pathway.

NMR Spectroscopy and Solution Structure Determination

Structure determination of nucleic acids by NMR is complementary to that by x-ray crystallography. Molecules are studied directly in solution without the need for crystallization. Solution conditions can be widely varied to determine the effects of counterions, temperature, small ligands, and proteins on conformation. Information pertaining to molecular motion can also be obtained. Kurt Wüthrich's landmark text, "NMR of Proteins and Nucleic Acids", elegantly outlines the fundamental approach to structure determination by NMR (*90*). Structures are determined by compiling constraints involving distances among protons, bond torsion angles, and hydrogen bonds in conjunction with molecular dynamics and energy minimization algorithms. The desire to accurately determine ever larger structures has driven the development of new NMR methods. Enrichment of RNA and DNA samples with ^{13}C and ^{15}N allows one to take advantage of the wide spectral dispersion of these nuclei to reduce spectral overlap between resonances. More accurate resonance assignments and larger sets of distance restraints can be obtained. In addition, backbone torsion angles can be determined more precisely via three-bond heteronuclear J-couplings (*91*).

Two nuclear spins in a biomolecule can relax one another by through-space magnetic dipole-dipole interactions (*92*). The efficiency of magnetization transfer depends on $1/r_{ij}^6$, where r_{ij} is the internuclear distance, which is measured by nuclear Overhauser effect experiments (NOE). As a first approximation, the efficiency of transfer between two nuclei separated by a fixed distance (e.g. the cytosine H5-H6) can be used as a ruler to estimate the distances between other pairs of protons whose NOEs have been measured (i.e. two-spin approximation). However, biomolecules do not consist of isolated pairs of spins. Rather, the many hydrogen nuclei in a biomolecule mutually relax one another leading to so-called "spin-diffusion" effects which distort the apparent distances obtained by the two-spin approximation. A solution to this problem is to measure initial rates of magnetization transfer in a transient NOE experiment (NOESY). The idea is illustrated in a three-spin system consisting of spins A, B, and C, whereby A is close to B and B is close to C while A and C are more removed from each other. After a short time interval (the mixing time in the two-dimensional NOESY experiment), magnetization transfers efficiently from spin A to spin B and from spin B to spin C, but inefficiently from spin A to spin C because of the long direct distance separating A from C and because more time is required for indirect transfer from A to B to C. It has been found, however, that for mixing times short enough to ignore spin diffusion (less than 40 msec), very little magnetization is transferred so that all signals are weak. This is problematic for biomolecules that have limited solubility, availability, and spectra exhibiting broad linewidths. Fortunately, approximate structures of nucleic acids can be used with complete relaxation matrix calculations to account for spin-diffusion effects. In a process analogous to crystallographic refinement, the observed NOESY spectrum is compared to a spectrum calculated from the atomic coordinates of an initial model. Modifications in the structure are then made in an iterative fashion using molecular mechanics techniques to minimize the residuals between experimental and calculated NOE interactions. Several contributions in this volume illustrate the use of these techniques and address ways of improving them.

Molecular motion and conformational flexibility are vital to biological functioning. Characterizing the whole range of molecular motions that occur on time scales of 10^{-12} to 10^{2} seconds is daunting, yet NMR is able to provide information over much of this range. For the purposes of NMR structure determination, it is the motions of large portions of the molecule on millisecond time scales that is most problematic and also most interesting for biological function.

NMR first revealed the 3D structures of stable, widely occurring hairpin loops found in large RNA structures, the UUCG (*93*) and the GNRA hairpins (*94*). The structures revealed the unique hydrogen bonding and base-stacking interactions that stabilize these loops and, in the case of GNRA, allow them to take part in specific tertiary structure interactions. The structure of a 29-nucleotide model of the α-sarcin loop from 28S ribosomal RNA (rRNA) (*95*) and the subsequent determination of a 41-nucleotide section of 5S rRNA (*96*) represent the current limits of RNAs that can be studied without isotope labeling. The use of NOEs involving exchangeable protons has played a key role in structure determination of RNA and gives direct information on hydrogen bonding. The development of methods for ^{13}C, ^{15}N, and ^{2}H isotope enrichment of RNA allows for routine assignment of resonances in small RNAs and more importantly has extended the size range of RNAs amenable to NMR approaches. The first RNA-protein complexes solved by NMR, which include tat-TAR from HIV-1 (*97*), rev-RRE (also from HIV) (*98*), and the U1A-snRNA structures (*99*), have provided information on how RNA, with only four bases, can specifically recognize many different proteins and other ligands. Recent simulation studies by Varani and co-workers have demonstrated that detailed structures of RNAs >40 nucleotides can be determined by current NMR methods with precision and accuracy comparable to similar-sized proteins (*100*). Studies of even larger RNAs present unique challenges for the future. The main problems, which are exacerbated by increasing molecular weight, are broad linewidths and low efficiency of COSY type magnetization transfers for torsion angle determinations. It appears that uniform and selective deuteration procedures help to sharpen linewidths and improve proton detection. We can also look forward to the sensitivity improvements promised by the introduction of super-conducting probes and by the development of higher static magnetic fields.

Reduced Models

Nucleic acids are modeled over the complete range of atomic detail from representations in which all atoms, including solvent and counterions, are represented explicitly to mechanical models in which double stranded DNA is modeled as a thin, isotropic, elastic rod. Intermediate levels of modeling range from all-atom models of the macromolecule, in which solvent is treated as a dielectric continuum, to reduced representations, in which relatively rigid atomic groupings are treated as single units. The choice of model depends on the problem that is being addressed. For example, much understanding of the supercoiling of DNA has been achieved using the elastic rod model (*101*). Supercoiling is observed in circular DNA molecules which are too long to be amenable to high-resolution methods such as X-ray crystallography and NMR. The dynamic behavior of supercoiled DNA relates directly to biological function -- winding and unwinding of DNA during replication and transcription. DNA supercoiling was discovered using electron microscopy (EM) (*102*). Although EM gives a two-dimensional view of what in solution is a 3D structure, EM has provided the detailed information about the dependence and range of supercoiling conformations on solution conditions and the supercoiling density (*103*). It was observed that the parameter which varied most among different EM studies was the degree of branching. Insight into the reason for this was obtained by computer simulations (*101*). Mechanical analysis aims to identify the one conformation of a supercoiled DNA that has the least elastic energy. However, in solution, DNA adopts many conformations, and so DNA supercoiling must be treated using statistical mechanics. Computer simulation of the equilibrium distribution of DNA conformations as a function of relevant parameters (ionic strength, supercoiling density, and chain length) can be effectively carried out using the Monte Carlo approach, in which a random set of conformations is generated (usually with the Metropolis procedure (*104*)) based on a given DNA model. A simple model which produces quantitatively accurate descriptions of DNA supercoiling represents the DNA as a closed chain of rigid cylinders. This model has only three parameters: the effective diameter of the cylindrical segments (which increases markedly at low ionic strength due to electrostatic repulsions, the torsional rigidity constant between the cylinders), and the persistence length which is a measure of the stiffness of the DNA chain (*105*). The motions of this so-called "wormlike chain" model of DNA have also been analyzed analytically (*106*). The Monte Carlo approach allows one to rapidly generate an ensemble of representative structures at equilibrium. Also of interest from a biological point of view is the way structures evolve in time. For example, one would like insight into the motions of the DNA strand as super-coiling is induced by topoisomerases. Appropriate molecular dynamics approaches have been developed to model this behavior (*107*).

Conclusions

Molecular models of nucleic acids are useful insofar as they provide insight into biological function and suggest fruitful directions for devising new experiments. The fruitful interplay of theory and experiment, so crucial for the first model of the DNA double helix, can be expected to continue. Recently, pure RNA oligonucleotides became available for crystallization. The result has been a virtual boom in RNA single-crystal crystallography (see for example the contribution of Holbrook et al. in this volume). Besides the crystal structure of the Group I intron, two structures of the hammerhead ribozyme recently appeared (*108,109*), the first natural RNAs solved crystallographically since tRNA more than 20 years ago. X-ray crystallography and NMR spectroscopy reveal new secondary and tertiary structure motifs, most of which, like the Hoogsteen base pair and the Z-DNA helix, were unforeseen by purely theoretical considerations. These new structures augment and enrich the repertoire in the molecular modeler's "Lego construction set" of 3D motifs.

Phylogenetic analysis has also played an important role in identifying new structural motifs. For example, the pseudo-knot was first discovered by comparative sequence analysis (*110*). This structure may be considered either a secondary or a tertiary interaction since it involves base-pairing and helix formation, but often serves to bring together two domains distant in the primary sequence. The detailed structures of pseudo-knots have been studied by NMR spectroscopy, and now they are standard building blocks in the RNA Lego set. The power of the phylogenetic approach in turn increases with the knowledge of new structural motifs, as for example the identification of GNRA hairpin loops and their receptors (*111,112*).

Molecular modeling of nucleic acids involves the application of diverse tools, each appropriate for a particular level of analysis: Quantum mechanics is required for a precise description of the covalent structure and electronic properties of the building blocks and their covalent connections. Empirical force fields are appropriate for analysis of the conformational properties of polynucleotides and for a description of non-covalent interactions leading to secondary and tertiary structure. Accurate methods for calculating electrostatic forces are key to their successful realization. Reduced models are appropriate for large-scale structures. As the capabilities of digital computers increase, it becomes possible to employ more detailed models on larger structures. The same applies to modeling the dynamics of phenomena that occur on different time scales. The dynamics of chemical reactions and electronic excitation require quantum mechanical analysis, molecular dynamics simulation employing Newtonian mechanics currently provides access to phenomena occurring in sub-picosecond to nanosecond time frames, whereas Langevin approaches are needed to gain access to longer time scales, such as those involved in transient base-pair opening and fraying (*82*).

In modeling nucleic acids, physics and chemistry encounter biology. Nucleic acids are molecules and require the methods of physics and chemistry to understand their structures and dynamics. But one cannot forget that nucleic acids are the product of biological evolution and contain within their sequences and in their 3D structures a molecular record of the evolutionary history of the organism in which they are found (*113*). It is through application of the methods and thought-patterns of all three disciplines that further progress can be anticipated.

Acknowledgments

The support of NIH Grant 1-R15-GM/OD55898-01 and ACS-PRF Grant 31427-B4 to NBL is acknowledged.

Literature Cited

1. Watson, J. D.; Crick, F. H. C. *Nature* **1953**, *171*, 737-738.
2. Wilkins, M. H. F.; Stokes, A. R.; Wilson, H. R. *Nature* **1953**, *171*, 738-740.
3. Franklin, R. E.; Gosling, R. G. *Nature* **1953**, *171*, 740-741.
4. Crick, F. H. C.; Watson, J. D. *Proc. Royal Soc. A* **1954**, *223*, 80-96.
5. Sadron, C. *Prog. Biophys.* **1953**, *3*, 237-304.
6. Williams, R. C. *Biochim. Biophys. Acta* **1952**, *9*, 237.
7. Zamenhof, S.; Brawerman, G.; Chargoff, E. *Biochim. et Biophys. Acta* **1952**, *9*, 402.
8. Furberg, S. *Acta Cryst.* **1950**, *3*, 325.
9. Donohue, J. *J. Phys. Chem.* **1952**, *56*, 502-510.
10. Cochran, W. *Acta Cryst.* **1951**, *4*, 81-92.
11. Gulland, J. M. *Cold Spring Harbor Symp. Quant. Biol.* **1947**, *12*, 95-104.
12. Langridge, R.; Marvin, D. A.; Seeds, W. E.; Wilson, H. R. *J. Mol. Biol.* **1960**, *2*, 38-64.
13. Hoogsteen, K. *Acta Cryst.* **1959**, *12*, 822-823.
14. Arnott, S.; Wonacott, A. J. *Polymer* **1966**, *7*, 157-166.
15. Jack, A.; Ladner, J. E.; Klug, A. *J. Mol. Biol.* **1976**, *108*, 619-649.
16. Arnott, S.; Hukins, D. W. L. *J. Mol. Biol.* **1973**, *81*, 93-105.
17. Rosenburg, J. M.; Seeman, N. C.; Kim, J. J. P.; Suddath, F. L.; Nicholas, H. B.; Rich, A. *Nature* **1973**, *243*, 150-154.
18. Wang, A. H.-J.; Quigley, G. J.; Kolpak, F. J.; Crawford, J. L.; Boom, J. H. v.; Marel, G. v. d.; Rich, A. *Nature* **1979**, *282*, 680-686.
19. Wing, R.; Drew, H.; Takano, T.; Broka, C.; Tanaka, S.; Itakura, K.; Dickerson, R. E. *Nature* **1980**, *287*, 755-758.
20. Yanagi, K.; Privé, G. G.; Dickerson, R. E. *J. Mol. Biol.* **1991**, *217*, 201-214.
21. Dickerson, R. E.; Goodsell, D. S.; Neidle, S. *Proc. Natl. Acad. Sci. U.S.A.* **1994**, *91*, 3579-3583.
22. Sauer, R. T.; Harrison, S. C. *Curr. Opin. Struct. Biol.* **1996**, *6*, 51-52.
23. Seeman, N. C.; Rosenburg, J. M.; Rich, A. *Proc. Natl. Acad. Sci. U.S.A.* **1976**, *73*, 804-808.
24. Dickerson, R. E. *Methods Enzymol.* **1992**, *211*, 67-111.
25. Dickerson, R. E.; et al. *EMBO J.* **1989**, *8*, 1-4.
26. Babcock, M. S.; Pednault, E. P. D.; Olson, W. K. *J. Biomol. Struct. Dynam.* **1993**, *11*, 597-628.
27. Lavery, R.; Sklenar, H. *J. Biomol. Struct. Dynam.* **1988**, *6*, 63-91, 655-667.
28. Hunter, C. A.; Lu, X.-J. *J. Mol. Biol.* **1997**, *265*, 603-619.

29. Bansal, M.; Sasiekharan, V. *Molecular Model-Building of DNA: Constraints and Restraints*; Bansal, M.; Sasiekharan, V., Ed.; Elsevier: New York, 1986, pp 127-214.
30. Pauling, L.; Corey, R. B.; Branson, H. R. *Proc. Natl. Acad. Sci. U.S.A.* **1951**, *37*, 205-211.
31. Ramachandran, G. N.; Ramakrishnan, C.; Sasisekharan, V. *J. Mol. Biol.* **1963**, *7*, 95.
32. Donohue, J.; Trueblood, K. N. *J. Mol. Biol.* **1960**, *2*, 363-371.
33. Yathinda, N.; Sundaralingam, M. *Biopolymers* **1973**, *12*, 297-314.
34. IUPAC-IUB, *Eur. J. Biochem.* **1983**, *131*, 9-15.
35. Sundaralingam, M. *Biopolymers* **1969**, *7*, 821-860.
36. Saenger, W. *Principles of Nucleic Acid Structure*; Springer Verlag: New York, 1984.
37. Kilpatrick, J. E.; Pitzer, K. S.; Spitzer, R. *J. Am. Chem. Soc.* **1947**, *69*, 2483-2488.
38. Altona, C.; Sundaralingam, M. *J. Am. Chem. Soc.* **1972**, *94*, 8205-8212.
39. Gierer, A. *Nature* **1957**, *179*, 1297-1299.
40. Doty, P.; Boedtker, H.; Fresco, J. R.; Haselkorn, R.; Litt, M. *Proc. Natl. Acad. Sci. U.S.A.* **1959**, *45*, 482-499.
41. Fresco, J. R.; Alberts, B. M.; Doty, P. *Nature* **1960**, *188*, 98-101.
42. Fontana, W.; Konings, D. A. M.; Stadler, P. F.; Schuster, P. *Biopolymers* **1993**, *33*, 1389-1404.
43. Holley, R. W.; Apgar, J.; Everett, G. A.; Madison, J. T.; Marquisee, M.; Merrill, S. H.; Penswick, J. R.; Zamir, A. *Science* **1965**, *147*, 1462-1465.
44. Zachau, H. G.; Dütting, D.; Feldmann, H.; Melchers, F.; Karau, W. *Cold Spring Harbor Symp. Quant. Biol.* **1966**, *31*, 417-424.
45. Woese, C. R.; Pace, N. R. *Probing RNA Structure, Function and History by Comparative Analysis*; Woese, C. R.; Pace, N. R., Ed.; Cold Spring Harbor Laboratory Press: Cold Spring Harbor, NY, 1993.
46. James, B. D.; Olsen, G. J.; Pace, N. R. *Meth. Enzymol.* **1989**, *180*, 227-239.
47. Waterman, M. S.; Jones, R. *Methods Enzymol.* **1990**, *183*, 221-237.
48. Turner, D. H.; Sugimoto, N.; Freier, S. M. *Ann. Rev. Biophys. Biophys. Chem.* **1988**, *17*, 167-192.
49. Zuker, M.; Jaeger, J. A.; Turner, D. H. *Nucleic Acids Res.* **1991**, *19*, 2707-14.
50. Arnott, S.; Wilkins, M. H. F.; Fuller, W.; Langridge, R. *J. Mol. Biol.* **1967**, *27*, 535.
51. Fuller, W.; Hutchinson, F.; Spencer, M.; Wilkins, M. H. F. *J. Mol. Biol.* **1967**, *27*, 507-524.
52. Robertus, J. D.; Ladner, J. E.; Finch, J. T.; Rhodes, D.; Brown, R. S.; Clark, B. F. C.; Klug, A. *Nature* **1974**, *250*, 546.
53. Kim, S.-H.; Suddath, F. L.; Quigley, G. J.; McPherson, A.; Sussman, J. L.; Wang, A. H.-J.; Seeman, N. C.; Rich, A. *Science* **1974**, *185*, 435-439.
54. Westhof, E.; Jaeger, L.; Dumas, P.; Michel, F. *Modeling the Architecture of Large RNA Molecules: A Three-dimensional Model for Group I Ribozymes*; Westhof, E.; Jaeger, L.; Dumas, P.; Michel, F., Ed.; Oxford University Press: Oxford, 1991.
55. Cramer, F. *Prog. Nucl. Acid Res. Mol.* **1971**, *11*, 391-417.
56. Levitt, M. *Nature* **1969**, *224*, 759-763.
57. Levitt, M.; Lifson, S. *J. Mol. Biol.* **1969**, *46*, 269-279.
58. Fuller, W.; Hodgson, A. *Nature* **1967**, *215*, 817-821.
59. Holbrook, S. R.; Sussman, J. L.; Warrant, R. W.; Kim, S.-H. *J. Mol. Biol.* **1978**, *123*, 631-660.
60. Major, F.; Turcotte, M.; Gautheret, D.; Lapalme, G.; Fillion, E.; Cedergren, R. *Science* **1991**, *253*, 1255-1260.

61. Michel, F.; Westhof, E. *Science* **1996**, *273*, 1676-1677.
62. Brion, P.; Westhof, E. *Annu. Rev. Biophys. Biomol. Struct.* **1997**, *26*, 113-137.
63. Brehm, S. L.; Cech, T. R. *Biochemistry* **1983**, *22*, 2390-97.
64. Michel, F.; Westhof, E. *J. Mol. Biol.* **1990**, *216*, 585-610.
65. Cate, J. H.; Gooding, A. R.; Podell, E.; Zhou, K.; Golden, B. L.; Kundrot, C. E.; Cech, T. R.; Doudna, J. A. *Science* **1996**, *273*, 1678-1685.
66. Berman, H. M.; Gelbin, A.; Westbrook, J. *Prog. Biophys. Molec. Biol.* **1996**, *66*, 255-288.
67. Berman, H. M.; Olson, W. K.; Beveridge, D. L.; Westbrook, J.; Gelbin, A.; Demeny, T.; Hsieh, S.-H.; Srinivasan, A. R.; Schneider, B. *Biophys. J.* **1992**, *63*, 751-759.
68. Grzeskowiak, K.; Yanagi, K.; Prive, G. G.; Dickerson, R. E. *J. Biol. Chem.* **1991**, *266*, 8861-8883.
69. Ravishankar, G.; Swaminathan, S.; Beveridge, D. L.; Lavery, R.; Sklenar, H. *J. Biomol. Struct. Dyn.* **1989**, *6*, 669-699.
70. Hehre, W. J.; Radom, L.; Schleyer, P. V. R.; Pople, J. A. *Ab Initio Molecular Orbital Theory*; John Wiley & Sons: New York, 1986.
71. Cornell, W. D.; Cieplak, P.; Bayly, C. I.; Gould, I. R.; Merz Jr., K. M.; Ferguson, D. M.; Spellmeyer, D. C.; Fox, T.; Caldwell, J. W.; Kollman, P. A. *J. Amer. Chem. Soc.* **1995**, *117*, 5179-5197.
72. MacKerell, A. D.; Wiorkiewicz-Kuczera, J.; Karplus, M. *J. Am. Chem. Soc.* **1995**, *117*, 11946-11975.
73. Pullman, B.; Pullman, A. *Quantum Biochemistry*; Wiley (Interscience): New York, 1963.
74. Ladik, J. J. *Adv. Quant. Chem.* **1973**, *7*.
75. Sponer, J.; Leszczynski, J.; Hobza, P. *J. Phys. Chem.* **1996**, *100*, 1965-1974.
76. Sponer, J.; Leszczynski, J.; Hobza, P. *J. Biomol. Struct. Dyn.* **1996**, *14*, 117-135.
77. Sponer, J.; Hobza, P. *J. Phys. Chem.* **1994**, *98*, 3161-3164.
78. Sponer, J.; Leszczynski, J.; Hobza, P. *J. Phys. Chem.* **1996**, *100*, 5590-5596.
79. Pearlman, D. A.; Kim, S.-H. *J. Mol. Biol* **1990**, *211*, 171-187.
80. Auffinger, P.; Beveridge, D. L. *Chem. Phys. Lett.* **1995**, *234*, 413-415.
81. van Gunsteren, W. F.; Berendsen, H. J. C. *Angew. Chem. Int. Ed. Engl.* **1990**, *29*, 992-1023.
82. McCammon, J. A.; Harvey, S. C. *Dynamics of Proteins and Nucleic Acids*; Cambridge University Press: Cambridge, 1987.
83. York, D. M.; Yang, W.; Lee, H.; Darden, T.; Pedersen, L. G. *J. Am. Chem. Soc.* **1995**, *117*, 5001-5002.
84. Jaeger, J. A.; SantaLucia, J., Jr.; Tinoco, I., Jr. *Annu. Rev. Biochem.* **1993**, *62*, 255-287.
85. Allawi, H. T.; SantaLucia, J., Jr. *Biochemistry* **1997**, *36*, 10581-10594.
86. Crothers, D. M.; Cole, P. E.; Hilbers, C. W.; Schulman, R. G. *J. Mol. Biol.* **1974**, *87*, 63-88.
87. Tinoco, I.; Borer, P. N.; Borer, P. N.; Dengler, B.; Levine, M. D.; Uhlenbeck, O. C.; Crothers, D. M.; Gralla, J. *Nature New Biology* **1973**, *246*, 40-41.
88. Jaeger, J. A.; Turner, D. H.; Zuker, M. *Proc. Natl. Acad. Sci. U.S.A.* **1989**, *86*, 7706-7710.
89. Serra, M. J.; Turner, D. H. *Methods Enzymol.* **1995**, *259*, 242-261.
90. Wuthrich, K. *NMR of Proteins and Nucleic Acids*; Wiley: New York, 1986.
91. Tinoco, I., Jr.; Cai, Z.; Hines, J. V.; Landry, S. M.; SantaLucia, J., Jr.; Shen, L. X.; Varani, G. in *Stable Isotope Applications in Biomolecular Structure and Mechanisms* , Trewhella, J., Cross, T. A., and Unkefer, C. J. Eds., Los Alamos National Laboratory, Los Alamos, 1994, pp 247-261.

92. Neuhaus, D.; Williamson, M. *The Nuclear Overhauser Effect in Structural and Conformational Analysis*; VCH: New York, 1989.
93. Varani, G.; Cheong, C.; Tinoco Jr., I. *Biochemistry* **1991**, *30*, 3280-3289.
94. Heus, H. A.; Pardi, A. *Science* **1991**, *253*, 191-194.
95. Szewczak, A. A.; Moore, P. B. *J. Mol. Biol.* **1995**, *247*, 81-98.
96. Dallas, A.; Rycyna, R.; Moore, P. B. *Biochem. Cell Biol.* **1995**, *73*, 887-897.
97. Aboula-ela, F.; Karn, J.; Varani, G. *J. Mol. Biol.* **1995**, *253*, 313-332.
98. Battiste, J. L.; Mao, H.; Rao, N. S.; Tan, R.; Muhandiram, D. R.; Kay, L. E.; Frankel, A. D.; Williamson, J. R. *Science* **1996**, *273*, 1547-1551.
99. Gubser, C. C.; Varani, G. *Biochemistry* **1996**, *35*, 2253-2267.
100. Allain, F. H.-T.; Varani, G. *J. Mol. Biol.* **1997**, *267*, 338-351.
101. Vologodskii, A. V.; Cozzarelli, N. R. *Annu. Rev. Biophys. Biomol. Struct.* **1994**, *23*, 609-643.
102. Vinograd, J.; Lebowitz, J.; Radloff, R.; Watson, R.; Laipis, P. *Proc. Natl. Acad. Sci. U.S.A.* **1965**, *53*, 4125-4129.
103. Boles, T. C.; White, J. H.; Cozzarelli, N. R. *J. Mol. Biol.* **1990**, *213*, 931-51.
104. Metropolis, N.; Rosenbluth, A. W.; Rosenbluth, M. N.; Teller, A. H.; Teller, E. *J. Chem. Phys.* **1953**, *21*, 1087-1092.
105. Hagerman, P. J. *Annu. Rev. Biophys. Biophys. Chem* **1988**, *17*, 265-286.
106. Barkley, M. D. *J. Chem. Phys.* **1979**, *70*, 2991-3007.
107. Schlick, T.; Olson, W. K. *J. Mol. Biol.* **1992**, *223*, 1089-1119.
108. Pley, H. W.; Flaherty, K. M.; McKay, D. B. *Nature* **1994**, *372*, 68-74.
109. Scott, W. G.; Finch, J. T.; Klug, A. *Cell* **1995**, *81*, 991-1002.
110. Woese, C. R.; Gutell, R.; Gupta, R.; Noller, H. F. *Microbiol. Rev.* **1983**, *47*, 621-669.
111. Jaeger, L.; Michel, F.; Westhof, E. *J. Mol. Biol.* **1994**, *236*, 1271-1276.
112. Massire, C.; Jaeger, L.; Westhof, E. *RNA* **1997**, *3*, 553-556.
113. Zuckerlandl, E.; Pauling, L. *J. Theoret. Biol.* **1965**, *8*, 357-366.

Quantum Mechanical Calculations and Empirical Force Field Parameterization

Chapter 2

The Energetics of Nucleotide Ionization in Water–Counterion Environments

Harshica Fernando, Nancy S. Kim, George A. Papadantonakis, and Pierre R. LeBreton

Department of Chemistry, The University of Illinois at Chicago, Chicago, IL 60607–7061

Results from self-consistent field (SCF) molecular orbital calculations, in combination with gas-phase photoelectron data and results from post-SCF calculations have provided a basis for descriptions of the valence electronic structure of gas-phase nucleotides and of nucleotides in water-counterion clusters. These descriptions contain values for 11 to 14 of the lowest energy ionization events in the DNA nucleotides 5'-dGMP$^-$, 5'-dAMP$^-$, 5'-dCMP$^-$and 5'-dTMP$^-$. When used with an evaluation of the difference between the Gibbs free energies of hydration for the initial and final states associated with ionization, this approach also describes the influence of hydration on the energetic ordering of ionization events in nucleotides.

Much of the biochemistry and biophysics of DNA relies on the electron donating properties of nucleotides, which, in the simplest sense, are reflected in ionization energies. For example, electron donation, as reflected in the susceptibility of nucleotides to electrophilic attack, plays a ubiquitous role in mechanisms of chemical mutagenesis and carcinogenesis (1, 2). Similarly, nucleotide ionization is an initiating step associated with radiation induced DNA strand scission (3-6). Nucleotide electron donation and ionization is also central to mechanisms responsible for electron transport in oligonucleotides (7).

Gas-phase appearance potentials for nucleotide bases were measured in early mass spectrometry experiments (8). In the first photoelectron (PE) probe of a nucleotide component, ionization potentials (IPs) of the valence manifold of π and lone-pair orbitals of uracil were measured (9). This was followed by numerous photoelectron investigations of other RNA and DNA bases (10-12), sugar model compounds (13, 14), phosphate esters (15, 16) and nucleoside analogues (17). Many of the PE investigations were accompanied by results from theoretical calculations of ionization potentials (17-19).

Theoretical and Experimental Ionization Potentials of Nucleotide Components

Figure 1 shows He(I) UV photoelectron spectra of water, and of the base and sugar model compounds, 1,9-dimethylguanine (1,9-Me_2G) and 3-hydroxytetrahydrofuran (3-OH-THF). In earlier investigations (13, 20, 21), the model compounds were employed in the evaluation of IPs for 5'-dGMP$^-$. The figure gives experimental energies and assignments associated with the 7 lowest energy vertical ionization potentials in 1,9-Me_2G and the two lowest energy IPs in 3-OH-THF. Figure 2 shows the PE spectrum and assignments for 9-methyladenine (9-MeA). The assignments for the π and lone pair IP's of 1,9-Me_2G, 3-OH-THF and 9-MeA were obtained from previous results (13, 22, 23). In addition to experimental IPs, Figures 1 and 2 also contain theoretical ionization potentials evaluated by employing Koopmans' theorem which, for closed-shell systems, equates vertical IPs to orbital energies (24). Here SCF molecular orbital calculations were carried out with the 3-21G basis set (25) and the Gaussian 94 program (26). The figures show diagrams for the 6 and 7 highest occupied orbitals in 9-MeA and 1,9-Me_2G, respectively, and for the 2 highest occupied orbitals in 3-OH-THF. The orbital diagrams were derived from the 3-21G SCF results using criteria described earlier (21). The results indicate that for 1,9-Me_2G and 9-MeA, calculated IPs of the highest occupied π orbitals differ from the experimental vertical IPs by less than 0.26 eV. The calculated lone-pair IPs are less accurate. For 3-OH-THF, the calculated lone-pair IPs are larger than the experimental vertical IPs by 1.19 and 1.16 eV.

The results in Figures 1 and 2 demonstrate that values of valence π and lone-pair ionization potentials of nucleotide components and model compounds, calculated at the SCF level, may differ significantly from experimental vertical IPs. The unreliability of the SCF results is also demonstrated by the results in the top and bottom panels of Figure 3, which contain computed and experimental vertical IPs for uracil, 6-methyluracil, 3-methyluracil, thymine (5-methyluracil), 1-methyluracil, and 1-methylthymine. The top panel shows IPs obtained from the application of Koopmans' theorem to 3-21G SCF results. The geometries used in the calculations were obtained by 3-21G SCF optimization of the heavy-atom bond lengths and bond angles, the H atom bond angles, and the torsional angles describing CH_3 rotation. The N-H bond lengths were 1.01 Å, and the C-H bond lengths of the ring, and of CH_3 were 1.08 and 1.09 Å, respectively. These values were obtained from X-ray data (27). The solid lines in the bottom panel give experimental IPs (28). A comparison of results in the top and bottom panels indicates that the energetic ordering of ionization events obtained from the 3-21G SCF calculations (top panel) is different from that which experiment provides. For example, experiment (9, 28) and results from post-SCF calculations (19) indicate that the second IP is associated with the removal of an electron from an oxygen-atom lone-pair orbital

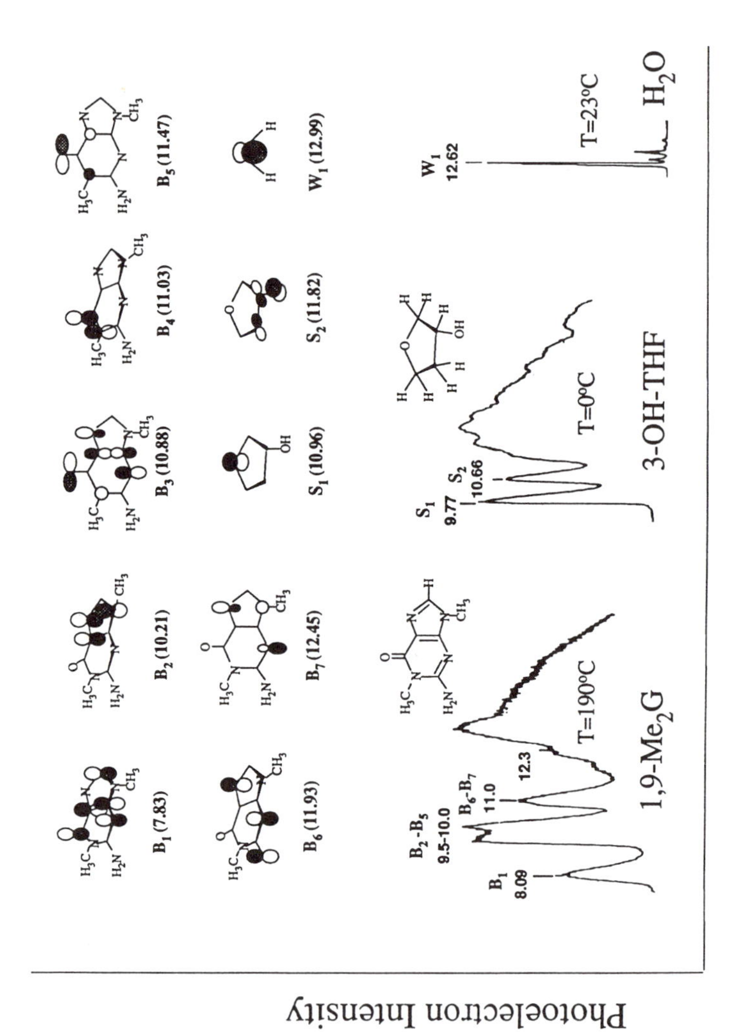

Figure 1. Photoelectron spectra and assignments for 1,9-dimethylguanine (1,9-Me$_2$G), 3-hydroxytetrahydrofuran (3-OH-THF), and water. Theoretical ionization potentials and molecular orbital diagrams obtained from 3-21G SCF calculations are also given. Reproduced with permission from ref. 21. Copyright American Chemical Society.

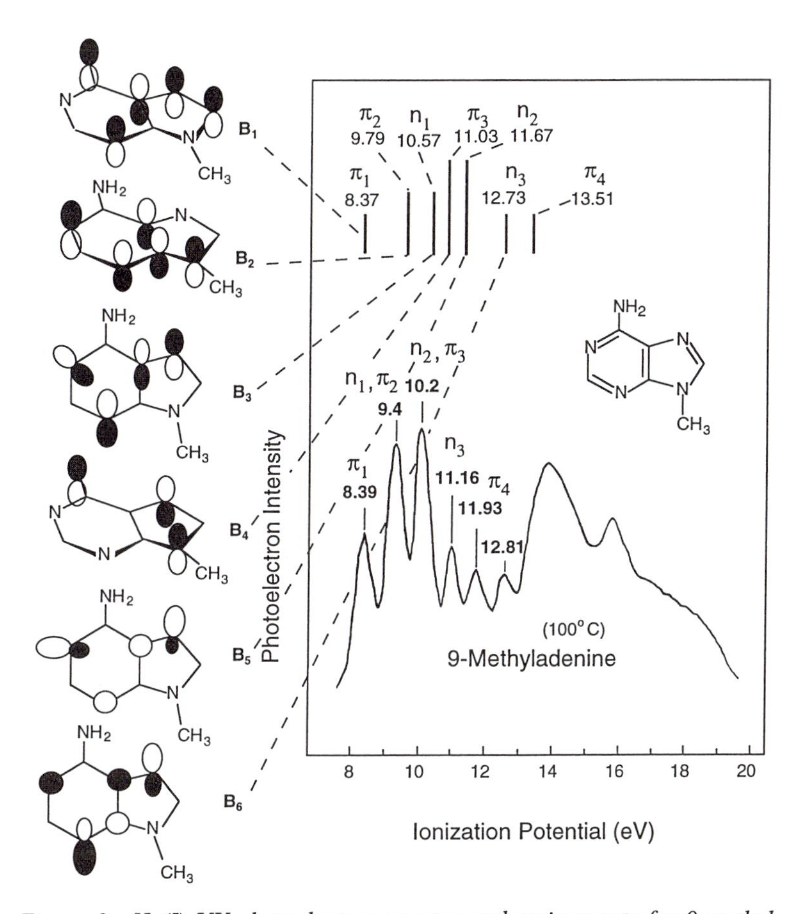

Figure 2. He(I) UV photoelectron spectra and assignments for 9-methyladenine (9-MeA). Molecular orbital diagrams and theoretical IPs obtained from 3-21G SCF calculations are also given. (Adapted with permission from reference 31. Copyright 1981 Wiley.)

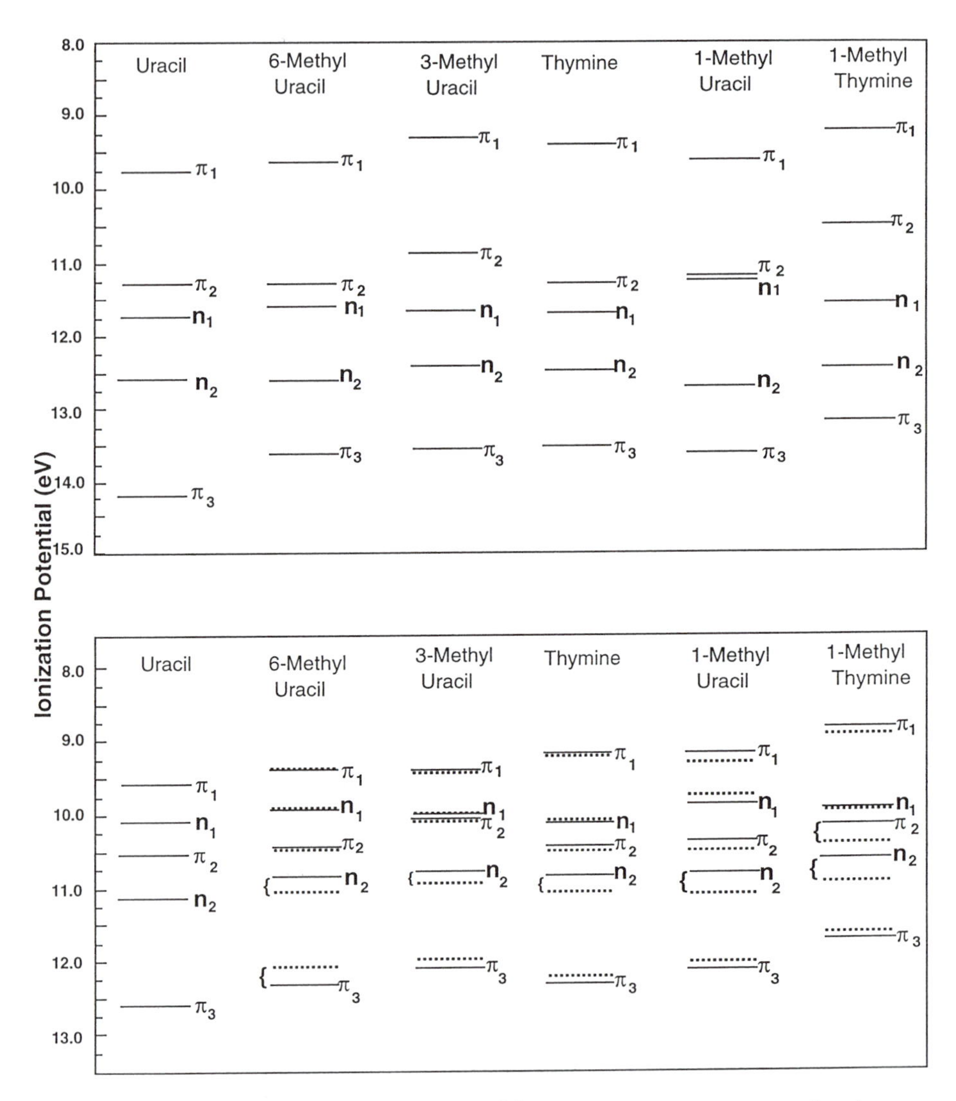

Figure 3. The five lowest energy π and lone-pair ionization potentials of uracil and methyl uracils obtained from application of Koopmans' theorem to results from SCF calculations with the 3-21G basis set (top panel). The figure also contains experimental vertical ionization potentials (solid lines, bottom panel) and, for the methyl uracils, 3-21G SCF ionization potentials (dashed lines, bottom panel) which were corrected using a method analogous to that described by eqs 1 and 2. Experimental IPs were taken from ref. 26.

(n_1) and the third IP is associated with a π orbital (π_2). The 3-21G SCF calculations predict that the π_2 ionization potential is smaller than n_1. The IPs of model compounds calculated at the ab initio SCF level are basis set dependent, and the energetic ordering varies. However, the general agreement between experimental values for the six or seven lowest energy ionization events does not significantly improve when the size of split-valence basis sets is increased (13, 20, 21).

A consideration of the application of PE spectroscopy and of computational approaches to obtain valence manifold IPs of intact nucleotides and larger DNA subunits reveals three impediments. Two are experimental. The first is the experimental difficulty associated with preparing gas-phase samples of anionic nucleotides at pressures sufficiently high to permit PE measurements. The second is the complex electronic structure of nucleotides, which contain a large number of orbitals with similar energies. This will give rise to PE spectra that are poorly resolved (17). In this regard, much of the advantage of PE spectroscopy, which provides as many as 7 valence IPs of nucleotide bases, is diminished when the method is applied to larger molecules. The third barrier is computational, and is also associated with the large size of nucleotide electronic systems. To date, the largest of these for which IPs have been evaluated contains 330 electrons (21). With readily available computational resources, it is not currently possible to calculate multiple ionization energies for systems of this size at a rigorous ab initio level.

These difficulties have been overcome by employing a strategy which relies on experimental photoelectron data, and post-SCF calculations to provide accurate valence π and lone-pair IPs for nucleotide components and component model compounds, together with less rigorous SCF calculations to provide perturbation energies associated with combining nucleotide components into larger units. With this approach, SCF calculations have also been employed to evaluate perturbations due to electrostatic interactions. The strategy, as applied to the evaluation of nucleotide IPs, is outlined in eqs 1 and 2.

$$IP_{corr}(i) = IP_{calc}(i) + \Delta IP(i) \quad (1)$$

$$\Delta IP(i) = IP(i) - IP'_{calc}(i) \quad (2)$$

In eq 1, $IP_{corr}(i)$ is the corrected IP associated with the i'th orbital in a nucleotide, and $IP_{calc}(i)$ is the IP obtained at the SCF level. When eq 2 is used to correct base or sugar IPs, IP(i) is the experimental IP associated with the i'th orbital in a base or sugar model compound, which most closely correlates with the i'th orbital in the isolated nucleotide. $IP'_{calc}(i)$ is the IP of the i'th orbital in a base or sugar model compound obtained from SCF results. In these investigations, 1-methylthymine (1-MeT), 1-methylcytosine, (1-MeC), 1,9-dimethylguanine (1,9-Me_2G) and 9-methyladenine (9-MeA), were employed as model compounds for the DNA bases, and 3-hydroxytetrahydrofuran (3-OH-THF) was used as a model compound for 2-deoxyribose. 1,9-Me_2G was chosen as a model compound for guanine, because the most stable gas-phase tautomeric structure of 1,9-Me_2G corresponds to that of the guanine structure which participates in Watson-Crick base pairing (22).

When eq 2 is used to correct IPs of the anionic phosphate group, IP(i) is the previously reported (29) ionization potential of $H_2PO_4^-$, which was obtained using a combination of post-SCF calculations. Here, the lowest energy IP was taken to be the difference between the ground-state energies of $H_2PO_4^-$ and of the $H_2PO_4\cdot$ radical. These energies were obtained from Möller Plesset second-order perturbation (MP2) calculations with a 6-31+G* basis set (30). The second through fifth IPs were obtained by adding excitation energies of $H_2PO_4\cdot$ to the lowest energy IP of $H_2PO_4^-$. These excitation energies were evaluated with a complete active space second-order perturbation (CASPT2) calculation using a complete active space SCF (CASSCF) reference wave function (31). For $H_2PO_4^-$, there is no experimental ionization potential data available. However, MP2/6-31+G* calculations yielded values of 1.51, 3.34, and 4.90 eV for the lowest energy IPs of the phosphorus and oxygen containing anions CH_3O^-, PO_2^-, and PO_3^-, respectively. These values agree well with the experimental values 1.57, 3.30 ± 0.2, and 4.90 ± 1.3 eV (13).

Figure 4 gives five of lowest energy IPs of $H_2PO_4^-$, obtained from 3-21G SCF calculations, along with orbital diagrams. The figure also gives IPs obtained from the combination of MP2 and CASPT2 calculations (29). An earlier comparison (21) of 3-21G SCF descriptions of the five lowest energy ionization potentials in $CH_3PO_4^-$ with descriptions obtained from a combination of MP2 and configuration interaction singles (CIS) calculations (32) indicated that the SCF descriptions of the changes in charge distributions associated with the ionization events were in qualitative agreement with the results from the MP2 and CIS calculations.

Results from a simple test of the strategy employed to obtain nucleotide IPs is provided in the bottom panel of Figure 3. Here, the dashed lines represent corrected IPs of the methyl uracils. These were obtained by applying eqs 1 and 2 to the results from the 3-21G SCF calculations. In this test, uracil was used as the model compound. After correction, the computational description of the perturbation pattern associated with methyl substitution is in good agreement with that obtained experimentally.

Gas-Phase Ionization Potentials of Nucleotides

Figure 5 contains a 3-21G SCF description of the 14 smallest ionization potentials of 5'-dGMP$^-$. The geometry is the same as that reported in an earlier investigation (21). The figure also contains orbital diagrams. The SCF results indicate that each orbital is largely located on either the base, sugar or phosphate groups, and that the upper occupied orbitals of the nucleotide correlate closely with corresponding orbitals in 1,9-Me_2G, 3-OH-THF and $H_2PO_4^-$.

Figure 6 shows ionization potentials of 5'-dGMP$^-$ after incorporation of the corrections described in eqs 1 and 2. For the base and sugar groups, the corrected IPs are the same as those reported earlier (21). For the phosphate group, the IPs have been revised. Here, the corrections of the 3-21G SCF values for the P_2 to P_5 ionization potentials are based on the CASPT2 results, described above, for $H_2PO_4^-$.

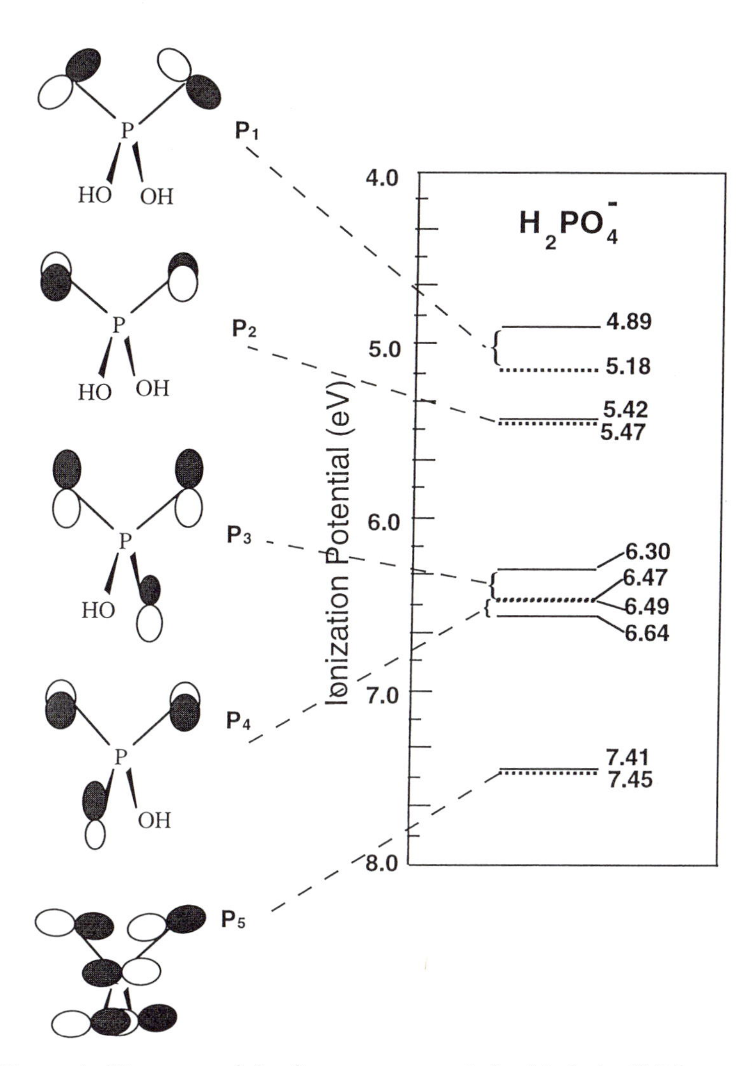

Figure 4. Diagrams of the five upper occupied orbitals in $H_2PO_4^-$, and ionization potentials (dashed lines) obtained from results of 3-21G SCF calculations. Solid lines show IPs of $H_2PO_4^-$ obtained from MP2/6-31+G and CASPT2 calculations. See ref. 27.*

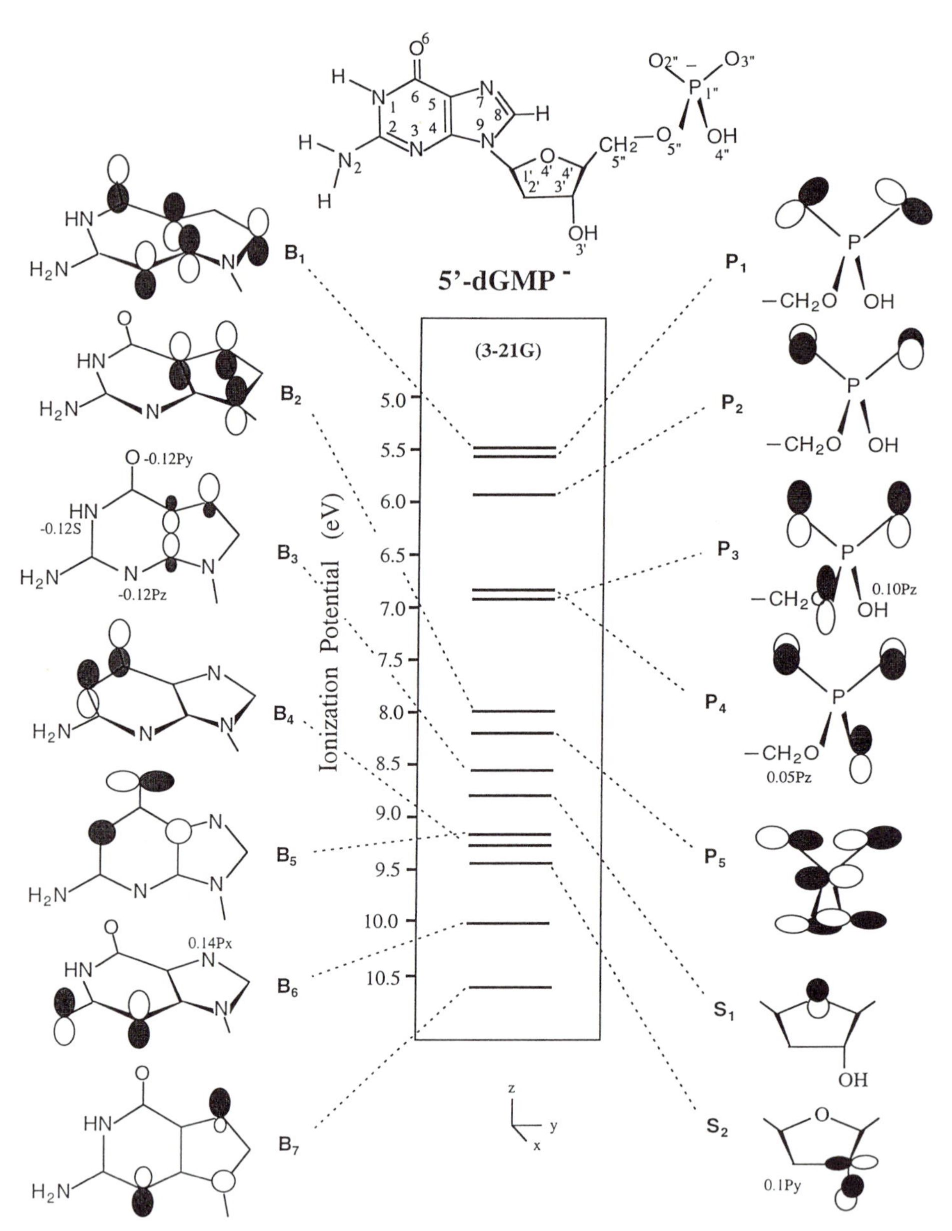

Figure 5. Ionization potentials and molecular orbital diagrams of 5'-dGMP⁻ obtained from 3-21G SCF calculations. The orbitals localized on the base, sugar, and phosphate groups are designated B, S and P, respectively. For the B_3 orbital, apparent differences between orbital diagrams in the intact nucleotide and in the in the model compound, 1,9-dimethylguanine (1,9-Me_2G), are exaggerated by the cut-off criterion. In this case molecular orbital coefficients are given. See ref. 21.

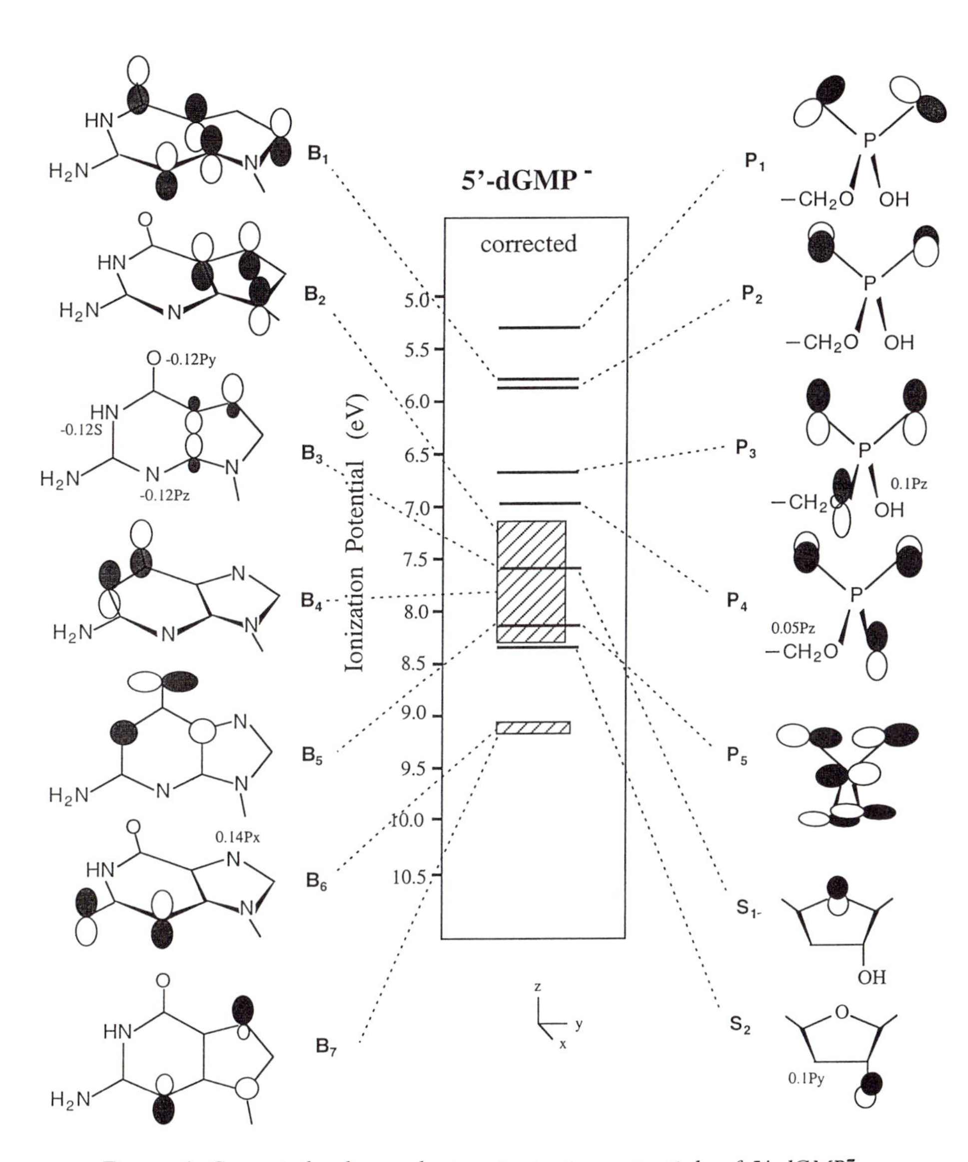

Figure 6. Corrected valence electron ionization potentials of 5'-dGMP⁻. The hatched area corresponds to an unresolved energy region in the PE spectrum of 1,9-Me₂G which contains overlapping bands. (Adapted from reference 21. Copyright 1996 American Chamical Society.)

A comparison of results in Figures 5 and 6 indicates that, in some cases, the values of the corrected IPs differ from the SCF values by more than 1.0 eV . This comparison also indicates that the energetic ordering of ionization potentials changes after correction. According to the 3-21G SCF results, the lowest energy IP is associated with the base. After correction, the lowest energy ionization is associated with the phosphate group. This difference in the energetic ordering of the base and phosphate IPs is due to the fact that the 3-21G SCF calculations predict phosphate lone-pair ionization potentials which are too large.

Figure 7 shows corrected gas-phase ionization potentials of 5'-dTMP$^-$, 5'-dCMP$^-$ and 5'-dAMP$^-$ obtained by applying eqs 1 and 2 to results from 3-21G SCF calculations. For 5'-dAMP$^-$, the results are the same as those reported earlier (33). For 5'-dCMP$^-$, the IPs in Figure 7, like the results for 5'-dGMP$^-$ in Figure 6, represent a revision of previously reported results (14). Here, again, the P_2 to P_5 ionization potentials were corrected using the CASPT2 results.

Figure 7 contains diagrams for the base orbitals in the nucleotides. For 5'-dTMP$^-$ and 5'-dAMP$^-$, all of the base orbitals correlate closely with corresponding orbitals in 1-methylthymine and 9-methyladenine. The sugar and phosphate orbitals are similar to the S_1, S_2 and P_1 to P_5 orbitals in Figures 1 and 4. For 5'-dCMP$^-$, the B_1 to B_5, S_2 and P_1 to P_5 orbitals are similar to corresponding orbitals in 1-methylcytosine (1-MeC), 3-hydroxytetrahydrofuran (3-OH-THF) and $H_2PO_4^-$. However, the S_1 orbital in 5'-dCMP$^-$ contains mixing of the S_1 , S_2 orbitals of 3-OH-THF with the B_4 orbital of 1-MeC. The delocalization of the S_1 orbital in 5'-dCMP$^-$ may be due to details of the 5'-dCMP$^-$ geometry used in the calculation. For 5'-dCMP$^-$, 3-21G SCF results also indicate the occurrence of an additional sugar lone-pair orbital (S_3) with a corrected gas-phase IP between those of S_2 and B_5. However, unlike the other orbitals of 5'-dCMP$^-$ which have been examined, the corrected IP of this orbital is strongly basis set dependent. For this reason, a description of S_3 in 5'-dCMP$^-$ has not been included in Figure 7.

For calculations on the model compounds 1-MeT and 1-MeC, which were used in evaluation of the 5'-dTMP$^-$ and 5'-dCMP$^-$ ionization potentials, the geometries were obtained in the same manner as for uracil and the methyl uracils of Figure 3. The geometries of 5'-dTMP$^-$ and 5'-dCMP$^-$ were obtained from crystallographic data on the B-DNA dodecamer, 1C-2G-3C-4G-5A-6A-7T-8T-9C-10G-11C-12G (34). The geometries of 5'-dTMP$^-$ and 5'-dCMP$^-$ were based on those associated with the 8 position in strand B, and with the 9 position in strand A, respectively. For 5'-dCMP$^-$, this represents a change in geometry from that examined earlier (14). For the nucleotides, the same N-H, C-H and O-H bond lengths were used as in the model compounds.

For the 5'-dTMP$^-$ and 5'-dCMP$^-$ geometries employed here, and for the geometry of 5'-dAMP$^-$ employed in an earlier investigation (33), small adjustments in the glycosidic and phosphate-ester dihedral angles ($< 5^o$), and in the angles describing sugar pucker ($< 2.2^o$) were introduced in order to enhance the localization of the valence orbitals, and to improve correlation with orbitals in the model compounds. However, in all cases these adjustments resulted in valence orbital energy changes of less than 0.1 eV, as calculated at the 3-21G SCF level.

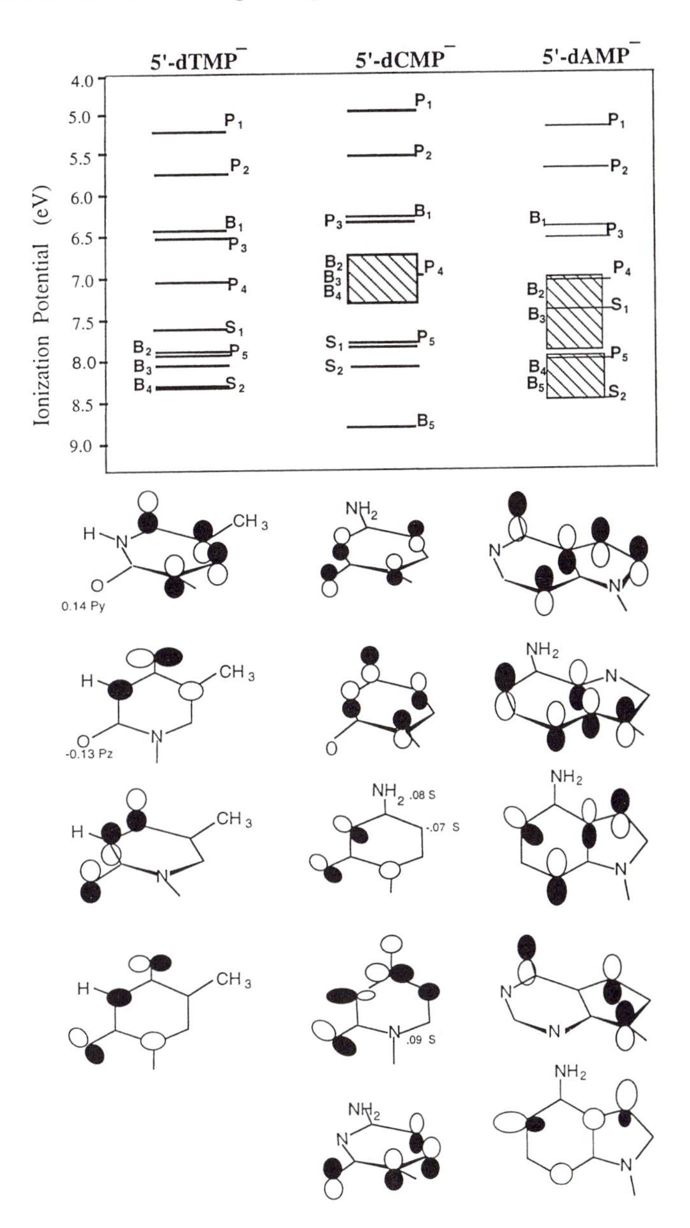

Figure 7. Corrected valence electron ionization potentials of 5'-dTMP⁻, 5'-dCMP⁻ and 5'-dAMP⁻. Molecular orbital diagrams for base orbitals which correlate with orbitals in the model compounds, 1-methylthymine (1-MeT), 1-methylcytosine (1-MeC) and 9-methyladenine (9-MeA), are shown below. For 5'-dCMP⁻ and 5'-dAMP⁻, hatched areas correspond to unresolved energy regions in the PE spectra of the 1-MeC and 9-MeA. See refs. 14 and 31.

The results in Figure 7 demonstrate that, like 5'-dGMP$^-$, the ionization potentials of 5'-dAMP$^-$, 5'-dTMP$^-$ and 5'-dCMP$^-$ increase in the order phosphate < base < sugar. The small IP associated with the phosphate group is consistent with the more negative charge on phosphate compared to charges on the base and sugar groups. According to the 3-21G SCF calculations, the net charge on phosphate is in the range -1.281 to -1.308 e, while the charges on the bases and the sugars are -0.400 to -0.425 e, and 0.693 to 0.708 e, respectively. These results are similar to earlier results from 6-31G SCF calculations on 5'-dGMP$^-$ (13). However, in Table II of ref. 13, a misprint occurs in the total 2'-deoxyribose charge listed. Here, the correct sign is positive. Most importantly, in all the nucleotides, negative charge decreases in the order phosphate > base > sugar, which is consistent with the ordering of IPs.

The results in Figures 6 and 7 indicate that the base IPs increase in the order guanine (5.76 eV) < cytosine (6.27 eV) < adenine (6.42 eV) < thymine (6.48 eV). This ordering is different from that associated with the model compounds in the gas phase, where the IPs decrease in the order 1,9-Me_2G (8.09 eV) < 9-MeA (8.39 eV) < 1-MeC (8.65 eV) < 1-MeT (8.79 eV) (9, 12, 33). The difference between the ordering of base IPs in the nucleotides versus the model compounds is, most likely, due to details of the nucleotide geometries.

This sensitivity of nucleotide gas-phase IPs to geometry is demonstrated by a consideration of B_1 ionization potentials of 5'-dCMP$^-$ for the geometries associated with position 3 in strand B (3C), and position 9 in strand A (9C) of the oligonucleotide described above (34). Here the B_1 ionization potentials (6.79 eV and 6.27 eV, for 3C and 9C, respectively) differ by 0.52 eV. This difference can be understood in terms of the distances between the base and phosphate groups. For 3C, the distances between the N1 atom of the base, and the P atom and the two negatively charged O atoms of phosphate are 6.11, 7.24, and 6.30 Å. For 9C, which has the smaller B_1 ionization potential, these distances are 5.47, 6.73, and 5.53 Å.

The Influence of Na^+ Counterions on Gas-Phase Ionization Potentials of 5'-dTMP$^-$ and 5'-dCMP$^-$

In aqueous solution, the description of DNA binding to small counterions, such as Na^+, is complicated by the fact that the binding is dynamic and occurs on a time scale of picoseconds (35-37). NMR results suggest that in a DNA solution (100 mg/ml) containing an equivalent of Na^+, about 90% of the Na^+ ions are within 7 Å of the DNA (38). X-ray data for dinucleotides in Watson-Crick base pairs (39, 40) indicates that most Na^+ binding occurs at the negatively charged phosphate O atoms. Theoretical results indicate that, in polymeric double-stranded B-DNA, binding of Na^+ also occurs with high probability in the major and minor grooves (37, 41, 42).

In this investigation, clusters of 5'-dTMP$^-$ and 5'-dCMP$^-$ were examined which contain Na^+ and 5 H_2O molecules, where the Na^+ ion is partially solvated and bound to the nucleotide phosphate group. Similar geometries were previously examined in clusters of 5'-dGMP$^-$ and 5'-dAMP$^-$ (20, 21, 33). The structures of the 5'-dTMP$^-$ and

5'-dCMP⁻ clusters are shown in Figure 8. As in earlier investigations (21, 33), the structures in Figure 8 were not fully optimized. These geometries were based on a combination of X-ray data and results from partial geometry optimization calculations (21, 33). The optimization involved the use of docking procedures and symmetry restrictions. For the geometries examined, the water-counterion interactions with 5'-dTMP⁻ and 5'-dCMP⁻ are favorable, and these interactions are expected to occur with high frequency for DNA in an aqueous solution containing Na^+ ions.

In Figure 8, water molecules labelled W(4) and W(5) make up part of the thymine and cytosine inner solvation shells. W(1) to W(3) make up part of a sodium hydration shell. The positions of W(4) and W(5) were chosen on the basis of X-ray data for water molecules interacting with B-DNA oligonucleotides (43). The distances between the O atom of W(4) and O4 of thymine, and between the O atom of W(5) and O2 of thymine are 3.60 and 3.00 Å, respectively. The C4-O4-O(W(4)) and C2-O2-O(W(5)) bond angles are 135.0° and 160.0°. The torsion angles (43) are -100.0° and 30.0°. For cytosine, the distances between the O atom of W(4) and N4 of cytosine, and between the O atom of W(5) and O2 of cytosine are 3.10 and 3.90 Å. The C4-N4-O(W(4)) and C2-O2-O(W(5)) bond angles are 120.0° and 143.0°. The torsion angles are -165.0° and 10.0°.

In both the 5'-dTMP⁻ and 5'-dCMP⁻ clusters, the orientation of W(1) to W(3) relative to Na^+ was obtained from 6-31G SCF optimization calculations (20). These water molecules make up part of an octahedral shell surrounding Na^+ and are coplanar with Na^+, the P atom and the negatively charged O atoms of the phosphate group. The distance (2.4 Å) between Na^+ and the O atoms of W(1) to W(3) was obtained from a separate 6-31G SCF optimization of an octahedral cluster containing Na^+ and six water molecules (20). In the clusters of Figure 8, Na^+ and W(1) to W(3) have a local C_{2v} symmetry. The distance between Na^+ and the P atom of the phosphate group is 5.60 Å. This distance was obtained from results of a previously reported 6-31G optimization calculation (20). Here, a cluster containing Na^+ and three water molecules, in the same relative orientation as Na^+ and W(1) to W(3) in Figure 8, was docked to the phosphate group of 5'-dGMP⁻ containing a water molecule bound to the O6 atom. In this optimization calculation, the geometries of the two individual clusters (Na^+ + 3 H_2O, and 5'-dGMP⁻ + H_2O) were fixed.

Table I lists the corrected IPs of the base, sugar, and phosphate groups in the 5'-dTMP⁻ and 5'-dCMP⁻ clusters shown in Figure 8. The table also lists IPs obtained in earlier investigations of clusters containing 5'-dAMP⁻ (cluster A in ref. 33) and 5'-dGMP⁻ (cluster B in ref. 20, 21) with Na^+ bound to phosphate. A comparison of the IPs of the nucleotide clusters in Table I, to the IPs for the isolated gas-phase nucleotides in Figures 6 and 7 demonstrates that the ionization potentials of the clusters are larger than corresponding IPs of the isolated nucleotides. The results indicate that the corrected lowest energy base, sugar and phosphate IPs for the clusters are larger than corresponding IPs for the isolated nucleotides by 1.47 - 2.07, 1.66 - 2.19, and 3.79 - 3.96 eV, respectively. For these clusters, interactions with H_2O have a smaller effect on ionization potentials than does electrostatic stabilization by Na^+. For 5'-dAMP⁻, this observation was supported by a supplementary calculation on a cluster containing only

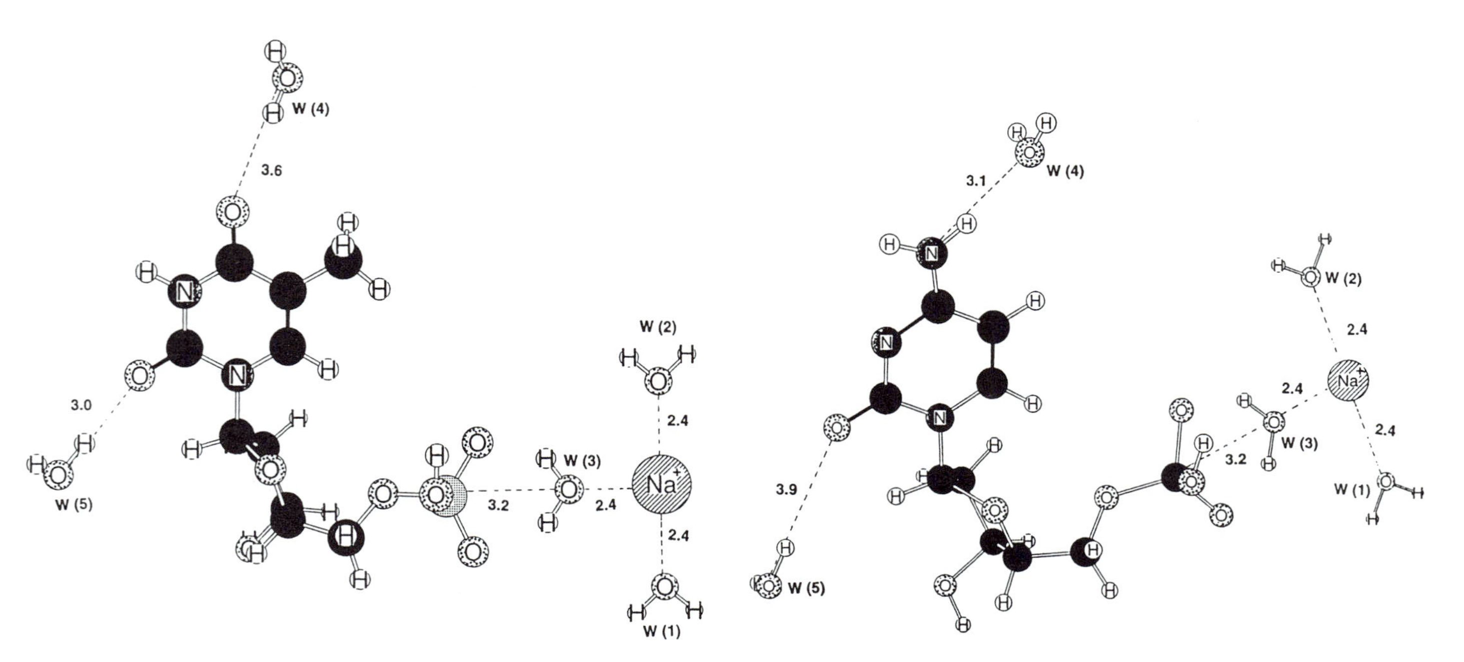

Figure 8. Structures of 5'-dTMP$^-$ and 5'-dCMP$^-$ clusters with 5 water molecules and Na^+ bound to the phosphate groups. In each of these clusters, there is part of a single solvation shell between Na^+ and the negatively charged O atoms of phosphate.

Table I. Corrected Nucleotide Ionization Potentials of 5'-dTMP$^-$, 5'-dCMP$^-$, 5'-dAMP$^-$, and 5'-dGMP$^-$, in Clusters Containing Na^+ Bound to Phosphate and H_2O[a].

	5'-dTMP^{-b}	5'-dCMP^{-b}	5'-dAMP$^{-c,d}$	5'-dGMP$^{-d,e}$
B_1	8.55	7.74	8.42 (7.04)	7.60 (6.44)
B_2	9.90	8.16-8.68	8.90-9.50	9.10-10.10
B_3	10.21	8.16-8.68	9.30-9.90	9.10-10.10
B_4	10.67	8.16-8.68	9.80-10.40	9.10-10.10
B_5	---	10.45	9.80-10.40	9.10-10.10
S_1	9.83	9.76	9.10 (7.80)	9.70 (8.23)
S_2	10.15	9.99	10.30	10.30
P_1	9.18	8.86	9.12 (8.15)	9.10 (8.53)
P_2	9.71	9.45	9.61	9.61
P_3	10.30	10.01	10.37	10.15
P_4	10.71	10.53	11.43	10.55
P_5	11.49	11.36	13.18	11.57

[a]All ionization potentials in eV.
[b]Evaluated for the clusters shown in Figure 8.
[c]Evaluated for cluster **A** of ref. 31. This cluster contains 5 water molecules.
[d]Energies in parentheses are values of ΔG_{ioniz}(solution) for the clusters obtained from eqs 3 to 5.
[e]Evaluated for cluster **B** of ref. 20. This cluster contains 4 water molecules. The B_1 and S_1 ionization potentials are the same as those reported in ref. 21. the P_1 to P_5 ionization potentials have been corrected using ionization potentials of $H_2PO_4^-$ evaluated from a combination results from MP2 and CASPT2 calculations. See text and ref. 27.

5'-dAMP$^-$ and Na^+, in which the nucleotide-counterion geometry is the same as that in a cluster containing 5'-dAMP$^-$, Na^+, and 5 H_2O molecules (33). In the cluster without H_2O, the increases in the lowest energy B_1, S_1 and P_1 corrected ionization potentials, compared to isolated 5'-dAMP$^-$, represent more than 60% of the corresponding increases found in the cluster with H_2O.

The results in Table I, and Figures 6 and 7 indicate that, when Na^+ binds to the phosphate group, the nucleotide IPs not only increase, but the energetic ordering also changes. For the isolated nucleotides, the IPs increase in the order $P_1 < B_1 < S_1$. In the clusters, the stronger interaction between Na^+ and the phosphate group, compared to the base, causes a greater increase in the phosphate IPs. Here the ordering is $B_1 < P_1 < S_1$. In earlier investigations of 5'-dAMP$^-$ clusters (33) it is found that the ordering is sensitive to the cluster geometry. In 5'-dAMP$^-$ clusters, the ordering is $B_1 < P_1 < S_1$, when Na^+ is bound to phosphate, and $P_1 < B_1 < S_1$ when Na^+ is bound to the base.

The Influence of Aqueous Bulk Solvation on Nucleotide Gibbs Free Energies of Ionization

To determine the influence of solvation on the ionization of nucleotides and nucleotide clusters, values of Gibbs free energy of ionization in solution, which do not include the electron solvation energy were evaluated. The Gibbs free energies of ionization in aqueous solution (ΔG_{ioniz}(solution)) were obtained by employing the gas phase IPs and the differences ($\Delta\Delta G_{hyd}$) between the Gibbs free energies of hydration for the nucleotides and clusters before and after ionization. Here, ΔG_{ioniz}(solution) is given by eq 3, where ΔG_{ioniz}(gas) is the free energy of ionization in the gas phase, and $\Delta\Delta G_{hyd}$ is given by eq 4. In eq 4, $\Delta G_{hyd}(1)$ and $\Delta G_{hyd}(2)$ are the free energies of hydration before and after ionization, respectively. The rationale for eq 3 is based on a previously described thermodynamic cycle (21).

$$\Delta G_{ioniz}(\text{solution}) = \Delta G_{ioniz}(\text{gas}) + \Delta\Delta G_{hyd} \quad (3)$$
$$\Delta\Delta G_{hyd} = \Delta G_{hyd}(2) - \Delta G_{hyd}(1) \quad (4)$$
$$\Delta G_{ioniz}(\text{gas}) = \Delta H_{ioniz}(\text{gas}) - T\Delta S_{ioniz}(\text{gas}) \approx IP \quad (5)$$

In eq 5, ΔH_{ioniz}(gas) is the corrected gas phase ionization potential of the nucleotide, or cluster, and ΔS_{ioniz}(gas) is the change in entropy associated with ionization in the gas phase. Compared to ΔH_{ioniz}(gas), $T\Delta S_{ioniz}$(gas) is negligible, so that $\Delta G_{ioniz}(\text{gas}) \approx \Delta H_{ioniz}(\text{gas})$. This approximation is supported by the observation that, by electron convention Fermi-Dirac statistics, $T\Delta S$ for the ionization of a hydrogen atom at room temperature is 0.05 eV (44). In this investigation, the values of IP employed in eq 5 were the values of $IP_{corr}(i)$ obtained via eq 1.

Gibbs free energies of hydration, before and after ionization ($\Delta G_{hyd}(1)$ and $\Delta G_{hyd}(2)$) were obtained by employing the Langevin dipole relaxation method (45-47) incorporated in the Polaris 3.2 program (46). Before and after ionization, solvent relaxation is modelled by evaluating the relaxation of discrete dipoles distributed on a lattice surrounding each nucleotide. Here the nucleotide is described by a distribution of point-charges, where each charge is located at the nucleus of one of the nucleotide atoms. The method used to determine the nucleotide charge distributions before and after ionization at the base, sugar or phosphate groups was described earlier (21, 33).

Figure 9 shows aqueous Gibbs free energies of ionization, ΔG_{ioniz}(solution), for 5'-dGMP$^-$ and 5'-dAMP$^-$ obtained using eq 3. A comparison of the gas-phase IPs of 5'-dGMP$^-$ and 5'-dAMP$^-$ in Figures 6 and 7 with values of ΔG_{ioniz}(solution) in Figure 9 indicates that bulk hydration increases the ionization energies of nucleotides without Na^+. According to the results in Figures 6 and 7, the lowest energy phosphate, base, and sugar IPs of the gas-phase nucleotides (P_1, B_1, and S_1) lie in the range 4.95 to 5.31, 5.76 to 6.48, and 7.44 to 7.86 eV, respectively. The corresponding values of ΔG_{ioniz}(sol ution) are 2.00 to 2.30, 0.35 to 0.78 and 0.64 to 0.75 eV larger.

Bulk hydration has the opposite effect on the ionization energies of nucleotide clusters with Na^+. This is demonstrated by the results in Table I which compares P_1, B_1, and S_1 ionization potentials for the gas-phase clusters of 5'-dGMP$^-$ and 5'-dAMP$^-$ with values of ΔG_{ioniz}(solution). The P_1, B_1, and S_1 ionization potentials of the gas-phase clusters lie in the range 8.86 to 9.18, 7.60 to 8.55, and 9.10 to 9.83 eV, respectively. The corresponding values of ΔG_{ioniz}(solution) are 0.97 to 1.08, 1.14 to 1.38, and 1.30 to 1.42 eV smaller.

The difference which bulk hydration has on the IPs of the nucleotides with and without Na^+ can be understood in terms of charge associated with the nucleotide system before and after ionization. For nucleotides without Na^+, the initial state, prior to ionization, is charged, while the final states, after ionization, are uncharged. Here bulk hydration favors the initial state causing an increase in the ionization energies. For nucleotides with Na^+, the opposite is true. Here, the initial state has a net charge of zero, while the final states have a net positive charge. In this case, bulk hydration favors the final states causing a decrease in the ionization energies.

The results in Figures 6 and 7, and Table I also indicate that bulk hydration reduces the influence of Na^+ on the nucleotide ionization energies. The large differences noted above (Table 1, Figures 6 and 7) between corresponding P_1, B_1, and S_1 ionization potentials of the nucleotides with and without Na^+, result in much smaller differences between corresponding values of ΔG_{ioniz}(solution). For example, for 5'-dGMP$^-$ and 5'-dAMP$^-$, the results of Table I and Figure 9 indicate that corresponding P_1, B_1, and S_1 values of ΔG_{ioniz}(solution), obtained with and without Na^+, differ by only 0.66 to 0.71, 0.10 to 0.26 eV and 0.09 to 0.24 eV, respectively.

Finally, the results in Figures 6, 7 and 9 demonstrate that bulk hydration influences the energetic ordering of ionization events. In the gas phase, the lowest energy ionization is associated with the phosphate group. In aqueous solution, the lowest energy ionization is associated with the base. This change in the energetic ordering also depends on charge. Here, it is important to consider the changes in nucleotide charge distributions associated with ionization of the phosphate versus base groups. For example, for nucleotides without Na^+, the formal charge before ionization (minus one) is localized on the phosphate group and gives rise to a large negative Gibbs free energy of hydration. According to Polaris 3.2 calculations, the hydration energies of 5'-dGMP$^-$

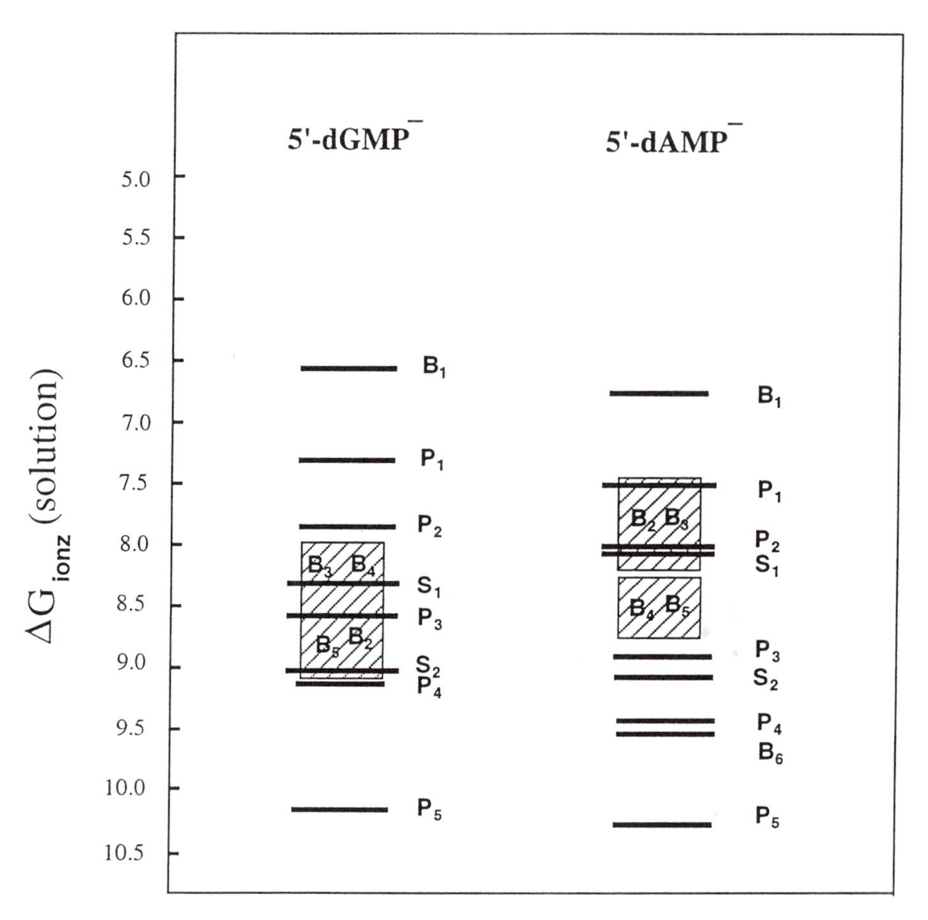

Figure 9. Free energies of ionization in solution (ΔG_{ioniz}(solution)) for 5'-dGMP$^-$ and 5'-dAMP$^-$. Orbital descriptions are the same as those given in the text and in Figures 5, 6 and 7. As in Figures 6 and 7, the hatched areas corresponds to energy regions containing orbitals in 1,9-dimethylguanine (1,9-Me_2G) and 9-methyladenine (9-MeA) that give rise to unresolved PE bands.

and 5'-dAMP$^-$ are -3.60 and -3.49 eV, respectively. Ionization at phosphate gives rise to a phosphate group which is formally neutral. Here, the final state free energy of hydration is significantly less negative. The Polaris 3.2 calculations indicate that the hydration energies for 5'-dGMP$^\cdot$ and 5'-dAMP$^\cdot$, formed via phosphate ionization, are -1.61 and -1.19 eV.

In contrast, ionization of the base results in formation of a zwitterion radical, with a negatively charged phosphate group and a positively charged base. In this case, the final state, which has significant charge separation, has a free energy of hydration that is significantly larger than that of the neutral radical formed via phosphate ionization. According to Polaris 3.2 calculations, the zwitterion radicals formed via base ionization of 5'-dGMP$^-$ and 5'-dAMP$^-$ have hydration energies of -2.82 and -3.15 eV. It is the large negative hydration energies associated with the zwitterions formed via base ionization that is responsible for the smaller base ionization energies versus phosphate ionization energies in aqueous solution.

It is interesting to consider the results for 5'-dGMP$^-$ and 5'-dAMP$^-$ in Figure 9 and Table I, in the light of experimental analyses of DNA damage induced by ionizing radiation. The results in Figure 9 and Table I indicating that for the nucleotides, both with and without an Na^+ counterion, the values of ΔG_{ioniz}(solution) are smaller for base ionization than for sugar or phosphate ionization is consistent with ESR data indicating that, in an aqueous environment at 77°K and neutral pH, γ-irradiated 5'-dGMP$^-$ and 5'-dAMP$^-$ give rise to highly stable radicals in which guanine or adenine are positively charged (48, 49). Furthermore, this finding and the observation that, at low temperatures, water geometries are largely frozen, provide evidence that electronic relaxation of water strongly influences the energetics of nucleotide ionization in aqueous solution.

In consideration of an experimental evaluation of the ionization energies of nucleotides in aqueous solution it is important to recall that the values of ΔG_{ioniz}(solution) reported in Figure 9 do not include the Gibbs free energy associated with the change-of-state in which a gas-phase electron makes a transition to the aqueous conduction band edge or to a fully solvated state. Furthermore, in considering future experiments, it is important to note that the values of ΔG_{ioniz}(solution) reported in Figure 9 have been evaluated under conditions where bulk solvent relaxation occurs. Since the time scale for solvent relaxation is significantly longer than the time scale for photoionization, (50) the experimental energetic influence of hydration, will be smaller than the $\Delta\Delta G_{hyd}$ values employed here. Nevertheless, we expect important qualitative aspects of these results to be verified by experiment.

Conclusions

The most important results to emerge from this investigation are:

1. When described at the ab initio SCF level with a split-valence basis set the valence orbital electronic structure of 5'-dTMP$^-$ is localized like those of the previously examined nucleotides 5'-dCMP$^-$, 5'-dAMP$^-$ and 5'-dGMP$^-$. This localized orbital structure of nucleotides makes it possible to employ PE data and results from post-SCF calculations on model anions to correct nucleotide valence IPs obtained from SCF

calculations. With this method, the 11 lowest energy base, sugar and phosphate IPs of 5'-dTMP⁻ and of a 5'-dTMP⁻ cluster containing five H_2O molecules and an Na^+ ion bound to phosphate have been evaluated.

2. In the gas-phase, the lowest energy ionization event in all four of the DNA nucleotides is associated with removal of an electron from phosphate. This is consistent with the observation that the phosphate group is more negatively charged than the sugar or base groups.

3. In nucleotide counterion clusters, the ionization energies are significantly different from those associated with the isolated nucleotides. The electrostatic interaction of the counterion increases the IPs of all of the highest occupied base, sugar and phosphate orbitals. Depending on the geometry of the cluster, the counterion also strongly influences the energetic ordering of the nucleotide IPs. For the 5'-dTMP⁻ and 5'-dCMP⁻ clusters examined here, in which Na^+ is bound to phosphate, the lowest energy IPs are no longer associated with the phosphate group but with the base.

4. In aqueous solution, hydration effects significantly modulate the ionization energies of nucleotides, with and without counterions. For both 5'-dGMP⁻ and 5'-dAMP⁻, with and without Na^+ bound to phosphate, the lowest energy ionization events in aqueous solution are associated with removal of an electron from the base groups. This is consistent with experimental analysis of products formed when 5'-dGMP⁻ and 5'-dAMP⁻ interact with ionizing radiation in an aqueous medium.

Acknowledgments

Support of this work by the American Cancer Society (Grant #CN-37E) is gratefully acknowledged. Computer access time has been provided by the Computer Center of the University of Illinois at Chicago, the Cornell National Supercomputer Facility, and the National Center for Supercomputing Applications, at the University of Illinois at Urbana-Champaign.

Literature Cited

1. Swenson, D. H.; Lawley, P. D. *Biochem. J.* **1978**, *171*, 575.
2. Singer, B.; Grunberger, D. *Molecular Biology of Mutagens and Carcinogens;* Plenum Press: New York, NY, 1983; pp 97-190.
3. Melvin, T.; Botchway, S. W.; Parker, A. W.; O'Neill, P. *J. Am. Chem. Soc.* **1996**, *118*, 10031.
4. Malone, M. E.; Cullis, P.M.; Symons, M. C. R.; Parker, A. W. *J. Am. Chem. Soc.* **1995**, *99*, 9299.
5. Gorner, H.; Gurzadyan, G. G. *J. Photochem. Photobiol. A: Chem.* **1993**, *71*, 155.
6. Schulte-Fronhlinde, D.; Simic, M. G.; Gorner, H. *Photochem. Photobiol.* **1990**, *52*, 1137.
7. Dandliker, P. J.; Holmlin, R. E.; Barton, J. K. *Science* **1997**, *275*, 1465.

8. Lifschitz, C.; Bergmann, E. D.; Pullman, B. *Tetrahedron Lett.* **1967**, *46*, 4583.
9. Padva, A; LeBreton, P. R.; Dinerstein, R. J.; Ridyard, J. N. A. *Biochem. Biophys. Res. Commun.* **1974**, *60*, 1262.
10. Hush, N. S.; Cheung, A. S. *Chem. Phys.* **1975**, *34*, 11.
11. Orlov, V. M.; Smirnov, A. N.; Varshavsky, Ya. M. *Tetra. Lett.* **1976**, *48*, 4377.
12. Urano, S.; Yang, X.; LeBreton, P. *J. Mol. Struct.* **1989**, *214*, 315 and refs. therein.
13. Kim, H.S.; Yu, M.; Jiang, Q.; LeBreton, P. R. *J. Am. Chem. Soc.* **1993**, *115*, 6169.
14. Tasaki, K.; Yang, X.; Urano, S.; Fetzer, S.; LeBreton, P. R. *J. Am. Chem. Soc.* **1990**, *112*, 538.
15. Cowley, A. H.; Lattman, M.; Montag, R. A.; Verade, J. E. *Inorg. Chim. Acta.* **1977**, *25*, L151.
16. Chattopadhyay, S.; Findley, G. L.; McGlynn, S. P. *J. Electron. Spectrosc. Relat. Phenom.* **1981**, *6*, 27.
17. Yu, C.; O'Donnell, T. J.; LeBreton, P. R. *J. Phys. Chem.* **1981**, *85*, 3851.
18. Heilbronner, E.; Hornung, V.; Bock, H.; Alt, H. *Angew. Chem. Int. Ed. Engl.* **1969**, *8*, 524.
19. O'Donnell, T. J.; LeBreton, P. R.; Petke, P. R.; Shipman, J. D. *J. Phys. Chem.* **1980**, *84*, 1975.
20. Kim, H.S.; LeBreton, P. R. *Proc. Natl. Acad. Sci.* **1994**, *91*, 3725.
21. Kim, H. S.; LeBreton, P. R. *J. Am. Chem. Soc.* **1996**, *118*, 3694.
22. LeBreton, P. R.; Yang, X.; Urano, S.; Fetzer, S.; Yu, M.; Leonard, N. J.; Kumar, S. *J. Am. Chem. Soc.* **1990**, *112*, 2138.
23. Lin, J.; Yu, C.; Peng, S.; Li, L.-K.; LeBreton, P. R. *J. Am. Chem. Soc.* **1980**, *102*, 4627.
24. Koopmans, T. *Physica* **1934**, *1*, 104.
25. Clark, T.; Chandrasekhar, J.; Spitznagel, G. W.; Schleyer, P. von R. *J. Compu. Chem.* **1983**, *4*, 294.
26. Frisch, M. J.; Trucks, G. W.; Schlegel, H. B.; Gill, P. M. W.; Johnson, B. G.; Robb, M. A; Cheeseman, J. R.; Keith, T. A.; Petersson, G. A.; Montgomery, J. A.; Raghavachari, K.; Al-Laham, M. A.; Ortiz, J. V.; Foresman, J. B.; Cioslowski, J.; Stefanov, B. B.; Nanayakkara, A.; Challacombe, M.; Peng, C. Y.; Ayala, P. Y.; Chen, W.; Wong, M. W.; Andres, J. L.; Replogle, E. S.; Gomperts, R.; Martin, R. L.; Fox, D. J.; Binkley, J. S.; Defrees, D. J.; Baker, J.; Stewart, J. P.; Head-Gordon, M.; Gonzalez, C.; Pople, J. A. *Gaussian 94*, Gaussian, Inc.: Pittsburgh, PA, 1995.
27. Bowen, H. J. M.; Donohue, J.; Jenkin, D. G.; Kennard, O.; Wheatley, P. J.; Wiffen, D. H. In *Tables of Interatomic Distances and Configuration in Molecules and Ions;* Sutton, L. E.; Jenkins, D. G.; Mitchell, A.D.; Cross, L. C., Eds.; The Chemical Society: London, 1958, pp M67, S7, S15-S17.
28. Padva, A.; O'Donnell, T. J.; LeBreton, P. R. *Chem. Phys. Letters.* **1976**, *41*, 278.
29. Fetzer, S. M.; LeBreton, P. R.; Rohmer, M. -M.; Veillard A. *Int. J. Quantum Chem. Quantum Biol. Symp.*, in press.
30. Hehre, W. J.; Random, L.; Schleyer, P. V. R.; Pople, J. A. *Ab Initio Molecular Orbital Theory;* Wiley: New York, NY, 1986; pp 38-40, 79, 86.

31. Siegbahn, P. E. M.; Almlof, J.; Heiberg, A.; Roos, B. O. *J. Chem. Phys.* **1981**, *74*, 2384.
32. Foresman, J.; Head-Gordon, M.; Pople, J. A.; Frisch, M. J. *J. Phys. Chem.* **1992**, *96*, 135.
33. Kim, N. S.; LeBreton, P. R. *Biospectroscopy* **1997**, *3*, 1.
34. Westhof, E. *J. Biomol. Structure and Dynamics.* **1987**, *5*, 581.
35. Seibel, G. L.; Singh, U. C.; Kollman, P.A. *Proc. Natl. Acad. Sci. USA.* **1985**, *82*, 6537.
36. York, D. M.; Darden, T.; Deerfield, D.; Pedersen, L. G. *Int. J. Quantum Chem.: Quantum Biol. Symp.* **1992**, *19*, 145.
37. Guldbrand, L. E.; Forester, T. R.; Lynden-Bell, R. M. *Molec. Phys.* **1989**, *67*, 473.
38. Strzelecka T. E.; Rill, R. L. *J. Phys. Chem.* **1992**, *96*, 7796.
39. Seeman, N. C.; Rosenberg, J. M.; Suddath, F. L.; Kim, J. J. P.; Rich, A. *J. Mol. Biol.* **1976**, *104*, 109.
40. Rosenberg, J. M., Seeman, N. C.; Day, R. A.; Rich, A. *J. Mol. Biol.* **1976**, *104*, 145.
41. Clementi E.; Corongiu, G. *Biopolymers* **1982**, *21*, 763.
42. Lee, W. K.; Gao, Y.; Prohofsky, E. W. *Biopolymers* **1984**, *23*, 257.
43. Schneider, B.; Cohen, D.; Berman, B. *Biopolymers* **1992**, *32*, 725.
44. Bartmess, H. J. M. *J. Phys. Chem.* **1994**, *98*, 6420.
45. Atkins, P. W., *Physical Chemistry*, 5th ed., W. H. Freeman and Company: New York, NY, 1994, p. 758.
46. Lee, F. S.; Chu, Z. T.; Warshel, A. *J. Compu. Chem.* **1993**, *14*, 161.
47. A. Warshel A.; Åqvist, J. *Ann. Rev. Biophys. Biophys. Chem.* **1991**, *20*, 267.
48. Gregoli, S.; Olast, M.; Bertinchamps, A. *Radiat. Res.* **1974**, *60*, 388.
49. Gregoli, S.; Olast, M.; Bertinchamps, A. *Radiat. Res.* **1977**, *72*, 201.
50. Bernas, A.; Grand, D. *J. Phys. Chem.* **1994**, *98*, 3440.

Chapter 3

Parameterization and Simulation of the Physical Properties of Phosphorothioate Nucleic Acids

Kenneth E. Lind[1], Luke D. Sherlin[1], Venkatraman Mohan[2], Richard H. Griffey[2], and David M. Ferguson[1]

[1]Department of Medicinal Chemistry, University of Minnesota, Minneapolis, MN 55455
[2]ISIS Pharmaceuticals, 2922 Faraday Avenue, Carlsbad, CA 92008

The physical properties of nucleic acid complexes containing phosphorothioate backbone modifications are studied using molecular mechanics and dynamics calculations. Parameters for the phosphorothioate oligonucleotide are derived from *ab initio* calculations in a manner consistent with the AMBER 4.1 force field database. The force field is applied to simulate the structural properties of hybrid DNA:RNA duplexes starting in both the A- and B-form geometries. The results show the phosphorothioate-DNA:RNA complex has an overall A-form geometry with minor groove widths between A- and B-form. Although model compound calculations indicate the sulfur substitution increases torsional flexibility around the phosphorous, molecular dynamics simulations show the modification does not have a great effect on backbone geometry. The results are also compared with previous studies of standard DNA:RNA hybrid structures. While a wide range of sugar puckers are typically associated with the DNA strand of hybrid duplexes, the average structures reported here show C3'-endo puckering, suggesting phosphorothioate substitutions may influence sugar conformation.

Background

Over the past few years there have been many advances in the design and characterization of antisense oligonucleotides for the treatment of various human diseases. These short, exogenous nucleic acid strands are designed to bind complementary messenger ribonucleic acids (mRNA) with high affinity and selectivity within the cell, thereby halting translation and/or promoting degradation of the mRNA strand by ribonuclease H (RNase H) (*1-3*). Although naturally occurring nucleic acids are effective templates for the design of antisense drugs, their utility is limited by several factors. These include poor penetration through cellular membranes, degradation by naturally occurring nuclease enzymes, and low affinity for the target mRNA sequence (*4,5*). To overcome these problems, synthetic modifications typically are proposed to alter the nucleotide (*4*). Common strategies include modifications to the phosphate backbone, alteration of the sugar ring (especially the 2' position), and base substitutions.

Although a wide variety of modified nucleic acids are now available that show high affinity for RNA as well as nuclease resistance, most cause a loss in RNase H activity toward the target mRNA strand (*6,7*). This appears to be linked to the structure of the modified-DNA:RNA hybrid. Several studies have shown that RNase H most likely recognizes the unique structural features of naturally occurring DNA:RNA hybrids (*8-13*). These duplexes tend to have an overall A-form geometry, with unique differences in each strand. The RNA strand has characteristics of A-form geometry, such as a northern sugar pucker, while the DNA strand has a unique form that is a mixture of both A- and B-type geometries. The shape of the minor groove may be especially critical, lying somewhere between ideal A- and B-form geometries, as seen by the positions of the phosphate groups and the width of the groove. It is hypothesized that these unique features of the hybrid duplex allow RNase H to distinguish DNA:RNA from pure B-form DNA:DNA and A-form RNA:RNA geometries (*8,13*). In fact, it does appear that the enzyme recognizes and binds to the minor groove of the hybrid duplex.

Some oligonucleotide modifications, however, do support RNase H activity. One of the first generation of antisense drugs developed, phosphorothioates (shown in Figure 1), falls in this category (*14*). While this

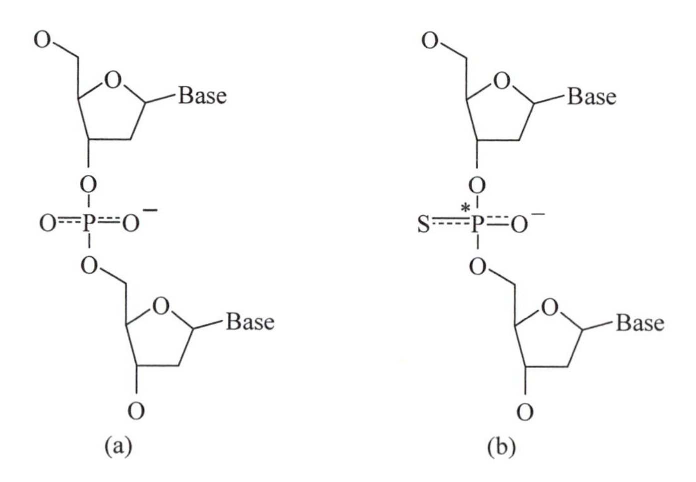

Figure 1. (a) Phosphate and (b) Phosphorothioate backbones.

change commonly is thought to have only a minor effect on the structure of the nucleic acid, phosphorothioates show a reduced binding affinity to complementary RNA (*15*). In addition, the affinity depends on the chirality of the phosphorothioate center. The *R* configuration (where the sulfur atom points into the major groove) has a higher affinity for the target RNA as compared to the *S* configuration (where the sulfur points outward from the duplex) (*16*).

Although phosphorothioates hold significant promise for the development of antisense therapies, the physical properties of these oligonucleotides, especially as they pertain to structure and function, are not well defined. Experimental structural data is simply not yet available for phosphorothioate-DNA:RNA hybrids. In the present study, we examine the structural properties of phosphorothioate oligonucleotides using a combination of *ab initio*, molecular mechanics, and dynamics techniques. Model compounds are used to parameterize the modified backbone consistent with the AMBER 4.1 force field database for simulating proteins and nucleic acids. The new force field is applied to simulate the average properties of a nucleic acid sequence taken from a library of antisense targets. Comparisons are made between phosphorothioate-DNA:RNA simulations started in both A- and B-form geometries. The results are further compared to experimental and theoretical studies of standard-DNA:RNA hybrids to identify structural features that may be induced by the modified oligonucleotide.

Parameterization

The model compounds used to parameterize the phosphorothioate backbone are shown in Figure 2. For completeness here, we have also re-evaluated the analogous phosphate fragment using the protocols reported by Cornell et al. in the development of the AMBER 4.1 force field database (*17*). Partial atomic charges for the phosphorothioate fragment were initially derived from *ab initio* calculations at the HF/3-21G* level (*18*). These values were further evaluated by fitting the 6-31G* electrostatic potential with the RESP algorithm (*19*). The original set (performed before the availability of the RESP program) was virtually identical to the RESP-fit set which is not surprising considering the size of the fragments. The partial charges derived for the fragments are shown in Figure 2. These have been included with the standard residue charges given in the AMBER 4.1 database. No other changes were made to the existing partial charges in the database. As might be expected, the sulfur substitution reduces the partial charge on the phosphorous center, slightly reducing the P-O dipole moments of the modified center.

The van der Waals parameters for the phosphate sp^2 oxygen atom (standard type O2) and phosphorothioate sulfur atom (new type SD) were modified based on water docking studies. The HF/6-31G* optimized *gauche-/gauche-* fragment was aligned with a water molecule such that the PO_2 or POS angle was bifurcated as shown in Figure 3. The water molecule was then manually adjusted along the water oxygen to phosphate non-bond vector while the rest of the molecule was held fixed to determine the energy as a function of distance. The

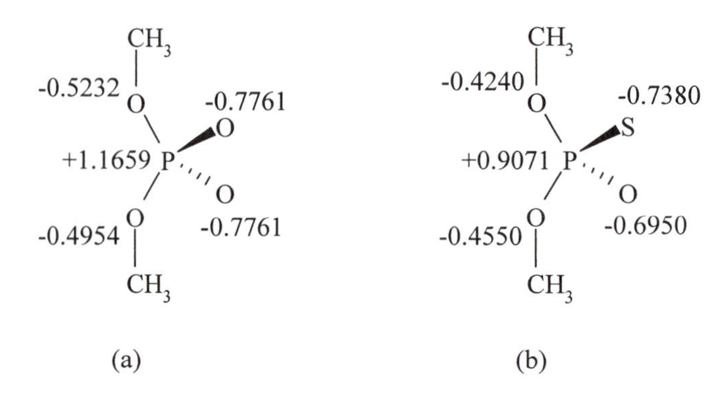

Figure 2. Model compounds used to parameterize the AMBER 4.1 force field database. Partial atomic charges for (a) current AMBER phosphate, and (b) new phosphorothioate for inclusion with existing nucleic acids.

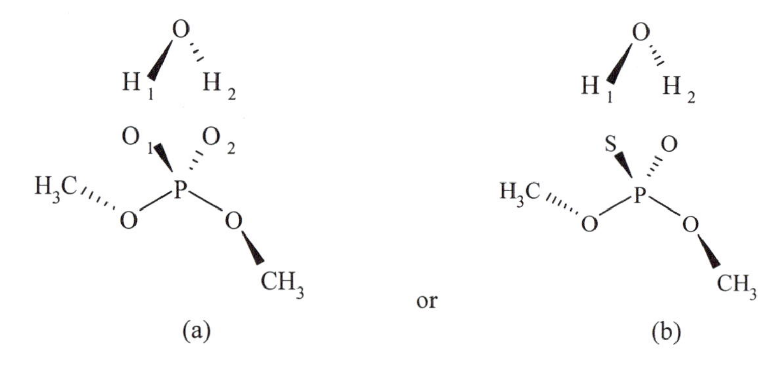

Figure 3. Water docking setup for (a) dimethyl phosphate and (b) dimethyl phosphorothioate models.

energy profile was then calculated using MP2/6-31G* single points and corrected to account for the monomer energies. This set of calculations was repeated using molecular mechanics with default van der Waals parameters taken from the AMBER 4.1 database. To reproduce the profiles, the R* and epsilon values for O2 (of the phosphate fragment) were increased from 1.66 to 1.768 Å and from 0.21 to 0.25 kcal/mol, respectively. The SD R* and epsilon were started from the values for atom type S of the AMBER 4.1 database and increased from 2.00 to 2.385 Å and from 0.25 to 0.50 kcal/mol, respectively. The resulting energy profiles are compared in Figure 4.

These non-bonded parameters were subsequently applied to evaluate the torsional energy profile of the two model compounds. Following the methodology outlined by Cornell et al., the low energy conformers of the

conformational energies calculated at the MP2/6-31G* level. This proved to be more difficult than we anticipated due to the asymmetry of the phosphorothioate fragment. In addition, the phosphorothioate energy surface is significantly flatter than the standard phosphate energy surface making direct comparisons of coefficients complex. The final energies are summarized in Table I.

For all model compound calculations, bond stretching and angle bending terms were taken from the AMBER 4.1 force field database. The missing terms involving the S-P bond and S-P-X angles were taken from the *ab initio* optimized dimethylester shown in Figure 2 were built and evaluated using both *ab initio* and molecular mechanics calculations. The four low energy forms, *trans/trans* (180°/180°), *trans/gauche+* (180°/60°), *trans/gauche-* (180°/300°), and *gauche-/gauche-* (300°/300°) are shown in Figure 5. The conformers were optimized at the HF/6-31G* level. The evaluated MP2/6-31G* single point energies are reported in Table I. Although these values show slight differences with those reported by Cornell et al. for the dimethyl phosphate conformers, the trends are the same, with the *gauche/gauche* conformer in the lowest energy form. Since we modified the O2 van der Waals parameters, the OS-P-OS-CT torsional coefficients for the standard phosphate backbone also required reparameterization. Based on the *ab initio* energies, the V1, V2, and V3 coefficients were adjusted starting from the default parameters taken from the AMBER 4.1 database. In

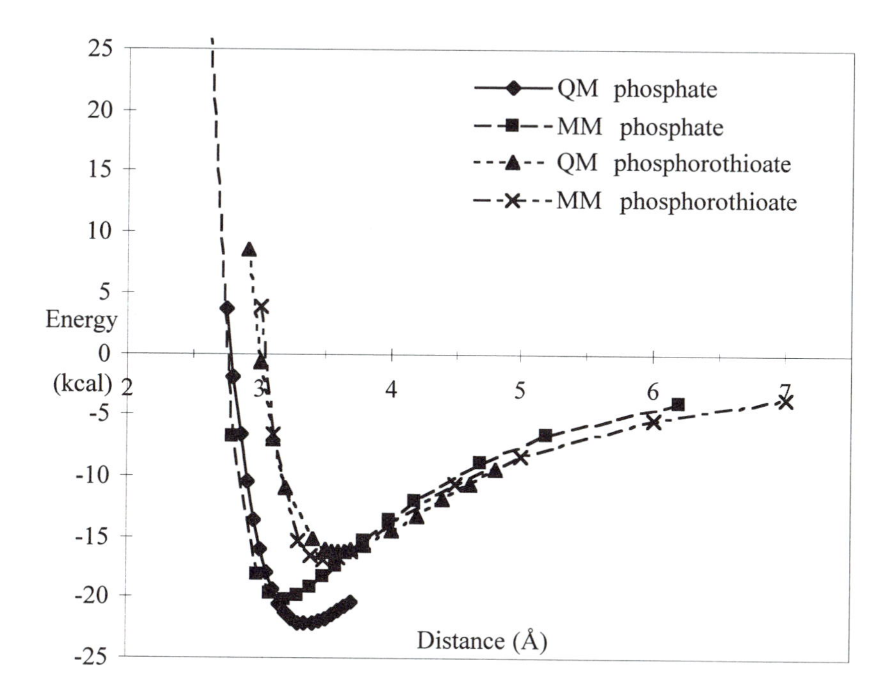

Figure 4. MP2/6-31G* (QM) interaction energy and new molecular mechanics (MM) interaction energy for dimethyl phosphate and dimethyl phosphorothioate with a single water molecule.

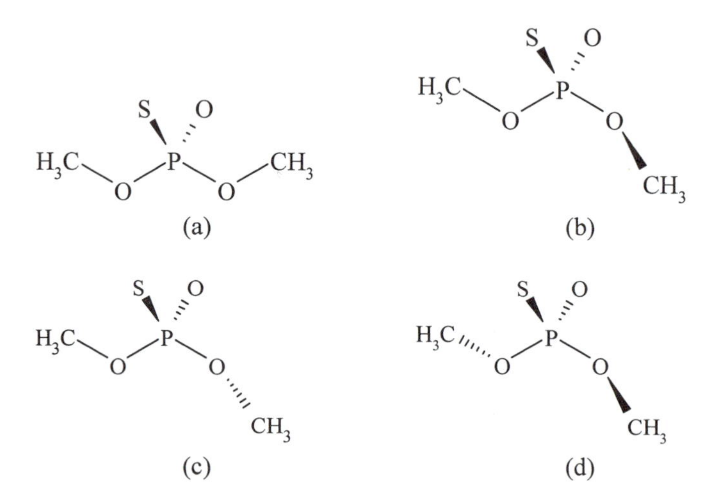

Figure 5. Dimethyl phosphorothioate model fragments: (a) *trans/trans* conformation, (b) *trans/gauche+* conformation, (c) *trans/gauche-* conformation, and (d) *gauche-/gauche-* conformation. Phosphate analogues of (b) and (c) conformers are identical due to symmetry.

Table I: Conformational energies for dimethyl phosphate and dimethyl phosphorothioate fragments.[a]

Conformation	HF/ 6-31G*	MP2/ 6-31G*	AMBER New	AMBER Original	HF/ 6-31G*	MP2/ 6-31G*	AMBER New
		Phosphate				Phosphorothioate	
trans/trans	0.00	0.00	0.00	0.00	0.00	0.00	0.00
trans/gauche+	-1.28	-1.74	-1.60	-1.41	-0.02	-0.27	-0.30
trans/gauche-		-- [b]			-0.59	-1.03	-1.04
gauche/gauche	-1.57	-2.85	-2.86	-2.83	-0.46	-0.89	-0.82

[a] kcal. [b] *trans/gauche+* and *trans/gauche-* values identical for phosphate due to symmetry of $PO_2(OCH_3)_2^-$.

contrast to the parameterization of the phosphate, the torsional coefficients for the phosphorothioate were initially set to zero, and fit using dihedral driver calculations with the SPASMS module of AMBER. An iterative process was performed in which we first measured the torsional profile due to van der Waals and electrostatics, and second adjusted V1, V2, and V3 to reproduce the geometries. The force constants were estimated from the standard O2-P and O2-P-X parameters based on trends noted in ether and thiol parameters reported in the AMBER 4.1 database. We used a simple percent reduction in sulfur versus oxygen stretching and bending which worked surprisingly well. The values were checked using a normal mode analysis with the NMODE module of AMBER. The predicted experimental frequencies were compared with *ab initio* frequencies calculated at the HF/3-21G* level. Good agreement was obtained using a uniform scaling factor of 0.89 for the *ab initio* values.

The complete set of new parameters for the phosphate and phosphorothioate model compounds is given in Table II. The new atom type (SD) and parameters have been added to the AMBER 4.1 force field database for simulating proteins and nucleic acids. For completeness, we have also included a comparison of geometrical parameters taken from optimized model compound structures using the updated AMBER 4.1 force field and HF/6-31G* basis set calculations. These values are given in Table III.

Table II: New parameters for phosphate and phosphorothioate fragments.[a]

Parameter	Value		
van der Waals:	R*, Å	ε, kcal/mol	
O2	1.7680 (1.66)	0.2500 (0.21)	
SD[b]	2.3850 (2.00)	0.5000 (0.25)	
Bonds:	r_{eq}, Å	K_r, kcal/(mol Å^2)	
SD-P	1.960	420.0	
Angles:	θ_{eq}	K_θ, kcal/(mol rad^2)	
O2-P-SD	122.90°	112.0	
OS-P-SD	108.10°	80.0	
Torsions:	$V_n/2$ [c]	γ [d]	n [e]
OS-P -OS-CT	1.00 (0.25)	0.0	3
	0.25 (1.20)	0.0	2
	1.25 (0.00)	180.0	1
SD-P -OS-CT	0.15	0.0	3
	1.05	0.0	2
	0.25	0.0	1

[a] Original values from AMBER 4.1 force field (17) given in parentheses. [b] New sulfur compared to standard sulfur atom of AMBER. [c] Magnitude of torsion in kcal/mol. [d] Phase offset in deg. [e] Periodicity of the torsion.

Table III: Selected Geometrical Parameters.

	MM[a]	QM[b]
Dimethyl Phosphate		
r (OS-P) (Å)	1.62	1.64
r (O2-P) (Å)	1.48	1.47
θ (OS-P-O2)	108.14°	107.41°
θ (OS-P-OS)	103.12°	99.63°
θ (O2-P-O2)	119.92°	124.76°
Dimethyl Phosphorothioate		
r (OS-P) (Å)	1.62	1.63
r (O2-P) (Å)	1.48	1.47
r (SD-P) (Å)	1.96	1.98
θ (OS-P-O2)	107.02°	108.13°
θ (OS-P-SD)	108.27°	108.18°
θ (OS-P-OS)	103.44°	99.57°
θ (O2-P-SD)	121.42°	122.32°

[a] AMBER optimized geometry with new parameters. [b] HF/6-31G* optimized geometry.

Molecular Simulations

Methods

Starting A-form and B-form structures of the sequence d[CCTATAATCC]-r[GGAUUAUAGG] were model built using the NUCGEN module of the AMBER 4.1 package (*20*). The DNA strand was created with an *R* configuration phosphorothioate backbone (designated as ps-DNA) and the RNA strand with a standard phosphate backbone. This setup causes the sulfur atoms to point into the major groove of the duplex. These structures were minimized in vacuo to relax the hydrogen atoms after which counter ions were added in the EDIT module of AMBER to neutralize the net charge. The resulting structure was solvated with explicit TIP3P water molecules to surround the nucleic acid by approximately 9Å in each direction. The B-form structure contained 8210 atoms in a box approximately 53Å x 40Å x 40Å. The A-form structure was composed of 7721 atoms and was contained in a box approximately 56Å x 37Å x 37 Å.

Initially, the SANDER module of AMBER 4.1 was used to minimize the water and counter ions to an energy convergence of <0.1 kcal/mol*Å while the nucleic acid molecule was held fixed. The water-counter ion system was then allowed to equilibrate using molecular dynamics for 10 picoseconds. For this and subsequent simulations, the SANDER module was used with the SHAKE algorithm (*21*) applied to hydrogen atoms, a 1 femtosecond (fs) timestep, and a temperature of 300 K with the Berendsen temperature coupling algorithm (*22*). A distance cutoff of 8.0 Å was used for all non-bonded interactions, and the pairlist was updated every 25 steps.

The entire structure was minimized with 10.0 kcal/mol*Å^2 positional constraints on the nucleotide base nitrogen atoms to an energy gradient convergence within 0.1 kcal/mol*Å. These constraints were applied for 10 ps of dynamics and were reduced to 1 kcal/mol*Å^2 for 10 ps, and then to 0.1 kcal/mol*Å^2 for 10 ps. Simulations were carried out using the particle mesh Ewald (PME) (*23*) option of AMBER 4.1 for treatment of non-bond electrostatic interactions. The grid was constructed of cubes of approximately 1.0Å^3. The order of the B-spline interpolation was 3 and the direct sum tolerance was set to $1x10^{-5}$. Simulations have been carried out for 300 picoseconds over which trajectories were recorded every picosecond for analysis. The CARNAL module of AMBER 4.1 was used to analyze the trajectories.

Results and Discussion

Figure 6 reports the structural analyses performed over the 300 ps trajectory simulated for the two oligonucleotide starting conformations. The results show both structures converge rather quickly to a stable geometry (~50 ps). Analysis of RMS deviations indicates that the simulation starting with an A-form geometry maintains its structure for the duration of the simulation. In contrast, the B-form structure rapidly moves away from its initial structure, and then remains at a shape partially between A-form and B-form as shown in Figure 6a. Although the B-form structure appears to be progressing towards an A-form geometry, the structure still displays the physical properties more characteristic of B-form geometries. This is especially true of the groove width, which remains fairly close to the ideal B-form distance. It is important to point out, however, that DNA:RNA hybrids have been shown to lie somewhere between ideal A- and B-like geometries. NMR (*8-12*) and x-ray (*13*) structural studies of hybrid DNA:RNA duplexes indicate that the minor groove tends to have a width intermediate between a pure RNA:RNA A-form duplex and DNA:DNA B-form duplex. Since both of our simulations led to structures with this general characteristic, some caution should be applied in interpreting the results.

One of the most interesting results can be found in the sugar pucker data. Sugar puckering is, of course, the classic measure of A- and B-form geometries in nucleic acid studies (*24*). Duplexes with ribose sugars generally have a C3'-endo sugar pucker or northern pseudorotation while deoxyribose sugars tend to adopt a C2'-endo sugar pucker or southern pseudorotation which defines the B-form geometry. The simulations show characteristic C3'-endo puckering for both strands of the A-form structure throughout the trajectory. The B-form structure, however, displays C3'-endo puckering for the phosphorothioate strand and C2'-endo puckering for the RNA strand, precisely the opposite of what might be

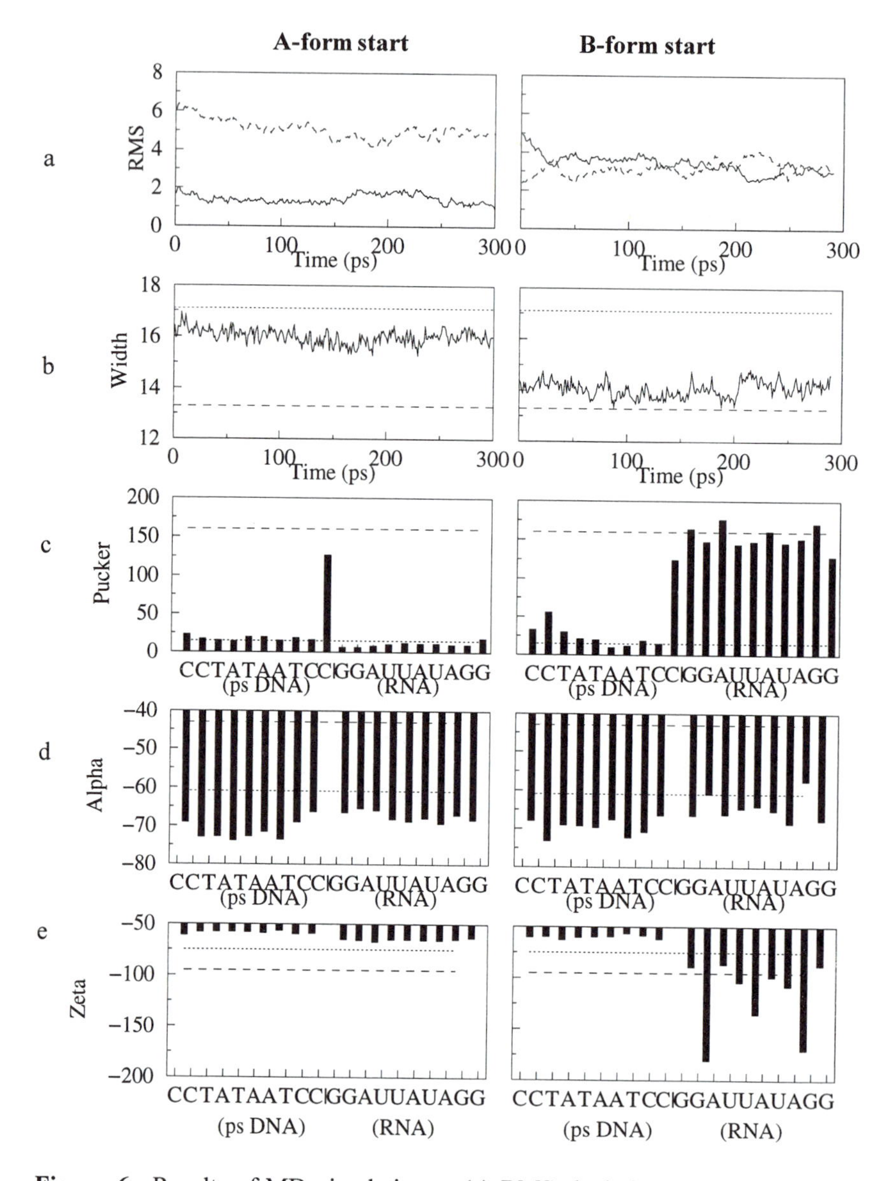

Figure 6. Results of MD simulations: (a) RMS deviations from canonical A-form geometry (solid line) and canonical B-form geometry (dashed line). (b) Average minor groove width during simulation in Å. (c) Average sugar pucker pseudorotation angle for each residue in deg. (d) Average alpha (C3-O3-P-O5) torsion angle between each residue in deg. (e) Average zeta (O3-P-O5-C5) torsion angle between each residue in deg. Dashed lines represent canonical B-form values; dotted lines represent canonical A-form values.

expected. This apparent anomaly can be partly explained by a closer examination of the starting geometries. When the RNA strand is started in canonical B-form, the geometrical constraints place the 2' hydroxyl into the major groove. To achieve the A-form structure, the hydroxyl group must flip into the minor groove. The flanking bases apparently hinder this transition. In the work reported by Cheatham and Kollman (*25*), such an interconversion was effected for RNA:RNA using large constraints (300 kcal/Å^2) to enforce A-form puckering values. Based on energetic differences for the repuckering of ribose versus deoxyribose rings, they estimated the interconversion time (without constraints) to be approximately an order of magnitude longer than current simulation time limits (~10-20 ns). Our results further support this hypothesis. Figure 7 shows a snapshot of the B-form trajectory. The 2'-hydroxyl group starts in a geometry facing the major groove and remains in this conformation for the duration of the simulation. In marked contrast, the deoxyribose phosphorothioate strand has interconverted to all A-type puckering.

Figure 6 also reports the average torsion angles, alpha (C3'-O3'-P'-O5') and zeta (O3'-P-O5'-C5'), for the nucleic acid backbones. In standard A and B form duplexes, these dihedrals are both *gauche-* (approximately -60°). Although the model compound calculations indicate the phosphorothioate backbone may prefer to be in a more extended conformation, *trans/gauche-*, the simulations do not show this behavior, suggesting the oligonucleotide structure imposes some conformational constraints on the backbone (albeit indirectly). The torsional profile about the phosphorous may have changed in the nucleotide as well (as compared to the model compound) due to the neighboring sugar atoms and proximity of the bases. These polar groups clearly have an effect on the

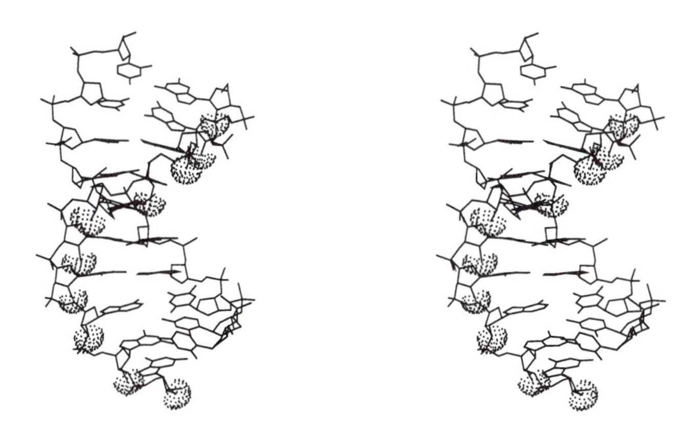

Figure 7. Protrusion of 2' hydroxyl groups (shown as van der Waals surface) into the major groove of the B-form starting structure after 300 ps simulation.

electrostatic energy of the backbone. Oddly, the only torsions that show *trans* configurations occur in the RNA strand of the B-form simulation. This is most likely linked to the B- to A-form interconversion barrier alluded to above. The backbone may be flexing to accommodate strain in the system produced as the sugars attempt to repucker to C3'-endo geometries.

Concluding Remarks

This study has presented the development and application of force field parameters to study the structural properties of phosphorothioate nucleic acids. Overall, the results indicate phosphorothioate oligonucleotides display similar structural characteristics to standard DNA. In comparing the sugar puckering of the phosphorothioate strand to experimental data from NMR and x-ray studies of DNA:RNA hybrids, however, the pseudorotation angles do not show the variable range observed for the DNA strand. Cheatham and Kollman have also reported a larger range of puckers for simulations of standard-DNA:RNA hybrids. This suggests that the phosphorothioate may have some influence on the sugar puckering which may, in part, be due to the differences noted in the energy surface of the backbone model compounds. The phosphorothioate backbone model does show a flatter energy surface than the standard phosphate suggesting the former may be more flexible. The result may also be a sequence specific effect or an artifact of the simulation time frame and starting geometry. Since structural data has not yet been reported for an all-phosphorothioate substituted duplex, no conclusive evidence is available to resolve this issue. Nevertheless, the result is provocative and suggests backbone substitutions may predispose oligonucleotides to particular conformations.

The results of this study may also help explain the physical basis for the lowered melting temperatures noted for phosphorothioate-DNA:RNA hybrids as compared to standard DNA:RNA hybrids. The differences in the energy surfaces reported here for the backbone models suggests the phosphorothioate single strand may have more configurations available, or configurational entropy than the phosphate single strand. Since the phosphorothioate-DNA:RNA simulations indicate the backbone shows fairly standard behavior, we would expect a net loss in free energy upon complexation as compared to the process for standard oligonucleotides. This, of course, does not account for differences in the hydration of the phosphorothioate, although one could argue this effect may be similar in both the single strand and duplex structures. It is important to note that the stability has also been linked to the chirality of the phosphorothioate center (*16*). Since we expect the configurational entropy to be approximately the same for the *R* and *S* stereoisomers, the differences noted are most likely due to changes in hydration or electrostatics when the sulfur points into (*R*) or away (*S*) from the minor groove. These contributions, however, appear to be minor since the melting point differences reported for *R* versus *S* configured oligonucleotides are small. Although this tends to support our hypothesis that configurational entropy may be a dominant thermodynamic force in destabilizing phosphorothioates, additional theoretical work is required to fully explore these properties.

Finally, the analysis of the B-form trajectory reported in Figure 6 provides further evidence that B-form geometries may "lock" RNA in C2'-endo sugar puckering. This behavior was first noted by Cheatham and Kollman in simulations of RNA:RNA duplex structures. Our results show only one strand of the DNA:RNA hybrid repuckering from C2' to C3'-endo, the DNA strand. Although it is fairly clear that the 2'-hydroxyl is responsible for the interconversion barrier noted in previous studies (and in our analysis here), the physical significance of this barrier is questionable. Since these "B-form" RNA structures do not appear to be observed experimentally, they are most likely an artifact of the simulation timeframe or force field. While the problem is easily avoided by careful system setup, the failure of 2'-substituted sugars to interconvert from B- to A-form may limit modeling experiments of non-standard structures. Additional work in this area is therefore necessary to quantify the barrier height and resolve the source of the anomalous behavior.

Acknowledgments

The authors thank the American Chemical Society for organizing this symposium series. Special thanks are due to Neocles Leontis and John SantaLucia, Jr. This work has been supported in part by a grant from ISIS Pharmaceuticals. K.E.L. would like to acknowledge support from NIH Training Grant GM07994. We would also like to thank the National Cancer Institute for allocation of computing time and staff support at the Frederick Biomedical Supercomputing Center.

Literature Cited

1. Crooke, S.T. *Med. Res. Rev.* **1996**, *16*, 319-344.
2. Bonham, M.A., Brown, S., Boyd, A.L., Brown, P.H., Bruckenstein, D.A., Hanvey, J.C., Thomson, S.A., Pipe, A, Hassman, F., Bisi, J.E., Froehler, B.C., Matteucci, M.D., Wagner, R.W., Noble, S.A., Babiss, L.E. *Nucleic Acids Res.* **1995**, *23*, 1197-1203.
3. Baker, B.F., Lot, S.L., Condon, T.P., Cheng-Flournoy, S., Lesnik, E.A., Sasmor, H.M., Bennett, C.F. *J. Biol. Chem.* **1997**, *272*, 11944-12000.
4. Mesmaeker, A.D.; Haner, R.; Martin, P.; Moser, H.E. *Acc. Chem. Res.* **1995**, *28*, 366-374.
5. Crooke, S.T. *Curr. Opin. Biotechnol.* **1992**, *3*, 656-661.
6. Lima, W.F.; Crooke, S.T. *Biochemistry* **1997**, *36*, 390-398.
7. Kanaya, S.; Ikehara, M. *Subcell. Biochem.* **1995**, *24*, 377-422.
8. Nakamura, H.; Yasushi, O.; Shigenori, I.; Inoue, H.; Ohtsuka, E.; Kanaya, S.; Kimura, S.; Katsuda, C.; Katayangi, K.; Mrikawa, K.; Miyashiro, H.; Ikehara, M. *Proc. Nat. Acad. Sci.* **1991**, *88*, 11535-11539.
9. Lane, A.N.; Ebel, S.; Brown, T. *Eur. J. Biochem.* **1993**, *215*, 297-306.
10. Fedoroff, O.Y.; Salazar, M.; Reid, B.R. *J. Mol. Biol.* **1993**, *233*, 509-523.
11. Iwai, S.; Kataoka, S.; Wakasa, M.; Ohtsuka, E.; Nakamura, H. *FEBS Lett.* **1995**, *368*, 315-320.

12. Gonzalez, C.; Stec, W.; Reynolds, M.A.; James, T.L. *Biochemistry* **1995**, *34*, 4969-4982.
13. Horton, N.C.; Finzel, B.C. *J. Mol. Biol.* **1996**, *264*, 521-533.
14. Matsukura, M.; Shinozuka, K.; Zon, G.; Mitsuya, H.; Reitz, M.; Cohen, J.S.; Broder, S. *Proc. Nat. Acad. Sci.* **1987**, *84*, 7706-7710.
15. Ghosh, M.K.; Ghosh, K.; Dahl, O.; Cohen, J.S. *Nucleic Acid Res.* **1993**, *21*, 5761-5197.
16. Koziolkiewicz, M.; Krakowiak, A.; Kwinkowski, M.; Boczkowska, M.; Stec, W.J. *Nucleic Acids Res.* **1995**, *23*, 5000-5005.
17. Cornell, W.D.; Cieplak, P.; Bayly, C.I.; Gould, I.R.; Merz, K.M.; Ferguson, D.M.; Spellmeyer, D.C.; Fox, T.; Caldwell, J.W.; Kollman, P.A. *J. Am. Chem. Soc.* **1995**, *117*, 5179-5197.
18. Frisch, M. J.; Trucks, G. W.; Schlegel, H. B.; Gill, P. M. W.; Johnson, B. G.; Wong, M. W.; Foresman, J. B.; Robb, M. A.; Head-Gordon, M.; Replogle, E. S.; Gomperts, R.; Andres, J. L.; Raghavachari, K.; Binkley, J. S.; Gonzalez, C.; Martin, R. L.; Fox, D. J.; Defrees, D. J.; Baker, J.; Stewart, J. J. P.; Pople, J. A. *Gaussian 92/DFT*, Revision G.4, Gaussian, Inc., Pittsburgh PA, 1993.
19. Cieplak, P.; Cornell, W.D.; Bayly, C.; Kollman, P.A. *J. Comp. Chem.* **1995**, *16*, 1357-1377.
20. Pearlman, D.A.; Case, D.A.; Caldwell, J.W.; Ross, W.R.; Cheatham, T.E. III; Ferguson, D.M.; Seibel, G.L.; Singh, U.C.; Weiner, P.K.; Kollman, P.A. *AMBER 4.1*; University of California, San Francisco, 1995.
21. Ryckaert, J.P.; Ciccotti, G.; Berendsen, H.J.C. *J. Comp. Phys.* **1977**, *23*, 327-341.
22. Berendsen, H.J.C.; Postma, J.P.M.; van Gunsteren, W.F.; DiNola, A.; Haak, J.R. *J. Comp. Phys.* **1984**, *81*, 3684-3690.
23. Essman, U.; Perera, L.; Berkowitz, M.L.; Darden, T.; Lee, H.; Pedersen, L.G. *J. Am. Chem. Soc.* **1995**, *103*, 8577-8593.
24. Saenger, W. *Principles of Nucleic Acid Structure*; Springer-Verlag;New York, 1984.
25. Cheatham, T.E., Kollman, P.A. *J. Am. Chem. Soc.* **1997**, *119*, 4805-4825.

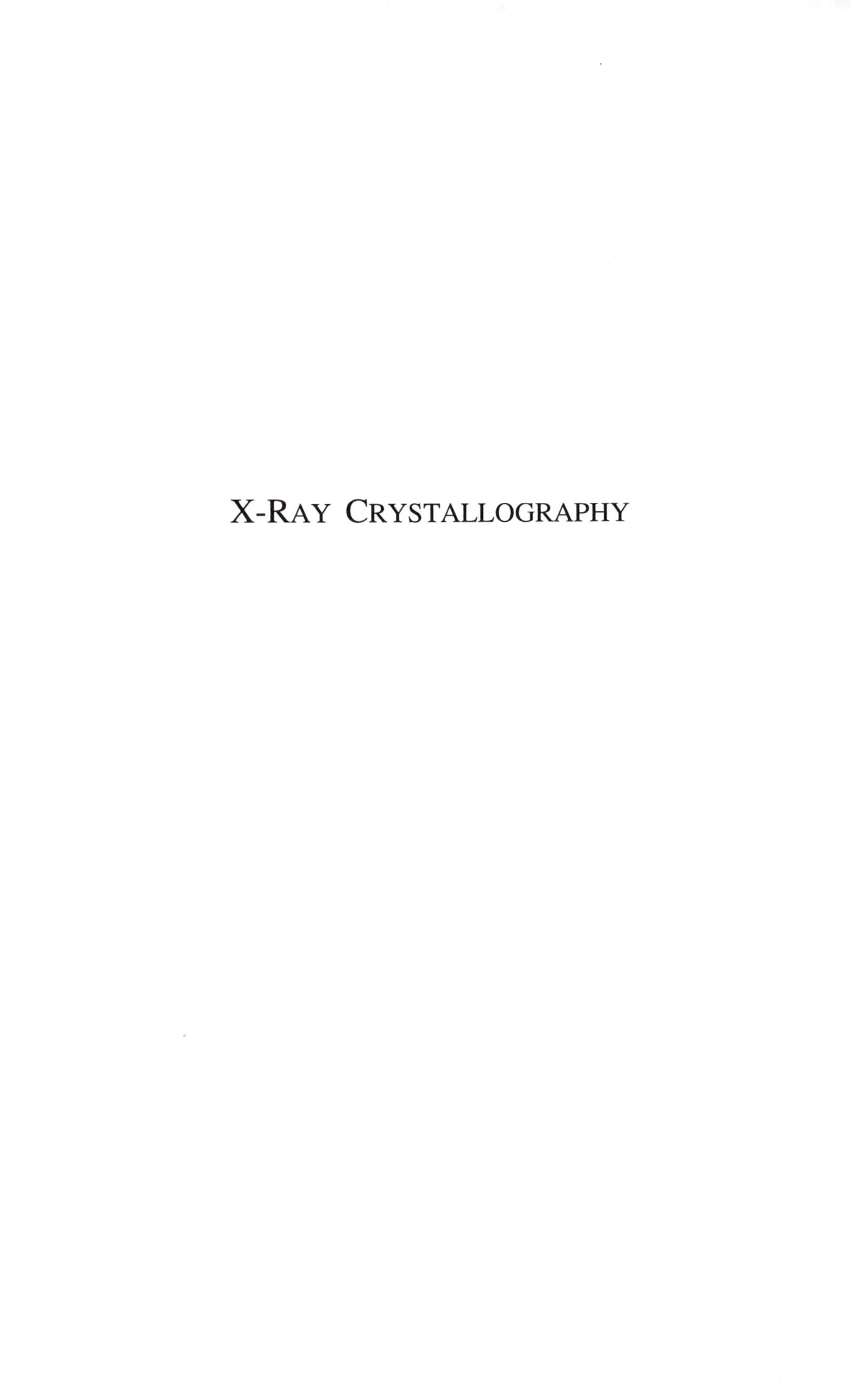

X-Ray Crystallography

Chapter 4

Crystallographic Studies of RNA Internal Loops

Stephen R. Holbrook

Structural Biology Division, Lawrence Berkeley National Laboratory, Melvin Calvin Building, University of California at Berkeley, Berkeley, CA 94720

Internal loops are common and ubiquitous elements of RNA secondary structure. They often form binding sites for metals, proteins, and other RNAs. To understand the effects of internal loops on RNA structure and function, we have determined the crystal and molecular structures of several RNA oligomers which include symmetric internal loops in the double helices formed in the crystal. The internal loops formed by these RNA oligomers generally continue the A-form double helices through formation of non-standard base pairs. Water and divalent metal ions are important for stabilization of some of these non-Watson-Crick base pairs. The distortion from a standard A-RNA helix both within the internal loop and propagated to the surrounding Watson-Crick duplexes varies dramatically depending on the internal loop sequence. The symmetric internal loops observed in these oligomers are compared to the asymmetric internal loops observed in the crystal structure of the P4-P6 domain of the group I catalytic intron. Our limited database is already suggesting improvements in RNA structure prediction algorithms and is providing the basis for comparative model building using phylogenetic information contained in the RNA sequence databases.

RNA double helices are frequently interrupted by short sequences of nucleotides on both strands which cannot form standard Watson-Crick base pairs. These regions which connect two Watson-Crick helices are historically referred to as 'internal loops' in analogy to hairpin loops which connect the ends of helices. A symmetric internal loop has an equal number of nucleotides on opposing strands, while an asymmetric internal loop has a different number of nucleotides on the two strands. Figure 1 schematically defines internal loops.

Internal loops are extremely common in biological RNA molecules. They are prevalent in ribosomal RNA, viroids, mRNA, retroviral RNA, snRNA, catalytic RNA and SELEX evolved RNAs. A large percentage of the secondary structure of ribosomal RNA, has been assigned as internal loop by comparative methods. In *E. coli* 16S rRNA, 11% of the total residues (167 nucleotides) are found in 50 internal loops, while human 16S rRNA has 56 internal loops containing 227 nucleotides (12% of the total). In 23S rRNA from *E. coli,* there are 97 internal loops containing 411 nucleotides or 14% of the total. Table I summarizes the occurrence and composition of internal loops in ribosomal RNA from diverse organisms. This Table shows that internal loops primarily consist of purine nucleotides, particularly adenosine. From 55% to 75% of the nucleotides in 16S rRNA and from 60% to 70% in 23S rRNA are purines. Strikingly, over 50% of the total internal loop nucleotides are adenosine in two of the five 16S rRNAs shown. The least represented nucleotide in internal loops is cytosine, representing only 6.7% of the nucleotides in the internal loops of one 16S rRNA. This uneven distribution of nucleotides in internal loops is probably due to structural as well as functional considerations and presents us with opportunities in terms of secondary and tertiary structure prediction.

Functions of RNA Internal Loops

The biological functions of RNA internal loops are as varied as the molecules in which they are found and are known for only a limited number of molecules. RNA internal loops serve as binding sites for regulatory proteins, are involved in intramolecular RNA-RNA interaction in the formation of tertiary structure, provide specific binding sites for metals and small molecule ligands such as antibiotics, and distort RNA helical structure by changing groove width, helical twist, and bending the RNA. Other functions of RNA internal loops, for example in those found in viroids, are not clearly understood.

Two well characterized internal loops that serve as protein recognition elements in an RNA mediated regulation are: the Iron Responsive Element, IRE, found in the 5'-untranslated region (UTR) of ferritin mRNA and the 3' UTR of transferrin receptor mRNA; and the Rev Response Element, RRE, found in the *env* gene of the HIV-1 retrovirus. These RNA elements are the binding sites of the IRP and Rev proteins respectively.

Small molecule binding to RNA internal loops has been studied in both biological systems and SELEX evolved RNA aptamers prepared *in vitro*. Examples include the A-site decoding region of 16S ribosomal RNA which is a binding site for

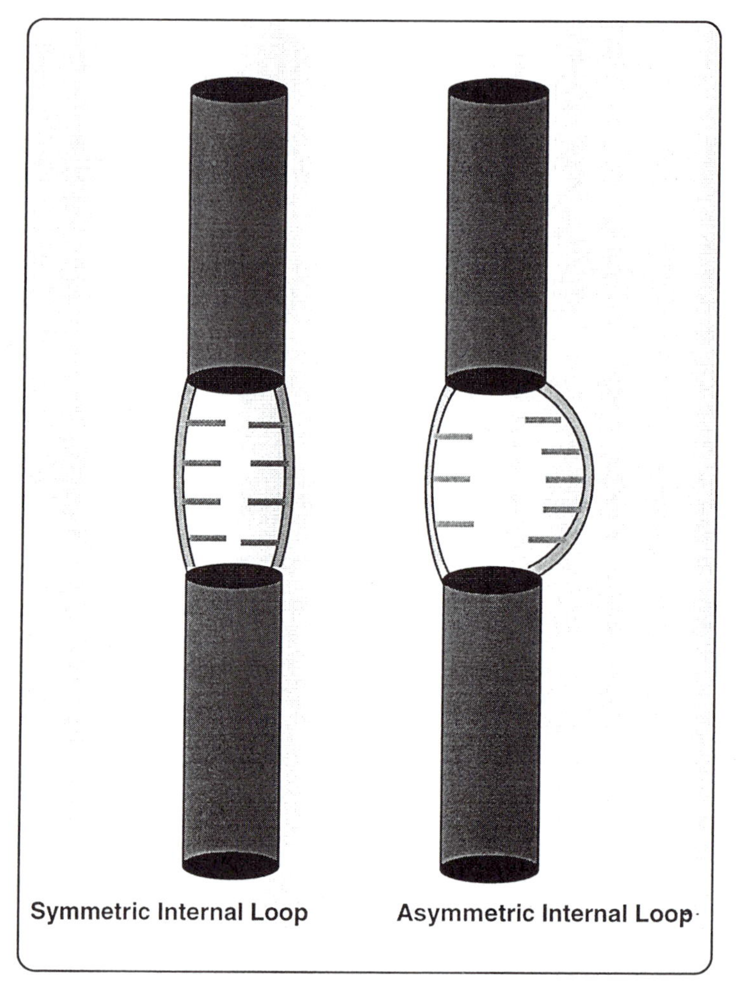

Figure 1. Definition of RNA Internal Loops. The cylindrical regions denote Watson-Crick base paired helices.

aminoglycoside antibiotics*(1)* and the ATP/AMP binding aptamer RNA identified by SELEX experiments*(2)*. Both of these molecules have been structurally studied by NMR methods and three-dimensional models are available*(3-6)*.

Although the role of RNA internal loops in the formation of RNA tertiary structure is only beginning to be understood, the crystal structure of the P4-P6 domain of the group I intron provides us with an instance of an internal loop which causes an approximately 180° bend between the two long helices which make up the three-dimensional structure and another internal loop which clamps down the two extended helices parallel to each other by forming a tertiary interaction with a tetraloop at the end of the other helix *(5)(7)*. The role of internal loops in stabilization of intermolecular RNA-RNA complexes has also been studied in several systems as discussed below.

X-ray Crystal Structures of Internal Loops

The x-ray crystal structures of several RNA oligomers incorporating symmetric internal loops have been determined. Four internal loop motifs are observed in the crystal structure of the P4-P6 domain of the group I intron. Three of these four internal loops are asymmetric. The x-ray crystal structures show the geometry of the non-Watson-Crick base pairs formed in the internal loops, the importance of bound water and metal ions, the distortion introduced into the RNA helix, and the mobility of the residues in and out of the internal loop.

Internal Loops in RNA Oligomer Crystals. The first crystal structure of an RNA internal loop was solved in 1991*(8)*. The dodecamer rGGACUUCGGUCC (internal loop underlined) forms a duplex in the crystal with an internal loop of consecutive U-G, U-C, C-U and G-U mismatches as shown in Figure 2. The chains of the double helix show two-fold symmetry in the crystal, thus the U-G and U-C pairs are identical to the C-U and G-U pairs. As can be seen, the internal loop generally continues the double helices which surround it by formation of non-Watson-Crick base pairs. The major groove of the A-form helix is opened with respect to a canonical RNA helix.

Subsequently, several other oligomer structures with symmetric internal loops have been determined and are illustrated and compared to canonical A-RNA in Figures 3 and 4. The dodecamer rGGACUUUGGUCC, which differs from the UUCG dodecamer described above by only a single nucleotide, crystallizes as a non-symmetric duplex with U-G and U-U base pairs in the internal loop *(9)*. The tandem U-U pairs in this structure form two hydrogen bonds between the imino nitrogens and carbonyl oxygens. The 5' uracil of each U-U pair is shifted toward the major groove to make these pairs. One U-U pair is highly twisted to allow formation of a single water bridge between the two U-U pairs. Interestingly, the double helix formed by this sequence does not have the expanded major groove seen in UUCG, but instead the minor groove width is diminished relative to A-RNA. The structures of these two internal loops have been verified by crystal structures of two other UUCG internal loops(a tridecamer rUGAGCUUCGGCUC*(11)* and a nonamer r(GCUUCGGC)d^{Br}U

Table I. Distribution of Nucleosides in rRNA Internal Loops

Organism	Internal Loops	% of rRNA nucleotides in int. loops	% of A in int. loops	% of G in int. loops	% of C in int. loops	% of U in int. loops
16S rRNA						
E. coli	50	10.8	50.9	25.7	11.4	12.0
P. occultum	50	9.0	47.8	27.6	10.4	14.2
H. sapiens	56	12.1	30.0	25.6	27.3	17.2
Z. mays (mt)	53	15.8	36.0	29.9	18.6	15.4
Euglena (cp)	51	10.0	51.3	25.3	6.7	16.7
23S rRNA						
E. coli	97	14.2	42.8	27.0	11.4	18.7
M. vanniellii	100	14.5	40.7	25.5	14.5	19.4
X. laevis	96	11.5	32.2	28.6	19.8	19.4
Z. mays (mt)	89	10.7	39.4	25.7	15.0	19.9
E. gracilis (cp)	97	14.2	43.8	21.4	12.4	22.4

mt = mitochondrial, cp = chloroplast

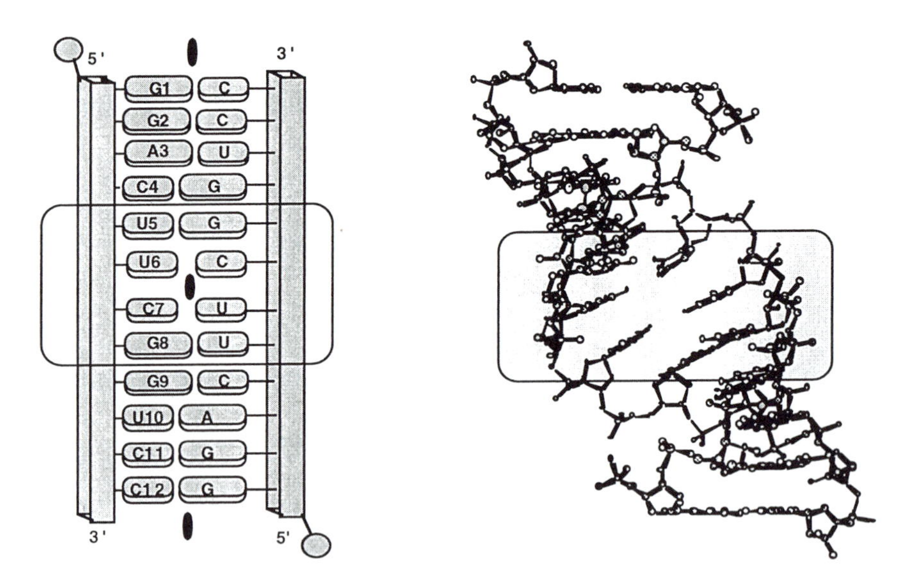

Figure 2. Structure of an RNA dodecamer duplex with a UUCG internal loop. Schematic diagram is shown on the left with the internal loop boxed. Darkened ellipsoids represent the crystallographic two-fold axis. The crystal structure is shown on the right with the internal loop boxed.

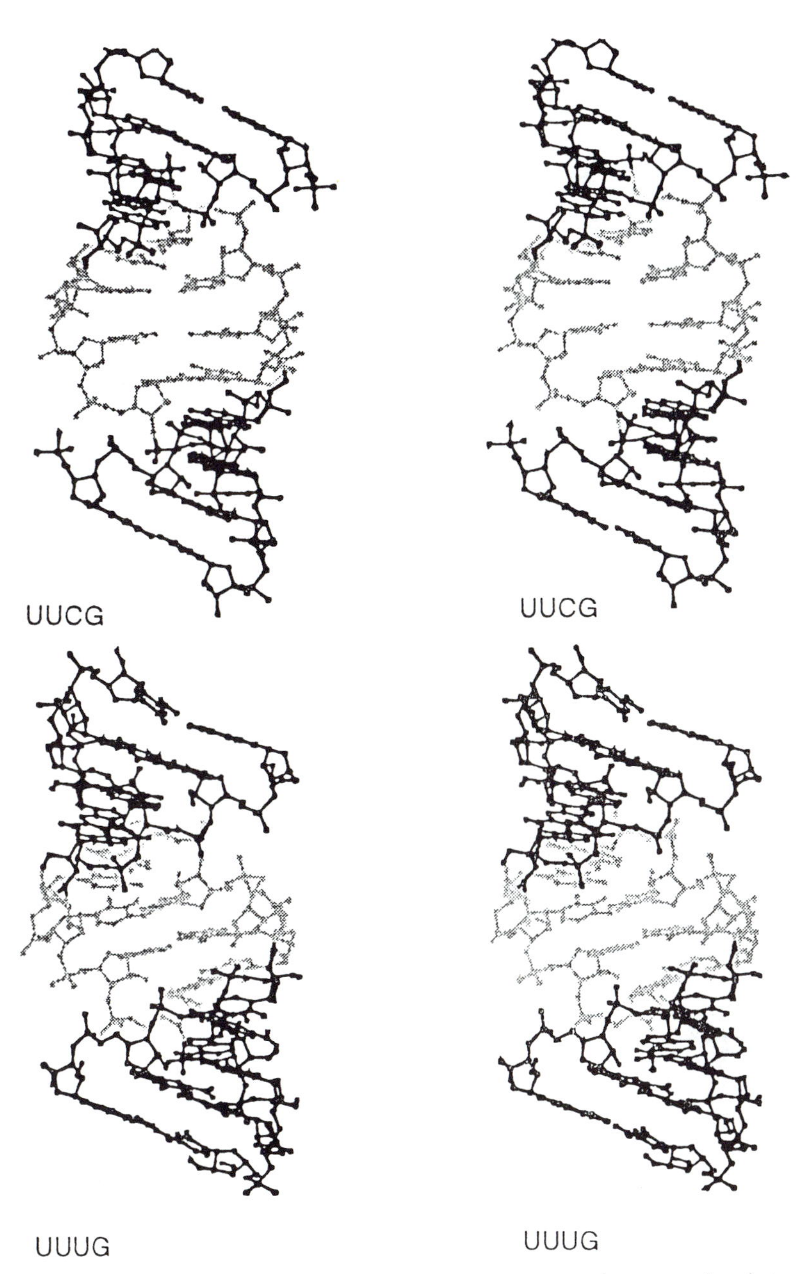

Figure 3. Stereoviews of four RNA oligomer structures incorporating internal loops compared to canonical A-RNA. The view is into the major groove. The internal loop is shaded grey. The labels under the molecules refer to the internal loop sequence as described in the text.

Continued on next page

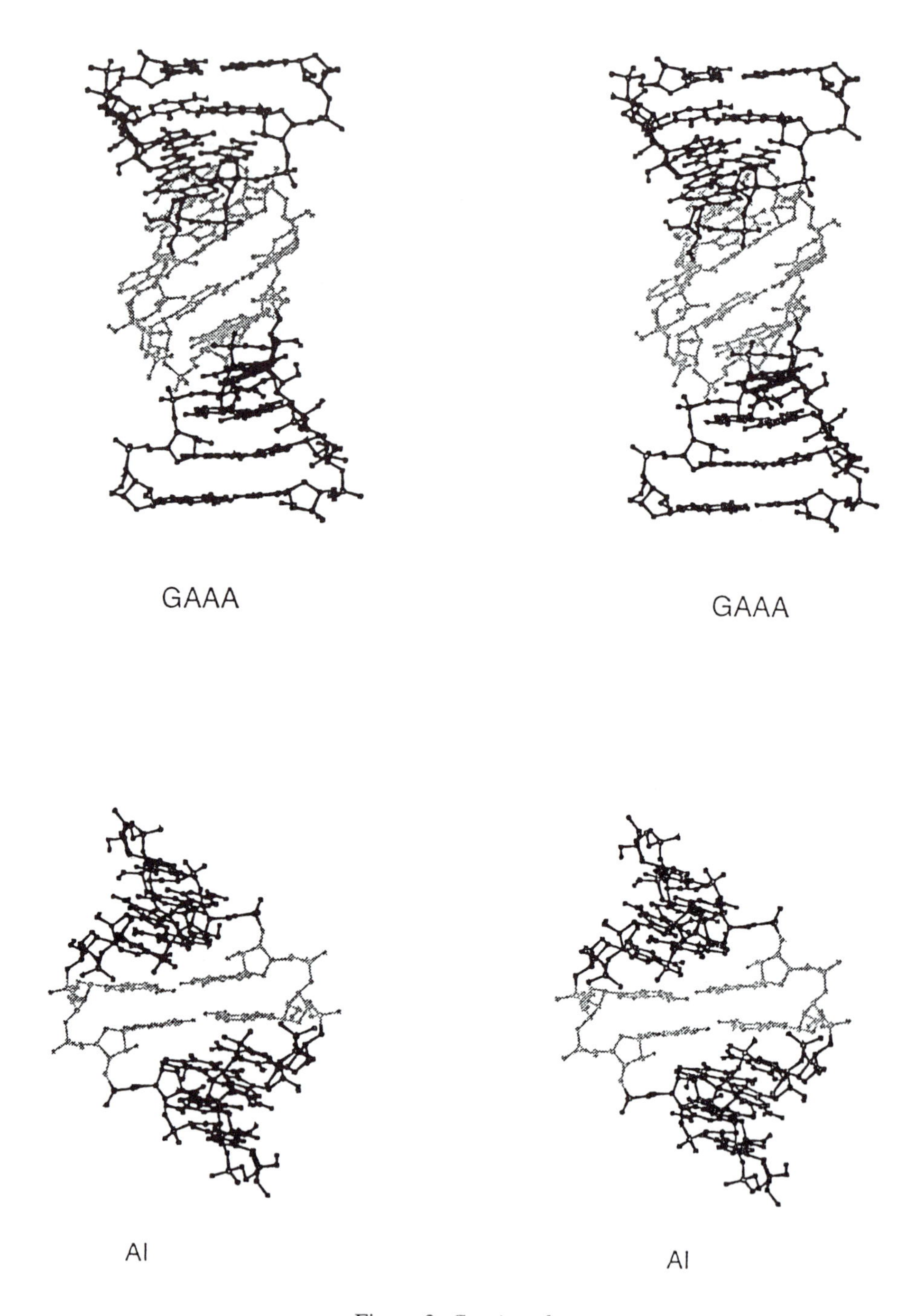

Figure 3. *Continued.*

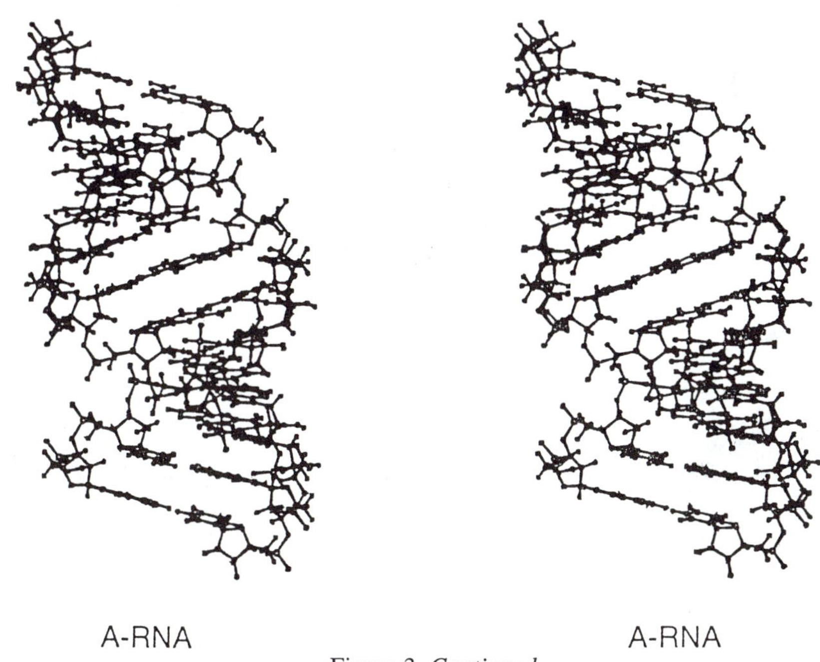

Figure 3. *Continued.*

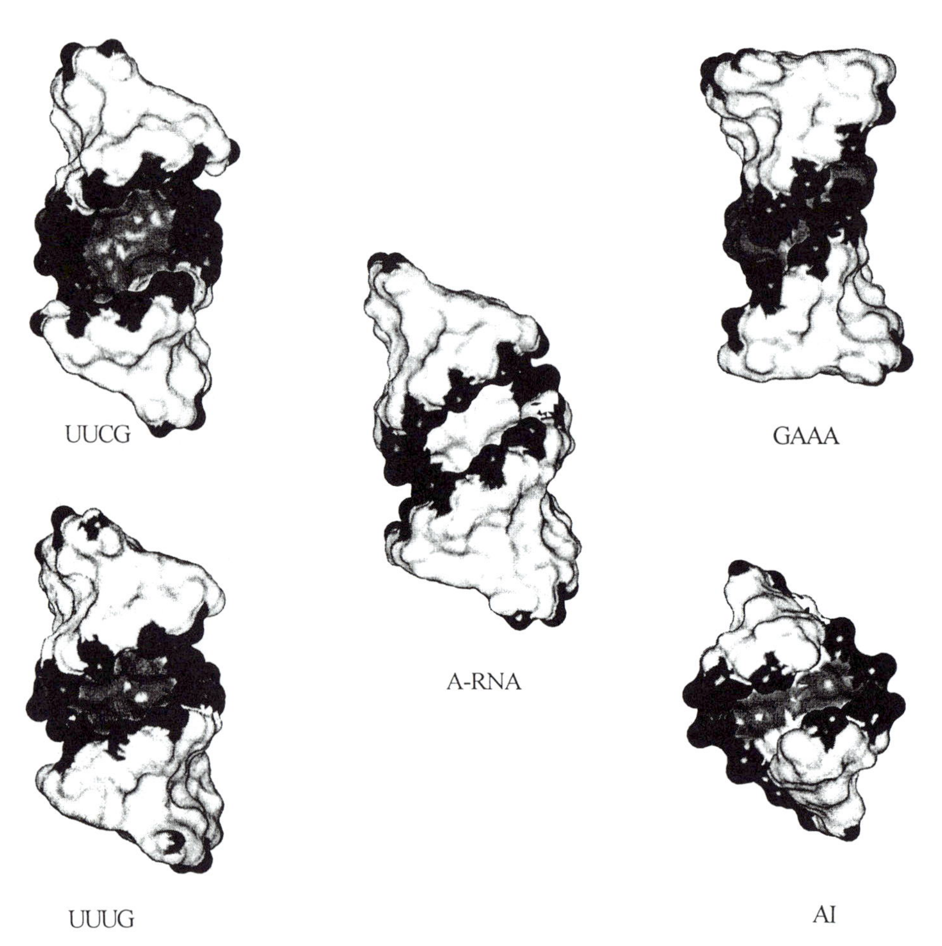

Figure 4. Solvent accessible surface representations of four RNA oligomers with symmetric internal loops compared to A-form RNA. The internal loops are shaded dark gray and the phosphodiester backbone is in black.

(10)) and a tandem UU internal loop (rGGCGCUUGCGUC *(12)*). The dodecamer structure, rGGCCGAAAGGCC, shown in Figure 3, also forms a symmetric duplex in the crystal with a run of G-A, A-A, A-A and A-G mispairs *(13)*. Of the oligomer structures incorporating symmetric internal loops, this is the most distorted from an A-form helix. The G-A pairs of this duplex are in a sheared conformation, also observed in the hammerhead catalytic RNA *(14)(15)* and in the NMR structure of a stable RNA tetraloop *(16)*. The RNA motif CG:AG observed in both the GAAA dodecamer and hammerhead catalytic RNA*(14)(15)* , contains a Mn^{+2} binding site which may provide additional stabilization. Recently, we have determined the structure of an RNA octamer, rGCCAIGGC, which forms a symmetric self-complementary duplex in the crystal (Carter and Holbrook, unpublished data). In contrast to the G-A base pairs discussed above, the A-I pairs are in a head-to-head (anti-anti) conformation. This type of base pairing produces significantly less distortion in the double helix than the sheared conformation. There is no divalent cation bound near the A-I pairs, such as that observed at the junction of the Watson-Crick and internal loop regions of the GAAA dodecamer. The G-A pairs in the RNA dodecamer rCGCGAAUUAGCG assume the same type of hydrogen bonding as that observed in the AI containing octamer. The non-standard base pairs observed in these RNA oligomers containing internal loops are illustrated in Figure 5.

Internal Loops in the Crystal Structure of the Group I Intron P4-P6 Domain. The crystal structure of the 160 nucleotide domain of the P4-P6 intron of *Tetrahymena thermophila* includes several internal loop regions *(5)*. Of the four internal loops (there is also a bulge), three are asymmetric in the number of nucleotides on the opposing strands. This provides a good comparison to the symmetric internal loops observed in oligomer structures. The four internal loops are shown in Figure 6.

One of the internal loops, UAAG_UAU (strand1_strand2), the tetraloop receptor, serves the important role of constraining the tertiary fold by interacting with the GAAA tetraloop at the end of the P5Bb extended helix. This internal loop, as well as the CAAG_UGCA internal loop, contains the adenosine platform structural motif formed by adjacent adenosines in the sequence that lie side-by-side and form a pseudo base pair.

Helical Distortion due to Internal Loops

The distortion of the helical axis induced by an internal loop on its surrounding helices relative to that of a continuous Watson-Crick duplex can be measured in many ways. One common approach is incorporated in the computer program CURVES *(17)*. This method, while quite useful for Watson-Crick duplexes, is somewhat ambiguous when considering non-canonical pairs or bulges which deviate significantly from the helix axis. Although these nucleotides may be excluded from the calculation, this choice affects the results greatly. Also, this approach produces a curved helical axis which follows a smooth path through the local oligomer itself, but does not assume continuation of the helical ends. Thus, it is more clearly called "internal curvature".

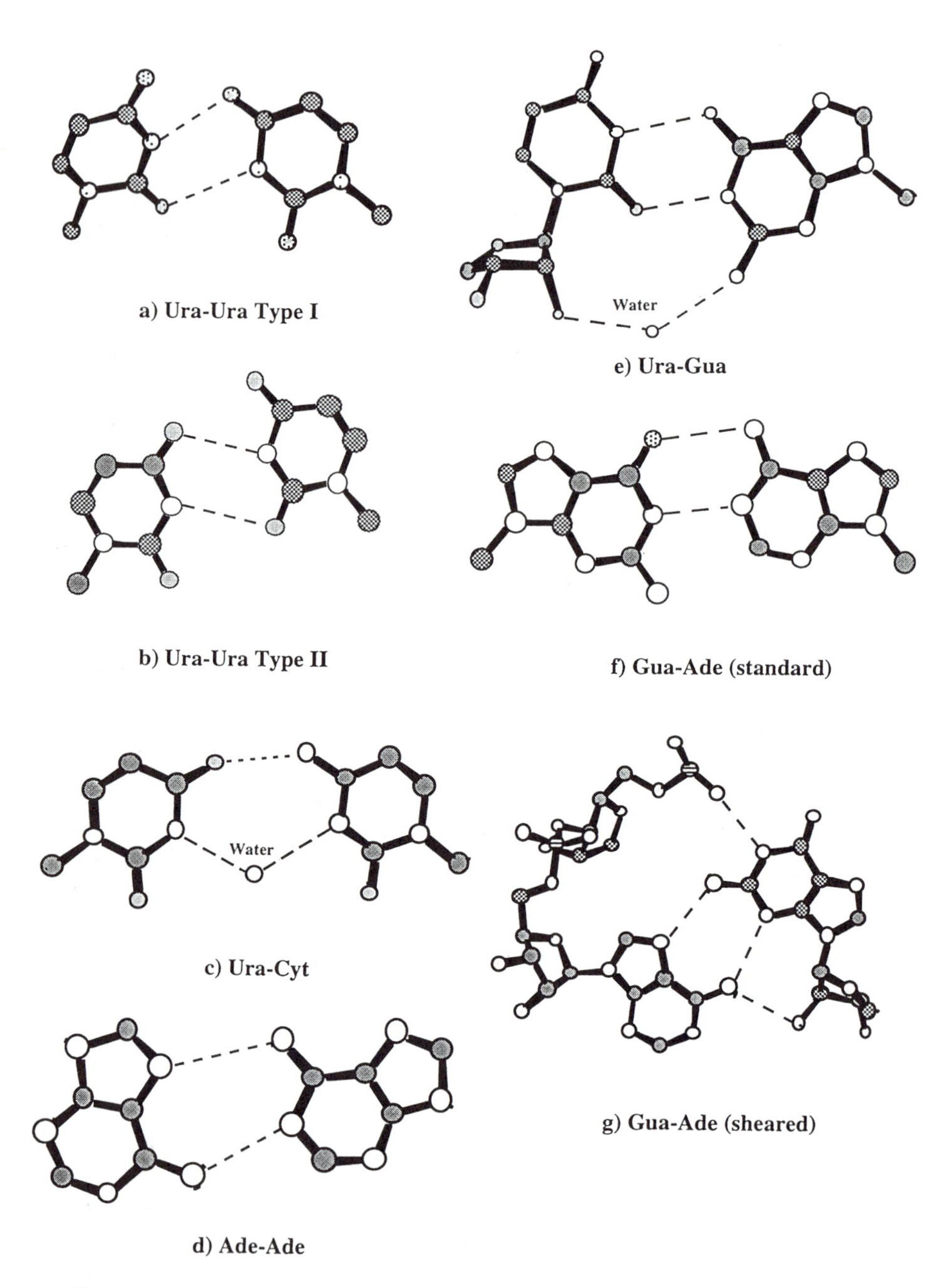

Figure 5 (a-g). Non-Watson-Crick base pairs found in the internal loops of RNA oligomers. Water molecules integral to the base pairing are indicated (5c, 5e). Hydrogen bonds are represented by dashed lines.

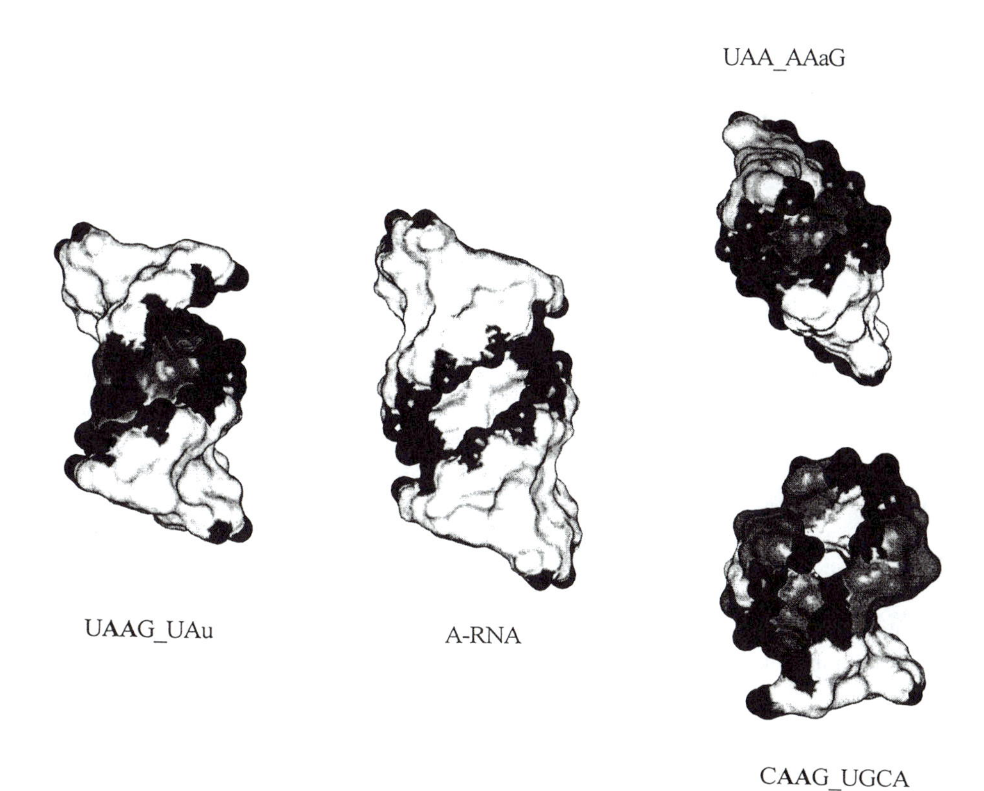

Figure 6. Internal loops found in the crystal structure of the P4-P6 domain of the group I intron. The solvent accessible surfaces are shown as in Figure 4. The labels under the molecules refer to the sequences of the internal loop.

In order to better characterize the distortion of the helical axes in the duplexes bounding the internal loop without consideration of an axis for the loop region itself, we have calculated the interhelical angle, translational displacement of the helical axes, and over(under)winding in helical turn relative to that of a standard A-form helix (32.7°/base pair). Since the Watson-Crick regions are quite short in the oligomers which have been studied, the first step is to superimpose canonical RNA-A helices onto the helical ends of the RNA oligomers by least squares fitting and then to calculate the angle and distance (shortest) between the axes of these standard helices. An example is shown in Figure 7. The winding angle is calculated by projecting C1'-C1' vectors from the Watson-Crick base pairs bounding the internal loop onto a common plane and calculating the angle between them. The results of these calculations for the helices surrounding internal loops as well as the curvature from the CURVES program, helical diameter and groove widths are summarized in Table II. The greatest distortions in the helices are emphasized by bolding. The greatest interhelical angle induced by symmetric internal loops is 16 degrees for the UUCG internal loop. Greater kinking of the helices is produced by asymmetric internal loops, with an extreme of 154.3°, which allows the two extended helices of the P4-P6 domain to lie side-by-side. An extremely large displacement, or translation, of the helical axes is induced by the GAAA internal loop. This oligomer is also narrowed in the internal loop to only 11.4Å, the smallest diameter of any of the internal loops observed so far. The greatest distortions in groove width are a widening of the major groove from 4.0 to 8.4Å in the UUCG internal loop and a narrowing of the minor groove from 11.1 to 8.7 and 7.6Å in the oligomers containing U-U mispairs (UUUG and UU).

Ligand Binding to Internal Loops

Localized binding of water, metal ions and other ligands is an important aspect of the structure (and likely the function) of RNA internal loops. The available crystal structures provide several examples of both directly coordinated divalent metal cations and water molecules integral to non-Watson-Crick base pairing. Figure 8 shows the U-G and U-C base pairs observed in the UUCG dodecamer structure *(10)* and other structures containing these same mismatches *(10)*. In both the U-G and U-C base pairs, water molecules are actually part of the base pair hydrogen bonding. In the U-G pair, a water bridges between the 2-amino group of guanosine and the 2'-hydroxyl of uridine on the opposite strand, while in the U-C pairs, a water links the N3 atoms of U and C opening the pyrimidine pair to more nearly conform to the helical diameter. Each of these water molecules integrated into the base pairs has a very high occupancy and low mobility as judged by the electron density and crystallographic temperature factors which are similar to those of the bases themselves. Additional waters seen in the major groove fulfill the hydrogen bond potential of the functional groups and form a stabilizing network. Resolution of the crystallographic data is crucial in the identification of bound water molecules. High resolution crystal structures typically show more ordered waters than lower resolution structures. For example, the 1.4Å structure of rUUCGCG has quite complete and well ordered water networks in both

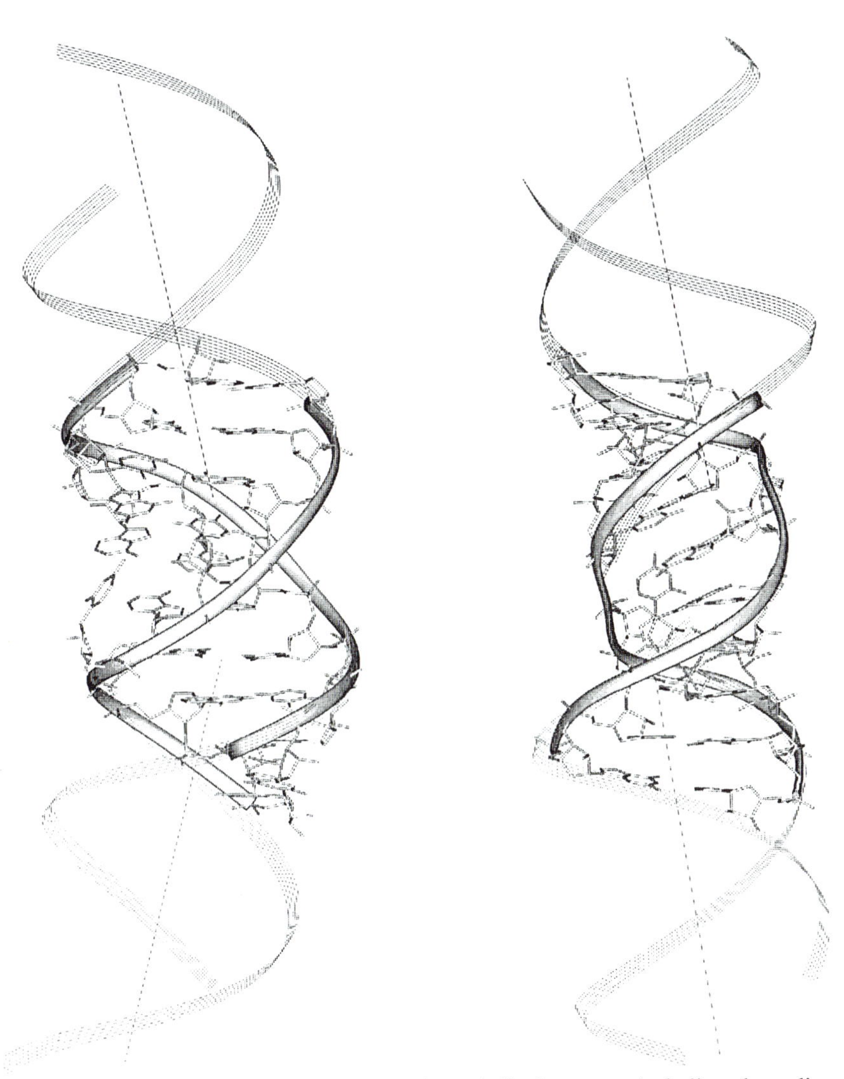

Figure 7. Calculation of interhelical angle and displacement in helices bounding an internal loop. An asymmetric internal loop from the P4-P6 domain of a group I intron is shown as an example. a) The helical axes for the Watson-Crick regions are indicated by straight lines. The angle is calculated as the dot product. b) The molecule is rotated by 90° to show the translational displacement between helix axes. The displacement is the minimum displacement between the skew lines.

Table II. Distortion of Helical Axes Surrounding Internal Loops

Internal Loop	Interhelical Angle (°)	Helix Axis Displacement (Å)	Over/Under-Winding (°)	Internal Curvature (°)	Helix Diameter (Min/Max)	Major Groove (Å)	Minor Groove (Å)
A-RNA	0.0	0.0	0.0	0.0	17.4	4.0	11.1
UUCG	**16.0**	0.1	**-10.2**	5.1	16.7-19.2	**8.4**	11.7
UUUG	2.1	0.5	9.2	15.5	17.2-18.6	4.9	**8.7**
UU	6.0	1.9	9.2	15.0	16.0-18.9	3.1	**7.6**
AI	0.2	0.1	0.7	15.2	19.0-19.3	n.a.	12.1
GAAA	0.4	**10.9**	**12.0**	10.4	**11.4**-20.2	**2.2**	13.0
CAAG_UGCA	**24.2**	**9.7**	**-16.7**	1.9	16.4-20.0	4.9	16.8
UAA_AAaG	6.0	1.3	-7.6	8.9	17.6-18.0	5.0	9.6
UAAG_UAu	**26.3**	6.4	-0.2	17.3	16.4-16.6	4.8	10.5
ACAG_ACA	**154.3**	2.1	n.a.	n.a.	n.a.	n.a.	n.a.

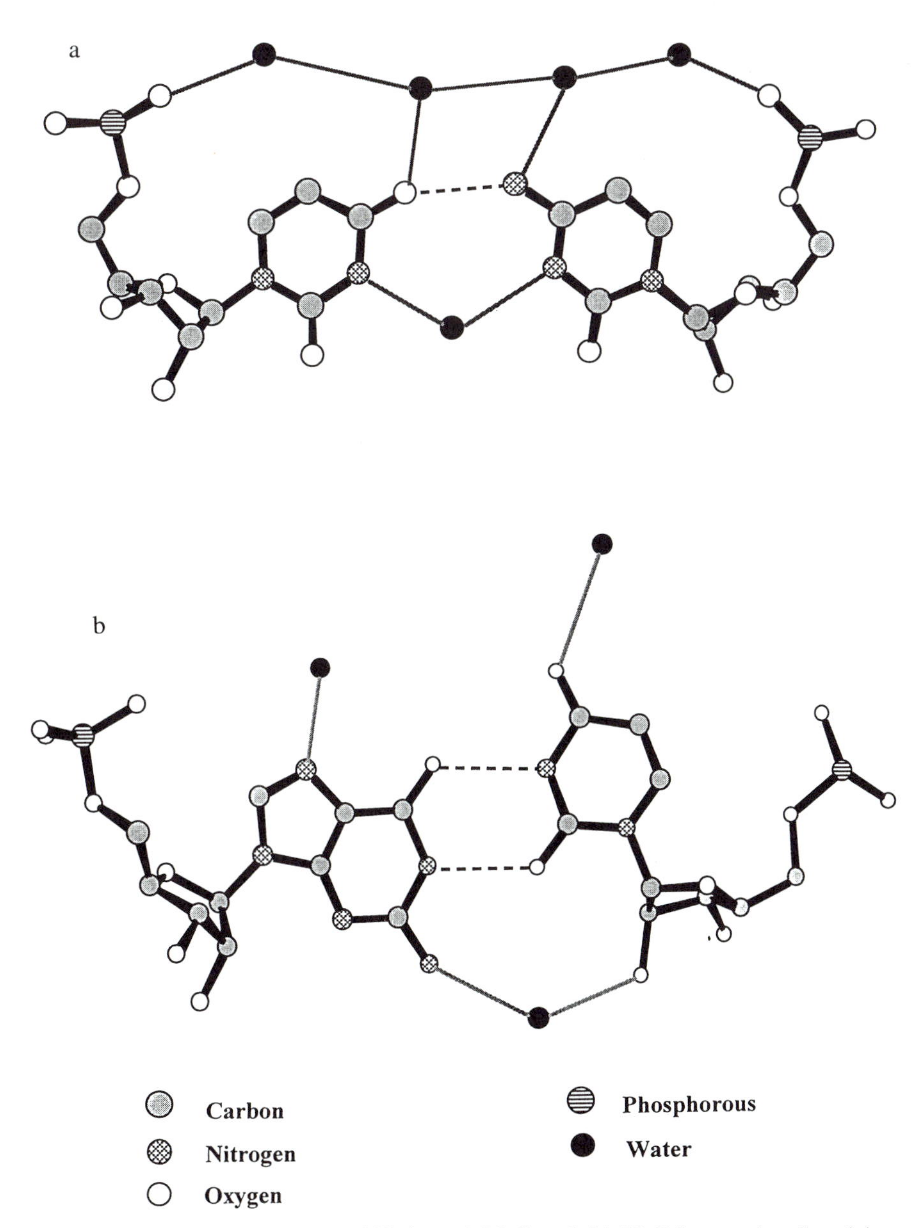

Figure 8. Water structure stabilizing (a) U-C and (b) U-G base pairs found in internal loops. Waters are denoted by blackened circles with hydrogen bonds indicated by grey lines. Atom type is shown in key.

grooves of the G-C pairs, linking the bases to the phosphate oxygens in the major groove and 2'-hydroxyls in the minor groove. These networks are not observed in lower resolution structures. Thus, resolution of the crystallographic data must be considered when examining the hydration of any RNA crystal structure.

Although metal ions, particularly divalent cations such as Mg^{+2}, are known to bind RNA and often be required for structure and function, localized metal ions are rarely observed in crystal structures of RNA oligomers. This may be due to the lack of specific, tight binding sites in regular helical RNA. More complex RNAs, such as tRNA, hammerhead catalytic RNA and the P4-P6 domain of the group I intron, display numerous specific metal binding sites in pockets formed by their tertiary structure. Certain internal loops, however, may form unique metal binding sites which may stabilize their structure or be available for a function. The dodecamer rGGCCGAAAGGCC has been crystallized in the presence of either Mn^{+2}, Co^{+2}, or Cd^{+2}, diffraction data collected, and each structure solved (Weng and Holbrook, unpublished data). In each of these structures, there are two specific, solvent binding sites of high electron density in the internal loop. These sites appear at identical positions in each of the three crystal structures, however, the relative occupancy of the two sites varies depending on which metal is bound. One of these sites clearly corresponds to a bound divalent cation, while the other "solvent" peak appears to be a tightly bound water. Figure 9 shows the binding sites for these ligands in the narrowed major groove of the internal loop of this dodecamer. The divalent metal ion sites coordinate directly to a guanosine N7 as well as to the RNA backbone and fill the coordination sphere with waters which hydrogen bond to functional groups of the RNA. Site I, at the junction of the Watson-Crick and internal loop regions, forms a motif which is also observed to bind divalent cations in hammerhead catalytic RNA*(14)(15)*. Since the divalent cation binds to the guanine of the C-G pair adjacent to the internal loop as well as the phosphate of the adenosine of the G-A pair, metal binding may explain why C-G, G-A sequences form sheared base pairs as in this structure, while G-C, G-A sequences which can not form this metal binding site make head-to-head G-A pairs. The Site II solvent binds between the two strands in the very narrow major groove (Table II).

Currently, twenty-four metals have been located in the 150-nucleotide model of the P4–P6 domain of the group I intron ribozyme*(18)*. Two magnesium ions are bound to and stabilize the A-rich corkscrew bulge motif bridging the two extended helices. One of these magnesium binding sites can coordinate Sm^{+3} in a similar manner. Other magnesium ions are located between phosphate ions of adjacent helices and serve the purposes of charge neutralization and connection. Crystals of the group I ribozyme domain were grown from a solution which included $Co(NH_3)_6$. The bound cobalt hexammine was replaced by soaking crystals with osmium hexammine triflate which was used for crystallographic phasing. One weak and three strong osmium hexammine sites were identified per molecule. The osmium hexammine molecules bind in the major groove near non-Watson-Crick base pairs at the ends of helices. Two of these sites involve consecutive G-U or U-G pairs. In the native RNA, two of these sites are occupied by fully-hydrated magnesium ions. Rhodium hexammine also bound

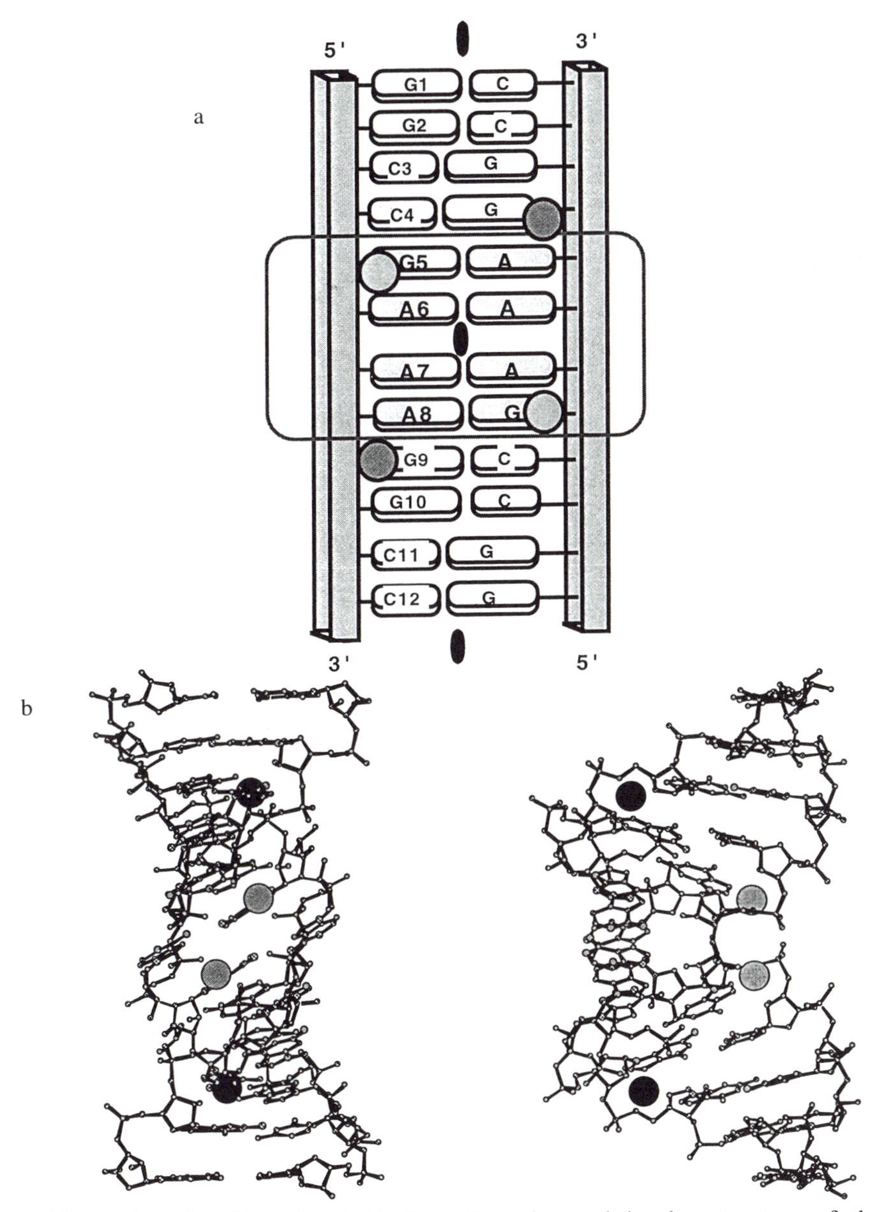

Figure 9. Specific solvent binding sites observed in the structure of the dodecamer incorporating a GAAA internal loop. a) Schematic diagram. Divalent metal sites corresponding to either Mn, Co, or Cd are indicated by black circles and tightly bound solvent molecules by gray circles. Internal loop mispairs are lightly shaded. b) Two views rotated by 90 degrees of the three dimensional structure of this dodecamer with the metals bound in the major groove indicated by black circles and the tightly bound solvents by gray circles.

to the guanosine O6 of the G-U pairs in the nonamer r(GCUUCGGC)dBrU duplex *(10)*, as well as to the O2C and phosphate of the C-U pairs.

Flexibility of RNA Internal Loops

The mobility of atoms within the crystal is inherent in the diffraction data and evaluated in crystallographic thermal parameters or B factors. These mobility parameters assume a spherical distribution of displacement from the equilibrium (average) position and are obtained through crystallographic least squares refinement of the model versus the observed data. A higher B factor for an atom or averaged over a residue indicates that it is more mobile or flexible than residues with a lower B factor and implies a less stable structure.

In order to understand the flexibility of RNA internal loops, we have examined the B factors obtained from crystallographic refinement of the existing structures. Generally, we can conclude that the flexibility of internal loops is no greater than the Watson-Crick double helical regions of the oligomer. For example, Figure 10 shows the average mobility parameters (B factors) for each residue in the dodecamer containing a UUCG internal loop. It is clear that the UUCG residues 5-8 forming U-G and U-C mismatches are approximately as rigid (average 9.9Å^2), as the other non-terminal residues of the molecule, with the greatest flexibility occuring at the 5'-terminus. This result is confirmed in other oligomers incorporating internal loops. As observed in other nucleic acid structures, the waters are the most mobile (average B=35.0Å^2), followed by the phosphates (15.5Å^2), the riboses (12.3Å^2) and finally the nucleotide bases (8.4Å^2).

Summary

Recent x-ray crystal structures have greatly clarified our understanding of the structure of RNA internal loops and how these structures can participate in RNA function. It is now apparent that non-Watson-Crick base pairing allows the double helical structure to continue through the internal loop, with distortion within and surrounding the internal loop dependent on the types of mismatches formed. While the oligomer structures studied so far are all symmetric with each base in the internal loop pairing with the opposing base in the other strand, the asymmetric internal loops observed in the P4-P6 domain structure have illustrated the possibility of same strand pairing (adenosine platforms) and looping out, or bulging of individual bases. Overall, the asymmetric loops can produce more distortion in the surrounding helices, although this is not strictly the case since some symmetric loops are more distorted than asymmetric ones. The presence of adenosine platforms and looped out bases also makes the traditional classification of symmetric and asymmetric loops less useful.

The dramatic differences in structure of internal loops of different sequence make it necessary to characterize the structures of many more sequence variations in order to understand and eventually model the structure and effect of internal loops in biological systems.

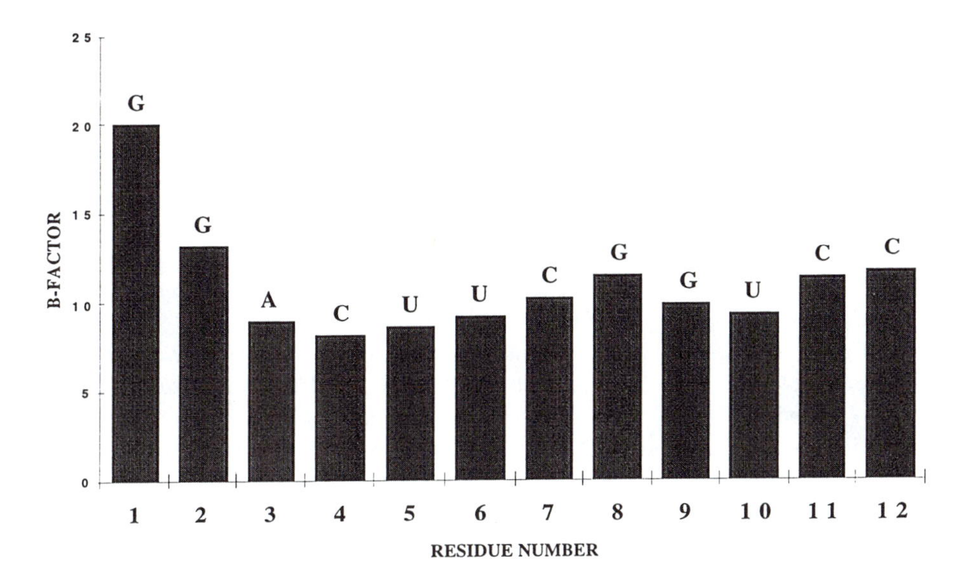

Figure 10. Distribution of crystallographic B factors averaged over the residues of the RNA dodecamer with UUCG internal loop. Larger B factors indicate regions of greater average mobility or flexibility.

Acknowledgements

The author acknowledges the support provided by U.S. Department of Energy Office of Health and Environmental Research (Contract DE-AC03-76SF0098) and the National Institutes of Health (NIGMS) GM 49215 both for his own research and his effort in compiling this chapter. The author is also grateful to Dr. Richard Carter and Dr. Xiangwei Weng for providing summaries of their studies prior to publication. Finally, special thanks are due Dr. Elizabeth Holbrook, Dr. Xiangwei Weng and Dr. Richard Carter for assistance in preparation of figures and critical proofreading of this manuscript.

Literature Cited

1. Recht, M. I., Fourmy, D., Blanchard, S. C., Dahlquist, D. D.,Puglisi, J. D. *J. Mol. Biol.* **1996**, *262*, 421-436.
2. Sassanfar, M.,Szostak, J. W. *Nature* **1993**, *364*, 550-553.
3. Fourmy, D., Recht, M. I., Blanchard, S. C.,Puglisi, J. D. *Science* **1996**, *274*, 1367-1371.
4. Jiang, F., Kumar, A. R., Jones, R. A.,Patel, D. J. *Nature* **1996**, *382*, 183-186.
5. Cate, J. H., Gooding, A. R., Podell, E., Zhou, K., Golden, B. L., Kundrot, C. E., Cech, T. R.,Doudna, J. A. *Science* **1996**, *273*, 1678-1685.
6. Dieckmann, T., Suzuki, E., Nakamura, G. K.,Feigon, J. *RNA* **1996**, *2*, 628-640.
7. Cate, J. H., Gooding, A. R., Podell, E., Zhou, K., Golden, B. L., Szewczak, A. A., Kundrot, C. E., Cech, T. L.,Doudna, J. A. *Science* **1996**, *273*, 1696-1699.
8. Holbrook, S. R., Cheong, C., Tinoco, I., Jr.,Kim, S.-H. *Nature* **1991**, *353*, 579-581.
9. Baeyens, K. J., DeBondt, H. L.,Holbrook, S. R. *Nature Str. Biol.* **1995**, *2*, 56-62.
10. Cruse, W. B. T., Saludijan, P., Biala, E., Strazewski, P., Prange, T.,Kennard, O. *Proc. Natl. Acad. Sci. USA* **1994**, *91*, 4160-4164.
11. Fujii, S., Tanaka, Y., Uesugi, S., Tanaka, T., Sakata, T.,Hiroaki, H. *Nucleic Acids Symposium Series* **1992**, *27*, 63-64.
12. Lietzke, S. E., Barnes, C. L., Berglund, J. A.,Kundrot, C. E. *Structure* **1996**, *4*, 917-930.
13. Baeyens, K. J., De Bondt, H. L., Pardi, A.,Holbrook, S. R. *Proc. Natl. Acad. Sci. (USA)* **1996**, *93*, 12851-12855.
14. Pley, H. W., Flaherty, K. M.,McKay, D. B. *Nature* **1994**, *372*, 68-74.
15. Scott, W. G., Finch, J. T.,Klug, A. *Cell* **1995**, *81*, 991-1002.
16. Heus, H. A.,Pardi, A. *Science* **1991**, *253*, 191-194.
18. Lavery, R.,Sklenar, H. *J. Biomol. Struct. Dynam.* **1989**, *6*, 655-667.
19. Cate, J. H.,Doudna, J. A. *Structure* **1996**, *4*, 1221-1229.

Chapter 5

Hydrogen-Bonding Patterns Observed inthe Base Pairs of Duplex Oligonucleotides

William N. Hunter[1,4], Gordon A. Leonard[2], and Tom Brown[3]

[1]Department of Biochemistry, University of Dundee, Dundee DD1 4HN, Scotland
[2]European Synchrotron Radiation Facility, F–38043 Grenoble, Cedex, France
[3]Department of Chemistry, University of Southampton, Highfield, Soughampton SO17–1BJ, United Kingdom

Watson-Crick base pairing in duplex structures utilises a unique combination of hydrogen bond donors and acceptors to stabilise the association. This stability contributes to the fidelity of replication. Non-Watson-Crick base pairs, mismatches or base pairs involving chemically modified bases adopt different patterns of hydrogen bond interactions sometimes utilising water molecules to stabilise the pairing and in some cases an argument can be put forward for C-H•••O interactions as a contributing factor in the pairing and not just in mismatches. The different hydrogen bonding patterns deduced on the basis of single crystal x-ray diffraction studies are described. These hydrogen bonding patterns have implications for protein-nucleic acid interactions and for the modelling of nucleotide structures.

Single crystal x-ray diffraction methods have been used to characterise numerous oligonucleotide structures enabling studies of the fine structure of DNA, oligonucleotide hydration, interactions with small molecule ligands and proteins. A particular focus has been on non-standard base associations and researchers have sought to characterise different non-Watson-Crick base pairs to further understanding of their influence on duplex DNA and RNA, and to investigate which structural features might be used in recognition and repair of these errors in DNA. Bases that can be chemically modified present distinct hydrogen bonding patterns and these too have been investigated. In this article we re-examine the hydrogen bonding patterns of the Watson-Crick pairing and survey some of the non-Watson-Crick base associations in duplex DNA and RNA.

Complementarity of the Watson-Crick pairs. Watson and Crick recognised that a complementary base pairing scheme in duplex DNA contributed to the fidelity of replication *(1)*. Purines interact with pyrimidines so that G pairs with C and A pairs with T to form what are termed Watson-Crick base pairs (Fig. 1). The very specific manner in which the Watson-Crick base pairs are formed contributes stability to an oligonucleotide structure and a particular arrangement of functional groups to interact with enzymes and proteins using specific hydrogen bonding patterns *(2)*. Since the human genome is estimated to contain around 10^9 base pairs it is not surprising that

[4]Corresponding author

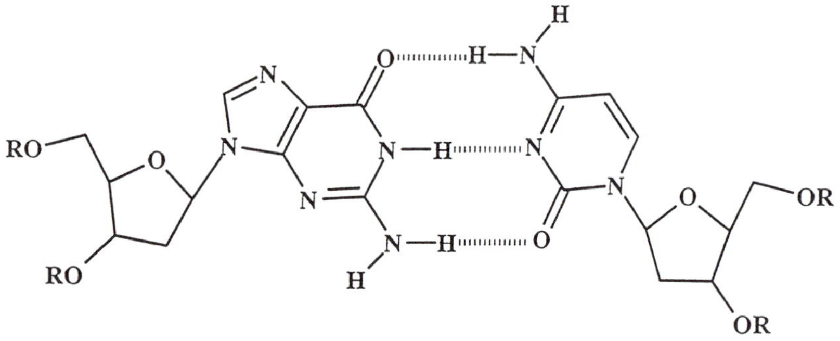

G.C

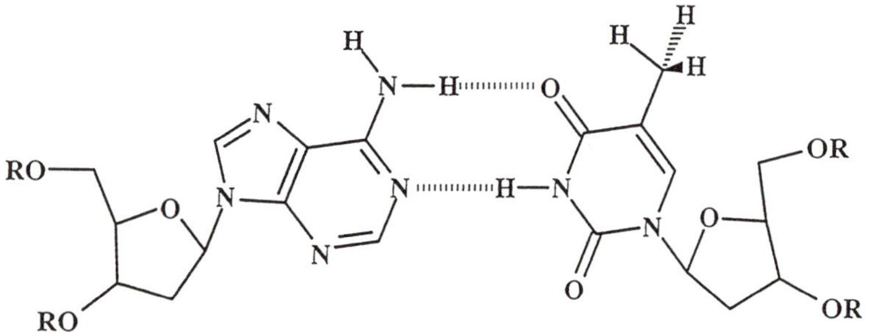

A.T

Figure 1. The G•C and A•T Watson-Crick base pairs. In this and all subsequent figures, dashed lines represent hydrogen bonding interactions.

mistakes occur during replication. A single error in a triplet can be carried through and eventually lead to a serious mutation in the gene product. Errors may be introduced via non-Watson-Crick base pairs termed mismatches or mispairs. Alternatively, damage to DNA can produce bases with altered chemical properties capable of scrambling the genetic code *(3)*.

Mistakes, if unchecked, can be deleterious and a complex protein recognition and repair system makes a contribution to maintaining the fidelity of replication *(4)*. Studies on these systems represent an exciting area of structural biology *(5)*. Crystallographic studies of mismatches and modified bases in DNA and RNA complement thermodynamic studies on stability of the mismatches or base pairs involving chemically modified components *(6)*. Our objective in this article is to reconsider the conventional view of hydrogen bonding in Watson-Crick base pairs and to describe some of the different hydrogen bonding patterns observed in non-Watson-Crick base associations in duplex structures. There are additional base-associations in larger RNA fragments, for example tRNA, which we will not cover and readers are directed elsewhere for coverage of those structures *(2, 7)*.

A Definition for the Hydrogen Bond. Before a description of the varied base pair hydrogen bonding patterns that have been observed in oligonucleotide structures is provided, it is instructive to consider what constitutes a hydrogen bond. Initially, it was thought that a hydrogen bond could only be formed if the donor-atom-acceptor atom distance was less than the sum of their van der Waals radii *(8)*. So, based on distances observed from single crystal X-ray diffraction studies hydrogen bonding could be assigned. It was later realised that this distance restraint was too strict and ignored the electrostatic arguments that the attractive energy diminishes linearly with increasing distance and so the assignment of hydrogen bonds should not be confined by such overly strict distance criteria *(9)*. The most appropriate definition, in our opinion, has been provided by Steiner and Saenger *(10)*. They defined a hydrogen bond as "any cohesive interaction X-H•••Y, where H carries a positive charge and Y a negative (partial or full) charge, and the charge on X is more negative than on H". Such a definition is attractive and can be widely applied to accurate X-ray structures or to less accurate modelling excercises *(11)*.

The conventional hydrogen bonds formed by the nucleotide bases involve N-H•••O and N-H•••N interactions. The existence of C-H•••O hydrogen bonds has in the past been a source of controversy yet now the role of this type of interaction in stabilising molecular assemblies is widely recognised *(12,13)*. Consideration of bond polarization effects suggests that some C-H•••O interactions satisfy the definition of the hydrogen bond as given above.

Watson-Crick pairing. The A•T and G•C pairs are shown in Figure 1. The conventional description of the hydrogen bonding is that the G•C is held together by three H-bonds whilst the A•T is held together by only two H-bonds. We had previously analysed the detailed geomety of the A•T pairing and suggested that there is in fact a third H-bond, a C-H•••O interaction on the minor groove side *(14)*. Recent theoretical studies have supported this idea *(15)*. This is not a strong stabilising interaction but rather it serves the purpose of alleviating the destabilising effects of having an unfulfilled hydrogen bond donor or acceptor group in the structure. The incorporation of such a weak interaction may be a useful consideration in molecular modelling of nucleic acids.

Mismatches. Watson-Crick A•T or G•C pairs have to compete with 8 non-Watson-Crick alternatives termed mispairs or mismatches. These are the purine-pyrimidine G•T and A•C pairings, the purine-purine G•G, A•A and G•A and the pyrimidine-pyrimidine C•C, T•T and C•T mismatches. Mutagenic pathways are divided into transition and transversion paths. The former invokes purine-pyrimidine mismatches, the latter purine-purine or pyrimidine-pyrimidine mispairs.

The incorporation of non-Watson-Crick base pairs in duplex DNA is the most common error occurring on replication. The theory of mispair formation was initially proposed by Watson and Crick *(16*, (extended by Topal and Fresco *(17)* and extensively reviewed by Strazewski and Tamm *(18)*) and postulated the involvement of rare tautomer forms of the bases. The mismatches involving these tautomers would be sterically equivalent to Watson-Crick base pairs and unlikely to distort the duplex into which they are formed. The crystallographic study of mispairs cannot give any information on the occurrence of rare tautomers during the replication process. The resolution to which most structures are determined does not allow for the precise location of hydrogen atoms and these have to be inferred using geometric considerations. One of the main conclusions from mismatch studies is that there is no need to invoke the presence of rare tautomers in mismatch formation and stability.

Crystallographic study of mismatches have concentrated on sequences known to crystallise into which the mispairs were engineered. A common framework has been the B-DNA Drew-Dickerson dodecamer duplex, d(CGCGAATTCGCG) *(19)*. Other templates have been A-form DNA octamers and Z-form hexamers *(20)*. In each case a duplex containing two mispairs has been formed. This approach maximises the chances of getting well ordered single crystals and means that there is a native Watson-Crick structure for comparative purposes.

Purine-pyrimidine base pairs. The G•T pair was the first mismatch to be characterised and this in an A-form octamer *(21, 22)*. The mispair was subsequently studied in different sequence environments and in different DNA forms *(23-25)*. This purine-pyrimidine pair adopts what is termed the wobble configuration that was first proposed by Crick to explain G•U pairing at the third codon position during codon-anticodon interactions *(26)*. In this mispair, the purine is shifted towards the DNA minor groove, the pyrimidine towards the major groove. The bases maintain the major tautomeric forms and create two inter-base hydrogen bonds (Fig. 2a). Solvent molecules bridge functional groups on the bases in both grooves and confer added stability. In addition, G•U^{Br} and G•U^{F} pairs (where uracil contains a bromine or fluorine at the 5 position) have been characterised in Z-form hexamers *(27, 28)* and wobble G•U pairs plus attendant solvent molecules observed in a fragment of 5S rRNA *(29)*.

Inosine (I) is a guanine analogue lacking the 2-amino group. It is commonly found in tRNA and is able to pair with A, C and U in codon-anticodon interactions. This is an important base since the ability to pair with three other bases contributes to the degeneracy of the genetic code. Inosine occurs rarely in DNA, as a result of deamination of deoxyguanosine, where it is potentially mutagenic. The I•T pair *(30)* assumes a similar structure to the G•T although the loss of N2 on the minor groove side of the duplex removes the possibility of a stabilising water bridge between the bases.

The A•C mismatch is similar to the G•T, but there are two arrangements that need to be considered to explain the formation of two H-bonds linking the bases *(31, 32)*, (Fig 2b, c). A solvent molecule can link the bases on the major groove side to aid stability but not on the minor groove side. The adenine is either protonated or in a rare tautomeric form. Energetic considerations support the former and biophysical characterisation of A•C mispairs using NMR and UV melting methods over a wide pH range were subsequently to prove this *(33)*. This pair should be denoted as A^{+}•C.

Purine-Purine Base Pairs. A•G and G•G pairs have both been characterised in duplex B-DNA. The A•A pairing has only been observed in RNA structures *(34, 7*

a

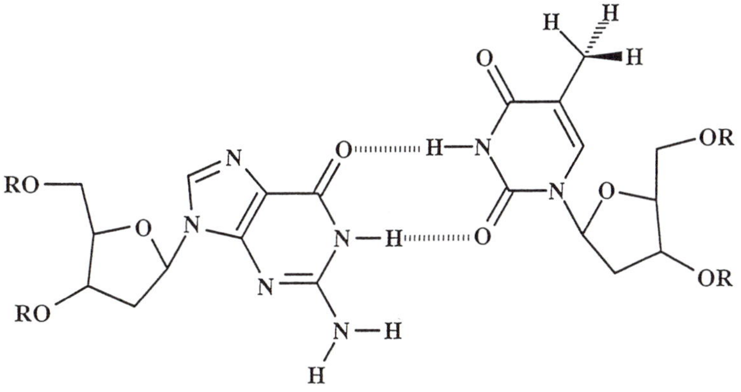

G.T

b

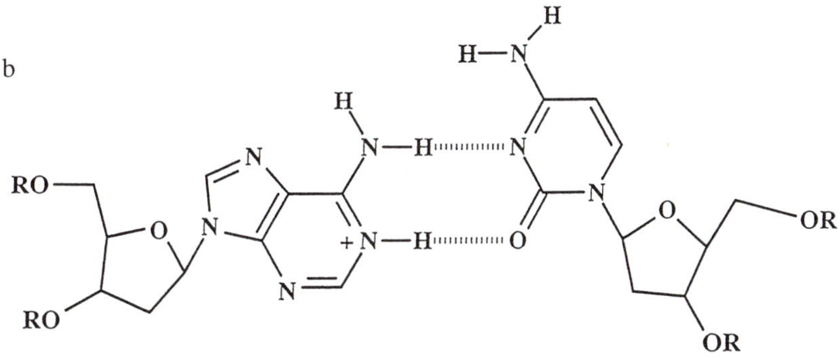

A^{+}.C

c

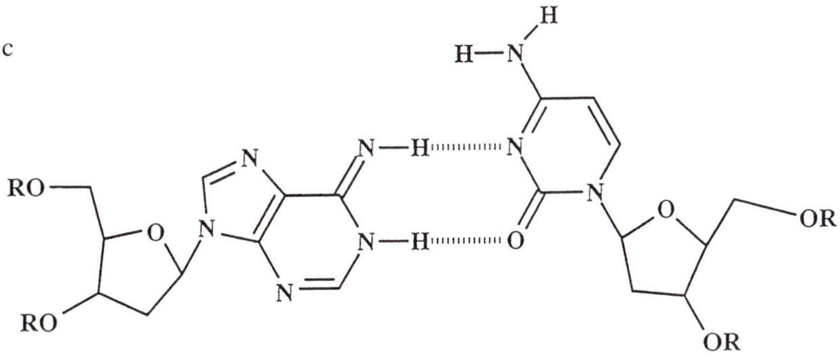

A(imino).C

Figure 2. (a) The wobble G•T (b) the A^{+}•C and (c) A•C with the adenine in the imino form.

and references theirin). Biochemical studies have indicated that A•G mismatches are repaired with much less efficiency than other mispairs *(35).*

Structural studies have identified four G•A configurations in DNA *(36-39),* (Fig. 3). The form of the mispair that is observed is dependent on the pH, salt concentration and in particular on the sequence environment. The sequence dependence of the G•A conformation can be rationalised by dipole-dipole interactions with adjacent bases *(40).* The possibility of forming a hydrogen bond using a functional group provided by an adjacent base can also be important. This is clear in the example of the G(*anti*)•A(*anti*) pairing where the presence of an inter-base-pair hydrogen bond with the amino N2 of guanine to the O2 of an adjacent thymine on the opposing strand has been noted *(36).* Without an O2 in this position some other G•A conformation might be preferred. The key point about studies on the G•A mispair is that the variablility of conformations that can be observed would present quite a challenge to an enzyme recognition and repair system and this may be the reason for low levels of G•A mismatch repair.

In the RNA duplex structure, r(CGCGAAUUAGCG) there are two A(*anti*)•G(*anti*) base pairs and evidence to suggest the same degree of variability observed in DNA *(38).* A careful investigation of the hydrogen bonding possibilities indicates that the A(*anti*)•G(*anti*) pairing uses a conventional hydrogen bond formed between N6 and O6 and what is termed a reverse, three-center hydrogen bond in which the lone pair on N1 is shared with the N-H groups of the guanine N1 and N2. This avoids the destabilising effects of having unsatisfied hydrogen bonding functional groups.

The structural variation of the G•A mismatch also applies to I•A pairs *(40-42)* and may help explain the mutagenicity of inosine. UV melting studies suggest that inosine containing mismatches are quite stable *(43)* whereas most other mispairs destabilise the DNA duplex and produce local melting effects that promote strand dissociation. Repair enzymes may utilise such a physical property of the mismatch duplex to recognise incorrect base pairing. Local destabilisation would also assist the flipping out of mismatched bases for excision. The phenomenon of base flipping as part of the protein recognition and repair process has been noted on the basis of crystallographic studies *(5).*

There has only been one duplex structure characterised which contains two homopurine G•G mismatchs. In each case, although a G(*anti*)•G(*syn*) arrangement is observed the details are slightly different for the two mispairs and two hydrogen bonding schemes have been put forward (Fig. 4). G•G transversion mismatches are readily repaired and in this case the authors note that the sugar-phosphate backbone is distorted at each mismatch site in comparison to the native duplex *(44).*

The dodecamer r(GGCCGAAAGGCC) displays both G•A and A•A associations each of which use two inter-base hydrogen bonds in pairing *(34).* The G•A in this case looks like that shown in Fig. 3d.

Pyrimidine-Pyrimidine Base Pairs. These mismatches have proven difficult to characterise when incorporated in duplex DNA but there are some examples of C•U and U•U associations in duplex RNA. The C•U mispair has been observed in r(GGACUUCGGUCC) with a single hydrogen bond between the bases involving CN4 and UO4 and a bridging solvent linking the two N3 groups *(45),* (Fig. 5).

The U•U pair is polymorphic. What are called *cis* U•U wobble pairs have been observed in two RNA dodecamer structures *(46,47).* The U•U pairs, each held together with two hydrogen bonds are formed (Fig. 6a) and although an ordered solvent is not observed in both crystal structures, this pair has what appears to be an attractive site to bring in a water molecule in both major and minor groove sides. This would be similar to the G•T mismatch discussed above. The nonameric

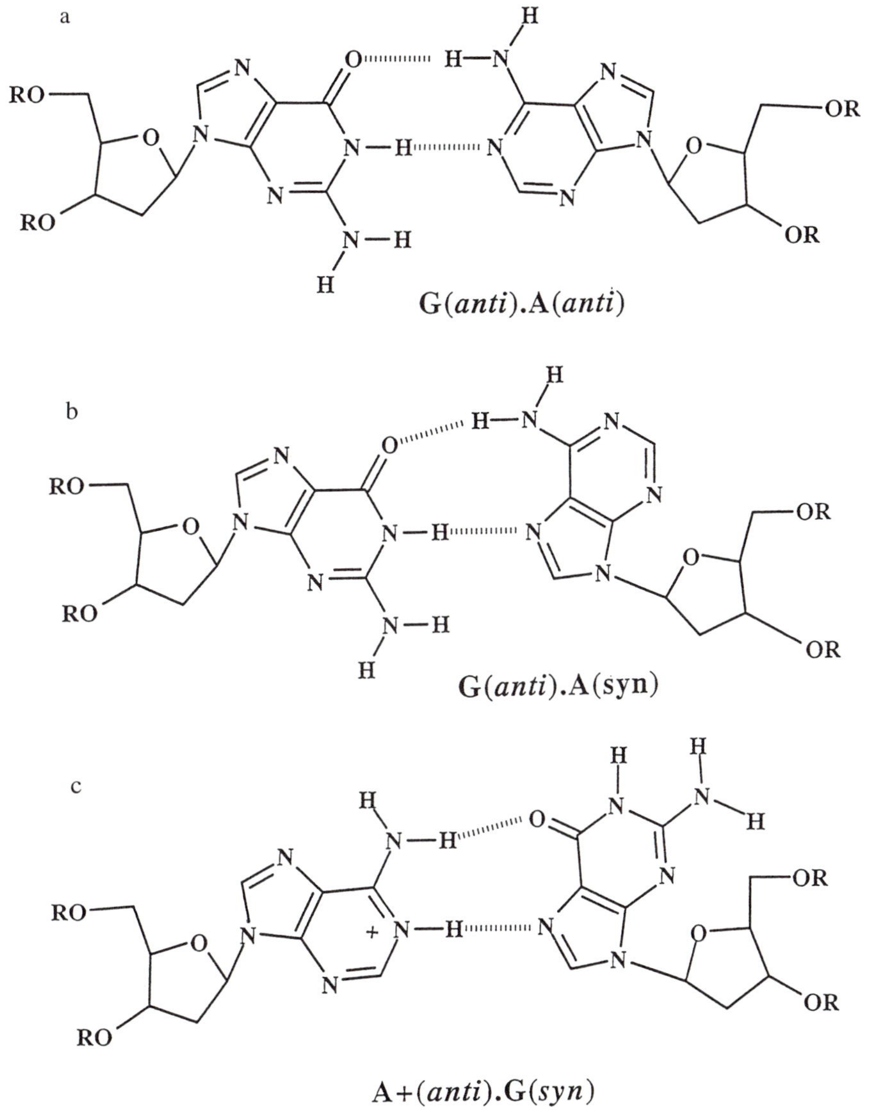

a

G(*anti*).A(*anti*)

b

G(*anti*).A(syn)

c

A+(*anti*).G(*syn*)

Figure 3. The variable G•A pairing (a) G(*anti*)•A(*anti*), (b) G(*anti*)•A(*syn*), (c) A^+(*anti*)•G(*syn*) and (d) G(*anti*)•A(*anti*) amino.

Continued on next page

d

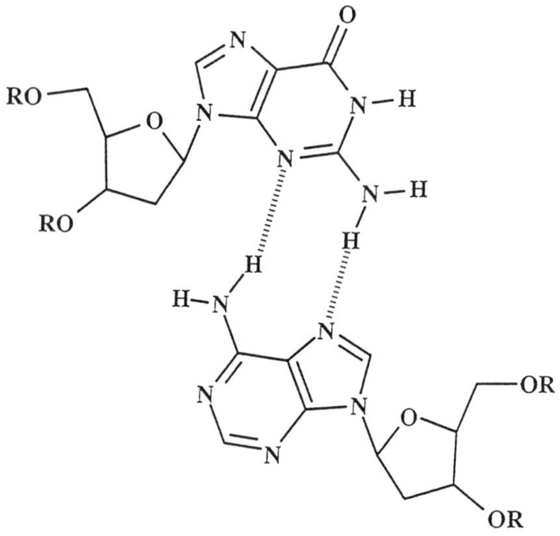

G.A(amino)

Figure 3. *Continued.*

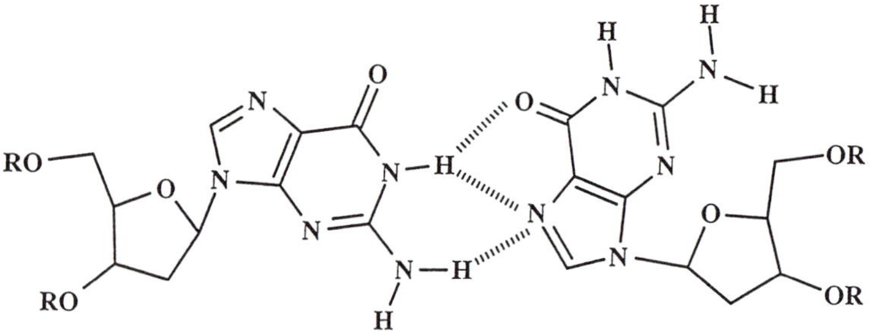

G(*anti*).G(*syn*)

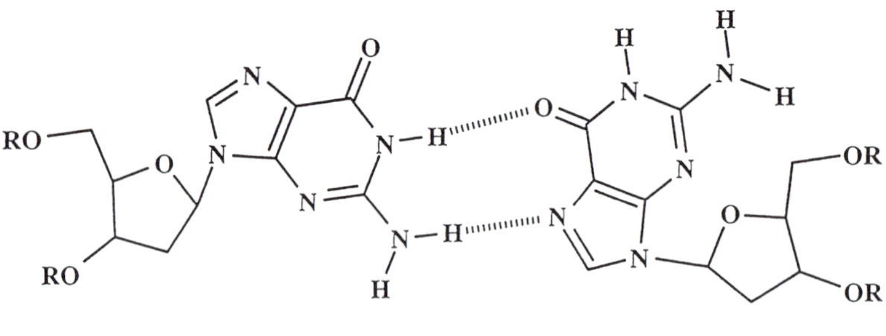

G(*anti*).G(*syn*)

Figure 4. Two slightly different G•G mismatches have been characterised in a B-form dodecamer duplex. Although they are both G(*anti*)•G(*syn*) the details of the assigned hydrogen bonding varies.

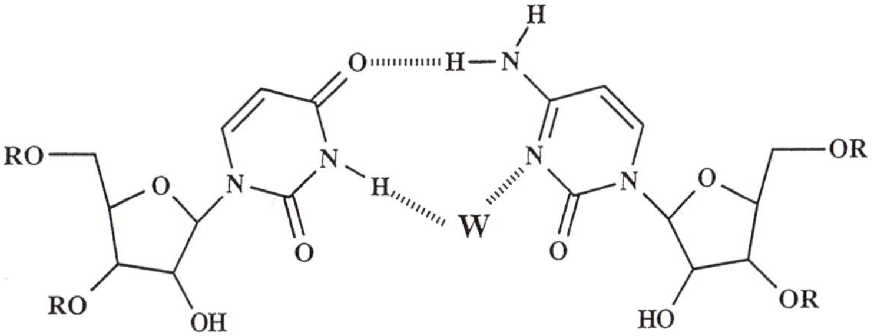

U.C

Figure 5. The U•C mispair observed in an RNA fragment. W represents the water molecule that bridges the bases.

a

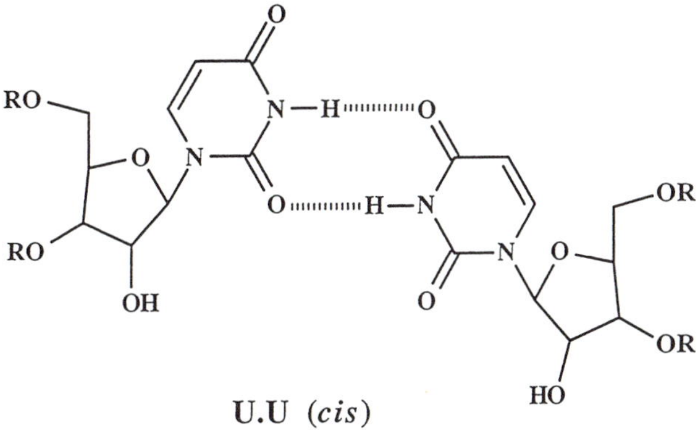

U.U (*cis*)

b

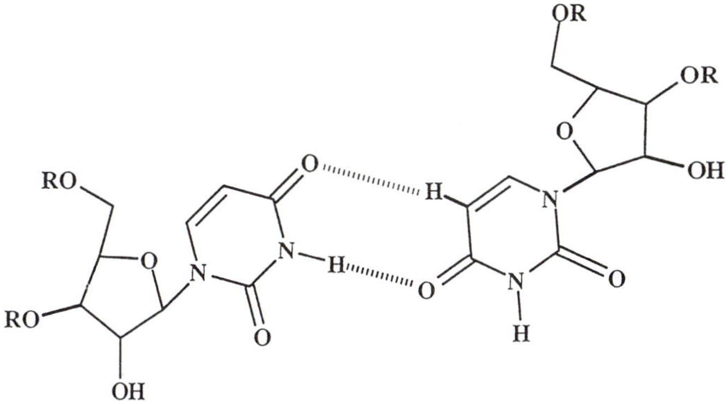

U.U (*trans*)

Figure 6. Two forms of the U•U pairing ((a)*cis* and (b)*trans*) as observed in RNA.

r(GCUUCGGC)d(U^{Br}) has a similar U•U pair at the end of one of the helices which is disordered *(48)* The hexanucleotide, r(UUCGCG) crystallises with a tetranucleotide duplex involving C•G pairs and two U•U pairs formed by the overhanging bases of neighbouring duplexes *(49)*. There is a conventional hydrogen bond between N3 to O4 but also a C-H••O hydrogen bond between C5 and O4 (Fig. 6b).

Pairings with Modified Bases. To compound the pressures of carrying out the replication of DNA involving many bases, the genetic code is constantly under attack from chemical and physical forces in the environment or that are generated in cells during the normal course of metabolism. Carcinogenic chemicals, ultraviolet light, ionising radiation and reactive oxygen species can all produce modifications to DNA *(3,4)*. Of particular interest are alterations to the purines and a number of examples are depicted in Figure 7.

Guanine reacts with alkylnitrosoureas to form O-6-methylguanine (O6MeG) which is potentially very damaging since it alters the hydrogen bonding potential of the base. The effect can be to promote G to A transition mutations. The O6MeG•T mispair could be selected during replication in preference to a O6MeG•C pair. The crystal structure of a O6MeG•C pair has been determined at physiological pH *(50-52)* and is shown to adopt a wobble conformation (Fig. 7a).

Chemical damage is not only induced by alkylating agents. Vinyl chloride, for example, reacts with adenine producing 1, N^6-ethenoadenosine (edA). The structure of the G•edA pairing has been determined *(53)* and the association is depicted in Figure 7d. There are two obvious hydrogen bonds and a C-H•••O hydrogen bond has been invoked between the 8H and O6 of G to alleviate the destabilisation of an unsatisfied hydrogen bond acceptor in the pair. Unlike many of the other non-Watson-Crick pairings that have been characterised, the G•edA pairing produces a significant distortion of the sugar-phosphate backbone. Alterations in the bond angles associated with the furanose-phosphate backbone lead to a bulge in the structure. Such perturbation might represent a signal for the recognition and repair of this modified base by 3-methyladenine-DNA glycosylase.

Purines undergo oxidation at the 8-position to produce 8-oxoadenine (O8A) and 8-oxoguanine (O8G) where the bases are predominantly in the *keto* form. Modification at the 8-position does not affect the hydrogen bonding patterns in G•C and A•T pairs but the presence of the 8-O and N7H does promote alternative possibilities and a *syn* conformation about the glycosidic bond. This has been observed in structures of O8G•A and O8A•G pairings *(54,55* Fig 7e, 7f). The highly mutagenic O8G in genomic DNA can facilitate the G to T transversion mutation via an intermediate O8G•A base pair. Such a pairing has been shown to be fairly stable. This property in combination with a psuedosymmetry about the glycosidic bonds suggests why this pair is not readily recognised by proof reading enzymes. O8A is not very mutagenic and the O8A•G pairing whilst again showing a *syn anti* pair is asymmetric about the glycosidic bonds, a structural feature that may assist recognition and repair. Four bifurcated hydrogen bonds resulting from two reverse three centred hydrogen bonding systems hold the bases in place. All functional groups participate in hydrogen bonds.

a

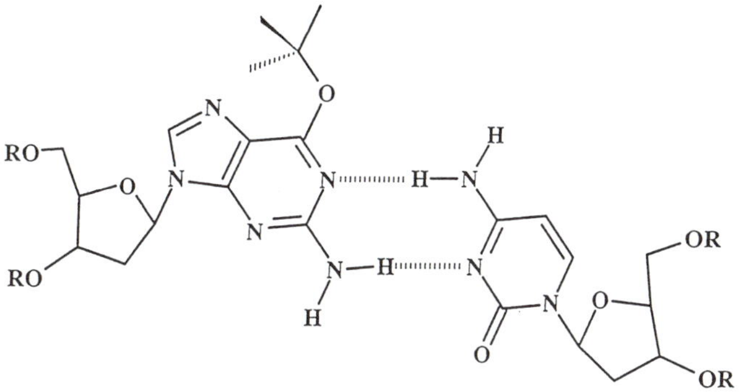

O(6)-MeG.C

b

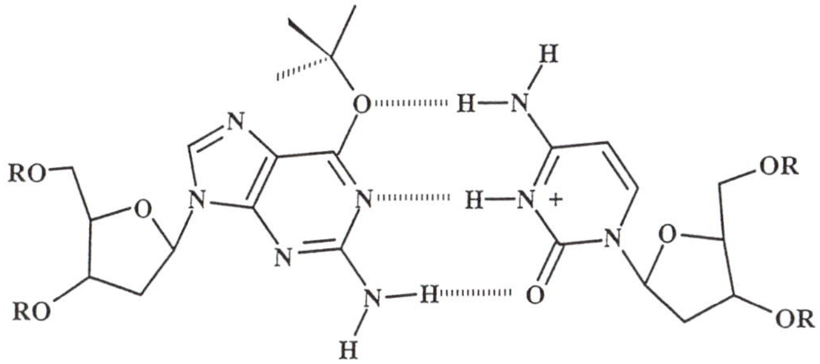

O(6)-MeG.C+

c

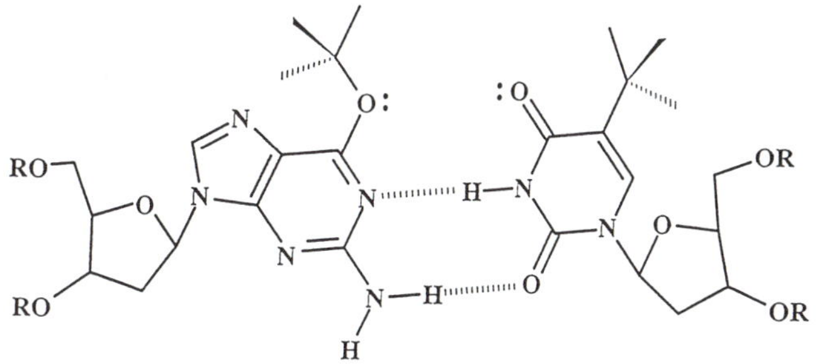

O(6)-MeG.T

Figure 7. (a) The O6 - MeG•C pair which mimics the G•T mismatch. (b) The O6 - MeG•C+ pairing which resembles a Watson - Crick base pair. (c) O6 - MeG•T mismatch which also resembles a Watson - Crick pair. (d) The G (*anti*)•edA pair where edA is ethenoA. (e) The A(*anti*)•O8G(*syn*) and (f) G(*anti*)•O8A(*syn*) pairings where O8G and O8A represent 8 - oxoG and 8 - oxoA respectively.

Continued on next page

d

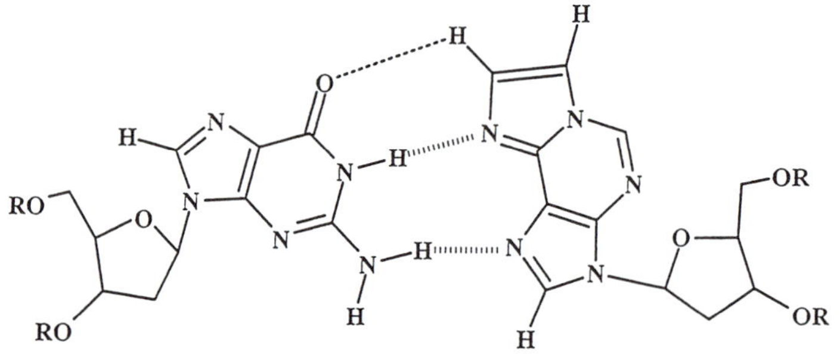

G(*anti*).ethenoA(syn)

e

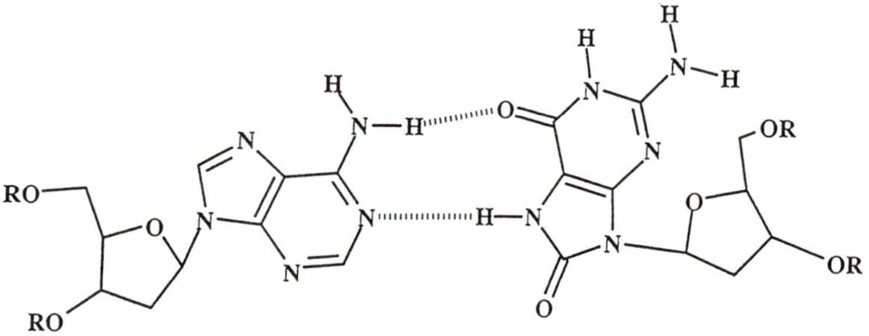

A(*anti*).8-oxoG(*syn*)

f

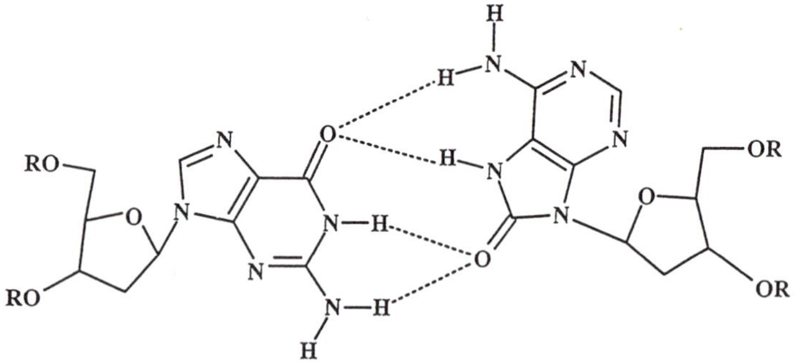

G(*anti*).8-oxoA(*syn*)

Figure 7. *Continued.*

Acknowledgments. The Wellcome Trust, the Biotechnology and Biochemistry Science Research Council, the Engineering and Physical Sciences Research Council have provided support for our studies. Dr T. Steiner is thanked for permission to reference unpublished material and for sharing his results with us.

Literature Cited.

(1) Watson, J.D. and Crick, F.H.C. **(1953)** *Nature 171*, 737-738.
(2) Saenger, W. **(1984)** *The Principles of Nucleic Acid Structure*, Springer-Verlag, NY.
(3) Loft, S. and Poulsen, H.E. **(1996)** *J. Mol. Med. 74*, 297-312.
(4) Modrich, P. (1987) *Ann. Rev. Biochem. 56,* 435-466.
(5) Pearl, L.H. and Savva, R. **(1995)** *TIBS 20*, 421-426.
(6) Brown, T., Hunter, W.N. and Leonard, G.A. **(1993)** *Chem. in Brit. 6*, 484-488
(7) Scott, W.G. and Klug, A. **(1996)** *TIBS 21*, 220-224.
(8) Hamilton, W.C. and Ibers, J.A. **(1968)** *Hydrogen bonding in solids*, W.A. Benjamin, New York, USA.
(9) Umeyama, H. and Morokuma, K.J. **(1977)** *J.Amer. Chem. Soc. 99*, 1316-1332.
(10) Steiner, T. and Saenger, W. **(1996)** *J. Amer. Chem. Soc. 114*, 10146-10154.
(11) Price, S.L. and Goodfellow, J.M. **(1992)** Ch 5. In *"Computer modelling of biomolecular procesess"*. Eds Goodfellow, J.M. and Moss, D.S. Ellis Howood Ltd, Chichester, UK.
(12) Desiraju, G.R. **(1996)** *Acc. Chem. Res. 29*, 441-450.
(13) Steiner, T. **(1996)** *Cryst. Rev. 6*, 1-57.
(14) Leonard, G.A., McAuley-Hecht, K., Brown, T. and W.N. Hunter., **(1995)** *Acta Cryst D51*, 136-139.
(15) Starikov, E.B. and Steiner, T. **(1997)** *Acta Cryst. Sect D* in press.
(16) Watson, J.D and Crick, F.H.C. **(1953)** *Nature 171*, 964-966.
(17) Topal, M.D. and Fresco, J.R. **(1976)** *Nature 263*, 290-293.
(18) Strazewski, P. and Tamm, C. **(1990)** *Angew. Chem. Intl. Edt. Engl. 29*, 36-57.
(19) Wing Drew H.R., Wing, R.M., Takano, T., Broka, C., Takana, S., Itakura, K. and Dickerson, R.E. **(1980)** *Nature 287*, 755-758.
(20) Kennard, O. and Hunter, W.N. **(1991)** *Angew. Chemie.* 30, 1254-1277.
(21) Brown, T., Kennard, O., Kneale, G. and Rabinovich, D. **(1985)** *Nature 315*, 604-606.
(22) Hunter, W.N.,Kneale, G., Brown, T., Rabinovich, D. and Kennard, O. **(1986)** *J. Mol. Biol. 190*, 605-618.
(23) Kneale, G., Brown, T., Kennard, O. and Rabinovich, D. **(1985)** *J. Mol. Biol. 186*, 805-814.
(24) Hunter, W.N., Brown, T., Kneale, G., Anand, N.N., Rabinovich, D and Kennard, O. **(1987)** *J. Biol. Chem.* ***262***, 9962-9970.
(25) Ho, P.S., Frederick, C.A., Quigley, G., van der Marel, G.A. van Boom, J.H., Wang, A.H-J. and Rich, A. **(1985)** *EMBO J. 4*, 3617-3623.
(26) Crick, F.H.C. **(1966)** *J. Mol. Biol, 19*, 548-555.
(27) Brown, T., Kneale, G., Hunter, W.N. and Kennard, O. **(1986)** *Nucleic Acids Res. 14*, 1801-1809.
(28) Coll, M., Saal, D., Frederick, C.A., Aymami, J., Rich, A., Wang, A.-H. J. **(1989)** *Nucleic Acids. Res. 17*, 911-923.
(29) Betzel, C., Lorenz, S., Furste, J.P., Bald, R., Zhang, M., Schneider, T., Wilson, K.S. and Erdmann, V.A. **(1994)** *FEBS Lett. 351*, 159-164.
(30) Cruse, W.B.T., Aymami, J., Kennard, O., Brown, T., Jack, A.G.C. and

Leonard, G.A. **(1989)** *Nucleic Acids Res. 17*, 55-72.
(31) Hunter, W.N., Brown, T., Anand, N.N. and Kennard, O. **(1986)** *Nature 320*, 552-555.
(32) Hunter, W.N., Brown, T. and Kennard, O. **(1987)** *Nucleic Acids Res. 15*, 6589-6606.
(33) Brown, T., Leonard, G.A., Booth, E.D. and Kneale, G. **(1990)***J. Mol. Biol. 221*, 437-440.
(34) Baeyens, K.J., De Bondt, H.L., Pardi, A. and Holbrook, S.R. **(1996)** *Proc. Natl. Acad. Sci. USA. 93,* 12851-12855.
(35) Fersht, A.R., Knill-Jones, J.W. and Tsui, W.C. **(1982)** *J. Mol. Biol. 156*, 37-51.
(36) Prive, G.G., Heinemann, U., Kan, L.S., Chandrasegaran, S., and Dickerson, R.E. **(1987)** *Science 238*, 498-504.
(37) Brown, T., Hunter, W.N., Kneale, G.G., and Kennard, O. **(1986)** *Proc. Natl. Acad. Sci. USA 83*, 2402-2406.
(38) Brown, T., Leonard, G.A., Booth, E.D. and Chambers, J. **(1989)** *J. Mol. Biol. 207*, 455-457.
(39) W.N. Hunter, T. Brown and O. Kennard. **(1986)** *J. Biomolecular Structure and Dynamics. 4*, 173-191.
(40) Leonard, G.A., McAuley-Hecht, K., Abel, S., Lough, D.M., Brown, T. and W.N Hunter, W.N. **(1994)** *Structure 2*, 483-494.
(41) Corfield, P.W.R., Hunter, W.N., Brown, T., Robinson, P and Kennard, O **(1987)** *Nucleic Acids Res. 15*, 7935-7949.
(42) Webster, G.D., Sanderson, M.R., Skelly, J.V., Neidle, S., Swann, P.F., Li, B.F. and Tickle, I. **(1990)** *Proc. Natl. Acad. Sci. USA. 87*, 6693-6697.
(43) G.Leonard, E.Booth, W.N. Hunter, T. Brown. **(1992)** *Nucleic Acids Res. 20*, 4753-4759.
(44) Skelly, J.V., Edwards, K.J., Jenkins, T.C. and Neidle, S. **(1993)** *Proc. Natl. Acad. Sci. USA. 90*, 804-808.
(45) Holbrook, S.R., Cheong, C., Tinoco Jr, I. and Kim, S. H. **(1991)** *Nature 353*, 579-581.
(46) Baeyens, K.J., De Bondt, H.L. and Holbrook, S.R. **(1995)** *Nature Structural Biology 2*, 56-62.
(47) Lietzke, S.E., Barne, C.L., Bergland, J.A., and Kundrot, C.E. **(1996)** *Structure 4*, 917-930.
(48) Cruse, W.B.T., Saludjian, P., Biala, E., Strazewski, P., Prange, T. & Kennard, O. *Proc. Natl. Acad. Sci.* **(1994)** *91*, 4160--4164.
(49) Wahl, M.C. Rao, S.T. and Sundaralingham, M. **(1996)** *Nature Structural Biology 3*, 24-30.
(50) Leonard, G.A., Thomson, J.B., Watson, W.P. and Brown, T. **(1990)** *Proc. Natl. Acad. Sci. USA 87*, 9573-9576.
(51) Ginell, S.L., Vojtechovsky, J., Gaffney, B., Jones, R. and Berman, H.M. **(1994)** *Biochemistry 33*, 3487-3493.
(52) Vojtechovsky, J., Eaton, M.D., Gaffney, B., Jones, R., Berman, H.M. **(1994)** *Biochemistry 34*, 16632-16640.
(53) Leonard, G.A., McAuley-Hecht, K.E., Gibson, N.J., Brown, T., Watson, W.P. and Hunter, W.N. **(1994)** *Biochemistry 33*, 4755-4761.
(54) Leonard, G.A., Guy, A., Brown, T., Teoule, R. and Hunter, W.N. **(1992)** *Biochemistry 31*, 8415-8420.
(55) McAuley-Hecht, K.E., Leonard, G.A., Gibson, N.J., Thomson, J.B., Watson, W.P., Hunter, W.N. and Brown, T. **(1994)** *Biochemistry 33*, 10266-10270.

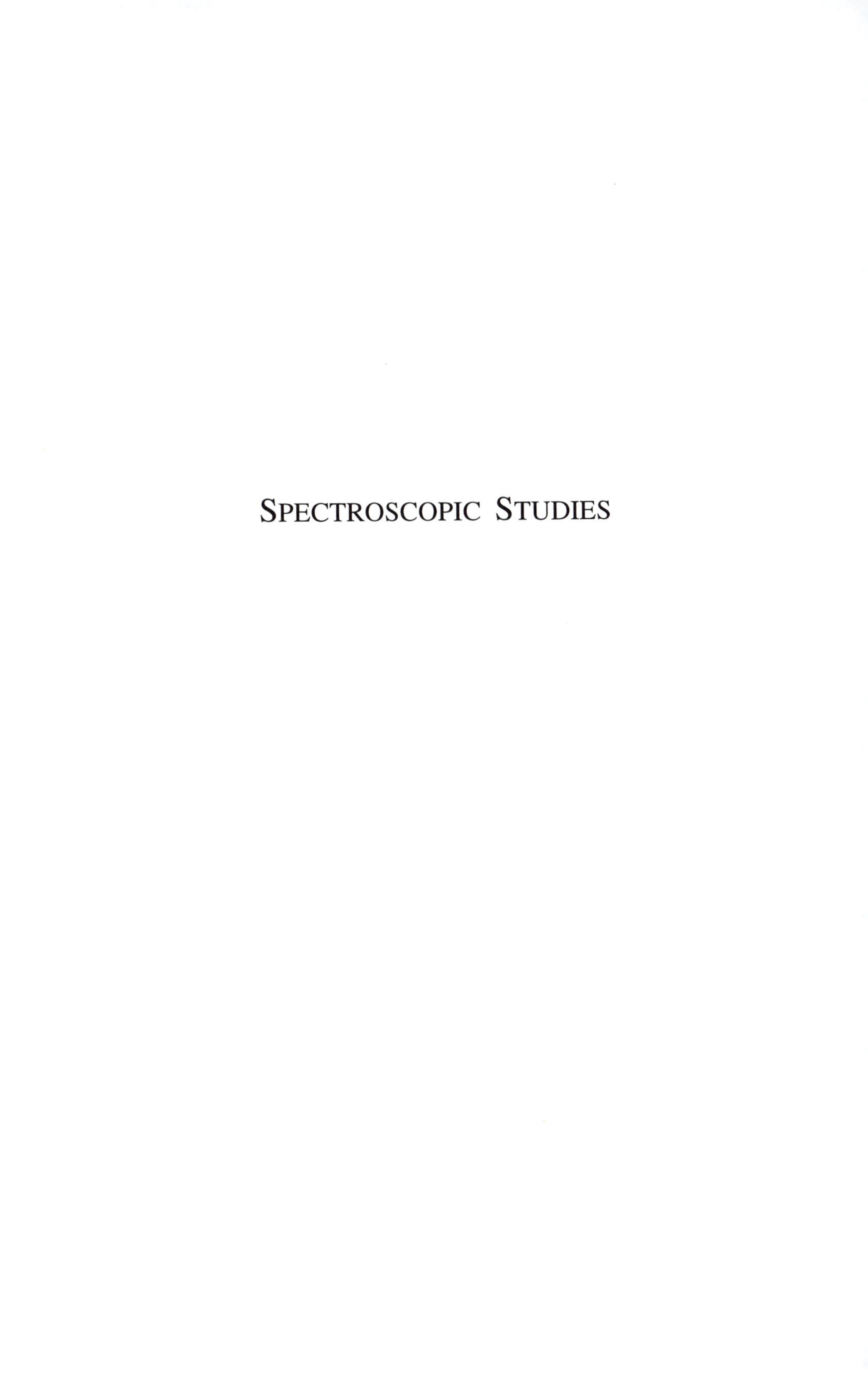

SPECTROSCOPIC STUDIES

Chapter 6

Structure and Stability of DNA Containing Inverted Anomeric Centers and Polarity Reversals

James M. Aramini[1], Johan H. van de Sande[2], and Markus W. Germann[1,3]

[1]**Department of Microbiology and Immunology, Kimmel Cancer Institute, Thomas Jefferson University, Philadelphia, PA 19107**
[2]**Department of Midical Biochemistry, University of Calgary, Calgary, Alberta T2N 4N1, Canada**

The promise of controlling gene expression in a specific and efficient manner has spurred a large research effort into antisense DNA therapy (*1*). To be effective as an antisense drug an oligonucleotide (ODN) must possess a number of properties including: i) nuclease resistance, ii) stable and specific complex formation with the target mRNA, iii) cellular uptake, and iv) the resulting hybrid duplex should be sensitive to cleavage by RNase H. In order to improve the antisense potential of natural ODNs numerous modifications to the phosphodiester backbone, bases and deoxyribose moieties have been explored (*2*).

Our strategy (*3*) and that of others (*4*,*5*) is to employ a combination of alpha anomeric nucleotides in conjunction with 3′-3′ and 5′-5′ linkages to generate a new class of antisense therapeutics which contain tracts of α- and β-anomeric DNA with improved properties (Figure 1). The 3′-3′ and 5′-5′ linkages allow the local inversion of the strand polarity of the α-anomeric tracts, enabling the parallel stranded α-anomeric component of the ODN to form Watson-Crick base pairs with the RNA target. Recent studies demonstrated that such an approach can lead to the design of ODNs that permit RNase H to destroy the RNA component of an ODN•RNA heteroduplex, a desired property that is not displayed by hybrids containing exclusively α-anomeric DNA (*6*,*7*). An attractive feature of this design is the possibility of generating ODNs that bind much more strongly to the target RNA than phosphorothioates, due to the enantiomeric purity of α-anomeric nucleotides and the inherent stability of α-ODN/β-RNA hybrids (*8*), and are also nuclease resistant and capable of activating and directing RNase H. These properties make oligonucleotides with polarity and anomeric center reversals viable candidates for antisense molecules.

In this chapter we give a synopsis of our recent enzymatic, thermodynamic, and spectroscopic investigations of ODNs containing α-anomeric nucleotides and polarity reversals (*7*,*9*,*10*).

RNase H Studies

Alpha Winged ODN/RNA Duplexes are Susceptible to Cleavage by RNase H. Mixed anomeric sequences with polarity reversals can activate RNase H. For example, we have investigated the RNase H sensitivity of constructs containing short tracts of β-anomeric nucleotides flanked by alternating α/β-anomeric wings (Figure

[3]To whom correspondence should be addressed: Tel (215) 503–4581; FAX (215) 923–2117; E-mail: MWG@bern.jci.tju.edu.

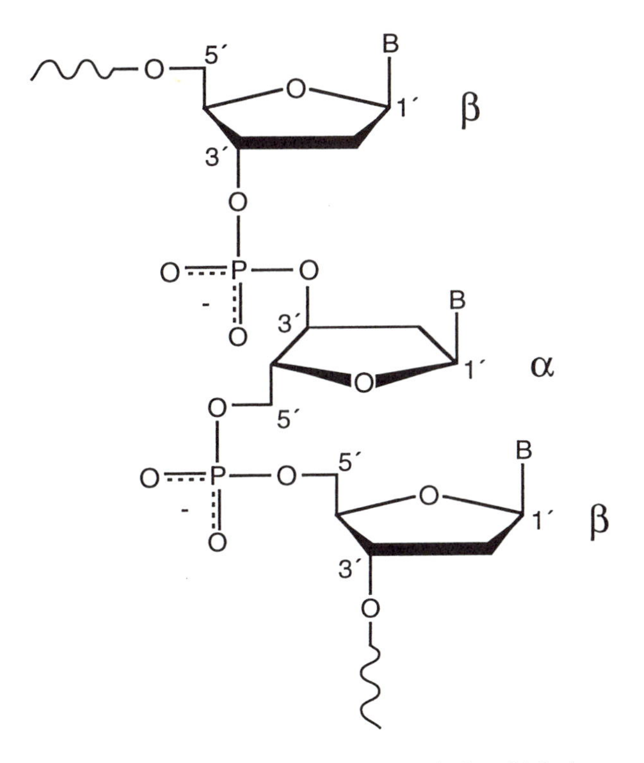

Figure 1. Schematic diagram of a β–α–β stretch, in which the α-anomeric nucleotide is fused to the flanking β-nucleotides via 3´-3´ and 5´-5´ phosphodiester linkages. The 1´, 3´ and 5´ furanose ring positions are labeled; B = nitrogenous base. α-Anomeric nucleotides differ from their natural β-counterparts by a single inversion of stereochemistry at the C1´ position.

2). The wings are composed of four sets of 3´-3´ and 5´-5´ linkages, and thus, are thermodynamically not optimal compared to flanking sequences containing purely α-anomeric nucleotides due to the large number of junction components. However, these wings convey nuclease resistance and, as intended, direct the action of RNase H (Figure 3). For both dA4 and dT4 we observe highly localized *E. coli* RNase H mediated RNA cleavage just outside the β-anomeric recognition window, directly across from the 3´-3´ linkage, resulting in the production of 8- and 12-mers, respectively. The low temperature (13 °C) digest reveals that the dA4 hybrid is more readily cleaved by the enzyme while at higher temperature (21 °C) both reactions went to completion.

α-Containing ODNs Activate RNase H and Destroy RNA Targets in Substoichiometric Amounts. The ability of alpha winged ODNs to activate RNase H was studied in more detail. Of particular interest from a biological point of view is whether or not these ODNs can sustain processive cleavage of RNA targets. Our data (Figure 4) show that a substoichiometric amount of the dA4 ODN (10%) results in the near complete digestion (80%) of a 10 fold larger amount of the complementary RNA strand. As the reaction nears completion very little RNA remains, with the consequence that the RNA-ODN duplex becomes progressively less stable; complete degradation was observed at a lower temperature (21°C). This demonstrates that RNase H cleavage of these duplexes is indeed catalytic, with each dA4 ODN being used an average of eight times. Furthermore, we did not detect any evidence for the degradation of the ODN (data not shown).

Spectroscopic Studies of α-Containing DNA Duplexes

Sequences. In an effort to gain insight into the thermodynamic and underlying structural effects of the introduction of α-anomeric nucleotides and polarity reversals into DNA in a systematic fashion, we have examined a self-complementary decamer model system based on the well-known Dickerson dodecamer (Figure 5). The series consists of an unmodified control, plus four constructs containing a single α-anomeric nucleotide (either αT, αC, αA, or αG) flanked by 3´-3´ and 5´-5´ phosphodiester linkages; the latter serve to reverse the polarity of the α-nucleotide within the predominantly β-anomeric sequence, thereby facilitating base pairing to its complement. As a secondary control, we have also investigated a sequence containing one α-anomeric residue but lacking polarity reversals (alphaT2). Finally, the core of the model sequence corresponds to the *Eco*RI consensus recognition sequence, which provides an additional built-in means of assaying the structural perturbations resulting from the unnatural moieties; our *Eco*RI enzymatic studies on this series of constructs are beyond the scope of this summary (*10*).

Thermodynamic Properties. UV thermal denaturation studies of the decamer family demonstrate that the insertion of two parallel stranded α-anomeric nucleotides in a decamer, along with four polarity reversal junctions, results in a relatively small penalty to the thermal stability of the duplex (Table I). On the basis of the T_m and $\Delta H°$ values, we conclude that duplex stability decreases from the control to alphaC in the following series: control > alphaT ≈ alphaA ≈ alphaG >> alphaC. Interestingly, we observed that alphaT is prone to hairpin formation under conditions of low concentration and elevated temperature. This is an inherent property of self-complementary DNA duplexes (*11*), and can be attributed to the modifications in the central portion of the helix, thereby destabilizing the duplex relative to the hairpin form. Such behavior is very dramatically observed for the alphaT2 sequence, in which the incorrect polarity of the α-nucleotide hampers base pair formation and results in the severe destabilization of the duplex form (see Studies of alphaT2 below).

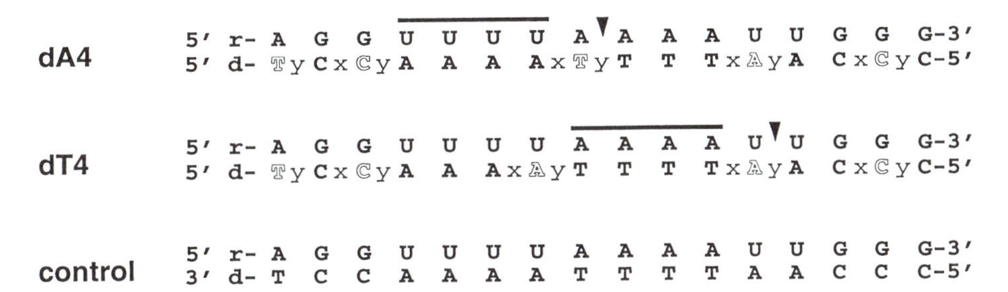

Figure 2. Sequences of the alpha-winged RNase H substrates and their control. In the DNA strand, y indicates a 3′-3′ linkage, x a 5′-5′ linkage, and α-anomeric nucleotides are depicted in outline type. In the dT4 and dA4 hybrids, RNase H sensitive windows are denoted by bars, and the major cleavage sites are marked by arrows.

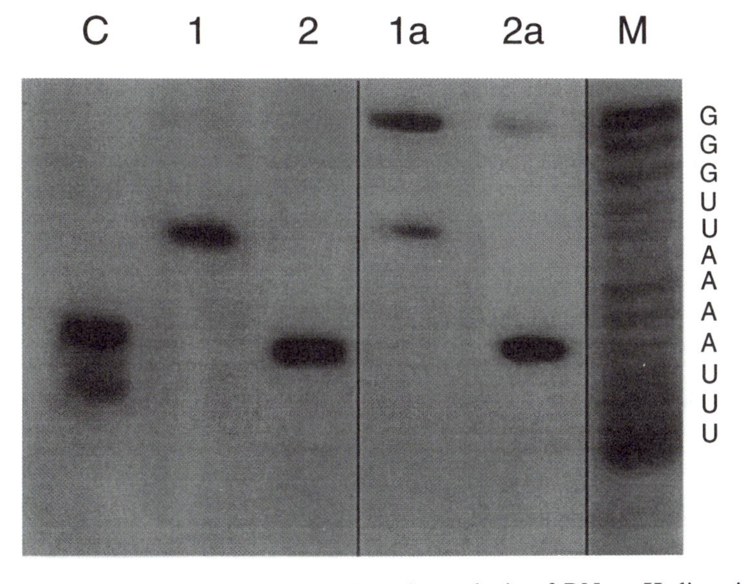

Figure 3. Denaturing gel electrophoretic analysis of RNase H digestion products of RNA•alpha winged ODN hybrids. C, control; 1, dT4; 2, dA4; M, Markers. The reaction temperature was 21 °C for lanes C, 1, and 2 and 13 °C for lanes 1a and 2a.

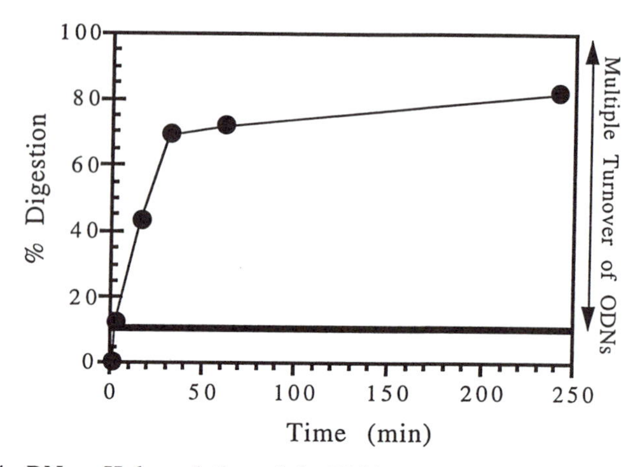

Figure 4. RNase H degradation of the RNA target (UUUU) in the presence of substoichiometric amounts of the dA4 ODN at 30° C. The target RNA was present in a 10 fold excess over ODN. RNA was ^{32}P labeled and quantified by scintillation counting after bands were excised from denaturing gels. The thick line in the plot represents a turnover number of unity.

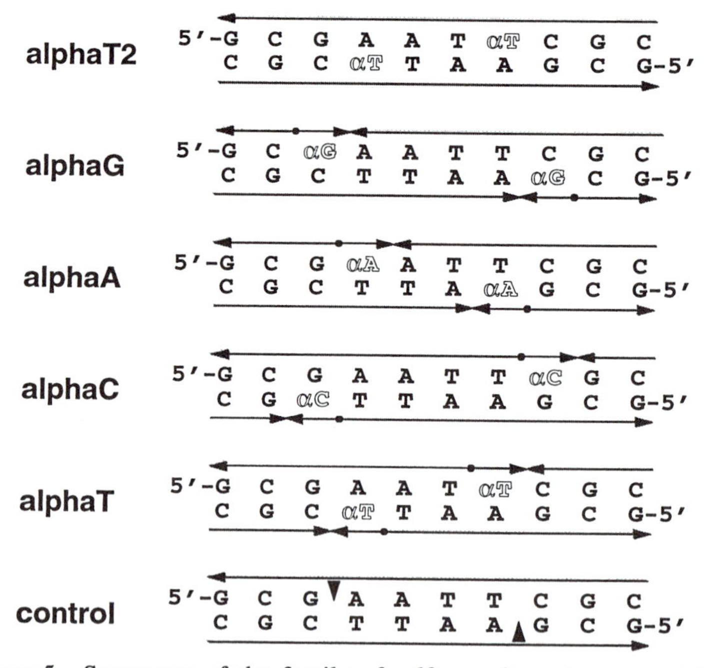

Figure 5. Sequences of the family of self-complementary α-containing decamer duplexes we have investigated and their control. Arrows indicate the strand polarity in the 3′→5′ direction; the 3′-3′ and 5′-5′ phosphodiester linkages are denoted by tail-to-tail and head-to-head junctions. α-Anomeric nucleotides are shown in outline type. Black arrows in the control duplex represent the *Eco*RI restriction enzyme cleavage sites.

Table I. Thermodynamic Data for the Control and α-Containing Decamer Duplexes[a]

duplex	T_m (°C) (C_T = 30 μM)	$\Delta H°$ (kJmol^{-1})	$\Delta S°$ (kJmol^{-1}K^{-1})
control	59.8 (0.1)	343 (12)	0.940 (0.033)
alphaT	54.7 (0.3)	330 (15)	0.946 (0.021)
alphaC	49.8 (0.3)	258 (7)	0.712 (0.022)
alphaA	54.4 (0.2)	323 (13)	0.898 (0.041)
alphaG	54.3 (0.3)	334 (14)	0.933 (0.043)

[a] All measurements were performed under the following buffer conditions: 400 mM NaCl, 10 mM Na_2HPO_4, 0.1 mM EDTA, pH 6.5. Thermodynamic data were obtained from analysis of melting curves using a six parameter fitting routine assuming a two-state (helix ↔ coil) transition and accounting for the temperature dependent absorbances of the helix and coil forms, as well as the concentration dependence of the T_m (*10,12*). Standard deviations are given in parentheses. (Reprinted from ref. 10. Copyright 1997 American Chemical Society).

NMR Studies. In the following sub-sections we present the salient NMR spectroscopic results which conclusively demonstrate that the α-containing duplexes are structurally analogous to the control (i.e., anti-parallel, right-handed B-DNA). Overall, our data indicate that structural perturbations in these species are localized to the regions encompassing the α-nucleotide and unnatural phosphodiester linkages, although subtle differences do arise depending on the nature of the α-nucleotide. The NMR data corroborates the results of other spectroscopic techniques, namely circular dichroism and UV hyperchromicity, which are not presented here.

Imino ^{1}H NMR. Imino ^{1}H NMR spectroscopy is a powerful direct probe for the presence and nature of the base pairing in nucleic acids. As shown in Figure 6, the imino ^{1}H spectra for the control and the four stable α-containing duplexes exhibit resonances for the five chemically distinct imino protons in each self-complementary decamer, indicative of stable base pair formation. In addition, temperature dependent experiments showed no evidence for pre-melting of the base pairs comprising the α-nucleotides. Virtually identical NMR spectra have been obtained for the α-duplexes at optical concentrations (3 to 6 μM duplex), demonstrating that the duplex form exists under the conditions used in the thermodynamic studies.

NOESY walks. For the assignment of non-labile ^{1}H NMR resonances in B-DNA it is standard practice to employ two major NOE pathways, specifically the sugar H1′ ↔ base H6/H8 and sugar H2′/H2′′ ↔ base H6/H8 pathways (*13*). In all of the α-containing decamers we observe that the H1′ network remains intact, as in the control. However, in each case the H2′/H2′′ pathway is broken between the α-nucleotide and the subsequent residue. This effect is due to a flip in the orientation of deoxyribose ring of the α-nucleotide, which places the H2′′ proton too far away from the base moiety in the following residue to make an NOE contact (Figure 7; also see Models below).

Deoxyribose ring conformation. Intraresidue spin-spin (*J*) coupling patterns are highly diagnostic of the type of sugar puckering adopted by deoxyribose rings (*14,15*). It is widely believed that the deoxyribose moiety in DNA undergoes a rapid conformational exchange between the so-called N- and S-type puckers, meaning that the observed *J*-coupling constants for vicinal protons in the ring are a weighted average of the couplings for the two inter-converting forms according to equation 1:

$$J_{XY\,obs} = (1 - f_S)J^N_{XY} + f_S J^S_{XY} \qquad (1)$$

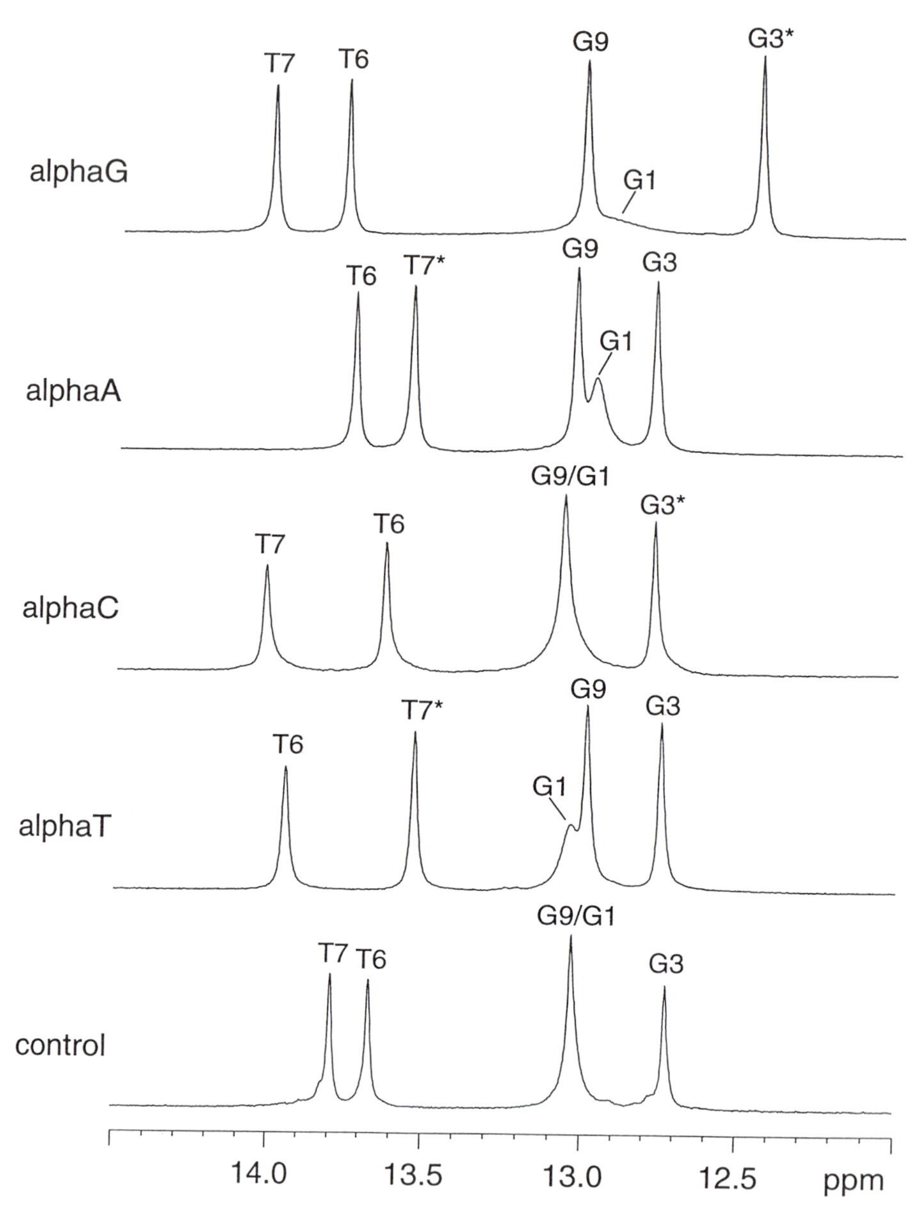

Figure 6. Imino proton region of the ^{1}H (600.1 MHz) NMR spectra of the control (2.6 mM), alphaT (6.8 mM), alphaC (2.4 mM), alphaA (2.9 mM), and alphaG (2.2 mM) duplexes in 90% H_2O/10% D_2O, 50 mM NaCl, 10 mM phosphate buffer, pH 6.5, 293 K. Note that the imino proton signals for G1 and G9 in the control and alphaC decamers are degenerate under these conditions. In each of the α-containing duplexes the imino proton within the base pair involving the α-nucleotide is denoted by an asterisk. (Reproduced from ref. 10. Copyright 1997 American Chemical Society).

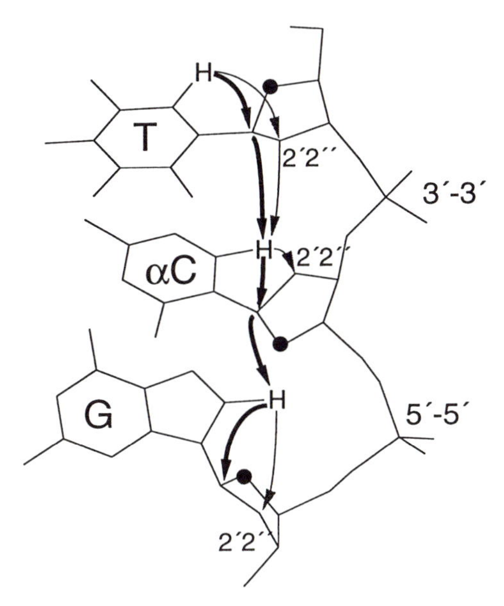

Figure 7. Schematic diagram showing the H1´ ↔ H6/H8 (thick arrows) and H2´/H2´´ ↔ H6/H8 pathways in the T-3´-3´-αC-5´-5´-G stretch of the alphaC decamer. Note the break in the H2´/H2´´ ↔ H6/H8 network between H2´/H2´´ of the α-nucleotide and H8 of the guanosine.

where f_S is the mole fraction of the S-pucker. On the basis of the sum of the H1´ coupling constants (Σ1´; Table II) and spectral simulations (*16,17*), illustrated in Figure 8, we conclude that the *J*-coupling data for all nucleotides in the control duplex, as well as the vast majority in the four α-containing species, are indicative of a high S-character (i.e., large Σ1´, $J_{1'2'} > J_{1'2''}$, very small $J_{2''3'}$); this is the normal situation in B-DNA. However, in three cases (alphaT, alphaC, and alphaA) we find that the conformational equilibrium of the deoxyribose ring in the nucleotide following the 5´-5´ phosphodiester linkage is drastically shifted away from a high S-character ($f_S \approx$ 40 to 60%; see Table II). In light of analogous decreases in f_S for deoxyribose groups within DNA/RNA hybrids reported in the literature (*18*), this effect may have biological ramifications. The sugar moiety in A4 of the alphaG decamer as well as those in all nucleotides preceding 3´-3´ linkages are not perturbed.

Table II. Sugar Puckering Data for the Nucleotides Following the 5´-5´ Phosphodiester Linkage in the α-Containing Decamer Duplexes[a]

	C8			G9	
decamer	Σ1´	f_S	decamer	Σ1´	f_S
control	15.1	0.90	control	15.7	1.00
alphaT	12.9	0.53	alphaC	13.5	0.63

	A5			A4	
decamer	Σ1´	f_S	decamer	Σ1´	f_S
control	15.7	1.00	control	16.2	1.00
alphaA	12.1	0.39	alphaG	14.5	0.80

[a] Data obtained from the H2´´(ω_1)/H1´(ω_2) multiplets along ω_2 in resolution enhanced DQF-COSY, P.E. COSY, or E. COSY spectra. Values of f_S were calculated using the following expression (*14*): $f_S = (\Sigma 1' - 9.8)/5.9$. (Adapted from ref. 10).

^{31}P NMR. Given our use of unusual phosphodiester linkages, the backbone conformation is of crucial importance in our studies of these duplexes. It is well-established that the chemical shift of the phosphorus atom is very sensitive to changes in backbone conformation in nucleic acids (*19*). ^{31}P NMR spectra of the control and the four α-containing duplexes are shown in Figure 10. With the exception of the large (± 1.1 to 1.5 ppm) chemical shift changes observed for the 3´-3´ and 5´-5´ phosphodiester linkages, the vast majority of phosphates resonate in a narrow window of the spectrum corresponding to that for the B-DNA control. The major exception to this is the phosphate directly across from the 3´-3´ linkage in alphaC (i.e., GpA), which experiences a +0.6 ppm shift compared to the control duplex. Downfield shifts of this magnitude have been repeatedly observed in duplex DNA containing mismatches (*20*), and are thought to be due to changes in the relative populations of the commonly observed B_I backbone conformation and the higher energy B_{II} form, toward the latter.

Models. The models of the alphaA and control decamer duplexes shown in Figure 9 are consistent with our spectroscopic data on several grounds. First, the models are in agreement with the finding that the local inversion of the strand polarity at the αA position enables this nucleotide to form a regular Watson-Crick base pair with the T residue on the complementary strand, fitting snugly into the helix while maintaining the base stacking. Second, the bases are in the anti orientation, with the exception of αA4. This residue is formally in the syn range; however, the H8-H1´ distance is similar to that expected for β-anomeric nucleotides in an anti conformation. Third, the observed H1´ ↔ H6/H8 and H2´/H2´´ ↔ H6/H8 NOE pathways are consistent

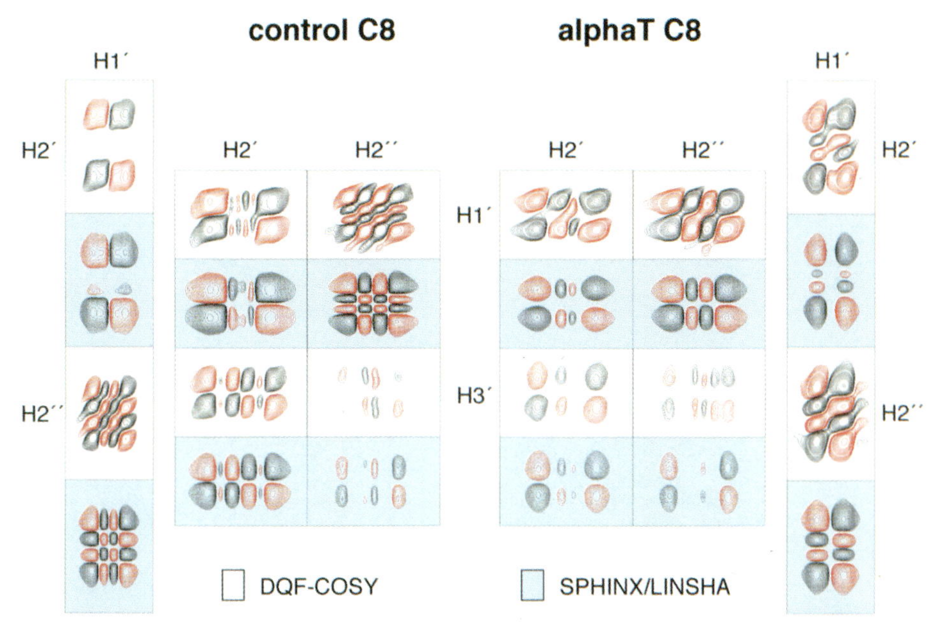

Figure 8. Expansions (60 x 40 Hz) of six DQF-COSY cross-peaks (white boxes) for the C8 residue in the control and alphaT duplexes (303 K) and their corresponding simulations (blue boxes). Simulations were obtained with the SPHINX and LINSHA routines (*16*) using the approach of the James group (*17*). Negative contours are shown in red; positive contours are black. The following coupling constants produced the best fits: control, $J_{1'2'} = 9.1$, $J_{1'2''} = 5.9$, $J_{2'3'} = 6.2$, $J_{2''3'} = 3.0$, $J_{2'2''} = -14$; alphaT, $J_{1'2'} = 6.0$, $J_{1'2''} = 6.5$, $J_{2'3'} = 6.3$, $J_{2''3'} = 5.3$, $J_{2'2''} = -14$. Sugar puckering results based on contour plots of the above couplings plus bounds ($J_{1'2'}$, $J_{1'2''}$, $J_{2'3'}$ $\pm$ 0.5 Hz; $J_{2''3'}$ $\pm$ 0.5 to 1.0 Hz) as a function of f_S and P_S assuming a two-state equilibrium in which $P_N = 9°$ and $\Phi_N = \Phi_S = 35°$ (*14*) are as follows: control, f_S = 81 - 93%; P_S = 145 - 185°; alphaT, f_S = 46 - 58%; P_S = 150 - 200°.

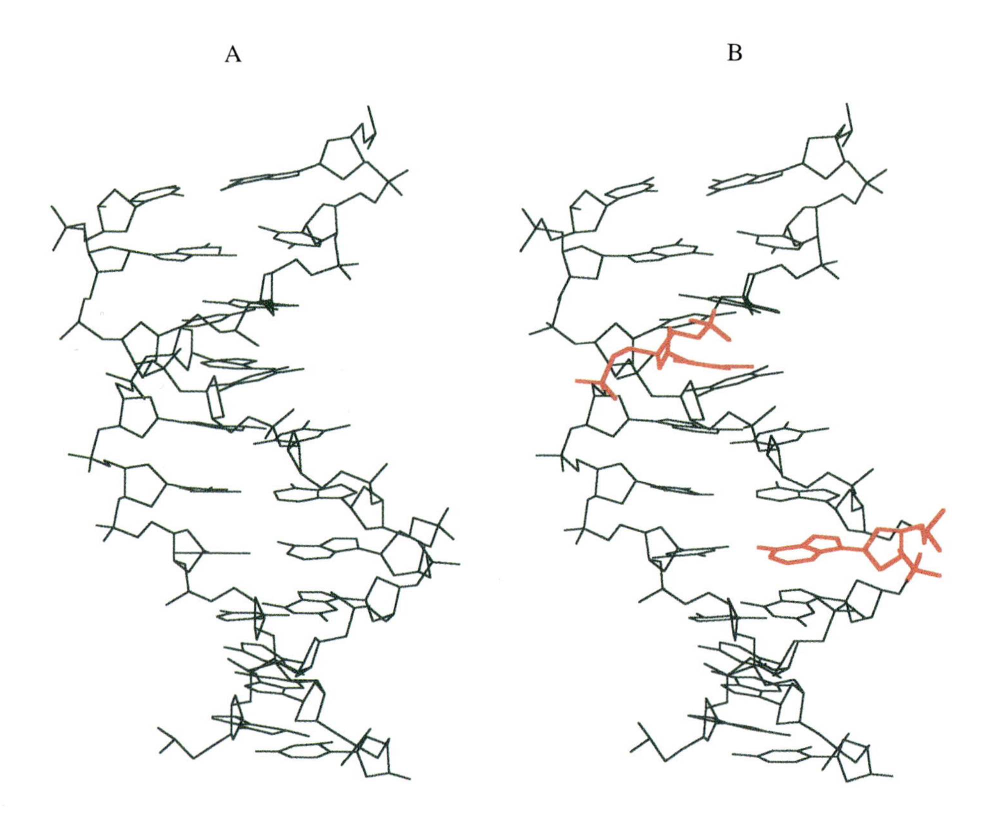

Figure 9. Models of the control (A) and alphaA (B) decamer duplexes. The initial models were based on the PDB coordinates of the Dickerson dodecamer (*22*), and energy minimized with AMMP, "Another Molecular Modeling Program" (*23*), as outlined previously (*9,10*). A dielectric screen with 5 Å correlation length and no cut-off for non-bonded interactions were used; also no NMR (i.e., NOE or *J*) restraints were incorporated into the calculations. α-Anomeric nucleotides are displayed in red. Both duplex models are nearly identical, with an RMSD of 0.32 Å (excluding the α-anomeric residues and their counterparts in the control), and total energies of -12419 $kJmol^{-1}$ for alphaA and -12461 $kJmol^{-1}$ for the control. The sugar puckering for each nucleotide in the models shown, including A5 in alphaA, is of the S-type.

with the corresponding distances in the models; specifically, the H1′ ↔ base NOESY walk can be followed throughout the entire sequence in both duplexes, whereas the H2′/H2″ ↔ base NOE pathway is disrupted between αA4 and A5 in alphaA, because of the increased distance between the H2′/H2″ of the α-nucleotide and the subsequent base. Similar conclusions have been drawn from the modeling and NMR studies of the remaining duplexes in this family.

Studies of alphaT2. The spectroscopic and modeling studies discussed above indicate that the linkage reversal (i.e. a local parallel strand disposition at the α-anomeric nucleotide) is a prerequisite for stable duplex formation. This is in agreement with the conclusions drawn by other groups based on UV melting studies (*4*). A recent molecular mechanics study proposed that an α-anomeric nucleotide (αA) would base pair and contribute to the overall duplex stability if it is present in an antiparallel strand orientation, thus obviating the need for the unusual 5′-5′ and 3′-3′ phosphodiester linkages (*24*).

We have investigated this possibility by studying a sequence analogous to the alphaT decamer (i.e. alphaT2 in Figure 5) which does not contain linkage inversions. The comparison of alphaT and alphaT2 assesses the energetic and structural requirements of the strand disposition of α-anomeric nucleotides on duplex stability. The alphaT2 sequence is characterized by a low melting temperature as well as a shallow melting profile. The melting temperature was found to be independent of the strand concentration at optical strand concentration, consistent with hairpin formation. This was confirmed by imino proton spectra recorded for a 12 μM alphaT2 sample at 283 K (Figure 11A). Under these conditions three GC base pairs are observed at $\delta \approx$ 13 ppm as well as the two T-loop imino protons at $\delta \approx$ 11.4 and 10.8 ppm, in agreement with previous studies on DNA hairpin structures with T-loops (*25*). These results are also consistent with the resonances observed in the methyl group region, where for either pure hairpin or pure duplex forms two methyl group signals are expected. Based on the imino and methyl group signals, a significant amount of hairpin is detected even up to 5.4 mM strand concentration (Figure 11B). Increasing the temperature to 303 K results in a shifting of the equilibrium to the hairpin form. The low thermodynamic stability and the strong tendency of alphaT2 to form hairpins demonstrate that a parallel strand disposition is indeed a requirement of α-anomeric nucleotides for the formation of stable duplex structures.

Conclusions and Future Prospects

Antisense ODNs containing a β-anomeric core flanked by nuclease resistant α-anomeric segments are capable of eliciting RNase H activity while retaining strong hybridization to the RNA target sequence. Such an ODN was shown to be able to mediate the degradation of an RNA target in substoichiometric amounts inescapably implying multiple turnover of the ODN. Ultimately any ODN must be eliminated from an organism. It is noted that endonucleolytic cleavage of the β-anomeric core of these ODNs will disable their antisense potential thereby minimizing the probability of nonspecific antisense activity of breakdown products.

Our systematic structural and thermodynamic analysis of a DNA decamer model system containing α-anomeric nucleotides has established that the α-anomeric components fit snugly into the double helix causing only local perturbations. In particular, base stacking and specific base pairing was retained for all α-anomeric nucleotides investigated, but only if they were provided with a local parallel stranded environment. However, we observed unique deoxyribose ring and backbone perturbations dependent on the nature and position of the α-nucleotide and polarity reversals in the sequence.

We are currently investigating strategies to enhance the stability of the α-containing sequences by alleviating constraints imposed by the unusual phosphodiester linkages.

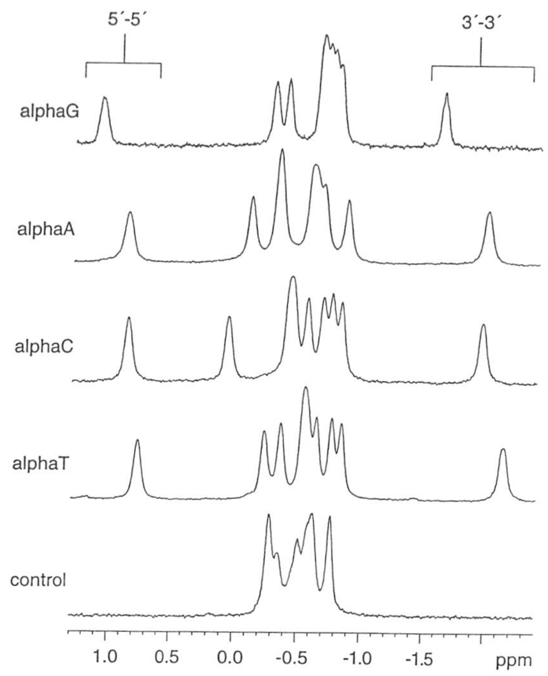

Figure 10. ^{31}P (242.9 MHz) NMR spectra of the control (2.6 mM), alphaT (6.8 mM), alphaC (5.9 mM), alphaA (5.1 mM), and alphaG (2.2 mM) duplexes in D_2O at 303 K. Resonance assignments were made using the ^{1}H-detected ^{31}P-^{1}H correlation technique of Sklenár et al. (*21*). (Reproduced from ref. 10. Copyright 1997 American Chemical Society).

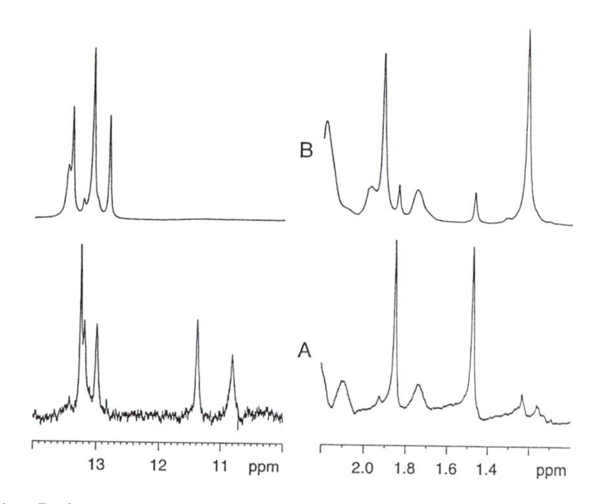

Figure 11. Imino proton and methyl group regions of the ^{1}H (600.1 MHz) NMR spectra of alphaT2 in 90% H_2O/10% D_2O, 50 mM NaCl, 10 mM phosphate buffer, pH 6.6, 283 K. A) Strand concentration 12 μM; B) strand concentration 5.4 mM.

Acknowledgment

This work was funded in part by a grant from the Natural Sciences and Engineering Research Council of Canada (JHvdS).

References

1. Wagner, R. W. *Nature* **1994**, *372*, 333-335.
2. De Mesmaeker, A.; Häner, R.; Martin, P.; Moser, H. E. *Acc. Chem. Res.* **1995**, *28*, 366-374.
3. van de Sande, J. H.; Kalisch, B. W.; Quong, B. Q.; Germann, M. W. **1994** Antisense Oligonucleotides with Polarity and Anomeric Center Reversals. First International Antisense Conference of Japan, Kyoto, Dec. 4 -7.
4. Debart, F.; Tosquellas, G.; Rayner, B.; Imbach, J.-L. *Bioorg. Med. Chem. Lett.* **1994**, *4*, 1041-1046.
5. Koga, M.; Wilk, A.; Moore, M. F.; Scremin, C. L.; Zhou, L.; Beaucage, S. L. *J. Org. Chem.* **1995**, *60*, 1520-1530.
6. Boiziau, C.; Debart, F.; Rayner, B.; Imbach, J.-L.; Toulmé, J.-J. *FEBS Lett.* **1995**, *361*, 41-45.
7. Germann, M. W.; Aramini, J. M.; Kalisch, B. W.; Pon, R. T.; van de Sande, J. H. *Nucleosides Nucleotides* **1997**, in press.
8. Paoletti, J.; Bazile, D.; Morvan, F.; Imbach, J.-L.; Paoletti, C. *Nucleic Acids Res.* **1989**, *17*, 2693-2704.
9. Aramini, J. M.; Kalisch, B. W.; Pon, R. T.; van de Sande, J. H.; Germann, M. W. *Biochemistry* **1996**, *35*, 9355-9365.
10. Aramini, J. M.; van de Sande, J. H.; Germann, M. W. *Biochemistry* **1997**, *36*, 9715-9725.
11. Marky, L. A.; Blumenfeld, K. S.; Kozlowski, S. A.; Breslauer, K. J. *Biopolymers* **1983**, *22*, 1247-1257.
12. Germann, M. W.; Kalisch, B. W.; van de Sande, J. H. *Biochemistry* **1988**, *27*, 8302-8306.
13. Wüthrich, K. *NMR of Proteins and Nucleic Acids*, John Wiley & Sons, Inc., New York, 1986.
14. Rinkel, L. J.; Altona, C. *J. Biomol. Struct. Dyn.* **1987**, *4*, 621-649.
15. van Wijk, J.; Huckriede, B. D.; Ippel, J. H.; Altona, C. *Methods Enzymol.* **1992**, *211*, 286-306.
16. Widmer, H.; Wüthrich, K. *J. Magn. Reson.* **1987**, *74*, 331-336.
17. Schmitz, U.; James, T. L. *Methods Enzymol.* **1995**, *261*, 3-44.
18. Gonzalez, C.; Stec, W.; Reynolds, M. A.; James, T. L. *Biochemistry* **1995**, *34*, 4969-4982.
19. Gorenstein, D. G. *Chem. Rev.* **1994**, *94*, 1315-1338.
20. Roongta, V. A.; Jones, C. R.; Gorenstein, D. G. *Biochemistry* **1990**, *29*, 5245-5258.
21. Sklenár, V.; Miyashiro, H.; Zon, G.; Miles, T.; Bax, A. *FEBS Lett.* **1986**, *208*, 94-98.
22. Westhof, E. *J. Biomol. Struct. Dyn.* **1987**, *5*, 581-600.
23. Harrison, R. W. *J. Comput. Chem.* **1993**, *14*, 1112-1122.
24. Ide, H.; Shimizu, H.; Kimura, Y.; Sakamoto, S.; Makino, K.; Glackin, M.; Wallace, S. S.; Nakamuta, H.; Sasaki, M.; Sugimoto, N. *Biochemistry* **1995**, *34*, 6947-6955.
25. Germann, M. W.; Kalisch, B. W.; Lundberg, P.; Vogel, H. J.; van de Sande, J. H. *Nucleic Acids Res.* **1990**, *18*, 1489-1498.

Chapter 7

Conformational Analysis of Nucleic Acids: Problems and Solutions

Andrew N. Lane

Division of Molecular Structure, National Institute for Medical Research, The Ridgeway, Mill Hill, London NW7 1AA, United Kingdom

There are numerous problems associated with the determination of nucleic acid structures in solution using NMR methods. These include spin diffusion, anisotropic rotation, conformational averaging, low density of experimental constraints and lack of long-range information. The difficulties arising from the first two problems can be treated reliably, but new experimental methods are needed for the last two. To deal effectively with conformational averaging, newer computational procedures are required. Approaches to these problems are described, and illustrated with an example of a DNA.RNA hybrid duplex.

Nucleic acids can adopt a wide variety of structures depending on composition, sequence and environmental conditions. This reflects the intrinsic plasticity of oligonucleotides. Not only are the nucleotide units themselves flexible, but also there is a large number of degrees of freedom in the phosphodiester linkage. In addition, in many nucleic acids the base-pairs interact only with their nearest neighbours. Polymers of this kind can be easily deformed by small anisotropic forces such as intercalation by small ligands, or formation of complexes with proteins as exemplified by CAP (*1*) and SRY(*2*). Further, one of the many interesting aspects of nucleic acids is the influence of base-pair mismatches on conformation. Studying mismatches is important in understanding DNA mutations, and in the phylogenetically conserved 'mismatches' found in many RNA species. As a rule, mismatches are thermodynamically destabilising (*3*), which is often associated with increased local flexibility in both DNA and RNA (*4,5*). Even ordinary duplexes can show significant conformational heterogeneity on the millisecond time scale (*6-9*). Such behaviour poses a serious challenge to describing the conformational properties of nucleic acids in solution. In this article, I will summarise the problems of determining solution structures from NMR data and molecular modelling, and outline some outstanding problems that may need to be addressed for a fuller understanding of their functional properties.

Information Content

If a nucleic acid is considered as a rigid body, then there is a fixed number of parameters needed to describe its structure. In terms of torsion angles (i.e. assuming that bond length and angles are known, as well as the chemical structure), then there are 8n-4 parameters per strand, or 16n-8 +6 parameters per duplex to be determined (cf. Structure I). Hence, for a decamer duplex, there would be 158 parameters. As a torsion angle cannot be described by a single distance, the minimum number of NOEs or distances needed is larger than this. Nevertheless, for a well resolved NMR spectrum, one can expect to be able to extract a sufficient number of experimental distance and dihedral constraints such that the problem is at least in principle overdetermined.

Structure I

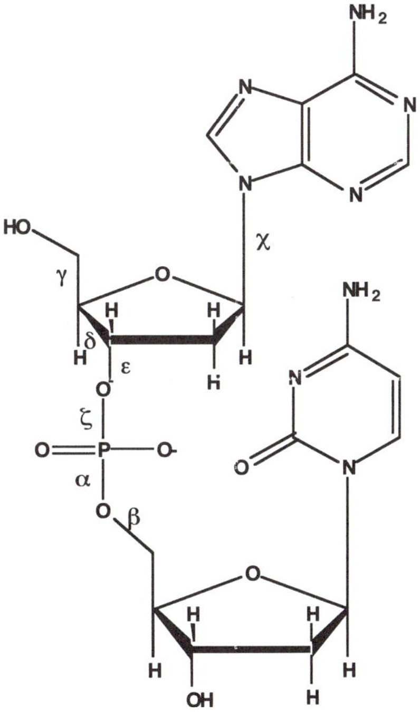

For example, the conformation of a rigid nucleotide can be specified by 4 parameters, namely the glycosidic torsion angle χ, the angles δ and γ, and a second endocyclic torsion, which together with δ defines the pseudorotation phase angle P and the maximum amplitude of the sugar pucker ϕ_m. The sugar pucker parameters can be determined from analysis of two or more coupling constants, which in favourable cases can be measured from various COSY variants (and see below). There are also two intraring proton-proton dipolar interactions in DNA that vary significantly over the pseudorotation phase cycle, namely H1'-H4', which has a maximum at P≈90° (O4'-endo), and H2"-H4' which has a maximum near P=18°(C3'-endo) and a minimum near P≈198° (C3'-exo). The remaining intraring proton-proton distances do not vary significantly with sugar conformation, and therefore provide no restraining power in distance-based algorithms. The base proton (H8 or H6)

interaction with several sugar protons is a strong function of the glycosidic torsion angle, especially the H2'-H8/H6. However, the H1'-H8/H6 NOE, perhaps the easiest to measure, is only weakly dependent on χ in the *anti* domain, and therefore for many nucleic acids it has minimal restraining power, though for *syn* nucleotides, this NOE provides the major source of information about χ.

To determine the helical conformation, there are again more experimental data available than the maximum number of degrees of freedom, i.e. three rotations and three translations for unlinked dinucleotides, or the torsions $\alpha,\beta,\varepsilon,\zeta$ for a linked dinucleotide (Structure I). 7-10 distances per dinucleotide (some of which are not completely independent) can be obtained from NOEs and in favourable cases, additional constraints on ε and β can be obtained from heteronuclear couplings, though in general these are rather loosely defined (10). The finite size of the constituent atoms makes some combinations of parameters physically impossible, so that they cannot be treated as independent, but rather conditionally dependent parameters.

The degree to which a structure is determined is a compromise between the number of constraints, their quality (which includes precision), and their sensitivity to variation of structural parameters (i.e. their restraining power). At a trivial level, a single exact constraint cannot determine the structure in any sensible way. Similarly, all possible distances known to very low precision (say between 0 and 10 Å), will also not specify the structure. The minimum number of constraints required is equal to the number of degrees of freedom in the system (within the context of conditional probabilities). At this constraint density, the precision of the structure will depend strongly on the quality of the constraints, and the accuracy of any force-field that is used for calculating structures. Weak restraints that do not further reduce the conformational space should not be included in a count of restraints. The quality of a restraint may also vary during refinement; once the gradient of a restraint with respect to a parameter (e.g. a torsion angle) approaches zero, that particular restraint no longer contains much information about the value of the parameter. Furthermore, a high density of constraints will only limit the available conformational space to a finite amount. Once the constraints are satisfied, any further convergence to a particular structure will be entirely determined by factors such as the algorithm used to generate the structures, the starting conditions, and the forcefield parameterisation. Ideally, a method is needed that is independent of calculation strategy, the starting point, and the force field.

Factors Which Affect the Number and Accuracy of Constraints

Coupling Constants. Coupling constants are extremely valuable for determining torsion angles. In practice there are several potential problems in extracting the information contained in the experimental splittings. These include the complexity of some multiplets, the large line-width of some resonances (particularly H2' and H2" in DNA), strong coupling, and the possibility of interference between dipolar and scalar coupling. Several groups, notably that of James (*11,12*) have shown that simulation of COSY cross-peaks can lead to improved estimates of individual coupling constants. However, for larger nucleic acid fragments, the influence of line-width becomes problematic, and accurate simulations require that the effective T_2 is known. It has also been shown that for reliable determination of coupling constants, good digitisation of the spectra is needed (*13*). Furthermore, dipolar interactions can affect the apparent splittings, which is probably most severe for coupling to H2' and H2" in DNA (*14-16*). To some extent, this problem can be alleviated by using the sums of the coupling constants, as they are less sensitive to the interference from dipolar effects, and the information content of three such sums is the same as that of three individual coupling constants (*17*). The extent of this problem remains unclear at present, though it may become significant for individual coupling constants when the effective correlation time exceeds 3 ns (ca. 12 base-pairs at 30°C). For correlation times much above 5-6 ns, the resonances become so broad that accurate coupling constants cannot be reliably retrieved in any case (*13*).

The parameterisation of the Karplus equation also limits the accuracy of the desired torsions that can be retrieved. The uncertainty of the Karplus curve may be in the range 0.5 to 1 Hz, depending on the exact values of P and ϕ_m. Furthermore, in DNA where the sugars are usually in the 'S' domain, there is only a weak dependence of the coupling constants on P: $^3J_{1'2'}$, $^3J_{1'2''}$ and $^3J_{2''3'}$ vary only slightly for $100° < P < 200°$. Only $^3J_{3'4'}$ and $^3J_{2'3'}$ vary strongly, and they are generally difficult to measure accurately (*17,18*). RNA presents additional problems as there is no H2", the splitting between H1' and H2' is usually too small to measure, and there is considerable chemical shift degeneracy among the H2', H3' and H4' resonances.

Spin Diffusion. Distances are not directly measured in solution NMR experiments, but rather are derived from an estimate of a cross-relaxation rate constant. This is limited by numerous considerations, including accuracy of integration, relaxation, and spin diffusion. Because spin diffusion is certainly important in nucleic acids (*19,20*), cross-relaxation rate constants estimated assuming isolated spin-pairs do not provide correct distances, and there is no 1:1 correspondence between the observed NOE and distance. The treatment of spin-diffusion in principle is straightforward, as the transfer of magnetisation is governed by linear first-order differential equations, and can therefore be treated by standard matrix algebra (*19*). These equations are also well behaved in numerical simulations, so that simple integration algorithms can be used (*20,21*). A complete set of NOEs measured as a function of mixing time completely specifies the relaxation matrix, R. For n dipolar coupled spins, there are n*n elements in R, of which n(n+1)/2 are independent, as follows:

$$R = \begin{bmatrix} \rho_1 \; \sigma_{12} \; \sigma_{13} \cdots \\ \sigma_{12} \; \rho_2 \; \sigma_{23} \cdots \\ \sigma_{13} \; \sigma_{23} \; \rho_3 \cdots \\ . \quad . \quad . \end{bmatrix} \qquad (1)$$

where ρ are the intrinsic spin-lattice relaxation rate constants and σ are the cross-relaxation rate constants. The peak volumes in a NOESY experiment are then given by:

$$v(t) = v(0)\exp(-Rt) \qquad (2)$$

where v(t) is the matrix of volumes at mixing time t, and v(0) the (diagonal) volumes at zero mixing time. v(0) is not necessarily equal to the equilibrium magnetisation of each spin; under conditions of incomplete relaxation between acquisitions, v(0) will be the steady state magnetisation, which depends on the spin-lattice relaxation properties of the system as a whole (i.e. non-exponential recovery). The appropriate saturation factors can usually be determined experimentally or by simulation (*22*). Only if values for all peak volumes are given can the relaxation matrix be constructed from a single time point. In general, NOE time-courses will be essential for detailed analysis. Using multiple time-points also improves the statistics of the analysis.

There are two main approaches to deriving a relaxation matrix, each having its pros and cons. In one method, the matrix of NOESY peak volumes is inverted (cf. Eq. (2)). This strictly applies only when the diagonal intensities are known in addition to the cross peak intensities, and the inversion of a large NOE matrix can be quite sensitive to missing values (due to overlap etc.). In practice, the experimental matrix of peak volumes has to be supplemented by model values (*19,23,24*). If the number of missing data is high, the inversion is dominated by the model data. The second approach uses numerical integration of the Bloch-Solomon equations, which is a very stable, if inelegant method, that does not require complete NOE data. Depending on the size of the spin system, the rates of magnetisation transfer, the accuracy required and the integration time, such methods can be either faster or slower than the matrix methods (*25*).

Assuming that a relaxation matrix can be derived (and there is always at least one such matrix that accounts for the data), the next problem is to determine distance constraints. If the molecule being studied has a single conformation, and there are no internal dynamics to consider, this is straightforward. However, if there are multiple conformations, internal motions or rotational anisotropy, the problem is much more complex.

Anisotropic and Rapid Internal Motions. The cross-relaxation rate constants depend not only on the internuclear separation but also the correlation time. Even for a spherically symmetric rotating body, each cross-relaxation rate constant depends on two parameters. However, for a rigid spherically symmetric rotor, there is a single unique correlation time, that can be determined by relaxation methods on X nuclei, by cross-relaxation between protons that have a known fixed separation or by non-NMR methods based on rotational diffusion.

Fast, small amplitude local oscillations scale down the correlation function, and therefore the spectral density function, decreasing cross-relaxation rate constants on average. Hence, even in a spherically symmetric rotor, each pair of spins may have different effective correlation functions. These kinds of motions are likely to have a second-order effect, though it does become important when the NOEs rather than distances are used in the refinement process. Fast motions, such as rotation of the thymine methyl group, may also lead to small fluctuations in the internuclear distance to other protons. This can be treated with sufficient accuracy (*26*). For very fast motions, the overall tumbling and internal motions are effectively decoupled, and the spectral density functions have a relatively simple form, albeit dependent on geometry. To a first approximation, these motions can be described using a generalised order parameter. Because it is the dipolar Hamiltonian that is averaged, the averaging is $\langle r^{-3} \rangle$, which is slightly different (depending on the geometry) from $\langle r^{-6} \rangle$. For small fluctuations in the distance, the difference is quite small. For example, in the sugar repuckering equilibrium between N and S states, the distance H2"-H4' varies between limits of about 2.6 and 3.7 Å. The $\langle r^{-6} \rangle$ average is 2.86 Å, and the $\langle r^{-3} \rangle$ average is 2.97 Å (assuming equal populations). However, if the Tropp model (*26*) is used, the average is 3.16 Å, which reflects the influence of the angular variation. This indicates an effective generalised order parameter, S^2 of around 0.6 for this vector. Recently it has been estimated that the transition rate constant for sugar repuckering in DNA is of the order 10^7 s^{-1} (*27*), which implies that $\langle r^{-6} \rangle$ averaging would be the most appropriate description of this equilibrium.

Because it is not possible to determine both the order parameter and the averaged distance for a given internuclear vector, one or the other must be known by some other means, such as molecular dynamics trajectories (*28*). Order parameters for some vectors of nearly fixed length have been estimated in DNA, and on the whole, they are not too far from unity ($S^2 > 0.7$) (*28-32*). The main exceptions were found for terminal residues, where $S^2 < 0.5$ (*31,32*). Low order parameters imply extensive conformational averaging, which presents a serious problem for the description of the dynamic conformation of terminal base-pairs.

Motions may also cause averaging of coupling constants, the main information that determines the conformation of the sugars in nucleotides. Although modest amplitude motions such as fluctuations within the S domain have rather small effects on the coupling constants (see above), interconversions between the S and N domains may have large effects on coupling constants (see below).

Nucleic acids > ca. 10 bp long are not spherically symmetric. To a good approximation they are equivalent to circular cylinders with a hydrodynamic diameter of 20-23 Å for DNA (*33-35*) and 25 Å for RNA (*35*). The correlation function for such symmetric top molecules consist of three exponentials, whose arguments are combinations only of the correlation time for end over end tumbling (τ_L) and for rotation about the principal symmetry axis (τ_S). Thus for anisotropic motion, two independent correlation times are needed to describe the rotational diffusion. The spectral density function also depends on the angle (θ) the interproton vector makes with the principal axis. J(0), and hence the cross-relaxation rate constant, varies as a function of this angle according to (*36*) :

$$J(0)=0.25(3\cos^2\theta-1)^2\tau_L + 6\cos^2\theta\sin^2\theta\ \tau_L\tau_S/(5\tau_S+\tau_L) + 2.25\ \sin^4\theta\ \tau_L\tau_S/(\tau_S+2\tau_L) \quad (3)$$

For a vector parallel to the long axis (θ=0), only the first term is important for relaxation, i.e. $J(0) = \tau_L$ whereas for a vector perpendicular to the helix axis (θ=90), only the first and the third term are important, and $J(0) = 0.25\tau_L + 2.25\tau_L\tau_S/(\tau_S+2\tau_L)$. For a vector oriented at the magic angle, the first term is zero, and at sufficiently high axial ratios, J(0) becomes very small compared with its value for vectors oriented at 0 or 90°. Note that the ratio of J(0) at 0 and 90° is smaller than τ_L / τ_S. If the axial ratio of the equivalent cylinder is known, then it is also possible to estimate the value of τ_L from a measurement of the cross-relaxation rate constant for Cyt (or Uri) H6-H5, as these vectors are perpendicular to the helix axis (*34*), and therefore the effect of rotational anisotropy on cross-relaxation for a given structure can be assessed in a straightforward manner.

For strongly anisotropic duplexes of say 20 bp (axial ratio ca. 3:1), the cross-relaxation rate constant for a pair of protons oriented parallel to the long axis is about twice that for the perpendicular orientation. This corresponds to a difference in distance for a vector oriented parallel and one antiparallel to the helix axis of about 12%, or 0.36Å at 3Å. Clearly for relaxation matrix analysis or direct refinements, such effects cannot be ignored.

Multiple Conformations: Ensembles. When multiple conformations are present, the problem rapidly becomes underdetermined. For example, if a nucleotide can exist in two major conformational states, with χ(N)P(N) and χ(S)P(S) where the glycosidic torsion angles in the N and S sugar states are not equal, each conformation will have its own characteristic NOEs and coupling constants. For n bases, there are 2^n conformations, and the contribution of each one to the ensemble will depend on its population. Even for a dinucleotide, there are at least 4 conformations (viz. SS, SN, NS, NN), which multiplies the number of structural parameters to be determined fourfold, plus four equilibrium constants. There are not enough independent NMR data to determine all of these parameters and the problem is underdetermined. In this situation, the best that can be hoped for is to derive a set of structures that in some way represent the ensemble of structures that is present in solution.

A fundamental problem (in addition to the lack of data), is that the experimental data are averages over a number of conformations which are not *a priori* known. It is therefore invalid to interpret the data as tight distance or torsion constraints, which would produce virtual structures. Thus the data must initially be interpreted as loose constraints, which decreases the quality of individual structures (see above). Hence, when conformational averaging occurs, the notion of single or mean structures has to be abandoned, and ensembles should be considered as the appropriate representation of the NMR data.

There are several approaches to this problem, but there is no general solution based on NMR data alone. One method is to use time-averaged distance constraints (*37-40*)., which relaxes the requirement implicit in distance-based structure algorithms that all constraints be simultaneously satisfied. Therefore a wider range of conformations can be sampled, and they are not necessarily treated as mutually exclusive. This approach gives improved agreement with the experimental data (*38,39*). A second approach is to generate ensembles of conformations which as a whole satisfy the experimental data (*41-44*). There is no direct indication of whether the ensemble is unique or not, nor can there be when the number of parameters exceeds the information content of the data. Cross-validation methods can be used only with rich data sets where the modelling includes only a small number of conformations (*45*), which is usually not the case for nucleic acids.

We have been experimenting with an intermediate approach, that progresses through stages of increasing complexity (*46*). First the nucleotides are assumed to be almost independent of one another, and can exist as a mixture of N and S states. This is partly a computational convenience, but is supported by the observation that coupling constants can often be accounted for by such a two-state equilibrium (*17*), and that DNA is a remarkably plastic molecule, such that large conformation changes in one nucleotide are not propagated along the helix. The mole fractions and phase angles of the N and S conformers can be determined for each nucleotide from analysis of the coupling constants, and the glycosidic torsion angles for the two-state model are determined from analysis of NOE time-courses. Rapid oscillations within either the N or S domain are accounted for by generalised order parameters. The conformations of the nucleotides are then treated as knowns, and are used to build up a (limited) picture of the conformational ensemble that as a whole must account for the experimental data. The number of structures generated is much smaller than the total suspected to be present (in a 10-mer, there would be ca. 10^6 structures), and therefore can only be a representative sample. To evaluate the quality of the structural ensemble, an R factor should be calculated such as:

$$R1 = (1/Nm)\sum\sum|NOE(c)-NOE(obs)|/(NOE(obs) \quad (4)$$

$$R6 = (1/Nm)\sum\sum[NOE(c)^{1/6}-NOE(obs)^{1/6}]/\sum NOE(obs)^{1/6} \quad (5)$$

where N is the number of NOEs per mixing time, m is the number of mixing times and the double sum is over all NOEs and mixing times.

In the following example, this method is used to demonstrate many of the problems described above, and to show how an ensemble can be discriminated from a unique 'intermediate' conformation.

An Example

DNA.RNA hybrids are of considerable biological interest and the DNA strands show a higher degree of flexibility in their sugar conformations than in similar all-DNA duplexes (*11,47,48*). The influence of such conformational averaging is likely to be at its most severe.

Several structures of the DNA.RNA hybrid d(CGAATTCG).r(CGAAUUCG) were generated within InsightII (MSI, San Diego). The first was the A conformation energy minimised using the full Amber forcefield (ε=4r). The second was the A form hybrid minimised with the deoxyriboses restrained to the C2'-endo conformation. The third structure was made in which the DNA sugars were constrained to the O4' -endo conformation. Finally, 8 structures were generated to represent an ensemble in which on average, each DNA nucleotide was 50% N and 50% S. All of the structures are stereochemically reasonable, and are globally in the A family of conformations, though they appear significantly different (Figure 1), and the rmsd values between the various structures are quite substantial (Table I). The potential energies, nucleotide parameters and helical parameters as determined using Curves 5.1 (*49*) are shown in Table II.

Table I. Rmsd values between structures

A(dN.rN), A(dE.rN), A(dS.rN) are the structures in which the DNA strand is C3'-endo, O4' endo and C2'-endo, respectively, and the RNA strand is C3'-endo. A(dNS,rN) is the ensemble of DNA duplexes with 4 N and 4 S deoxyriboses.

	A(dN.rN)	A(dE.rN)	A(dS.rN)	A(dNS.rN)
A(dN.rN)	-	0.78	1.37	0.88
A(dE.rN)		-	0.99	0.73
A(dS.rN)			-	0.8
A(dNS.rN)				0.64

Table II. Helical parameters of the DNA.RNA hybrids.

The nomenclature is the same as in Table I. A(dNS1.rN) is one member of an ensemble of structures in which 50% of deoxy sugars are constrained to C2'-endo and 50% to C3'-endo. A(dE.rN)r is the rMD refined structure using constraints derived from A(dE.rN). A(ens)r is the rMD refined structure using constraints derived from the ensemble of A(dNS1.rN) to A(dNS8.rN).

molecule	Upot kcal	twist deg	rise Å	inc^n. deg	prop. deg	dx Å	P deg	χ deg	groove width Å minor	major
A(dN.rN)	-137.0	32.5	2.61	12.2	-24.3	-4.62	21 20	-160 -160	9	19.6
A(dE.rN)	-133.3	31.9	2.72	9.63	-18.5	-4.4	96 20	-143 -161	8	21
A(dS.rN)	-138.3	34.5	2.90	5.83	-23.7	-3.41	137 19	-119 -161	7	20
A(dNS1.rN)	-133.7	32.9	2.79	7.65	-21.7	-3.77	140/27 19	-124/-150 -159	8.3	19.5
A(dE.rN)r	-118.7	31.5	2.60	12.3	-16.5	-4.46	100.5 29	-142 -155.4	9	21
A(ens)r	-97	31.3	2.76	11.5	-12.2	-4.3	78.5 28.5	-142 -158	9	21

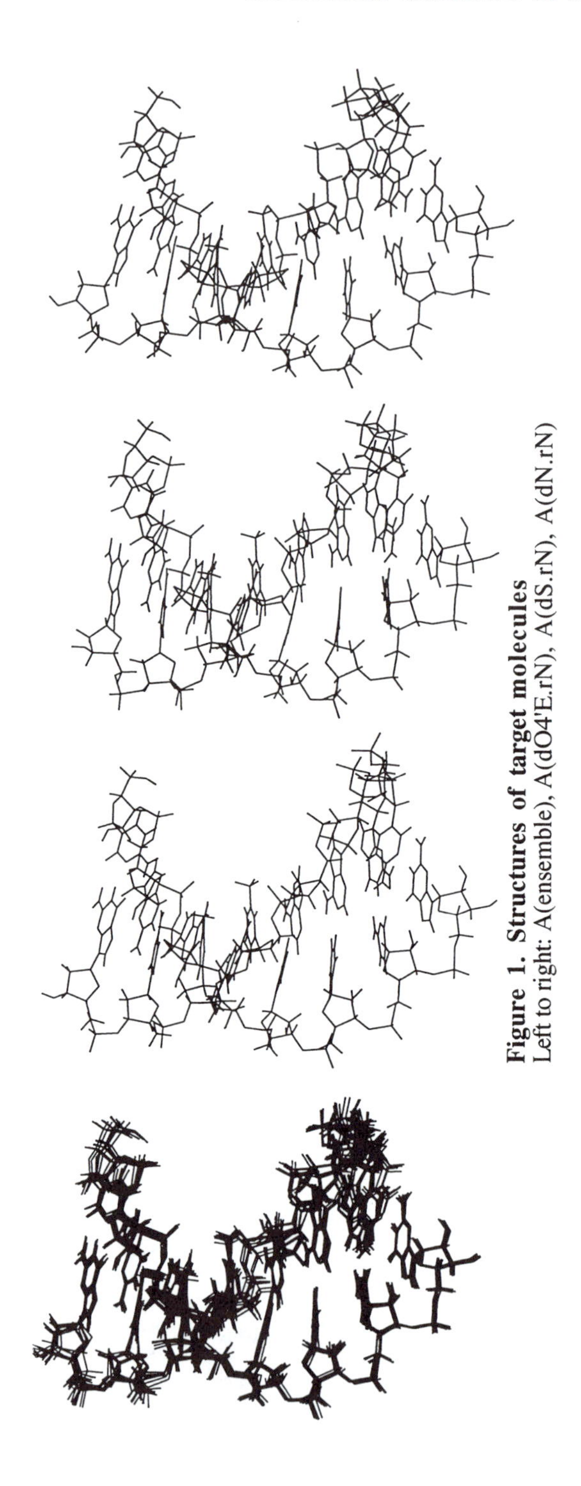

Figure 1. Structures of target molecules
Left to right: A(ensemble), A(dO4'E.rN), A(dS.rN), A(dN.rN)

The conformational adjustments required to convert a deoxy sugar from C3'-endo to C2'-endo are highly local (Fig. 2); the effect is not propagated beyond the nearest neighbours. This is because the backbone of DNA has a large number of degrees of freedom, allowing substantial local perturbations to be accommodated by small changes in a few local torsions (mainly χ, ζ and ε, Fig. 2). Fluctuations in ε should be observable in the ^{1}H-^{31}P scalar coupling, and such averaging has been observed in B_I<=>B_{II} transitions (*50*). Similarly, the helicoid parameters, notably the propeller twist, also show localised variation according to the conformation of the deoxy nucleotide (Fig. 2B). Furthermore, the conformation of the RNA strand seems to be largely unaffected by considerable variations in the DNA strand. This also *a posteriori* justifies the assumption of independent nucleotides. Figure 2 also indicates that some of the helicoid parameters, which are notoriously difficult to determine accurately even for a unique conformation (*21,51*) are likely to be even less well described in the presence of conformational averaging.

The NOEs or $<r^{-6}>$ for the ensemble are in most instances intermediate between those of the C2'-endo and C3'-endo DNA structures, and similar to those of the O4'-endo structure (Table III). The main exception is H1'-H4', for which the NOE is large only in the O4'-endo conformation, and is small in the ensemble average. It is notable though that the coupling constants, especially $\Sigma 1'$, which is the most straightforward coupling parameter to measure, are quite different for these structures, and would provide the primary information for distinguishing between them (Table IV).

Table III. Selected distances in the DNA.RNA hybrids.

Distances are shown for A4 (intranucleotide) and for the A3.A4 dinucleotide step. d(SN).rN refers to DNA strands in which different deoxy sugars were either S or N, and in particular where A3A4 are SN etc. $<r^{-6}>$ is the ensemble averaged distance as $<0.125*\Sigma 1/r_{ij}^{6}>^{-1/6}$. %NOE is the NOE at a mixing time of 100 ms assuming a correlation time of 2.5 ns, a spectral frequency of 600 MHz, and a recycle time of 3 seconds. NOE_{DR} refers to NOEs for A(dE.rN) and NOE_{ens} is for the ensemble.

	intranucleotide					sequential					
molecule	8/1'	8/2'	8/2"	8/3'	1'/4'	2"/4'	1'/8	2'/8	2"/8	3'/8	8/8
A(dN.rN)	3.83	3.89	4.71	2.88	3.16	2.55	4.79	2.25	3.94	3.53	4.67
A(dS.rN)	3.93	2.47	3.79	4.54	2.88	3.93	3.61	3.57	2.40	5.13	4.48
A(dE.rN)	3.88	3.02	4.35	4.25	2.50	3.38	4.12	2.57	2.60	4.64	4.43
$\%NOE_{DR}$	0.34	1.18	0.39	0.26	4.15	0.84	0.42	3.5	3.3	0.34	0.19
d(SN).rN	3.81	3.86	4.71	3.01	3.08	2.58	3.31	3.65	2.23	4.96	4.85
d(NS).rN	3.93	2.47	3.83	4.96	2.78	3.92	4.83	2.35	4.08	3.64	4.29
d(NN).rN	3.82	3.91	4.73	3.00	3.10	2.55	4.70	2.22	3.92	3.63	4.43
d(SS).rN	3.93	2.43	3.74	4.52	2.86	3.93	3.94	3.31	2.34	4.92	4.64
$<r^{-6}>$[a]	3.87	2.72	4.08	3.29	2.98	2.85	3.89	2.52	2.53	3.99	4.54
$\%NOE_{ens}$	0.32	1.95	0.53	0.81	1.39	1.52	0.43	3.45	3.35	0.38	0.16

[a] average over all 8 structures.

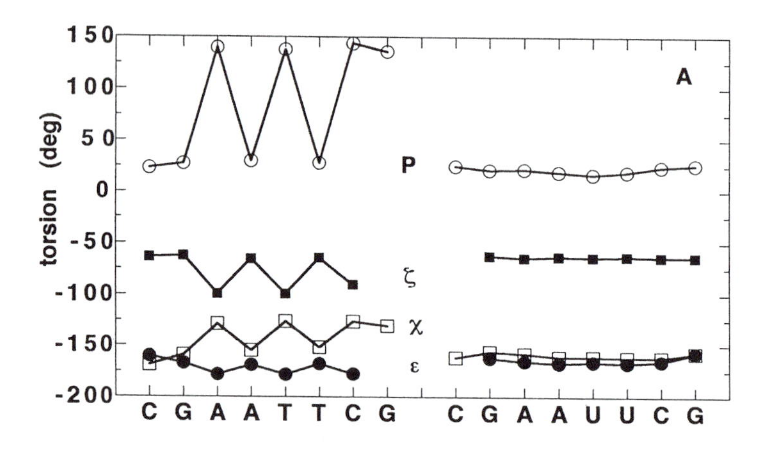

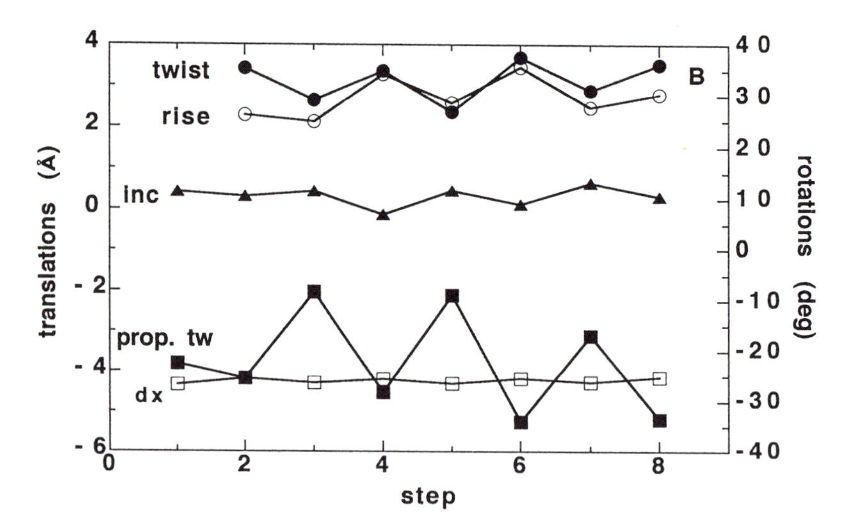

Figure 2. Variation of torsions and helicoid parameters
A. selected torsion angles B. selected helicoid parameters.

Table IV. Coupling constants

$|^2J_{2'2''}|$ = 14.2 Hz. øm=40°. In the ensemble, each deoxy sugar is on average 50% S and 50% N. Coupling constants were taken from (17).

Phase	$^3J_{1'2'}$	$^3J_{1'2''}$	$\Sigma_{1'}$	$^3J_{2'3'}$	$^3J_{2''3'}$	$\Sigma_{2'}$	$\Sigma_{2''}$	$^3J_{3'4'}$	$\Sigma_{3'}$
18	1.2	7.8	9.0	6.3	9.2	21.7	31.2	8.9	23.8
90	8.9	6.4	15.3	8.4	3.0	31.5	23.6	7.2	18.6
140	10.9	4.5	15.3	5.7	0.8	30.7	19.5	1.5	7.9
ensemble	6.3	6.5	12.8	6.5	5.7	27	26.4	5.5	17.7

Tight restraint sets were made for rMD refinement, to determine the differences between structures using restraints for a unique conformation, and restraints that would be derived from an ensemble. Thus, for the minimised O4'-endo structure, a restraint set was made in which the distances were ±0.1 Å for r<3 Å, ±0.2 Å for 3<r<4 Å and ±0.3 Å for 4<r<5Å. 208 such distances were used including those associated with H-bonding interactions (one for each H-bond). In addition, the torsions δ (±5°) and γ (±30°) were restrained, making a total of 240 restraints (15 per residue). A similar restraint set was made based on the ensemble averaged distances of the 8 individual structures. Calculations were started from the standard A structure, which was first partially randomised by heating to 500 K for 1.2 ps in the absence of any restraints, while the electrostatic potential was scaled down 5-fold, and the Lennard-Jones potential 2-fold. The restraints were then applied, briefly minimised (200 cycles of conjugate gradients), then refined using 10 ps rMD at 300 K followed by 5000 cycles of restrained energy minimisation. Finally the scaling factors were returned to their full values, and the potential energy of the system determined. Note that the refinement was carried out using a much weaker forcefield than that used to generate the target structures.

Unsurprisingly, the O4'-endo structure refined to a low pairwise rmsd (0.46 Å), especially if the terminal residues were ignored (these were less well determined by virtue of the lower constraint density) (rmsd = 0.35 Å, Table V). The accuracy was also quite good, showing on average an rmsd to the target structure of 0.67 Å. In contrast, when the ensemble-averaged constraints were refined assuming that there is a single structure, the final potential energy was much higher, and there were numerous small, but consistent restraint violations, even though the largest violation was only 0.08Å. Furthermore, the rmsd to the target structure was 0.81Å.

Table V. Statistics for rMD calculations

molecule	Upot	Uf	rmsd/Å		
	kcal mol^{-1}		d(O4'E).r(N)	d(ens).r(N)	model
d(O4'E).r(N)	-118.7	1.1	0.35±.11	1.07±.07	0.67±.01
d(ens).rN	-97.0	5.2	1.07±.07	0.45±0.1	0.81±.05

The restraints that were consistently violated were H2"-H4', H1'-H4' and H2'(i)-H8/H6(i+1). Further, the coupling constants calculated for this structure did not agree with the actual values for the ensemble (Table IV) and the R factors calculated for the ensemble NOEs versus the NOEs calculated for the O4'-endo structure were high (R1=0.2, R6=0.06). Furthermore, the R factor for the best fit single conformation against the ensemble NOEs was also high (R1=0.21, R6=0.047). The NOEs calculated for the best rMD structure are compared with the ensemble NOEs in Fig. 3. The best fit structure displays numerous systematic deviations from the input data, signalling a failure of the model. Hence, with good quality data, it should be possible to discriminate between a structure that is intermediate in many respects between two families of more extreme structures, and the ensemble of the families of structures, though this is a fairly demanding task.

If the data do show that the ensemble approach is more suitable than the single conformation method, then further refinement is necessary. As described, it is usually found that the ensemble does not always give as good agreement with the ensemble NOE data (not distances) as that obtained when fitting unique structures to model data calculated for a single conformation (*46*). This is due in part to the need at this stage to use relatively wide bounds on some of the constraints, and because the problem is then underdetermined. Individual structures would be partly determined by the nature of the force field used. The next stage would be to tighten the loose constraints, which within the context of this model may best be done by direct refinement of the assumed ensemble (*41,46*). A possible way forward would be to derive an accurate as possible relaxation matrix at this stage, and then refine directly against R using the given ensemble populations (*30*). The relaxation matrix will be:

$$\langle R \rangle = \Sigma p_i R_i \qquad (6)$$

where p_i are the mole fractions, and R_i are the relaxation matrices corresponding to each conformer. This immediately reduces the total computational burden at each cycle, as it is no longer necessary to integrate the Bloch-Solomon equations, or invert the experimental estimate of R. The model values of R_i will initially be equal to those found at the previous stage of the refinement, and become iteratively modified until agreement with $\langle R \rangle$ is found. Unfortunately, there can be no unique solution if the p_i are also allowed to vary, and the R_i have to incorporate some model of rapid small amplitude fluctuations (*28,31*). As the backbone must fluctuate to accommodate the changes in nucleotide conformations (see above), it may be possible to analyse the ^{1}H-^{31}P couplings in terms of the fraction of B_I and B_{II} conformations (*50*), and use this probability to reduce some of the underdetermination of the backbone conformation.

Discussion

Although it may be possible to provide plausible ensembles of conformations that are consistent with the data, the low amount of independent data certainly poses a limit on uniqueness. Thus, a description of the conformation of nucleic acids in solution will always include a substantial degree of modelling, and therefore be dependent on the quality of the force-fields used. A possible source of additional information that has received little attention in nucleic acids is the chemical shift. There is now a huge database of chemical shifts for a wide variety of conformations, and the chemical shifts are in the main determined by the ring currents of the bases. Detailed calculations of the ring current fields of the bases have been reported (*52*), and it should be possible to refine these against the data base to derive a consistent set of values. This would then add a large number of probably weak constraints throughout the entire molecule, which may well serve to reduce the amount of conformational space that is consistent with NOEs and coupling constants. The short range of all of the experimental parameters, and the fact that nucleic acids are

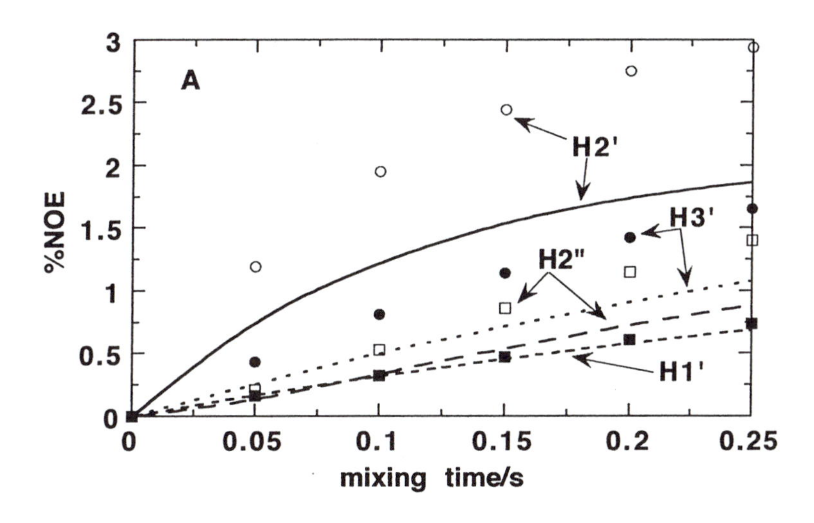

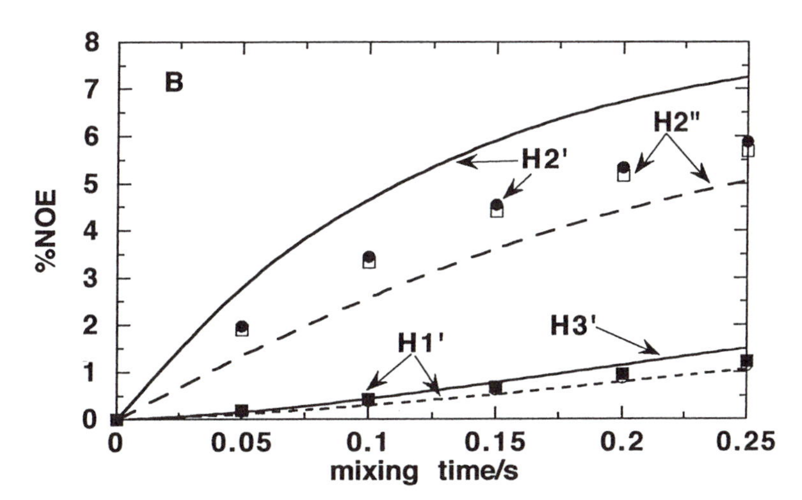

Figure 3. Ensemble NOEs
The points are the ensemble NOEs, and the continuous curves were calculated from the rMD refined structures assuming a single conformation.
A. Intraresidue NOEs from A4H8 B. Sequential NOEs from A4H8 to A3Hi.

determined mainly by nearest neighbour interactions means that long range order such as bending or even the global twist is not determined. Methods of supplying this information would be extremely valuable. One possibility is to use the slight ordering of a long DNA duplex in a strong magnetic field, which causes spin-spin splittings to be modulated by the residual angular dependence of the dipolar interaction (*53*). Measurements of the dipolar contribution of the splittings would provide information about orientation of vectors with respect to a global axis, and therefore provide some of this information.

Acknowledgments
This work was supported by the Medical Research Council of the UK. I am most grateful to my colleagues for discussions, and in particular to Drs. J.I. Gyi, T.A. Frenkiel and N.B. Leontis for careful reading of the manuscript.

Literature Cited

1 Steitz, T.A.*Q. Rev. Biophys.* **1990** *23,* 205-280
2 Werner, M.H.; Huth, J.R.; Gronenborn, A.M.; Clore, G.M. *Cell* **1995** *81,* 705-714.
3 Aboul-ela, F.; Koh, D.; Tinoco Jr., I. *Nucl. Acids Res.* **1985** *13,* 4811-4824
4 Lane, A.N.; Peck, P. *Eur. J. Biochem.* **1995** *230,* 1073-1087
5 Ebel, S.; Brown, T.; Lane, A.N. *Eur. J. Biochem* . **1994** *220,* 703-715
6 Lefèvre, J-F.; Lane, A.N. ; Jardetzky, O. *FEBS Lett.* **1985** *190,* 37-40
7 Schmitz, U.; Sethson, I. ; Egan, V.M.; James, T.L. *J. Mol. Biol.* **1992** *227,* 510-531
8 Kennedy, M.A.; Nuutero, S.T.; Davis, J.T.; Drobny G.P.; Reid B.R. *Biochemistry* **1993** *32,* 8022-8035
9 Lane, A.N.; Frenkiel, T.A. ; Bauer, C.J. *Eur. Biophys. J.* **1993** *21,* 425-431
10 Allain, F. H-T.; Varani, G. *J. Mol. Biol.* **1995** *250,* 333-353
11 Gonzalez, C. ; Stec, W.; Reynolds, M.; James, T.L. *Biochemistry* **1995** *34,* 4969-4982
12 Schmitz, U.; Zon, G.; James, T.L. *Biochemistry* **1990** *27,* 2537-2368
13 Conte, M.R.; Bauer, C.J.; Lane, A.N. *J. Biomolec. NMR.* **1996** *7,* 190-206
14 Harbison, G.S. *J. Am. Chem. Soc.* **1993** *115,* 3026-3027
15 Zhu, L.; Reid, B.R.;Kennedy, M. ; Drobny, G.P. *J. Magn. Reson. Series A.* **1994** *111,* 195-202.
16 Norwood, T.J. J. *Magn. Reson. A* **1995** *114,* 92-97
17 van Wijk, J.; Huckriede, B.D.; Ippel, J.H.; Altona, C. *Meths. Enzymol.* **1992** *211,* 286-306
18 Wijmenga, S.S. ; Mooren, M.N.W.; Hilbers, C.W. *NMR of Macromolecules. A Practical Approach* (G.C.K Roberts, Ed.) IRL Press, Oxford, 1993. Ch. 8.
19 Borgias; B.A.; Gochin, M.; Kerwood, D.J.; James, T.L. *Prog. NMR Spect.* **1990** *22,* 83-100
20 Lefèvre, J-F.; Lane, A.N. ; Jardetzky, O. *Biochemistry* **1987** *26,* 5076-5090
21 Lane, A.N. *Biochim. Biophys. Acta.* **1990** *1049,* 189-204
22 Lane, A.N. ; Fulcher, T. J. *Magn. Reson. B* **1995** *107,* 34-42
23 Boelens, R.; Koning, T.M.G.; Kaptein, R. *J. Mol. Sruct.* **1988** *173,* 299-308
24 Post, C.; Meadows, R.; Gorenstein, D.G. *J. Am. Chem. Soc.* **1990** *112,* 6796-6803
25 Forster, M.J. *J. Comp. Chem.* **1991** *12,* 292-298
26 Tropp, J. *J. Chem. Phys.* **1980** *72,* 6035-6043
27 Robinson, B.H.; Drobny, G.P. *Meths. Enzymol.* **261** *261, 451-512*
28 Koning, T.M.G.; Boelens, R.; van der Marel, G.A.; Kaptein, R. *Biochemistry* **1991** *30,* 3787-3797
29 Lane, A.N.; Forster, M.J. *Eur. Biophys. J.* **1989** *17,* 221-232

30 Koehl, P.; Lefèvre, J-F. *J. Magn. Reson.* **1990** *86,* 565-583
31 Lane, A.N. *Carbohydrate Research* **1991** *221,* 123-144
32 Borer, P.N.; LaPlante, S.R.; Kumar, A.; Zanatta, N.; Martin, A.; Levy, G.A. *Biochemistry* **1994** *33,* 2441-2450
33 Eimer, W.; Williamson, J.R.; Boxer, S.G.; Pecora, R. *Biochemistry* **1990** *29,* 799-811
34 Birchall, A.J; Lane, A.N. *Eur. Biophys. J.* **1990** *19,* 73-78
35 Bonifacio, G.F.; Conn, G.L.; Brown, T.; , Lane, A.N. *Biophys. J.* **1997** in press.
36 Woessner, D.E. *J. Chem. Phys.* **1962** *37,* 647-654
37 Torda, A.E.; Scheek, R.M.; van Gunsteren, W.F. *Chem. Phys. Lett.* **1989** *157,* 289-294
38 Torda, A.E.; Scheek, R.M.; van Gunsteren, W.F. *J. Mol. Biol.* **1990** *214,* 223-235
39 Pearlman, D.A.; Kollman, P.A. *J. Mol. Biol.* **1991** *220,* 457-479
40 Gonzalez, C.; Stec, W.; Kobylanska, A.; Hogrefe, R.I.; Reynolds, M.; James, T.L. *Biochemistry* **1994** *33,* 11062-11072
41 Ulyanov, N.B.; Schmitz, U.; James, T.L. (1993) J. Biomolec. NMR 3, 547-568
42 Ulyanov, N.B.; Schmitz, U.; Kumar, A.; James, T.L. *Biophys. J.* **1995** *68,* 13-24
43 Bonvin, A.M.J.J. ; Boelens, R.; Kaptein, R. *J. Biomolec. NMR* **1994** *4,* 143-149
44 Bonvin, A.M.J.J.; Brünger, A.T. *J. Mol. Biol.* **1995** *250,* 80-93
45 Bonvin, A.M.J.J., ; Brünger, A.T. *J. Biomolec. NMR* **1996** *7,* 72-76
46 Lane, A. N. *Magn. Res. Chem.* **1996** *34,* S3-S10
47 Gyi, J.I.; Conn, G.L.; Brown, T.; Lane, A.N. *Biochemistry* **1996** *35,* 12538-12548
48 Nishizaki, T.; Iwai, S.; Ohkubo, T.; Kojima, C.; Nakamura, H.; Kyogoku, Y.; Ohtsuka, E. *Biochemistry* **1996** *35,* 4016-4025
49 Lavery, R. ; Sklenar, H. **1996** Curves 5.1 Manual, CNRS, Paris.
50 Roongta, V.A.; Jones, C.R.; Gorenstein, D.A. *Biochemistry* **1990** *29,* 5245-5258
51 Metzler, W.J.; Wang, C.; Kitchen, D.B.; Levy, R.M.; Pardi, A. *J. Mol. Biol.* **1990** *214,* 711-736
52 Giessner-Prettre, C.; Pullman, B. *Q. Rev. Biophys.* **1987** *20,* 113-172
53 Kung, H.C.; Wang, K.Y.; Goljer, I.; Bolton, P.H. *J. Magn. Reson* **1995** *109B,* 323-325

Chapter 8

NMR Structure Determination of a 28-Nucleotide Signal Recognition Particle RNA with Complete Relaxation Matrix Methods Using Corrected Nuclear Overhauser Effect Intensities

Peter Lukavsky, Todd M. Billeci, Thomas L. James, and Uli Schmitz[1]

Department of Pharmaceutical Chemistry, University of California, San Francisco, CA 94143–0446

Abstract: Homonuclear NMR experiments have been used to obtain sequential assignments for the vast majority of exchangeable and non-exchangeable protons of a 28mer RNA representing part of the most conserved domain IV of SRP RNA. In the presence of magnesium cations, the SRP 28mer exhibits a remarkably stable structure even for an extended internal loop region. NOE crosspeak intensities for the non-exchangeable protons have been quantified and corrected for full relaxation and were then used for complete relaxation matrix calculations with the program MARDIGRAS to yield spin-diffusion free distance restraints. These restraints were used in restrained molecular dynamics calculations to refine the solution structure of the SRP 28mer. The preliminary NMR structure is presented.

List of Abbreviations: NMR, nuclear magnetic resonance; $O.D._{260}$, optical density at 260 nm at 25 °C; NOESY, nuclear Overhauser effect spectroscopy; DQF-COSY, double quantum filtered correlation spectroscopy; KP, potassium phosphate; T_m, UV-melting point; ER, endoplasmic reticulum; PAGE, polyacrylamide gel electrophoresis; TSP, trisilylpropionic acid sodium salt; fid, free induction decay; *ffh*, fifty-four-homologue; *ffhM*, methionine-rich domain of fifty-four-homologue.

[1]Corresponding author

Protein translocation across the ER membrane involves the interaction between ribosomes and the signal recognition particle (SRP). SRP is thought to recognize a signal sequence in the nascent protein which causes it to bind tightly to the active ribosome (1, 2). This in turn provokes translational arrest, followed by the targeting of the dormant ribosome/SRP complex to the SRP receptor in the ER membrane.

Mammalian SRP is comprised of a 300-nucleotide RNA and six distinct proteins (SRP 9, 14, 19, 54, 68, and 72) (1). SRP RNA homologues have been found for a wide range of species and phylogenetic comparisons (3, 4) suggest a conserved four domain structure for the RNA. Domain IV exhibits the largest conservation of both sequence and secondary structure, possibly indicating a key role for this region (see also the chapter of C. Zwieb in this volume). Indeed, domain IV interacts with SRP54, which is also instrumental in recognizing the peptide signal sequence. A SRP54 homologue was found in *E. coli*, termed *ffh*, (5, 6) which exhibits similarity in both sequence and domain structure compared with the mammalian system.

SRP54 may be separated proteolytically into two well-defined domains. The N-terminal domain (34 kD), whose crystal structure was solved recently (7), harbors GTPase activity, while the C-terminal domain (20kD) contains both the RNA binding and the signal sequence binding sites (8-11).

As a first part in the elucidation of the solution structure of the complex of SRP54 C-terminal domain and RNA domain IV, which is capable of recognizing signal sequences, we describe here the determination of the RNA motif and some of the methodological implications encountered.

The NMR constructs for domain IV RNA are depicted in Figure 1. Recent low resolution models (12, 13) suggest that domain IV folds back on domain II making tertiary interactions, which would require domain IV to make a U-turn. The flexibility for this turn could be associated with the asymmetrical internal loop. Besides determining the structure of the fairly large 43 nucleotide RNA, it seemed prudent to also study the structure of domain IV by isolating the two internal loops and thus their contribution to a potential bend.

Our recent NMR study of the 24mer hairpin motif comprising largely the symmetrical internal loop revealed that a unique and stable structure is formed only in the presence of Mg^{2+} ions, which seem to bind specifically in the internal loop region. Unfortunately, this construct had a large propensity for duplex formation and we did not consider the NMR data of sufficient quality to allow a high resolution structure determination. This particular problem could be circumvented by making the transition from the kinetically favored hairpin to the thermodynamically more stable duplex more difficult by adding two additional G:C basepairs. This 28mer hairpin (Fig. 1) is more stable than the 24mer as indicated by UV melting data and the 2D NOE spectra proved to be excellent for a full structure determination.

While we are applying recent heteronuclear NMR methods (14, 15) to ^{13}C, ^{15}N-labeled samples of the 43mer RNA, the 28mer offers a unique possibility to test the limits of well-established homonuclear NMR methods which work well for a molecular size under 20 nucleotides (16). High resolution structures of larger RNA fragments have been determined with either special spectral deconvolution techniques (17) or partial deuteration of RNA samples (18).

For the majority of NMR structures determined with natural abundance RNA samples, NMR-derived interproton distance information has been utilized in the form of semi-quantitative distance restraints. However, NOE-derived semi-quantitative distance information alone is not sufficient to precisely determine a high resolution structure. For most RNA structures (19), semiquantitative distance restraints are typically augmented with torsion angle information to restrain the ribose moieties and part of the backbone. However, well-defined backbone torsion angle restraints, most of which arise from ^{31}P-^{1}H correlated experiments, require largely complete assignments of the ^{31}P resonances and all ribose proton resonances including stereospecific assignments of H5' and H5" proton resonances. Although quite powerful for smaller systems, the aforementioned requirements for the torsion angle approach are hardly achievable for larger RNAs, such as the SRP 28mer.

An alternative appproach in cases where there is a paucity of accurate backbone torsion angle restraints, is the utilization of NOE-derived distances whose precision has been improved through accounting for multispin effects, commonly termed spin-diffusion (20). These complete relaxation matrix methods (21) have been established for a while now but have been applied mostly to DNA systems (22). With this approach, the theoretical NOE spectra including spin-diffusion effects can be calculated for a given structure with the assumption of a motional model. Conversely, this strategy can be used to compute accurate distances from NOE intensities for unknown structures using a hybrid matrix approach, where all experimentally unobserved NOEs are taken from a model structure that is similar to the target structure (23-25). The elements of the relaxation matrix are then varied until a consistent fit to the experimentally observed NOEs is obtained. Interproton distances are readily obtained from the converged relaxation matrix. This method, as it is implemented in our program MARDIGRAS(23), also allows the determination of error bounds for the distances which take into account errors arising from limitations in NOE volume accuracies due to peak overlap, spectral noise levels, distortion from solvent or diagonal peaks, exchange processes, incomplete proton relaxation or baseline problems.

To date, only two applications of complete relaxation matrix derived distance restraints have been reported for RNA. MARDIGRAS provided "precise" distance restraints for the structure determination of a 17mer RNA with a flexible loop (26) and a 31mer hairpin with a large internal loop (18). For the latter, assignments and NOE volume extraction was made feasible through deuteration of all cytosines.

This chapter describes the application of the complete relaxation matrix method to homonuclear NOE data of the SRP 28mer and the determination of a preliminary high fresolution structure of the latter.

Results and Discussion

Sequence Design: The conserved secondary structure for SRP RNA domain IV contains a distinct hairpin motif with two bulged regions separated by two to three canonical basepairs. The rationale for designing our model for domain IV, the SRP 43mer, has been described elsewhere (27). The SRP 28mer was prepared by *in vitro* transcription with T7 RNA polymerase (27). In comparison with our earlier work on the 24mer, the 28mer binds more tightly to the C-terminal part of *ffh*. On the other hand, the effect of Mg^{2+} ions on the thermodynamic stability as inferred from UV melting analysis is less pronounced. However, the NMR spectra of the two shorter SRP RNA constructs are virtually identical for the internal loop, top stem and tetraloop, such that we consider 24mer and 28mer structurally the same.

NMR Studies: In sharp difference to the 24mer, the 28mer can be trapped in pure hairpin conformation after snapcooling in the absence of additional cations, as indicated by gel filtration assays (data not shown).

1. Assignments of Nonexchangeable Protons. Our proton assignment strategy followed well-described methods utilizing largely 2D NOE data and homonuclear correlated experiments (16, 28).

In addition, the 28mer chemical shift assignments were largely simplified by comparison with the 24mer. The chemical shifts of exchangeable and non-exchangeable protons are summarized in Table 1.

Similar to the 24mer SRP RNA, 1D-spectra of the 28mer (in D_2O) in the presence of Mg^{2+} (5 mM) *versus* no Mg^{2+} or Na^+ (50 mM), clearly showed that Mg^{2+} stabilizes the internal loop as indicated by sharpening of several broad resonances and by leading to a greater chemical shift dispersion in the aromatic proton region (see Fig. 2).

The DQF-COSY spectrum allowed us to identify the ten pyrimidines in the 28mer sequence *via* the H5-H6 crosspeaks. The crosspeaks of one the two internal loop cytosines, C22, was significantly broadened compared to the other pyrimidines indicating different dynamic properties for C22. The DQF-COSY experiment also served to identify the H1'-H2' crosspeaks of ribose moieties that do not exhibit pure C3'-endo sugar puckers; two strong peaks in the H1'→ ribose region could be assigned to the loop residues G14 and A15. A weak peak in this region arose from the 3'-terminal C28.

The base → H1' region of the 2D NOE spectra at various mixing times (50, 150, 200 and 400ms) revealed only 24 of the 28 H1' resonances. The sequence of the 28mer loop is a member of the GnRA tetraloop family. One of the signatures of this tetraloop is the unusual strong upfield shift of the H1' proton which is situated under the baseplane of the adenine of the G:A closing base pair of the loop (29). This allowed to identify the H1' resonance of G17 (3.61 ppm). The last three missing H1' proton resonances are all located in the internal loop, namely G8, G21, and G24, respectively. Those resonances could be identified by acquisition of a 120-ms 2D NOESY at 10°C and a DQF-COSY without presaturation of the residual HOD resonance.

The DQF-COSY spectrum showed an additional strong H1'-H2' crosspeak on the water resonance line exhibiting a large active coupling constant of ≈12 Hz which indicates a ribose with a C2'-endo conformation. In the 2D NOESY spectrum at lower temperature weak H8 → H1' connectivities arising from this H1' resonance allowed to close the gap in the H6/H8 → H1' walk between G8 and G9 and to assign it for G8H1' (4.78ppm). Furthermore, the observation of the weak intraresidue H8 → H1' crosspeak excludes the possibility of a *syn* conformation for G8. Another interesting feature of this G8pG9 step is the relatively strong sequential G8H2'-G9H1' crosspeak which indicates a deviation from A-form RNA in this part of the internal loop.

The second H1' resonance that could be identified due to the downfield shift of the water resonance at 10°C was G21H1' (4.71 ppm): it showed a very weak intraresidue H8 → H1' crosspeak, no sequential H1' → H6 connectivity with C22 (possibly due to the increased linewidth of C22), and a strong A-form type sequential A20H2 → G21H1' crosspeak. A similar pattern could also be found for G24H1' (4.68 ppm): strong sequential A23H2 → G24H1' and weak intra and interresidue H1' → H8 connectivities. Those assignments were in accord with the observed intra and interresidue H2' → H6/H8 (the shortest sequential ribose → base distance in A-form is the H2' → H6/H8 distance)

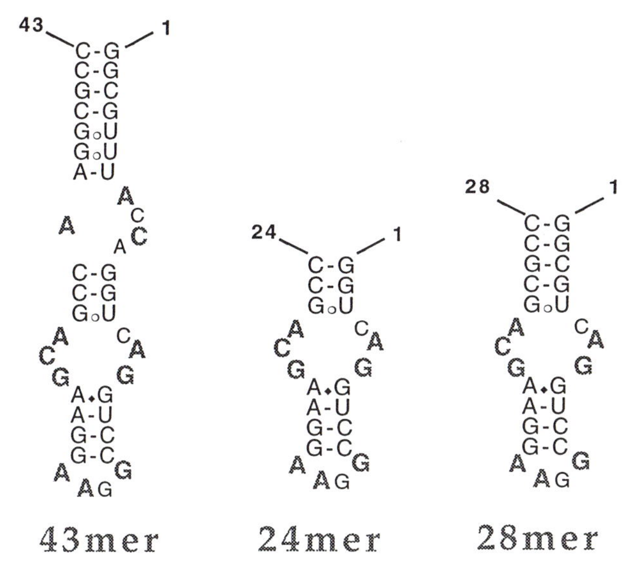

Figure 1. Sequences and phylogenetically-derived secondary structures of NMR constructs for SRP domain IV: 43mer, 24mer and 28mer (from left to right). Phylogenetically conserved bases are shown in bold.

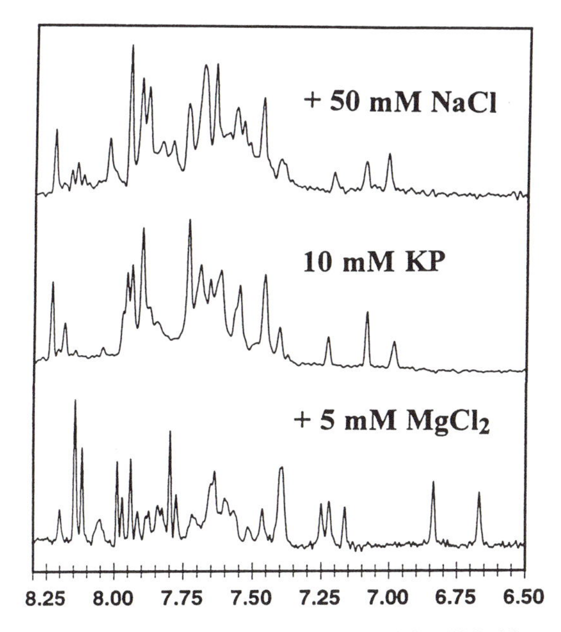

Figure 2. 1D NMR spectra of the aromatic region of the SRP 28mer showing the influence of different cations: (center) 10mM KP, pH=6.5; (top) 10mM KP, pH=6.5, 50mM NaCl; (bottom) 10mM KP, pH=6.5, 5mM $MgCl_2$; all spectra were taken at 25° C in D_2O.

Table 1: Proton Chemical Shift Assignments for SRP 28mer (ppm) and $^3J_{H1'H2'}$ coupling constants (Hz)[a]

nucleotide	H6/H8	H2/H5	H1'	H2'	H3'	H4'	H5'/H5''[b]	imino	amino	$^3J_{H1'H2'}$
G1	8.18	-[c]	5.93	4.50	4.66	4.38	n/a[d]	13.27	n/a	<3
G2	7.64	-	5.72	4.45	4.67	n/a	n/a	13.42	n/a	<3
C3	7.71	5.20	5.53	4.55	4.60	n/a	n/a	-	8.47/6.51	<3
G4	7.39	-	5.73	4.50	4.38	n/a	n/a	13.12	7.82/6.09	<3
U5	7.58	5.10	5.58	4.51	4.38	n/a	n/a	12.31	-	<3
C6	7.38	5.45	5.11	4.01	4.64	4.52	n/a	-	6.90	<3
A7	7.94	8.11	6.10	4.91	4.66	4.55	4.17/4.31	-	n/a	<3
G8	7.24	-	4.78	4.44	4.53	n/a	4.40	11.30	8.80/6.00	11-13
G9	7.84	-	5.39	4.34	n/a	n/a	n/a	13.60	5.98	<3
U10	7.39	5.42	5.52	4.36	4.45	4.07	n/a	12.40	-	<3
C11	7.88	5.62	5.44	4.12	4.50	n/a	n/a	-	8.29/6.98	<3
C12	7.57	5.31	5.48	4.41	n/a	4.50	n/a	-	8.13/6.47	<3
G13	7.52	-	5.64	4.28	4.67	4.41	4.11/4.34	10.57	n/a	<3
G14	7.98	-	5.46	4.72	n/a	4.18	3.88	12.74	6.15	4-6
A15	7.97	7.79	5.61	4.39	4.61	4.30	3.94	-	n/a	4-6
A16	8.19	8.16	5.99	4.65	n/a	4.45	n/a	-	n/a	<3
G17	7.93	-	3.61	4.44	4.25	4.17	4.25	12.38	7.78/6.27	<3
G18	7.17	-	5.72	4.47	4.45	n/a	n/a	12.40	7.79/6.31	<3
A19	7.78	6.83	5.80	4.30	4.45	n/a	n/a	-	8.49/6.54	<3
A20	7.82	8.14	6.08	4.65	4.42	4.51	n/a	-	8.80/5.98	<3
G21	7.46	-	4.71	3.89	4.38	4.34	n/a	10.31	6.16	<3
C22	8.04	5.91	5.85	4.51	4.43	4.76	4.16	-	8.54	<3
A23	7.60	6.67	5.94	4.63	4.39	4.53	n/a	-	n/a	<3
G24	7.21	-	4.62	4.54	4.37	4.27	n/a	10.59	6.69	<3
C25	7.64	5.39	5.37	4.29	4.55	4.30	n/a	-	8.42/6.73	<3
G26	7.63	-	5.72	4.46	n/a	n/a	n/a	12.96	8.27/6.12	<3
C27	7.64	5.20	5.52	4.30	n/a	n/a	n/a	-	8.56/6.74	<3
C28	7.64	5.40	5.71	4.00	4.17	n/a	n/a	-	8.62/6.75	4-6

[a] 10 mM potassium phosphate, pH 6.5, 5 mM $MgCl_2$, 25°C for non-exchangeable and 15°C for exchangeable protons. [b] not stereospecificly assigned. [c] not applicable. [d] not assigned.

connectivities which also confirmed some tentative assignments in the H6/H8 → H1' walk. The complete H6/H8 → H1' walk is shown in Figure 3.

Although we did not assume to get typical A-form NOE patterns within the internal loop region during the assignment process, the connectivities found did not reveal dramatic deviations from an A-form stacking pattern.

2. Assignments of Exchangeable Protons: 1D NMR spectra of the 28mer collected at 9:1 H_2O:D_2O at 10°C and 15°C also showed a significant structural stabilization of the internal loop in the presence of Mg^{2+} (5 mM) evident from the sharpening of several internal loop NH resonances.

The 2D SS-NOESY spectra (mixing time 120 ms and 200 ms, respectively) allowed identification of the regular Watson-Crick basepairs due to their typical NH → H5/H2/inter and intrastrand H1' NOE crosspeak patterns. The NH resonances in the upper stem, G17, G18, and U10 (one resonance cluster at 12.40 ppm), and the ones from the bottom stem, G4, G26, G2, and G1, could be assigned unambiguously. Furthermore, the quality of the spectra also permitted assignments of most NH_2 resonances in those stem regions.

The NH protons of the U5:G24 wobble pair could be easily identified through strong intra basepair NH → NH and weak inter basepair NH→NH crosspeaks. Further confirmation of this assignment arose from the NOE crosspeaks of the NH to inter and intrastrand H1' and H2'. However, an untypical, large exchange peak with water was observed for the G24NH (10.59ppm) resonance line, suggesting that another more solvent accessible NH resonates at this frequency as well. Indeed, weak crosspeaks to A16H8 and A15H8 as well as A15H2' and G13H2' assigned this NH resonance to G13NH which is involved in the closing G:A basepair (N7-amino, N3-amino) of the tetraloop.

All four remaining NH resonances showed exchange peaks with water, of which G14NH (12.74ppm) exhibits no NOE crosspeaks at all. The strong crosspeak of the most downfield shifted NH resonance and A20H2 not only assigned this NH resonance to G9NH (13.60 ppm) but also identified the G9:A20 mismatch as a basepair involving the Watson-Crick-face of both G9 and A20.

Thus, the most upfield NH resonance and the NH resonance at 11.3 ppm must be G8NH and G21NH. The absence of an NH → NH crosspeak clearly excluded the possibility of a G8:G21 NH-NH mismatch basepair. Furthermore, none of those NH resonances indicated any type of basepairing between G8:G21 involving the Watson-Crick-face of one of them and the Hoogsteen-face of the other one due to the absence of an NH → H8 crosspeak. The assignment of those resonances was finally achieved on the basis of the following assumptions: since the D_2O NOE data clearly indicated a stronger deviation of the G8pG9 step from A-form compared to the A20pG21 step, we assumed to observe

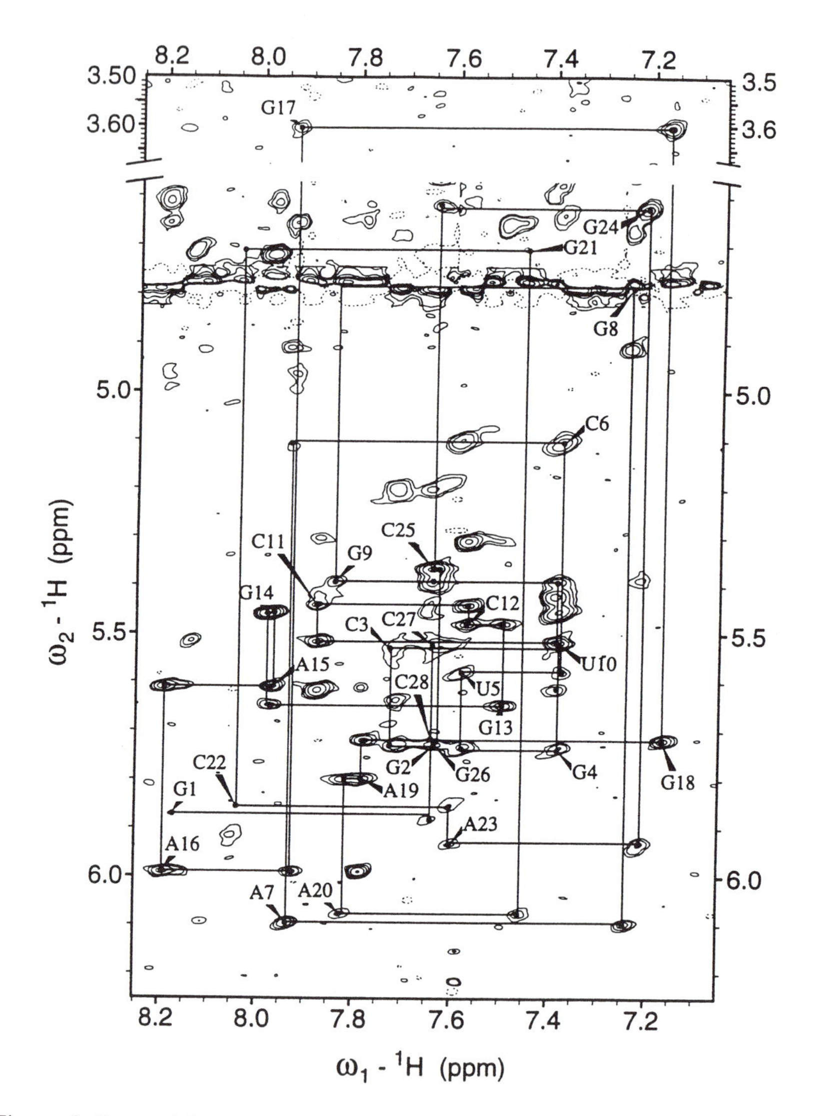

Figure 3. Sequential assignments for SRP 28mer: portion of the 400ms 2D NOESY showing the base→H1'/H5 proton crosspeaks. H6/8→H1' NOESY walk is traced: the respective intraresidue peak is labeled with the residue number. (Note the downfield shift of several H1' protons.)

more A-like NH connectivities for G21NH compared to G8NH. The G9NH resonance showed a weak crosspeak to G21H1' which also supported this assumption. G21NH was therefore assigned to the NH resonance at 10.31ppm which shows weak crosspeaks to C22H1' and G9H1' and a strong intraresidue NH → NH_2 crosspeak. The strongest crosspeak of the unusual G8NH → ribose connectivities could be assigned to G8NH → C6H2' due to the unique chemical shift of C6H2' (4.01ppm) and the presence of additional weak crosspeaks for G8NH → C6H1',G8NH → C6H3' and G8NH → C6H6.

Despite the fact that the base → NH_2 as well as H1' → NH_2 regions were reasonably well resolved in both 2D SS-NOESY spectra, only the NH_2 resonances of the two cytosines of the internal loop , C6 and C22, could be assigned due to their H5 → NH_2 crosspeaks. NH_2 resonances of A7 and A23 could not be observed. Furthermore, no additional crosspeaks of the NH_2 resonances of C6 and C22 could be detected which left us with no indication of the type of basepairing of the two mismatched A:C moieties, A7:C22 and C6:A23.

Comparison of the SS-NOESY spectra of the 28mer in the presence of Mg^{2+} (5 mM) *versus* no Mg^{2+} demonstrates again that the internal loop is significantly stabilized by Mg^{2+}; the U5:G24 wobble pair is still formed but the strong U5NH → C6H1' and G24 → C6H1' crosspeaks are not observed in the absence of Mg^{2+}. Furthermore, the mismatch NH protons G8, G9, and G21 can be identified only due to their exchange peaks with water. G13NH does not show any crosspeaks in the absence of Mg^{2+}, indicating increased flexibility within the tetraloop when no Mg^{2+} is present.

Structure Determination: Our strategy for the structure determination via complete relaxation matrix derived distance restraints is summarized in Figure 5. Assigned and integrated NOE crosspeak volumes are first corrected for the effects of partial relaxation and then used for complete relaxation matrix calculations (21) with the program MARDIGRAS (23, 30). The latter yields precise distances and error bounds which are subsequently employed in restrained MD calculations. Preliminary, refined structures are then carefully examined to exclude possible errors or ambiguities in the proton assignments. rMD structures obtained after this first round are then used as starting models for the complete relaxation matrix calculations to minimize the effect that the starting geometry might exert on the ensuing distances. The refinement process thus goes through several cycles of the entire procedure.

1. Generation of NOE-derived distance restraints: To obtain distances and their error bounds as accurately as possible, the recycling delay of the NOESY experiment should be long enough that the longitudinal magnetization of the protons is fully recovered.

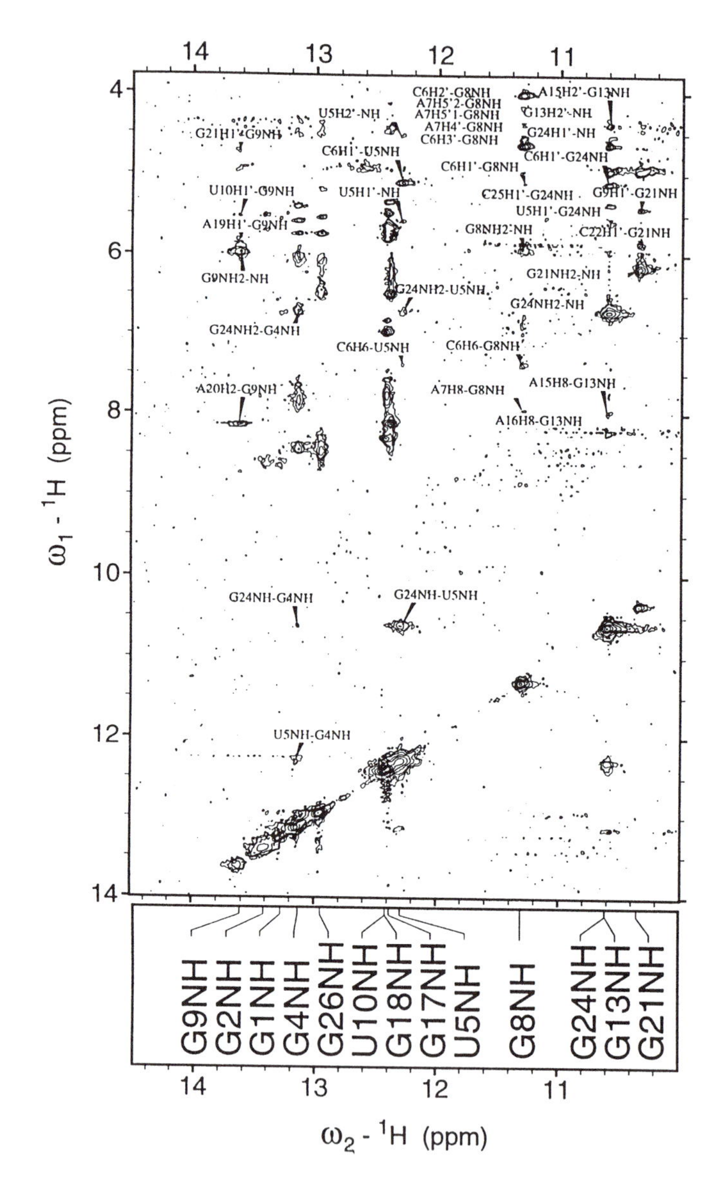

Figure 4. 2D NOE data (mixing time 120ms) for SRP 28mer in H_2O in 10mM potassium phosphate (pH=6.5, 10°C) (right) and with additional 5mM $MgCl_2$ (left). Shown is the portion with the crosspeaks involving imino protons.

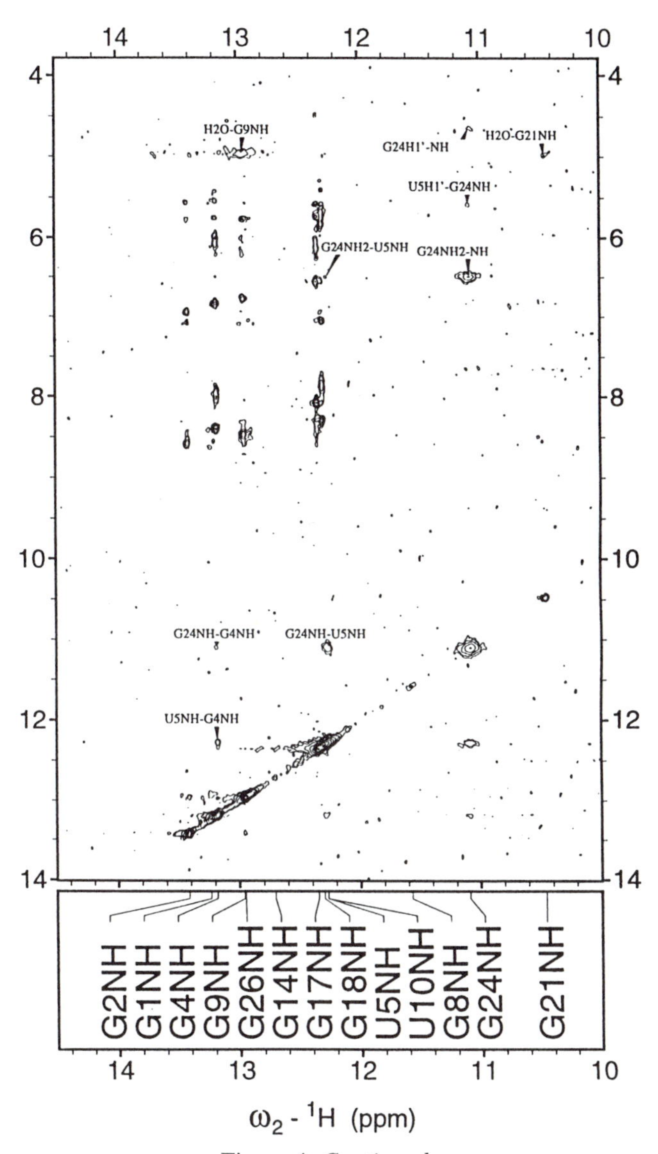

Figure 4. *Continued.*

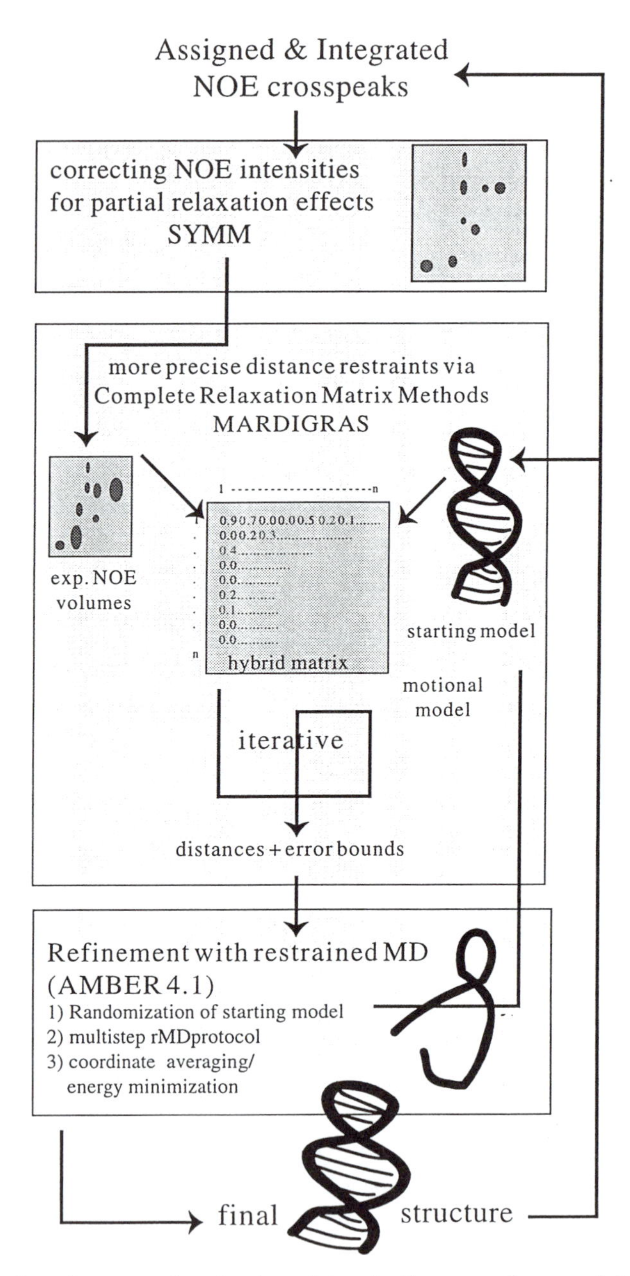

Figure 5. Overview over the structure determination procedure used for the SRP 28mer.

In our earlier work on short DNA duplexes (31), this was accomplished by using recycling delays of over 10 sec, which was at least three times the longest T_1-relaxation time. However, for larger RNA samples where T_1-relaxation times are significantly larger at least for certain groups of protons, sufficiently large recycling delays would make 2D NMR virtually impossible. For the SRP 28mer, measured T_1-relaxation times ranged from 6-9 sec (Tab. 2), which would require delays of over 20 sec between pulsing. However, it is possible to correct for the effects of truncated relaxation when the actual T_1-values and the recycle delay are known (32-35). A NOESY crosspeak volume I_{ij} is proportional to the longitudinal magnetization of spin j, $Mz_j(0)$, recovered before application of the NOESY pulses. For short repetition times where the spins are not fully relaxed, $Mz_j(0)$ is smaller then 1 and the intensities are scaled

$$I_{ij} = I_{ij}^{o}\, Mz_j(0) \qquad [1]$$

with

$$I_{ij}^{o} = [\exp(-\mathbf{R}t_m)]_{ij} \qquad [2]$$

and

$$Mz_j(0) = \sum_k [1 - \exp(-\mathbf{R}(t_d + t_{acq})]_{jk} \qquad [3]$$

where I_{ij}^{o} is the fully relaxed intensity, **R** is the relaxation rate matrix, t_m is the mixing time, t_d is the relaxation delay, t_{acq} is the acquisition time and the sum is over k nuclei. $T_r = t_d + t_{acq}$ is the repetition time of a NOESY experiment. For a given T_1 value, $Mz(0)$ values can be estimated using the following approximation:

$$Mz_j(0) \approx 1 - \exp[-T_r/T_{1j}]. \qquad [4]$$

Cross-relaxation rates and the ensuing interproton distances are determined by Eq.[2], which requires full relaxation. However, with Eq. [4] it is possible to extrapolate from partially relaxed NOE intensities to the fully relaxed quantities (32, 34). This approach, however, requires that accurate T_1 values are available for individual protons, which might be an obstacle in the case of macromolecules. Another possibility to correct for partial relaxation effects utilizes the ratio between above- and below-diagonal crosspeak intensities which in the case of a partially relaxed NOESY spectrum deviate significantly from 1. The details of this approach are beyond the scope of this chapter and have been described elsewhere (35). Both correction procedures have been implemented in our program SYMM (35), which we have used for the correction of the SRP 28mer NOESY data, which had been acquired with a typical, short repetition delay of 2.5 sec.

Table 2: T1-Relaxation Times (s) for SRP 28mer[a,b]

nucleotide	H6/H8	H2/H5	H1'[c]
G1	6.06	-[d]	7.21
G2	7.21	-	7.07
C3	6.20	6.92	6.92
G4	7.50	-	7.07
U5	7.79	7.79	8.08
C6	7.50	6.20	7.50
A7	8.37	7.93	7.79[d]
G8	7.50	-	[e]
G9	6.92	-	7.79
U10	7.50	7.50	7.21
C11	6.49	7.79	7.21
C12	7.21	7.36	7.50
G13	5.77	-	6.92
G14	6.35	-	6.35
A15	6.06	8.37	6.49
A16	5.92	8.08	6.20
G17	5.77	-	6.06
G18	6.06	-	7.07
A19	6.64	8.37	7.65
A20	7.21	8.08	7.50
G21	7.79	-	[e]
C22	6.92	7.79	7.65
A23	7.79	7.5	7.79
G24	7.50	-	[e]
C25	7.21	7.79	7.79
G26	7.21	-	7.07
C27	7.21	-	6.92
C28	7.21	-	7.07

[a] determined *via* Inversion Recovery Experiments. [b] for conditions see Table 1. [c] all other ribose protons exhibit T1-relaxation times in the range of 6.00-7.50 s. [d]A7H2': 7.21 s. [e] not determined.

We analyzed the practical effect of partial relaxation in NOESY spectra on distance restraints by comparing complete relaxation matrix derived distances and their error bounds for several 2D NOE data sets (mixing times 150, 200, 400ms). Both approaches, using measured T_1 values and intensities ratios for above- and below-diagonal peaks, lead to roughly the same distances. The method with direct T_1-values is limited by the fact that not all values are experimentally available, especially for ribose protons other than H1'. For all those protons we resorted to average T_1 values, defined by either the average of the actual T_1 values measured for a type of proton or the median of the range of T_1 values observed in the T_1 inversion recovery experiment (Tab. 2). Comparing intensity ratios above and below the diagonal works best when the crosspeak intensity difference is large and it is dominated by partial relaxation and not by other sources of error. This can be a serious limitation for weak peaks where the intensity differences are more due to noise. For the NOESY spectra (mixing times 150, 200 and 400 ms) of the SRP 28mer, approximately 500 NOE crosspeaks were integrated, of which 5 -8 % appeared only on one side of the diagonal and 20- 25% have such low intensities that the correction procedure is not reliable. However, it is possible to use both approaches together if proper normalization is performed before merging the results. (Note, the two approaches yield different absolute values, which is usually not important since for the complete relaxation matrix calculation only relative intensities are used.)

The effect of the partial relaxation correction appeared to be quite different for different groups of distances. The most dramatic effect was observed for the distances involving protons with longer T_1 values, namely adenine H2 protons. For example, for the three H2-H2 distances corrected distances were 2 - 3 Å shorter compared to using intensities without correction. In this case, the distance restraints would be shifted to the unreasonable range of ≈ 6 -8 Å. Figure 6 depicts the distribution of lower distance bounds and restraints widths for the group of restraints involving H2 or H8, respectively. For the restraints involving H2, the lower bounds are shifted dramatically toward shorter distances without increasing the restraint widths. Furthermore, during the MARDIGRAS calculations, a number of the very small, original intensities are rejected by the program since the unreasonably small values cause convergence problems. All these distances could be recovered with the partial relaxation correction. Even for a group of restraints involving mostly protons with shorter T_1 values, e.g., the peaks involving H8 protons, the intensity correction has a clear, although less pronounced effect (Fig. 6).

These effects are large enough to make it mandatory to correct NOE intensities before taking the ensuing complete relaxation matrix-derived distance restraints into structure refinement calculations. Naturally, this step is less important when NOE intensities are

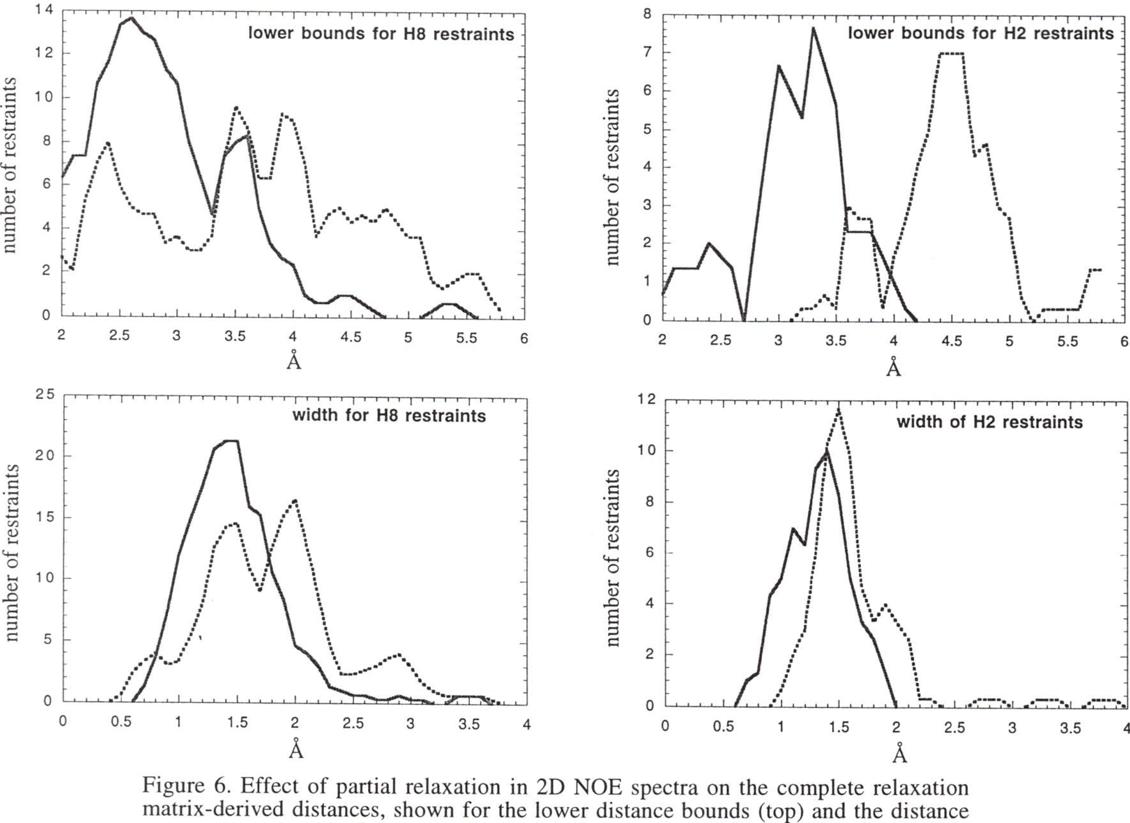

Figure 6. Effect of partial relaxation in 2D NOE spectra on the complete relaxation matrix-derived distances, shown for the lower distance bounds (top) and the distance restraint width (bottom) of H8 (left) and H2 (right) involving distances (For further explanation see text). (Corrected intensities, solid; partial relaxation intensities, dotted)

used only in a semiquantitative manner, where they are grouped in a few relative categories of, e.g., weak, medium and strong, corresponding to distances of 4-7, 3-5, 1.8-3.5 Å (36, 37). However, in the case of the 28mer, a significant number of peaks involving H2 protons would have been placed in the wrong category. Since distance restraints involving H2 protons are very powerful in structure refinement as they represent sparse cross-strand information, structural artifacts would have been inevitable.

Another important advantage of complete relaxation matrix methods for restraints generation lies in the fact that NOESY data acquired for long mixing times can be utilized easily since spin-diffusion effects are explicitly accounted for. For the 28mer, MARDIGRAS calculations were carried out for 150 ms, 200 ms and 400ms NOESY data using an isotropic motional model with a large window for the correlation time (2.5 - 5 sec). The latter was derived empirically by finding the best fit for the intensities of diagonal peaks and by obtaining the best match for fixed (e.g., H5-H6) and virtually fixed distances (e.g., H1'-H2' and H1'-H3' for N-type puckers). To fill in the gaps of the complete relaxation matrix we used a model structure that consisted of an A-form stem, comprising residues G1-G11 and G18-C28, topped with the C12-G17 segment in the coordinates of the GCAA-tetraloop NMR structure (29). The model was energy-minimized to alleviate steric clashes of the purine-purine basepairs. Our complete relaxation matrix calculations were carried out with the "randmardi" procedure which has been demonstrated as essential to get accurate error bounds (38). Multiple MARDIGRAS calculations are carried out with variations over a specific intensity set, where the variations are achieved through modeling systematic and individual errors in the crosspeak volumes. A distance restraint set was built using the MARDIGRAS results of all calculations for the 150 ms NOESY dataset as the core, which was supplemented with the distances that arose only from the 400ms NOESY dataset. The distance error bounds obtained from 400ms NOESY data are naturally larger than those for 150 ms data. Ultimately, we obtained a distance restraint set with 320 entries and an average restraint width of 1.4 Å which was used for the first round of structure refinement. Subsequently, the MARDIGRAS calculations were repeated with the preliminary rMD structure of the SRP 28mer and a slightly smaller window for the correlation time which led to an average restraint width of 1.0 Å.

Despite the improved precision for the distance restraints involving non-exchangeable protons, the much less precise information from exchangeable imino and amino protons is still essential for the determination of unusual RNA structures. The 120 ms SS-NOESY data acquired in H_2O yielded almost 200 additional restraints for exchangeable protons. As can be seen in Figure 4, some imino protons, especially G8NH, show connectivities with a number of ribose protons which turned out to be critical for defining the final structure.

Although it is possible to obtain complete relaxation matrix derived distances for exchangeable proton NOEs intensities if the exchange rate with the solvent is known (39), we used a conservative semiquantitative implementation for the restraints involving exchangeable protons using no lower bounds and upper bounds of 4, 5 and 6 Å for strong, medium and weak peaks, respectively.

2. Structure refinement with restrained molecular dynamics: Beyond the NOE-derived distance restraints we also used torsion angle restraints to constrain the ribose ring moieties based on the observed $J_{H1'H2'}$ coupling constants. All ribose were constrained to the C3'-endo conformation, except for residue G8 which was set to C2'-endo. Although riboses of G14, A15 and C28 were found to assume intermediate conformations by NMR standards, the NMR structure as well as the crystal structures of the GNAA-tetraloops (29, 40) exhibited conformations close to C3'-endo. Since the NOE-derived distances of our GGAA-tetraloop are a perfect match for the GAAA-crystal structure we used the corresponding pucker restraints in addition to loose backbone torsion angle restraints (width 40 - 60°) for residues C12-G17. Backbone torsion angle restraints were also used to keep the bottom stem close to A-form RNA since the two terminal basepairs were not well-defined by the NMR data due to peak overlap and typical rMD modeling procedures *in vacuo* yield artificial conformations for the helix termini. To enforce proper basepairing we also used hydrogen bonding restraints for all Watson Crick basepairs in the sequence. Note that for the internal loop of the 28mer we used only complete relaxation matrix-derived distance restraints, semiquantitative distance restraints for exchangeable protons and ribose torsion angle restraints without any further assumptions especially for the backbone.

Since coupling constant-derived torsion anlges constrain the ribose moieties more precisely and efficiently, all intra-ribose distance restraints were omitted in the refinement calculations. On a per-nucleotide basis, our list contained 7.5 MARDIGRAS distance restraints and 4.4 semiquantitative distance restraints. For the internal loop, comprising residues U5-G9 and A20-G24, the list had 7.3 MARDIGRAS distance restraints and 5.3 semiquantitative distance restraints.

All refinement calculations were carried out with the SANDER module of AMBER 4.1 (41), building on rMD protocols that have been established earlier for DNA fragments (31). For the 28mer RNA, a typical rMD protocol starts with the randomization of the energy-minimized A-form/GCAA-tetraloop model described above, which involves heating of the system *in vacuo* to 1500K for 3-5 ps of unrestrained MD, followed by unrestrained energy-minimization of the last frame.These conditions lead to sufficiently randomized starting geometries. (Note that this conformational melting step does not go to completely randomized structures since it basically melts an A-form hairpin from the open end toward the loop, which results in the lower stem having a larger conformational envelope than the upper stem and loop. Average atomic r.m.s. deviations for the bottom stem were 8 to 10 Å, 6 to 8 Å for the internal loop and below 5 Å for the top stem.)

For the conformational search, a multistep rMD protocol is invoked which besides varying temperature and force field terms uses different weights for different groups of restraints. In order to avoid structures getting trapped in local minima, it was crucial to have a period in the rMD protocol with low weights for electrostatic and van der Waals terms (10%), high temperature (300-600K) in combination with high restraint weights for all the restraints except those that involved the exchangeable protons. The following phase used the same weights for all distance restraints and full weights for the force field terms still under the condition of high temperature. At the end of the high temperature period we also implemented a short period with a five fold weight for the van der Waals which helps to find stacking interactions between bases. Finally, the system was cooled to 160K with regular weights for all force field components and force constants of 20 kcal/mol Å^2 for all distance restraints and 50 kcal/mol rad^2 for all torsion angle restraints. The last two picoseconds of a rMD simulation are averaged and subjected to restrained energy-minimization with distance restraints only (10 kcal/mol Å^2). The typical average deviation from the distance target of a final structure is between 0.17 -0.2 Å.

Although we consider the structural details obtained so far preliminary, we can say that our protocol in combination with the NMR-derived restraints converged, despite the fact that the overall atomic R.M.S.D. for 5 final structures is 2.6 Å. This high value is due to a hinge effect associated with the internal loop. When only the top stem and tetraloop, bottom stem or internal bulge residues are superimposed, average R.M.S.D values are 0.7 Å, 0.8Å and 1.3Å, respectively (Fig. 7). This demonstrates that even for the internal loop comprising 10 residues, complete relaxation matrix derived distance restraints in combination with relatively loose restraints involving exchangeable protons can define a high resolution RNA structure reasonably well.

The preliminary NMR structure of SRP 28mer is shown in Figure 8. We are currently performing the final rMD refinement simulations with improved MARDIGRAS-derived distance restraints where the preliminary NMR structure was used as a starting model for the MARDIGRAS calculations. Therefore, we refrain from a detailed discussion of the structure of the SRP 28mer in this article and refer to future publications.

Nevertheless, a few comments are in order. Besides the expected result that the loop part of the hairpin (C11-G18) is very similar to the GAAA-loop in the previously determined ribozyme crystal structure, we were surprised how well-structured the internal

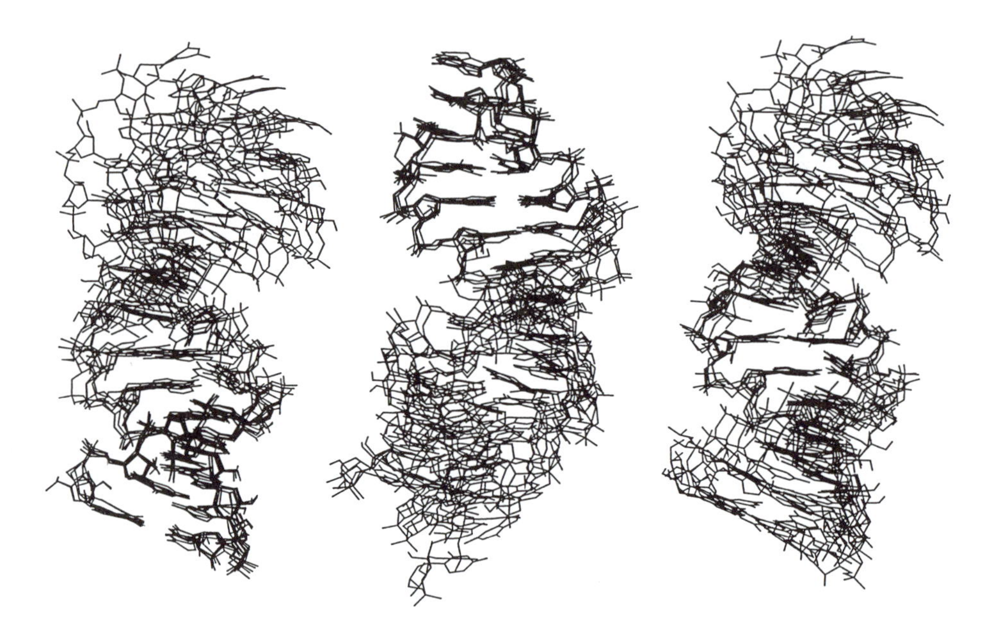

Figure 7. Convergence of restrained MD refinement procedure for SRP 28mer. Five rMD structures are superimposed using just the five basepairs of the bottom stem (left), the top stem and tetraloop (middle) and the six residues of the internal loop (right).

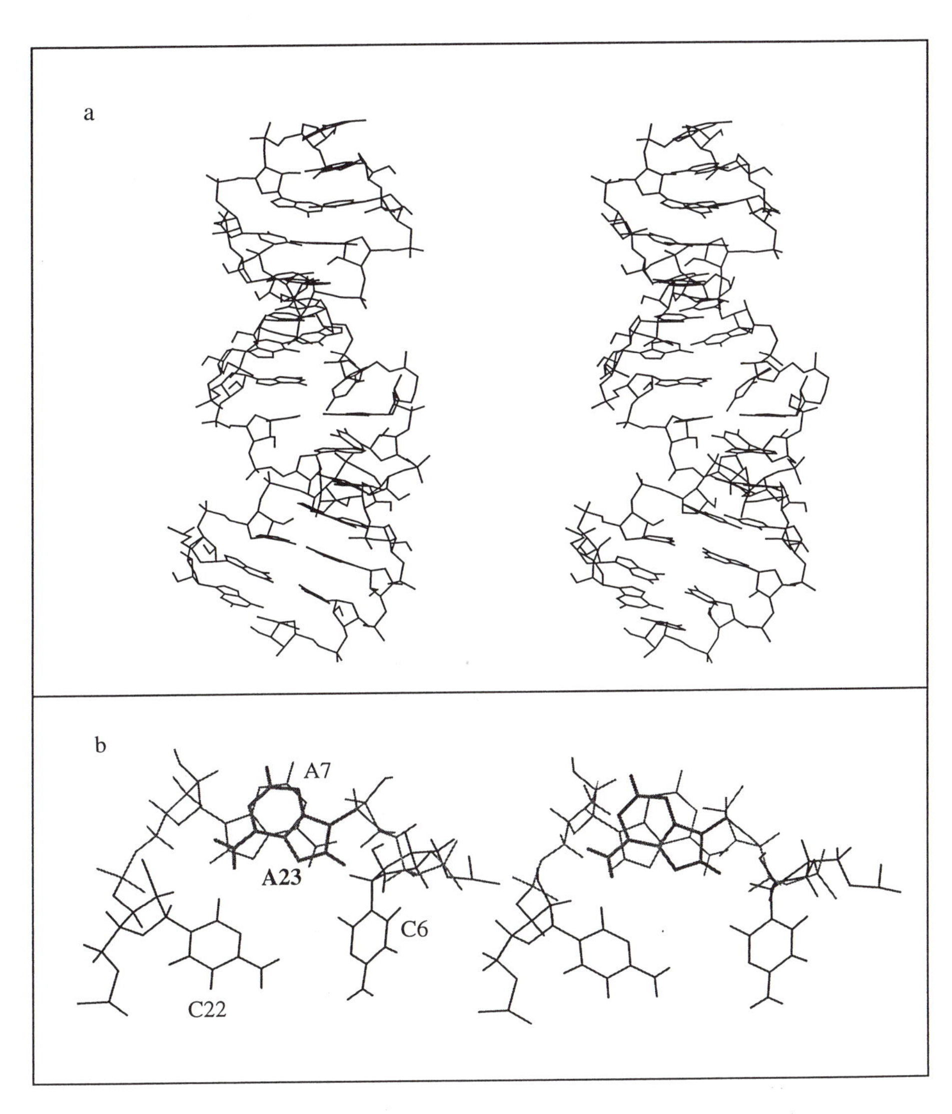

Figure 8. Stereoviews of final rMD structure of SRP 28mer: a) overview with 5'-G1 at left in back; b) part of the internal loop with A7-A23 cross-strand stacking (A23 is on top and in bold). Hydrogen atoms are not shown.

loop actually is. Flanked by a G:U wobble pair and a G:A pair (G-imino to A N1, A-amino to G-carbonyl), one of the most obvious features is the cross-strand stacking of the two adenines A23 and A7 (Fig.8b). The two guanines G8 and G21 are not involved in any known G:G basepairing, although a number of discrete hydrogen bonds could be found in the NMR structure for the central part of the internal bulge. Since the unusual stability and thus also the unique structure involve Mg^{2+} ions, we expect a Mg^{2+} binding site in the center of the internal loop which would explain the unique positioning of residues G8 and C6. Also, the global helix axes of bottom and top strand are not parallel, showing that a significant bend arises from the internal loop. However, a more accurate description of the bend requires helical analysis of a larger ensemble of final structures, since the internal loop bottom stem junction is less well defined than most other parts of the molecule.

Conclusions

We have demonstrated that even for a relatively large RNA fragment of 28 nucleotides assignments of all the non-exchangeable proton resonances except H5',H5" and a few H3' and H4' protons can be obtained with conventional homonuclear 2D NOE spectroscopy. Assignment of the 28mer resonances was greatly aided by the prior assignments of the closely related24mer. Roughly 250 peaks could be integrated on one or both sides of the spectral diagonal which were subsequently used in complete relaxation matrix calculations to yield distance restraints with higher precision since spin-diffusion has been accounted for. However, it was crucial to correct the NOE intensities for partial relaxation effects, due to repetition delays in our NOE experiments that are several times shorter than the longitudinal relaxation times of the protons. Since most 2D NOE data of larger RNAs most likely have not and will not be acquired allowing full relaxation, it is important to assess the effects of this on the structural results. Most importantly, our restraint set comprising only NOE-derived distance restraints and pucker restraints for the internal loop region of ten nucleotides was sufficient for the restrained MD refinement starting from largely randomized starting models to produce converged structures. Although the relative orientation of the two stems is not well-defined due to a hinge effect of the internal loop, a clear picture could be obtained for the internal loop including the unusual cross-strand stacking of the two adenine residues.

Materials and Methods

RNA synthesis and purification. The SRP 28mer was synthesized by *in vitro* transcription using T7 RNA polymerase on a synthetic DNA template according to the method of Milligan *et al.* (42, 43) The transcription reaction was optimized to yield 5 O.D./mL reaction mixture using final conditions of 0.4 M Tris-HCl (pH 8.1), 36 mM $MgCl_2$, 50 mM dithiothreitol, 10 mM spermidine, 4 mM each ATP, CTP, GTP, and UTP, 0.1 % Triton X-100, 800 nM DNA template, and 5 µg/ml T7 RNA polymerase. After 4 hours at 37°C the 25 ml reaction was precipitated with ethanol. The RNA dissolved in 4 ml 7 M urea and purified on two 20 % polyacrylamide/7 M urea gels. The bands of the full-length RNA and the "n+1"-product were excised, electroeluted, precipitated with ethanol, and finally dialyzed against a cascade of buffers [10 mM potassium phoshpate, 5 mM EDTA, 100 mM NaCl; 10 mM, 0.1 mM EDTA, 100 mM NaCl; 10 mM, 50 mM NaCl; water] for 24 hours each at 4 °C, pH 6.5, using a membrane with a 3500 molecular weight cut-off.

NMR sample preparation and NMR Experiments. The SRP 28mer was lyophilized several times from D_2O for analysis of nonexchangeable protons. The sample was annealed immediately prior to use by heating at 85°C for 10 min and snap-cooling on ice for 30 min. To minimize the formation of dimeric species, Mg^{2+}-salts were added after annealing, immediately prior to conducting NMR experiments. Note that the SRP 28mer was subjected to the complete procedure of dialysis, snap-cooling and addition of Mg^{2+} before each 2D NMR experiment. Final conditions for the 28mer were 0.25 mM RNA, 5 mM $MgCl_2$, 10 mM potassium phosphate at pH 6.5.(Note, the low RNA concentration helped to reduce line-broadening which is a common effect of even low Mg^{2+}-concentrations). For analysis of exchangeable protons, the sample was lyophilized and dissolved in 9:1 H_2O/D_2O for the same conditions as above. For acquisition of 2D NMR data we used a 8 mm heteronuclear probe (sample volume 870 µl) on a VARIAN UNITYPLUS 600MHz spectrometer.

All 2D NOE spectra (τ_m = 50, 150, 200 and 400 ms) in D_2O were acquired in the hypercomplex mode (44) at 25 °C, using a spectral width of 5999 Hz in both dimensions with the carrier frequency set to the HDO resonance frequency. A total of 400 t_1 values were recorded for each fid. 32 scans with a repetition time of 2.5 s including the t_2-acquisition time were recorded for each t_1 increment with 2K data points in f_2. DQF COSY spectra were acquired under the same conditions. All 2D NOE experiments in H_2O were collected at 10 °C or 15 °C using the SS-NOESY pulse sequence (45) with a spectral width of 12999 Hz using a symmetrically-shifted S-pulse with pulse width of 88.8 µs for water suppression. The excitation maximum occurs at ± 7576 Hz from the carrier frequency.

Mixing times were 120 and 200 ms. Longitudinal relaxation experiments were carried out using the inversion-recovery method.

2D datasets were transferred to a SUN SPARCstation 2 and were processed using Striker and Sparky (46).

Restraint generation and refinement procedures: NOE crosspeak intensities were corrected for partial relaxation effects with the program SYMM (35), which is available with the latest version of our program CORMA (47). Corrected intensities were used for all complete relaxation matrix calculations with the program MARDIGRAS (30). The randmardi procedure with 50 repetitions (38) was invoked for three NOESY datasets (150, 200, 400ms) and five correlation times (2.5, 3.0, 3.5, 4.0, 4.5 and 5.0 sec). Final upper and lower distance bounds were calculated by adding and subtracting the standard deviation of all runs (750 values for most distances) to the average upper and lower bounds from individual randmardi runs (15).

All models were generated with the TLEAP module of the AMBER 4.1 program suite (41) with the Cornell et al. force field (48), using "fat" sodium ions for the neutralization of the phosphate charges. The 5'-end was modeled without the triphosphate group which is present in the real sample. All structural analysis was carried out with the CARNAL module.

A typical refinement protocol consisted of (i) 3 ps free MD with heating from 0 to 1500K and tight Berendsen temperature coupling (49), (ii) 1000 steps of conjugate gradient energy-minimization, (iii) 3 ps restrained MD with temperature and restraint weight ramping to 500K and 50 kcal/mol·Å^2 for distance and 250 kcal/mol·rad^2 for torsion angle restraints, respectively, (iv) continuation of restrained MD for 4 ps under high temperature and restraint weight conditions, (v) rescaling temperature and force-constants to 200K and 20 kcal/mol·Å^2 for distance and 100 kcal/mol·rad^2 for torsion angle restraints within 0.5 ps, after which the simulation is run out to 10 ps. The time-step was 1 fs. For the first 5 ps of the restrained MD period weight of the van der Waals term was set to 0.1, ramped to 5 for 1 ps and rescaled to 1 for the remainder of the run. The weight of the electrostatics term was kept at 0 for the first 7 ps, ramped to 1 for the rest of the time. The first 5 ps of restrained MD did not contain the restraints for the exchangeable protons. For the second half of the run, all distance restraints had the same weight.

Acknowledgments. This research was supported by a UC Academic Senate Research Grant and a NSF grant (MCB-9513214) to US. We thank A. Pardi for making the coordinates of the GCAA tetraloop available at an early stage. We gratefully acknowledge the Frederick Biomedical Supercomputing Center and the UCSF Computer Graphics laboratory supported by NIH grant RR01081.

References

(1) Walter, P. & Johnson, A. E., *Ann. Rev. Cell Biol.* **1994**, 10, pp. 87-119.

(2) Lütcke, H., *Eur. J. Biochem.* **1995**, 228, pp. 531-550.

(3) Larsen, N. & Zwieb, C., *Nuc. Acids Res.* **1993**, 21, pp. 3019-3020.

(4) Althoff, S., Selinger, D. & Wise, J. A., *Nuc. Acids Res.* **1994**, 22, pp. 1933-1947.

(5) Bernstein, H. D., Poritz, M. A., Strub, K., Hoben, P. J., Brenner, S. & Walter, P., *Nature* **1989**, 340, pp. 482-486.

(6) Römisch, K., Webb, J., Herz, J., Prehn, S., Frank, R., Vingron, M. & Dobberstein, B., *Nature* **1989**, 340, pp. 478-482.

(7) Freymann, D. M., Keenan, R. J., Stroud, R. M. & Walter, P., *Nature* **1997**, 385, pp. 361-4.

(8) Römisch, K., Webb, J., Lingelbach, K., Gausepohl, H. & Dobberstein, B., *J. Cell Biol.* **1990**, 111, pp. 1793-1802.

(9) Zopf, D., Bernstein, H. D., Johnson, A. E. & Walter, P., *EMBO J.* **1990**, 9, pp. 4511-4517.

(10) Zopf, D., Bernstein, H. D. & Walter, P., *J. Cell Biol.* **1993**, 120, pp. 1113-21.

(11) High, S., Flint, N. & Dobberstein, B., *J. Cell Biol.* **1991**, 113, pp. 25-34.

(12) Zwieb, C., Müller, F. & Larsen, N., *Fold. Des.* **1996**, 1, pp. 315-324.

(13) Gowda, K., Chittenden, K. & Zwieb, C., *Nuc. Acids Res.* **1997**, 25, pp. 388-94.

(14) Nikonowicz, E. P. & Pardi, A., *J. Mol. Biol.* **1993**, 232, pp. 1141-1156.

(15) Pardi, A., Meth. Enzymol. **1995**; ed. James, T. L. (Academic Press, New York, NY), Vol. 261, pp. 350-382.

(16) Varani, G. & Tinoco, I., *Q. Rev. Biophys.* **1991**, 24, pp. 479-532.

(17) Borer, P. N., Lin, Y., Wang, S., Roggenbuck, M. W., Gott, J. M., Uhlenbeck, O. C. & Pelczer, I., *Biochemistry* **1995**, 34, pp. 6488-6503.

(18) Glemarec, C., Kufel, J., Foldesi, A., Maltseva, T., Sandstrom, A., Kirsebom, L. & Chattopadhyaya, J., *Nucl. Acids Res.* **1996**, 24, pp. 2022-35.

(19) Shen, L. X., Cai, Z. & Tinoco, I., *FASEB J.* **1995**, 9, pp. 1023-1033.

(20) Borgias, B. A. & James, T. L., Meth. Enzymol. **1989**; , Vol. 176, pp. 169-183.

(21) Keepers, J. W. & James, T. L., *J. Magn. Reson.* **1984**, 57, pp. 404-426.

(22) Schmitz, U., Blocker, F. J. B. & James, T. L., *Standard DNA duplexes and RNA•DNA hybrids in solution* **1997,** ed.^eds. Neidle, S. (Oxford University Press, Oxford).

(23) Borgias, B. A. & James, T. L., *J. Magn. Reson.* **1990**, 87, pp. 475-487.

(24) Post, C. B., Meadows, R. P. & Gorenstein, D. G., *J. Am. Chem. Soc* **1990**, 112, pp. 6796-6803.

(25) Boelens, R., Koning, T. M. G., van der Marel, G. A., van Boom, J. H. & Kaptein, R., *J. Magn. Reson.* **1989**, 82, pp. 290-308.

(26) Yao, L., James, T. L., Kealey, J., Santi, D. V. & Schmitz, U., *J. Biomol. NMR* **1997**, 9, pp. 229-44.

(27) Schmitz, U., Freymann, D. M., James, T. L., Keenan, R. J., Vinayak, R. & Walter, P., *RNA* **1996**, 2, pp. 1213-1227.

(28) Wijmenga, S. S., Mooren, M. M. W. & Hilbers, C. W., NMR of Macromolecules **1994**; ed. Roberts, G. C. K. (Oxford University Press, Oxford), Vol. 134, pp. 217-288.

(29) Heus, H. A. & Pardi, A., *Science* **1991**, 253, pp. 191-4.

(30) Liu, H., Borgias, B., Kumar, A. & James, T. L., *MARDIGRAS* **1990, 1994,** (University of California, San Francisco).

(31) Schmitz, U. & James, T. L., Meth. Enzymol. **1995**; ed. James, T. L. (Academic Press, New York), Vol. 261, pp. 1-43.

(32) Geppert, T., Kock, M., Reggelin, M. & Griesinger, C., *J. Magn. Reson. B* **1995**, 107, pp. 91-93.

(33) Kock, M. & Griesinger, C., *Angew. Chem. -Int. Edt.* **1994**, 33, pp. 332-334.

(34) Zhu, L. & Reid, B. R., *J. Magn. Reson. B* **1995**, 106, pp. 227-35.

(35) Liu, H., Tonelli, M. & James, T., *J. Magn. Reson. Ser. B* **1996**, 111, pp. 85-9.

(36) Wüthrich, K., *NMR of Proteins and Nucleic Acids* **1986** (Wiley, New York).

(37) Jaeger, J. A. & Tinoco, I., Jr., *Biochemistry* **1993**, 32, pp. 12522-30.

(38) Liu, H., Spielmann, H. P., Ulyanov, N. B., Wemmer, D. E. & James, T. L., *J. Biomol. NMR* **1995**, 6, pp. 390-402.

(39) Liu, H., Kumar, A., Weisz, K., Schmitz, U., Bishop, K. D. & James, T. L., *J. Am. Chem. Soc.* **1993**, 115, pp. 1590-91.

(40) Pley, H. W., Flaherty, K. M. & McKay, D. B., *Nature* **1994**, 372, pp. 68-74.

(41) Pearlman, D. A., Case, D. A., Caldwell, J. C., Ross, W. S., Cheatham III, T. E., Ferguson, D. N., Seibel, G. L., Singh, U. C., Weiner, P. K. & Kollman, P. A., *AMBER version 4.1* **1995,** (University of San Francisco, San Francisco).

(42) Milligan, J. F. & Uhlenbeck, O. C., Meth. Enzym. **1989**; ed. Dahlberg, J. E. & Abelson, J. N. (Academic Press, New York), Vol. 180, pp. 51-62.

(43) Torda, A. E., Scheek, R. M. & van Gunsteren, W. F., *J. Mol. Biol.* **1990**, 214, pp. 223-235.

(44) States, D. J., Haberkorn, R. A. & Ruben, D. J., *J. Magn. Reson.* **1982**, 48, pp. 286-292.

(45) Scalfi-Happ, C., Happ, E., Nilges, M., Gronenborn, A. M. & Clore, G. M., *Biochemistry* **1988**, 27, pp. 1735-1743.

(46) Kneller, D. G., *Sparky, NMR display and processing program* **1992,** (copyright 1992 University of California, San Francisco.

(47) Liu, H., Borgias, B., Kumar, A. & James, T. L., *CORMA-Complete Relaxation Matrix Analysis Version 31* **1992,** (University of California, San Francisco).

(48) Cornell, W., Cieplak, P., Bayly, C. I., Gould, I. R. & Kollman, P. A., *J. Am. Chem. Soc.* **1996**, 118, pp. 2309-2309.

(49) Berendsen, H. J. C., Postma, J. P. M., van Gunsteren, W. F., Di Nola, A. & Haak, J. R., *J. Chem. Phys.* **1984**, 81, pp. 3684-3690.

Chapter 9

Molecular Modeling of DNA Using Raman and NMR Data, and the Nuclease Activity of 1,10-Phenanthroline–Copper Ion

W. L. Peticolas[1], M. Ghomi[2], A. Spassky[3], E. M. Evertsz[1], and T. S. Rush III[1]

[1]Department of Chemistry and the Institute of Chemical Physics, University of Oregon, Eugene, OR 97403
[2]Laboratoire de Physicochimie Biomoléculaire et Cellulaire, Unité de Recherche Associée, Centre National de la Recherche Scientifique 2056, Université Pierre et Marie Curie Case 138, 4 Place Jussieu, 75252 Paris Cedex 05, France
[3]Laboratoire de Chimie et Biochimie Pharmacologiques et Toxicologiques, 45 rue des Saints-Pères, 75270 Paris Cedex 06, France

Molecular modeling studies have been carried out on two completely different types of bent or kinked DNA. The first of these involves a DNA that contains alternating tracts of A and T bases that cause a curvature [1-4]. The second study characterizes a kinked DNA that contains a T tract--CG tract--A tract motif that creates a pair of B/Z junctions. Raman and/or NMR data were used to constrain these models. Duplexes of the sequence, 5'-d$(A5T5)_n$ and poly(dA)•poly(dT) are considered to be in the B' conformation. Both of these DNA show a pair of Raman bands at 817±3 cm^{-1} and 840±5 cm^{-1}. The latter band is well known to be a B form DNA marker band. The 817±3 cm^{-1} band is weaker or non-existent in non-A tract DNA. It is now interpreted as being due to furanose rings with a value of the pseudo-rotational angle, P = 115°±15°. This assignment is necessary to make the Raman data consistent with published NMR results and normal mode calculations. It is shown how the introduction of this ring pucker can lead to curvature of the DNA.

Raman spectra were taken of duplex oligomers with the sequences, [5'-d$(T_N(CG)_8A_N)]_2$ in aqueous 6M NaCl solutions. These oligomers showed the presence of two B/Z junctions. Molecular modeling was used to show that the oligomer conformation went abruptly from the B to the Z form and from the Z back to the B form inside the $(CG)_8$ tract. It was found that there were no base pairs in the junctions. This model was strikingly verified when it was found that the nuclease activity of 1,10-phenanthroline-Cu^+ ion occurred only at the sugar phosphate link between the last G in the B form and the first C in the Z form. In every case the cuts were made at exactly the sites predicted by the molecular model that had been published previously.

In our molecular modeling of DNA we use Raman spectroscopy and NMR to build the DNA models using the Biograf modeling program. We started with A tract DNA , i.e., DNA that has repeating sequences of two or more adenines which is a well studied sequence motif [1-4]. Crystal structures of different duplexes containing these A tracts indicate that they adopt a non-canonical B form that is characterized by a high propeller twist, possible bifurcated hydrogen bonds, absence of significant tilt, and a narrow minor groove [5-7]. Several models for sequence directed curved DNA have been suggested [1-18]. One of these models is the A form-B form hybrid model that was first suggested by Peticolas and Thomas [16] and Arnott *et al.*[17-18]. It was proposed by Peticolas and Thomas because two different vibrational frequencies were observed for a phosphate-deoxyribose ring vibration in poly(dA)• poly(dT). One occurs at 817±3 cm^{-1} and one at 841±3 cm^{-1}.The existence of two Raman bands in this region instead of one was a very original discovery and definitely indicated the existence of two different furanose conformations. A tract DNA shows these two bands much more strongly than native DNA or synthetic DNA containing both GC and AT base pairs. As will be shown below, these bands are the most intense in poly(dA)•poly(dT), that has an infinite A tract. It was suggested [16] (correctly) that the 840±5 cm^{-1} band was due to C2'-endo sugar pucker where the pseudo-rotational angle, P = 165°±15°. [For a review of the assignment of the Raman bands of DNA see references 19 and 20. For a definition and discussion of P, the pseudo-rotational angle see references 21 and 22]. It was suggested[16] (erroneously) that the 817±3 cm^{-1} band in A tract DNA was due to C3'-endo sugar puckers [where P = 20°±10°]. These assignments were interpreted to mean that one of the strands in poly(dA)• poly(dT) was more A-like in character than the other strand. Since the number of base pairs per turn are different for these two types of DNA, it was suggested that matching the A strand with the different length B strand would cause a curvature in the DNA. The diffraction data of Arnott *et al.* [17], seemed to support this conformation of polydA•polydT. Recent X-ray diffraction studies on crystals of A tract containing oligomers [5-7] and fibers of poly(dA)• poly(dT) [23] show no evidence of any C3'-endo sugar puckering. Alexeev *et al.* [23] reinterpreted the diffraction data of Arnott *et al.* [17]. They claim that Arnott *et al.*'s data is consistent with a modified B-type geometry with an unusually high propeller twist. No evidence for A-type C3'-endo ring pucker was found from the assignments of NMR signals from short A tract sequences in solution [24-26]. It would seem that the early 817±3 cm^{-1} band assignment was made in error [16, 27]. As we will show, the band at 817±3 cm^{-1} is now assigned to a furanose ring vibration when the furanose ring as has P value of 115°±15. We have found that models of alternating A tract--T tracts motif with some of the furanose rings in the P = 115°±15° conformation leads to a bent DNA with a large propeller twist and a narrow minor groove.

Molecular models have been made of the sequences $dT_N(CG)_M$ and $dT_N(CG)_MA_N$.that predict that the B/Z junctions always occur in the CG tracts and never in the A or T tracts. In these models there are no base pairs in the junction. We have used these models to predict the unique sites for the for nuclease attack by 1,10-phenathroline-Cu^+. In every such oligonucleotide the B form is continued from the adenine tracts two to four bases into the CG tract where the B/Z junctions and bending occur. Since no bases are in the junction which is formed by an abrupt change from the B to Z helix there is only one site for nuclease attack --the junction between the last G in the B form and the first C in the Z form. Gel studies show that predicted sites for nuclease attack are the only sites observed to be cut.

Part I: The Modeling of Curved A-Track DNA and B' DNA

Measurement and Assignments of the Raman Bands that Aid the Modeling of A-Track DNA Deoxyoligonucleotides and polymers containing only adenine (A) and thymine (T) were purchased from Pharmacia and were desalted using Centricon-3 filters (Amicon) and then concentrated to dryness by rotary evaporation. All spectra were taken in 0.5 M NaCl, pH 6.8 at 5° C unless otherwise noted. Some of the oligomers have a base sequence that conceivably could lead to hairpin or triplex or quadriplex formation. To show that under our conditions oligomer only formed a simple duplex, the gel was analyzed at 0.1 M salt on a non-denaturing acrylamide gel. It was found to run as a single band in the correct position between duplexes of 15 and 30 base pairs. No evidence was found for either hairpins or higher aggregates. Oligomers prepared in this manner give excellent Raman spectra.

In order to compare the behavior of the oligomers with that of the polymers, we used a salt concentration of 0.5 M NaCl that significantly raises the melting point, T_m, of the oligomers and tends to prevent fraying of the ends. Salt concentration does not seem to have much effect on the bending behavior. Diekmann[28] reports that measurements of the bending *versus* NaCl concentration showed no change in the polyacrylamide gel pattern for salt concentrations varying between 20 mM and 100 mM. To see the effect of raising the salt concentration from 0.1 to 0.5 M on a fully formed duplex structure, the Raman spectrum of the $[d(A_5T_5)_2]_2$ duplex was run at 5° C at both salt concentrations and the spectra were found to be identical. From this we concluded that there was no significant change in conformation between 0.1 and 0.5 M salt for the fully formed duplex. In order to obtain the Raman bands of the thymine carbonyl groups, free of the strong, broad water bending mode at 1640 cm^{-1}, some Raman spectra were taken in D_2O. These samples were prepared by repeated solution washes in D_2O using the centricon-3 apparatus, before drying them and redissolving them in D_2O, 0.5 M NaCl, pD 6.8. The samples were then heated to 80° C and slowly cooled to insure complete exchange of C8 protons.

The Raman equipment has been described previously [16]. Spectra were recorded from 500-1800 cm^{-1} and 2500 to 3000 cm^{-1} using the 514.5 nm argon excitation line with slit setting corresponding to a band pass of 2 cm^{-1}. Analysis of the spectra consisted of first subtracting the water or D_2O spectra by using the high wave number region of water or D_2O to monitor the subtraction. The data are fit to a series of Lorenztian curves using a least square procedure [29] Spectra are compared with others by subtraction, using the 1016 cm^{-1} band for normalization. This band is conformationaly insensitive to both melting and the B to A transitions [30,31]. It was chosen as the normalization band because the more commonly used 1095 cm^{-1} band appears to change with temperature in the premelting studies. It is important to note that there is no change in the low temperature difference spectra whether the band at 1016 cm^{-1} or 1095 cm^{-1} is used for normalization.

Interpretation of the Raman bands at 817±3 cm^{-1} and 840±5 cm^{-1}. Figure 1a shows the Raman spectra of three DNA sequences in the 600 to 1160 cm^{-1} frequency region. The top spectrum is that of $[d(A_5T_5)_2]_2$, the middle spectrum is that of $[d(TTATTATAATATTATAATAA)]_2$, and the bottom spectrum is that of the duplex polydA•polydT. All of these three samples contain some A tract DNA. The least amount of A tract DNA is in the middle sample that only has runs of two A's or two T's but it does have 6 ApT junctions in each chain. These may be explicitly shown as TTApTTApTAApTApTTApTAApTAA where only the phophates in the ApT junctions are shown. As we will discuss below, there is a tendency for ApT junctions to have an unusual ring pucker. All of these samples show a strong Raman band at 842 cm^{-1} which is well known to be a B form marker band [30,31]. In order to determine the structure of A tract DNA, we need to be able to assign the 817±3 cm^{-1} Raman band that is so conspicuous in the Raman spectrum of this DNA and absent in native DNA and synthetic DNA that does not contain A tracts. A clue to the correct assignment of the 817±3 cm^{-1} band is found in the NMR data of Wüthrich et al. [26] on the oligomer $[d(A_5T_5)]_2$. It was found that the deoxyribose rings attached to the

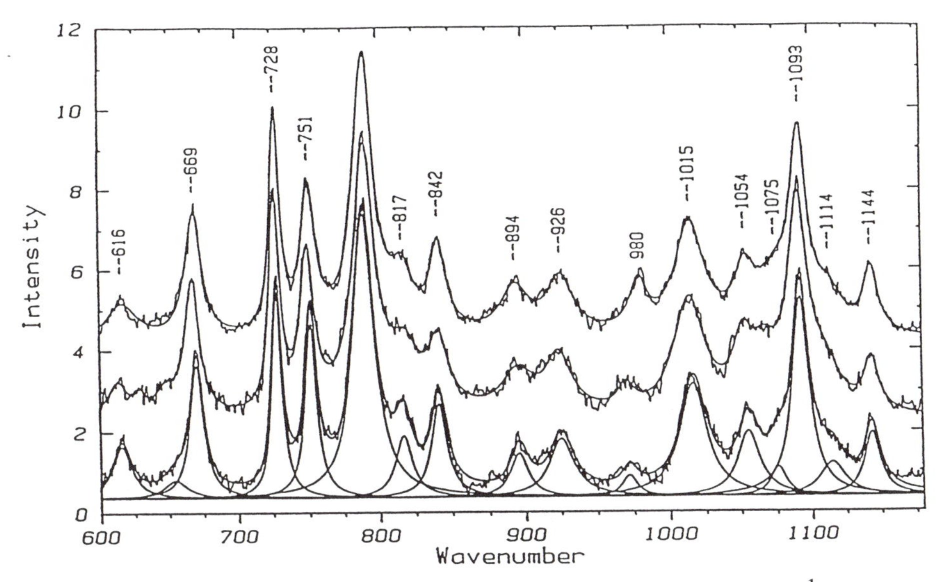

Figure 1a. Raman spectra of three DNA sequences in the 600-1160 cm^{-1} frequency region. .Middle: $[d(TTATTATAATATTATAATAA)]_2$, Bottom: poly(dA)poly(dT). Top: $[d(A_5T_5)_2]_2$,. The overlapping smooth curves are sums of Lorenzians that were fit to the spectra using a least squares techniques.Spectra were taken at 5° C in 0.5 M NaCl, (pH 6.8).

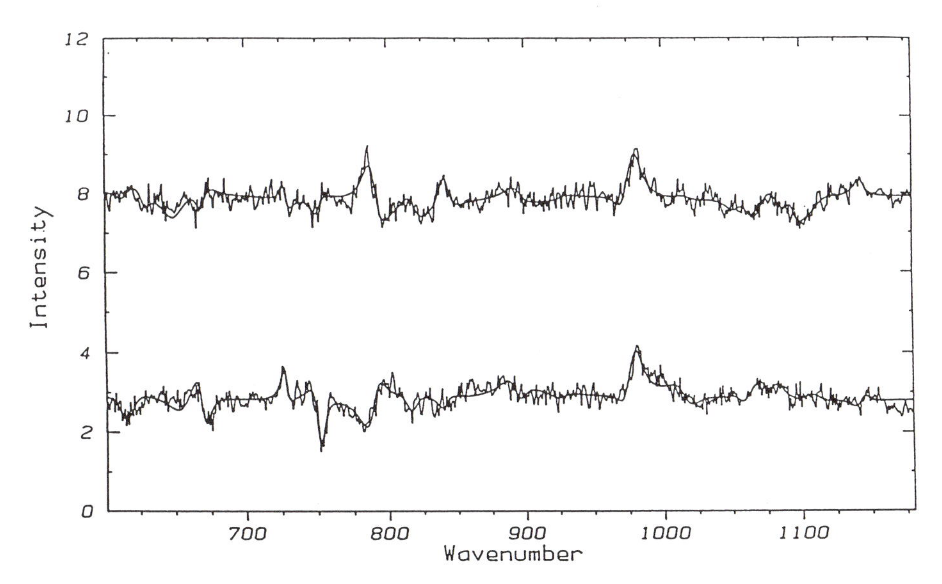

Figure 1b Raman difference spectra that were obtained from the Raman spectral data from Figure 1a. Data were subtracted using the 1016 cm^{-1} band for normalization. Top: $[d(A_5T_5)_2]_2$ minus $[d(TTATTATAATATTATAATAA)]_2$ Bottom: $[d(A_5T_5)_2]_2$ minus poly(dA)•poly(dT)].

two thymines on each side of the ApT junction have unusual P values of 115°±15°. This corresponds to the East (strictly the East-Southeast) direction of the pseudo-rotational angle conformational wheel[21,22] and is rarely observed in solutions or in crystals because the potential energy of this conformation is higher than in the regions corresponding to the C2'-endo (South) and C3'-endo (North) regions [21,22]. It seems that A tract DNA tends to have an unusually high fraction of the furnanose rings in the E form and this can account for the presence of two Raman bands--the usual Raman band at 840±5 cm^{-1} that corresponds to the usual Southern values of P and the Raman band at 817±3 cm^{-1} corresponding to the unusual Eastern values found by the NMR work [26]. The band at 807°±3° found in A-form DNA at low water activity is due to the furanose ring in the Northern part of the pseudo-rotational conformational wheel [30]. There appears to be a correlation between the frequency of this Raman band which goes from 807±3 cm^{-1} to 840±5 cm^{-1} and the pseudo-rotational angle P that goes from the North at P = 20±10° to the South where P = 165°±15°.

Normal mode calculations of nucleosides have been performed by Taillandier, Ghomi *et al.* [32] as a function of P, the pseudo-rotational angle. These calculations indicate that there is a conformationaly dependent furanose-phosphate vibrational mode that ranges in frequency from 818 cm^{-1} to 843 cm^{-1} as P goes from P=0° to P=180°. However, their calculated value of 818 cm^{-1} cannot be correct for the P = 0° conformation. For example, the polymer, polydG•polydC, and oligomers of the sequence, dG_N•dC_N have a strong tendency to go into the A form. These DNA exhibit this band at the frequency of 804 cm^{-1} in the A form [33] that is well below the calculated frequency [32] of 818 cm^{-1}. This computational error can be corrected by scaling the calculated results, based on an empirical force field to fit the experimental values of 807±3 cm^{-1} for P=20±10° and 840±5 cm^{-1} for P=165°±15°. When this is done, the 817±3 cm^{-1} band of A tract DNA now falls on the curve corresponding to a P value of about 115°±15°. Since A tract DNA is the only DNA to show both the East side furanose ring pucker in NMR studies [26] and the presence of the Raman band at 817±3 cm^{-1} it is reasonable to assume that the Raman band at 817±3 cm^{-1} is due to the presence of this Eastern conformation where P = 115°±15°.

These new spectra/structure correlations explain very nicely the following observations: (1) the band at 840±5 cm^{-1} is always present in native DNA because the vast majority of the furanose ring puckers have values of P = 165°±15° (ordinary B form or C2'-endo DNA). (2) Fibers of native DNA, oligomers of $d(C)_N$•$d(G)_N$, and poly(dG)• poly(dC) readily go into the A form at low (thermodynamic) water activity. The low activity of water can be induced in fibers at low humidities, in crystals by the organic cosolvents, and in concentrated salt solutions [34]. These DNA samples have P = 20°±10° and the band at 807±3 cm^{-1} is very strong while the band at 840±5 cm^{-1} is absent [19,30,31]. The 817±3 cm^{-1} band never occurs alone in any DNA but always in addition to the B form marker band at 840±5 cm^{-1} and only occurs in DNA containing a preponderance of A-T base pairs. Thus the 817±3 cm^{-1} is always found in a certain kind of B form DNA since it always accompanies the B form marker band at 840±5 cm^{-1}. We now assign this Raman band at 817±3 cm^{-1} as a marker band for B' DNA that is strongest in A tract DNA, and is due to furanose rings in the conformation corresponding to P = 115°±15°. This last observation is also supported by the work of Benevides *et al.* [35]. These authors have found bands at 820 cm^{-1} and 842 cm^{-1} in crystals of d(CGAAATTTCG). The 842 cm^{-1} band would predict the presence of high furanose P values at 165°±15° typical of B DNA. Although these authors do not attempt to assign the Raman band at 820 cm^{-1}, we may note that it falls in the 817±3 cm^{-1} region and predicts that this crystal contains furanose rings with P values in the 115°±15° region. This prediction can be verified since the X-ray determined coordinates [5] now deposited in the Brookhaven data bank are known. They show that P values for the thymidines of this sequence fall in the 90-130° range, while the P values for the adenines fall in the 165°±15° range just as we would predict from the published Raman spectrum of this crystal! As noted above, DNA containing tracts of guanine goes easily into the A or Northern form[31] where P = 20°±10° . On the other hand, A tract DNA cannot be induced into the Northern part of the pseudo-rotational conformational wheel under any conditions [34, 36]. Apparently, the P values for G tracts go easily from P = 165°±15° at high water activity to P = 20°±10° at lower water activity such as in concentrated salt solutions, crystals or fibers under low (75%) relative humidity [34], but the P value for A tracts stops at the East side around P= 150°± 15° as the humidity is lowered and will decrease no further [36]. As the activity of the water is reduced, a fraction but not all of the furanose rings in A tract DNA goes from South to East. On the other hand DNA containing G tracts such as $d(C)_n \bullet d(G)_n$ or alternating $d(AT)_n$ goes from South to North as the water activity is reduced. Alternating CG tracts go into the Z form with reduced water activity but that transition is complicated by the fact that the position of the base relative to the furanose ring changes.

Molecular Modeling of Alternating A Tract--T Tract (Bent) DNA

Energy minimizations were performed using a modified Dreiding II force field [37] in the Biograf molecular mechanics program, on a Silicon Graphics Unix Indigo workstation. The effect of counterions was taken into account by reducing the charge on the phosphate group to -0.5 e, and the effect of solvent was taken into account by scaling the dielectric "constant" linearly with distance from 0-10 Å and then exponentially down to 0 within 1 Å. The minimization procedure consists of the following steps: (1) Build an initial geometry of the sequence in a purely B form using the Build routine of the Biograf program. This program automatically puts every furanose ring in the P = 151° conformation. Although this value falls in the P = 165°±15° range typical of B form DNA it is rather small. (2) The local geometrical information obtained from the Raman spectra or NMR data is then used to change the P values. For the duplex, $[d(A_5T_5)]_2$ we changed the values of P of four furanose rings (the two that are attached to the thymines on each side of the ApT junction) to 100°. This falls in the range the value that is found in the NMR data [26], P = 115°±15°. This is done using the formula, $\tau_i = \tau_m \mathrm{Cos}[P + (i - 2)4\pi/5]$ where $\tau_m = 40°$ and i = 0, 1, 2 3, 4. The angle τ_o is the torsion about the O1'-C1' bond, τ_1 is the torsion about the C1'-C2' bond, τ_2 is the torsion about the C2'-C3' bond and has the simple formula, $\tau_2 = 40°\mathrm{Cos}[P]$. The angles τ_3 and τ_4 are the torsions about the C3'-C4' and C4'-O1' bonds respectively. The latter bond closes the furanose ring [21, 22]. (2) The energy minimization routine in the Biograf program is then carried out while constraining these P conformations and allowing all other internal coordinates the freedom to move during the minimization, (3) Running the minimization routine again to relieve any stress caused by the rough placement of atoms during the building procedure, (4) Removing the constraints, (5) Obtaining the structure of minimum energy without any restraints, and (6) Performing quenched dynamical runs to sample the conformational space immediately surrounding this minimized structure.

The simple oligonucleotide, $[d(A_5T_5)]_2$ was modeled using the procedure given above. The final energy minimized distribution of the P values are given in the bottom circular graph of Figure 2. In these graphs, P = 0° is at the top of the circle (North) while P = 180° is at the bottom of the circle (South). The dashed lines are the P values for the deoxyribose rings attached to the thymine bases and the solid lines correspond to the deoxyribose rings attached to the adenine bases. For comparison, the P values for two crystal structures containing short A tracts are also given. In the top graph is shown the X-ray diffraction determined values of P for the deoxyribose rings attached to the A's and T's in $[d(CGCGAATTCGCG)]_2$ as reported by Holbrook *et al.* [38]. Note that even in this short tract of two A's followed by two T's, the P values for the furanose rings attached to the T's are consistently smaller than the values for those attached to the A's. Even so this sequence must be regarded as a B form DNA since there are no T's in the Eastern part of the conformational wheel and the sequence does not show a strong Raman band at 817±3 cm^{-1} [xxx]. The middle graph shows the P values for the furanose rings attached to the A's and the T's for the oligomer, $[d(CGCAAATTTGCG)]_2$ as reported by Coll *et al.* [5]. Here we see that some of the T's have P values lying on the East side of the conformational wheel representation of the furanose ring pucker. These East side T values give rise to the 817±3 cm^{-1} Raman band observed from this crystal as discussed above. We consider this oligomer to be a B' form DNA because of the presence of the Raman band at 817±3 cm^{-1} denoting the presence of the Eastern P values that are observed in the X-ray determined structure. It is clear that the P values for the thymidines in $[d(A_5T_5)]_2$ tend to be lower than those for the adenosines, on the average. This is consistent with the NMR results of Wuthrich et al. [26]. This seems to be a structural characteristic of $[d(..AAAApTTTT..)]_2$ tracts which may partially account for the tendency of these sequences to bend. The top two dashed lines correspond to P values of approximately 100° and are due to the furanose rings attached to each side of the thymines on each side of the in agreement with the NMR results of Wuthrich *et al.*

In order to model a longer A tract--T tract chain, we started with $[d(A_5T_5)_2]_2$ using the Biograf program. First, the sequence was input using standard geometry of the B DNA as given in the build subroutine of the Biograf program. Then the sugar rings of the two thymines on each side of the ApT steps in each of the two chains were changed to have a P value of 100°. These sugar puckers were then constrained to the 100° value and the initial minimization was performed followed by a minimization without these constraints but with the quenched dynamics runs. The P values for the furanose rings attached to the T's in the ApT junctions in the energy minimized structure fall between 100° and 135°; the T sugar puckers are significantly higher than the A sugar pucker values. Figure 3 shows the model of the 40mer, $[d(A_5T_5)_4]_2$ which is made by connecting two of the 20mers, $[d(A_5T_5)_2]_2$. The prononced curvature in this 40mer is evident.

There are several geometrical features in our model of $[d(A_5T_5)_2]_2$ that are consistent with other experimental evidence besides the sugar pucker conformations and the fact that this new minimized structure is bent. Most notably our model shows a narrow minor groove characterized by a short distance between phosphates and a large propeller twist of the base pairs. Table I gives a listing of the propeller twists for this oligomer. The published experimental evidence that the A tract sequences contain large propeller twists includes three X-ray diffraction studies. These are the investigations of Alexeev *et al.* [23] on fibers of poly(dA)• poly(dT), Nelson *et al.*[6] on single crystals of d(CGCAAAAAAGCG), and Coll *et al.* [5] on single crystals of the oligomer, d(CGCAAATTTGCG). Table I gives a listing of the calculated propeller twists of the 20 base pairs in the duplex $[d(A_5T_5)_2]_2$. As seen in Table I, which was compiled with the aid of R.E. Dickerson's New Helix '93 program, the propeller twisting values for the $[d(A_5T_5)_2]_2$ model are found to be above the average B-DNA values of 11.7±4.8°. Also easily observed from the table is the fact that the bases are especially propeller twisted in the region of the ApT steps but have much smaller

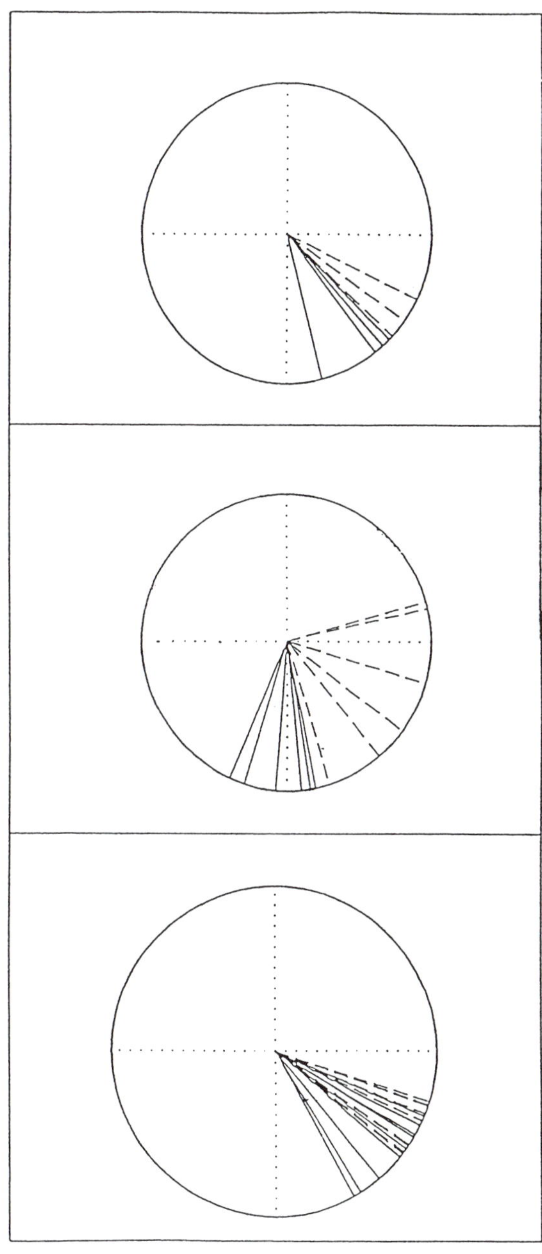

Figure 2. Pseudo-rotational angle plots for the crystal conformation of $[d(CGCGAATTCGCG)]_2$ (top), and $[d(CGCAAATTTGCG)]_2$ (middle), and the energy minimized "solution" conformation of $[d(A_5T_5)]_2$ [bottom]. The P values go from P=0° at the top of the wheel,[North] and increases clockwise to the right (East side) and then to the bottom of the wheel [South] where P = 180°.

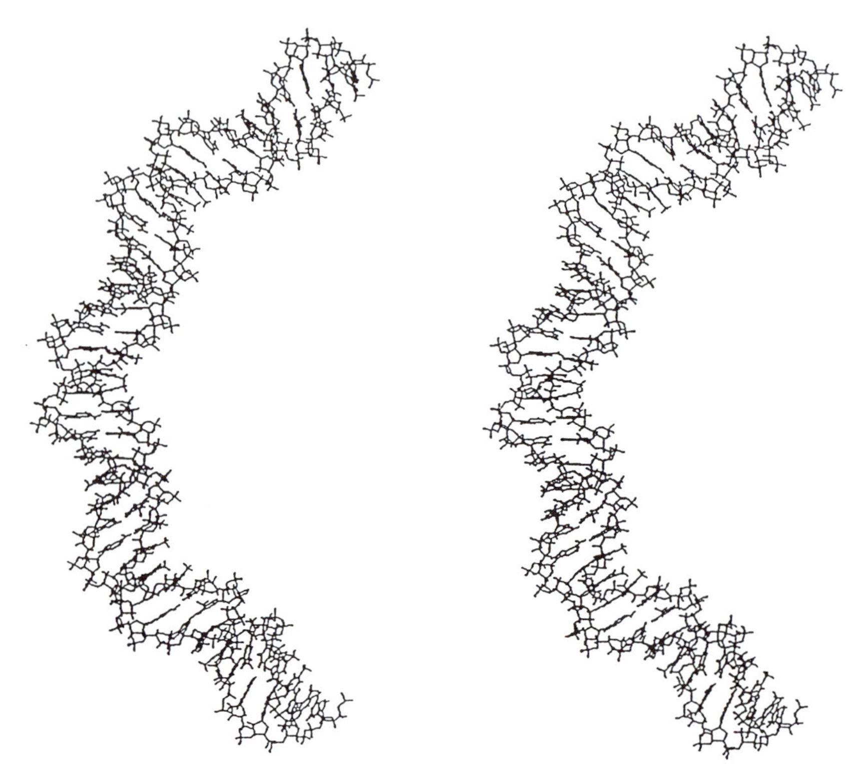

Figure 3. Stereo view of the energy minimized structure of $[d(A_5T_5)_4]_2$.

Table I. Propeller Twist of the Base Pairs in $[d(A_5T_5A_5T_5)]_2$

Base Pair	Propeller Twist	Base Pair	Propeller Twist
A1-T40	11.53	A11-T30	13.64
A2-T39	21.86	A12-T29	16.76
A3-T38	21.89	A13-T28	23.12
A4-T37	25.54	A14-T27	29.23
A5-T36	21.94	A15-T26	23.23
T6-A35	16.39	T16-A25	18.55
T7-A34	27.40	T17-A24	27.99
T8-A33	23.18	T18-A23	19.07
T9-A32	14.80	T19-A22	14.49
T10-A31	12.32	T20-A21	14.08

propeller twists in the TpA steps. It should be noted that these large propeller twists were not put into our model explicitly but came out of the energy minimized structure after constraining the P values of the furanose rings attached to the ApT junctions as described above.

There have been several experimental investigations that indicate that A tract DNA has a narrow minor groove. These include X-ray analyses [6,23], and acrylamide gel studies employing base analogs [24,41]. From the X-ray diffraction analyses of fibers of poly(dA)•poly(dT) by Alexeeve et al. [23] it was found that the polymer had a shortest interstrand separation between phosphates across the minor groove of 9.2 Å. Single crystals studies have yielded similar results [6]. NMR studies of A tract DNA in solution also provide evidence of a narrow minor groove even in solution [39,40]. As noted above, the energy minimized model we have made also shows a narrow minor groove even though, again, this feature was not explicitly put into the model. It arises as a result of the P values chosen to agree with the NMR and Raman measurements.

Raman Measurements of the Carbonyl Vibrations Show No Evidence for Bifurcated Hydrogen Bonds.

In order to observe possible differences in the carbonyl environment of bent and straight DNA, Raman spectra were taken in the 1550-1750 cm-1 region in D_2O. Figure 4 shows the Raman spectrum of the three DNA fully deuterium exchanged in D_2O. The top spectrum (A) is from the bent $[d(A_5T_5)_2]_2$; spectrum (B) is from the straight $[d(TTATTATAATATTATAATAA)]_2$ and the spectrum (C) is from poly(dA)•poly(dT). There are two difference spectra: (D) is spectrum (A) minus spectrum (B); spectrum (E) is spectrum (A) minus spectrum (C). The large band at 1660±5 cm-1 is assigned to strongly coupled vibrations of C2=O and C4=O in the thymines. This band has been shown to be sensitive to the environment [19,20]. Small changes in the environment of the carbonyl group can cause large changes in position, width and intensity of this band. As can be seen in Figure 4 there is almost no difference between the spectra of the curved and straight oligomers. From this we conclude that there is no evidence for bifurcated hydrogen bond to these carbonyls in aqueous solution since bifurcation should lead to a shift or splitting of the carbonyl frequency. The bifurcated hydrogen bonds seen in crystals of similar A tract oligonucleotides may be due to the lower water activity in the crystals.

Part II. Location of the B-Z Junctions in T tract-CG tract-A tract Deoxyoligonucleotides of the Form d-5'-$T_N(CG)_8A_N$-3' Using Raman Spectroscopy and the Nuclease Activity of 1,10-Phenanthroline-Cu+ Ion.

Just as the alternating tracts of five A's and five T's lead to a non-straight DNA, so do alternating tracts of A (or T) and $(CG)_n$ at least in concentrated salt solutions where a B/Z junction may occur. It now seems apparent that Z DNA requires an alternating pyrimidine-purine sequence to form easily, and that tracts of alternating cytosine and guanine (CG tracts) are the tracts that most readily support the Z form. (Since Z DNA must be a minor part of the DNA in the cell, each side of any Z form sequence must contain a B/Z junction leading to a B/Z/B conformational sequence.) The location of these junctions may be of importance in understanding the biological activity of these structures. A recent model for B/Z/B junctions has been presented for duplexes of deoxyoligonucleotides that are of the form 5'-d(T tract)--(CG tract)--(A tract)-3' or more precisely, 5'-d($T_N(CG)_8A_N$)-3'. Note that sequences of this type contain two types of junctions: the sequence junctions and the conformational junctions. There are two sequence junctions in these oligomers. These occur between the T tract on the left and the CG tract in the middle (starting at the 5' end and going to the 3' end), and between the CG and A tracts, i.e. at the positions: 5' d...TTTCGCG... and

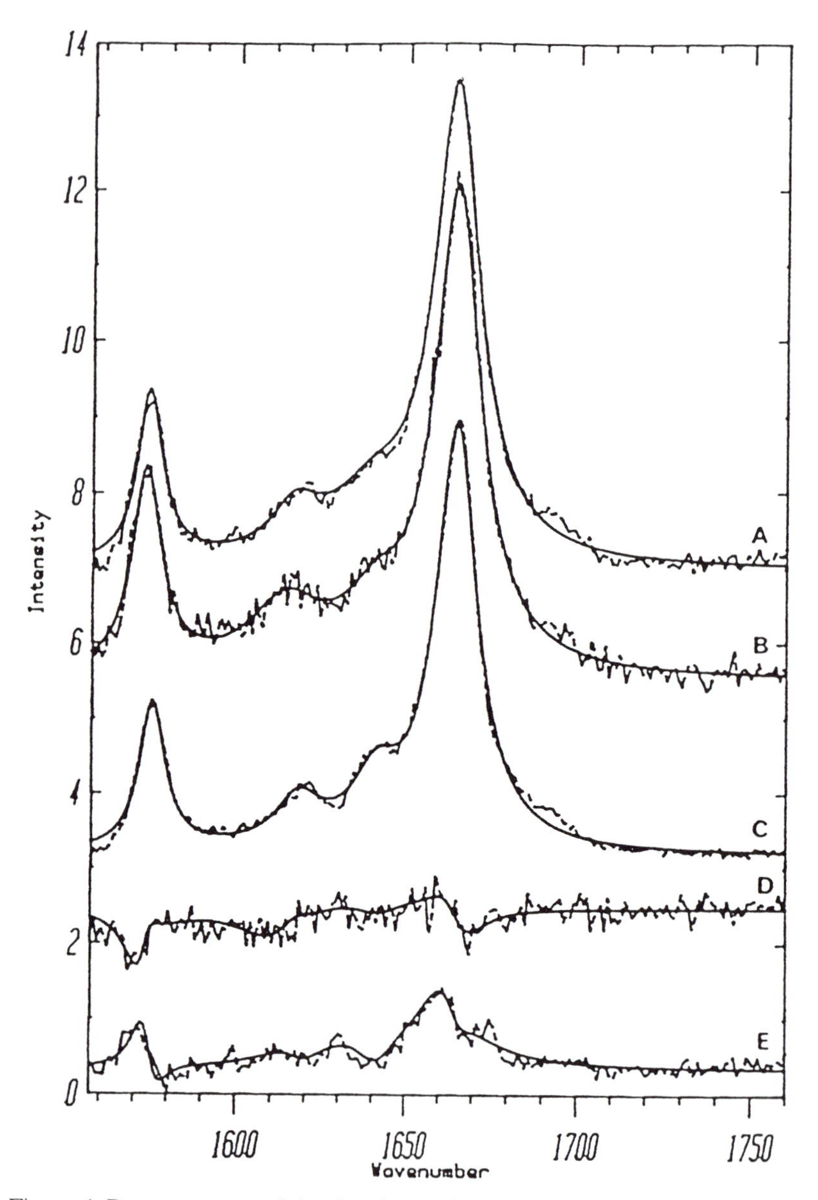

Figure 4. Raman spectra of the thymine carbonyl vibrations the three DNA fully deuterium exchanged in D_2O. The spectrum (A) is from $[d(A_5T_5)_2]_2$; the spectrum (B) is from $[d(TTATTATAATATTATAATAA)]_2$ and the spectrum (C) is from poly(dA)•poly(dT). Spectra are taken at 5° C in 0.5 M NaCl (pH 6.8). The difference spectra are (D) is spectrum (A) minus spectrum (B) and spectrum (E) is spectrum (A) minus spectrum (C).

...CGCGAAA....-3' There there are two conformational junctions; the first occurs between the B tract on the left and the Z tract in the middle. The second occurs between the Z tract in the middle and the B tract at the right (3') end. One might naively think that the conformational junction would occur exactly at the sequence junction so that the last Watson-Crick base pair in the B form is a TA base pair while the first Watson-Crick base pair in the Z form is a CG base pair. However, as we shall see below this is not the case. It will be shown that the last Watson-Crick base pair in the B form is a CG base pair and the last base pair in Z form is a GC base pair so that both B/Z junctions occur in the CG tract. This is to be expected since the CG tract sustains the B form easily in aqueous solutions at low salt concentrations while the A and T tracts will not go into the Z form even in saturated NaCl solutions. The B form is the more stable form and is projected from the T and A tracts into the CG tracts.

To show where the B/Z junction occurs in these sequences we used the method of Dai et al. [42] that is based on the following rules. DNA containing only alternating CG base pairs (CG tracts) will go into the Z form at high (4-6 M) salt concentrations. At low salt concentrations where the conformation is purely B form, there exists a strong Raman band at 680 cm^{-1} that is the frequency of a conformationally dependent guanine-backbone vibration that is present when the DNA is in the B form. At 4-6M salt concentrations, when the DNA is completely in the Z form, this band occurs with the exact same intensity at the frequency 625 cm^{-1}. If one measures the intensities of the Z-form band at 625 cm^{-1} and the B-form band at 680 cm^{-1} as a function of salt concentration, one finds that the sum of the intensities of these two bands remains constant during the B to Z transition. This is evidence of a transition from one state (the B state) to a second state (the Z state) with no intermediate state between the B and Z forms. By taking the ratio of the intensity of the band at 625 cm^{-1} to the sum of the intensities of the bands at 680 cm^{-1} and 625 cm^{-1} one can obtain the fraction of the CG bases in the sample that are in the Z form. When an oligonucleotide contains a CG tract contiguous to an A or T tract all of the CG base pairs will not go into the Z form even at saturated salt conditions. By measuring the intensities of the 625 cm-1 and 680 cm^{-1} bands one can measure the fraction of the CG base pairs in the Z form and the fraction of base pairs in the B form. Assuming that the B form CG tract is contiguous with the B form T and A tracts one can locate the exact position of the B/Z junctions in the oligonucleotide and build a model showing the exact position of the junctions [42]. In our published models there are no base pairs in the junctions. This is a result of the energy minimization part of the Biograf program that produces this B helix flip Z helix motif. Figure 5 shows a stereo view of $[5'\text{-d}(T_2(CG)_4A_2)\text{-}3']_2$.[42].

The rules that we devised for locating the B/Z junction in oligodeoxyribonucleotides of the form, d-5'-$T_N(CG)_8A_N$-3' in concentrated NaCl solutions are very simple [42]: (1) The AT base pairs are always in the B form. (2) The B/Z junction always occurs in the CG tract. The evidence for this rule is the observation that the Z-form guanine Raman band at 625 cm^{-1} is always present along with the 680 cm^{-1} band due to guanine in the B form. This means that some of the guanine is in the B form and some is in the Z form so there must be a junction between the two forms inside the CG tract. (3) If N = 2, the B form is projected from the A and T tracts on each side of the oligomer exactly one longitudinal (not Watson-Crick) CG base pair into the CG tract. (4) If n = 4 or more, the B form is projected two longitudinal CG base pairs into the CG tract. Note that each base in a longitudinal base pair forms a lateral base pair by Watson-Crick hydrogen bonding to its lateral partner.

In this study we chose three duplexes of the form, $[5'\text{-d}(T_N(CG)_8A_N)\text{-}3']_2$, with N = 2, 4, and 6. Based on the published rules reviewed above, the B/Z junctions of these duplex oligonucleotides have been located and published as follows [42]:

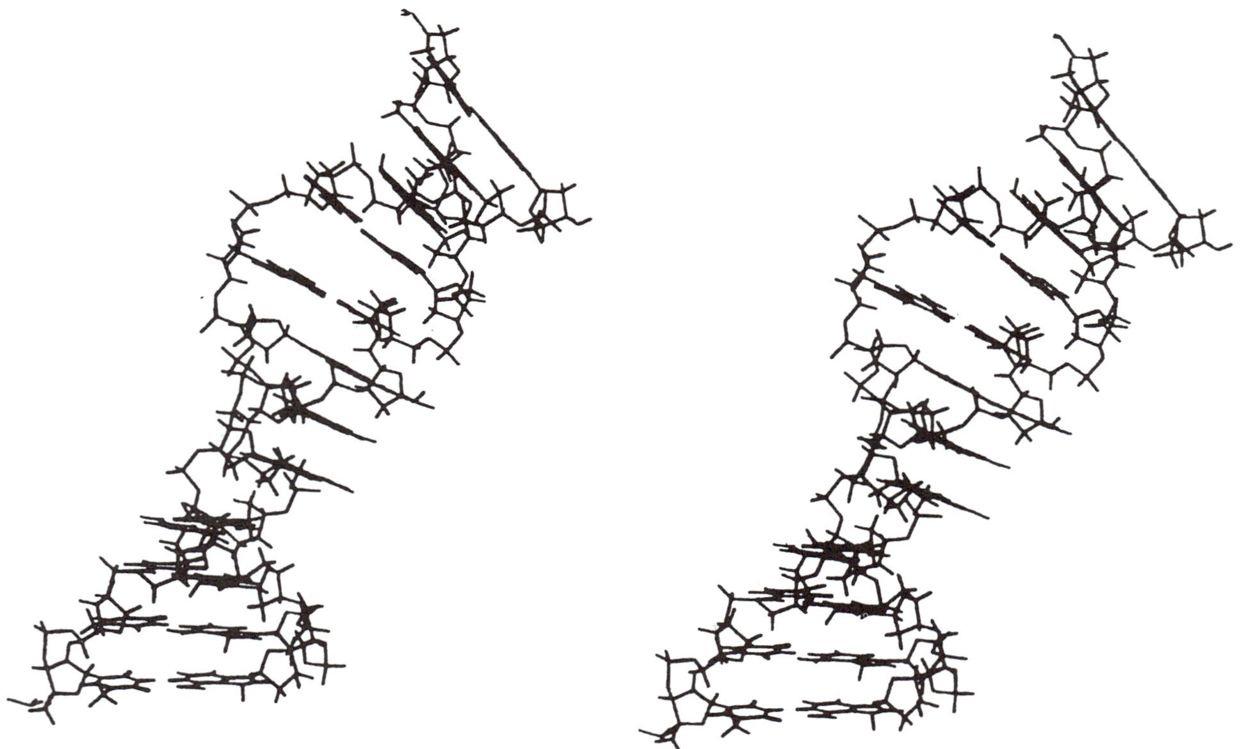

Figure 5. A stereoview of the double B/Z junction in the duplex of the self-complementary sequence d(TTCG/CGCG/CGAA) showing the location of the two B/Z junctions. The reagent, 1,10-phenanthroline-Cu^+ ion cuts such sequences exactly at the B/Z junctions.

```
      BBBB/ZZZZZZZZZZZZ/BBBB
5'-d(TTCG/CGCGCGCGCGCG/CGAA)-3'
3'-d(AAGC/GCGCGCGCGCGC/GCTT)-5'
```

I

```
      BBBBBBBB/ZZZZZZZZ/BBBBBBBB
5'-d(TTTTCGCG/CGCGCGCG/CGCGAAAA)-3'
3'-d(AAAAGCGC/GCGCGCGC/GCGCTTTT)-5'
```

II

```
      BBBBBBBBBB/ZZZZZZZZ/BBBBBBBBBB
5'-d(TTTTTTCGCG/CGCGCGCG/CGCGAAAAAA)-3'
3'-d(AAAAAAGCGC/GCGCGCGC/GCGCTTTTTT)-5'
```

III

Prof. David S. Sigman at UCLA and Prof. Annick Spassky at the University of Paris have shown that 1,10-orthophenanthroline-cuprous complex [$(OP)_2Cu^+$] with hydrogen peroxide as a coreactant is an artificial nuclease that cleaves DNA by an oxidative mechanism under physiological conditions [43-45]. They have shown that this nuclease activity depends on the secondary structure of the DNA being cut. They found DNA is cleaved by the interaction in the minor groove when the minor groove is more open. It is plain to see from our models of the B/Z junction that the minor groove is wide open because of the sudden change in conformation. (See Figure 5). Thus we predicted that [$(OP)_2Cu^+$] would cut our oligomers containing B/Z junctions exactly where we modeled the change from B to Z form to occur. This prediction has been strikingly verified.

Experimental procedures The deoxynucleotides (I, II & III) were labeled at the 5" end with [$\gamma^{32}P$]ATP under the reaction conditions described elsewhere [43-45]. The digestion of these oligonucleotides by $(OP)_2Cu+$ was carried out in the same manner as previous work on other oligonucleotides [43-45]. The resulting scission products were then dissolved in a loading buffer and electrophoresed on a 20% denaturing sequencing gel. The electrophoresis was carried out as previously described [43-45]. Exposure times were adjusted to allow visualization of relevant bands. The scission of each of the oligonucleotides gave rise to two products--one from the labeled 5' end to the nearest junction and one from the 5' end to the furthest junction. The 20mer (I) gave two products, one, containing four bases corresponding to the 5'-d(TTGG) and one containing 16 bases corresponding to (I) with the last four bases (CGAA) cut off. the 3' end. This corresponds exactly to the position of the B/Z junctions described in our previous publication [42]. Similarly the 24mer (II) gave two bands--one containing eight bases and one containing 16 bases while (III) gave two bands corresponding to 10 bases and one containing 18 bases, In our view, this work shows how molecular modeling can be used to make predictions about the sites for nuclease attack by [$(OP)_2Cu^+$], and the combination of the nuclease attack and molecular modeling is a more powerful tool for understanding DNA structure than either method alone.

Conclusions: The purpose of these studies is to show how molecular modeling combined with selected experimental data such as may be obtained from Raman spectroscopy, NMR or nuclease attack by [$(OP)_2Cu^+$] can be used to develop an understanding of the structure of DNA in solution. The purpose of the first study has been to provide insight into the solution structure of A tract DNA in an aqueous environment, away from crystal packing constraints. DNA containing A tracts has been shown to have a mixture of furanose ring puckers. A newer type of DNA called B' DNA is characterized by the appearance of two furanose conformations. The majority of the furanose rings are in the Southern part of the P conformational wheel as is the case for the classic B DNA. In the B' conformation, a substantial fraction of the furanose rings are in the Eastern part of the P conformational wheel giving rise to a Raman band at the intermediate frequency of 817 ± 3 cm^{-1} in addition to the normal B type DNA Raman band at 840 ± 5 cm^{-1}.

It is shown that the method developed previously in this laboratory for determining the position of the B/Z junctions in DNA has been strikingly verified by using 1,10-phenanthroline-Cu+ as a nuclease that cuts DNA where the minor groove is more open. Since the minor groove is much more open in the junction between the B and Z helices the reagent cuts exactly at this site. It is suggested that a good way to test molecular models of DNA is to locate the sites on the model where the minor groove is more open and see if the model is confirmed by nuclease by 1,10-phenanthroline-Cu+ ion.

References:

1. Trifonov, E. N. *CRC Critical Reviews Biochemistry* **1985,** 19:89-106.
2. Diekmann, S. and v Kitzing, E., in *Structure and Expression,* vol 3. eds. W.K. Olson, M.H. Sarma, R.H. Sarma and M. Sundaralingam Adenine Press Albany, N.Y. **1988** pp 57-67.
3. Hagerman, P. J., *Ann. Rev. Biochem.* **1990,** 755-781; Hagerman, P.J. in *Unusual DNA Structures,* R.D. Wells and S.C. Harvey, eds. **1988,** Springer Verlag, New york pp. 225-2326.
4. Sanghani, S.R., Zakrzewska, K., Harvey, S.C., and Lavery, R., *Nucleic Acids Research* **1996,** 24, 1632-1637.
5. Coll, M., Frederick, C.A., Wang, J. and Rich, A., *Proc. Natl. Acad. Sci. USA* **1987,** 84, 8385.
6. Nelson, H. C. M., Finch, J. T., Bonaventure, R. L., Klug, A., *Nature* **1987,** 330, 221-226.
7. Di Gabriele, A. D., Sanderson, M.R., Steiz, T.A., *Proc. Natl. Acad. Sci. USA* **1989,** 86, 1816.
8. Trifonov, E. N., and Sussman, J. L., *Proc. Natl. Acad. Sci. USA* **1980,** 77, 3816.
9. Levene, S. D., and Crothers, D. M., *J. Biomol. Struct. & Dyn.* **1983,** 1, 429.
10 Bolshoy, A., McNamara, P., Harrington, R.E., and Trifonov, E.N., *Proc. Natl. Acad. Sci. USA* **1991,** 88, 2312-2316.
11. Cacchione, S., De Santis, P. Foti, D.P. , Palleschi, A. and Savino, M., *Biochemistry* **1988,** 28, 8706-8713.

12. Callandine, C.R., Drew, H.R. and McCall, M.J., *J. Mol. Biol.* **1988,** 201, 127-137.
13. Koo, H.-S. and Crothers, D. M., *Proc. Natl. Acad. Sci. USA* **1988,** 85, 1763-1767.
14. Satchwell, S. C., Drew H.R., and Travers, A. A., *J. Mol. Biol.* **1986,** 191, 127-137.
15 Goodsell, D.S. and Dickerson, R.E., *Nucleic Acids Resarch* **1994,** 22, 5497-5503.
16. Thomas, G. A. and W. L. Peticolas, *J. Am. Chem. Soc.* **1983,** 105, 993-996.
17. Arnott, S., Chandrasekaran, R., Hall, I.H. and Puigjaner, L.C., *Nucleic Acids Research* **1983,** 11, 4141-4155.
18. von Kitzing, E. and Diekmann, S., *Eur. Biophys. J.* **1987,** 15, 13-26.
19. Peticolas, W.L., and Evertsz, E., in *Methods in Enzymology* Vol. 211 D.M.J. Lilley, J.E. Dahlberg eds.**1992,** pp.335-352; Peticolas, W. L., in *Methods in Enzymology* Vol 246 K. Sauer ed. **1995,** pp. 389-416.
20. Thomas Jr., G.J. and Tsuboi, M. *Adv. Biophys. Chem.* **1993,** 3, 1.
21a. Levitt, M. and Warshel, A. *J. Am. Chem. Soc.* **1978,** 100, 2607-2613.
21b. Olson, W.K. and Sussman, J. L. *J. Am. Chem. Soc.* **1982,** 104, 270.
22. Van Wijk, J., Huckriede, B.D., Ippel, J.H., and Atona, C. in *Methods in Enzymology*, Vol 211, D.M.J. Lilley, & J.E. Dahlberg, eds. **1992,** pp. 286-306.
23. Alexeev, D.G., Lipanov, A.A., and Skuratovskii, I.Y. *J. Biomol. Struct. & Dyn.* **1987,** 4, 989-1012.
24. Behling, R. W. & Kearns D. R. *Biochemistry* **1986,** 25, 3335-3346.
25. Behling, R. W., Rao, S. N., Kollman, P., & Kearns, D. R. *Biochemistry* **1987,** 26, 4674-4681.
26. Celda, B., Widmer, H., Leupin, W., Chazin, W.J., Denny, W.A., and Wüthrich, K. *Biochemistry* **1989,** 28, 1462-1471.
27. Katahira, M., Y. Nishimura, Y., Tsuboi, M., Tomohiro, S., Mitsui, Y., Iitaka, Yu., *Biochem. Biophys. Acta* **1986,** 867, 256-267.
28. Diekmann, S. *Nucleic Acids and Molecular Biology*, New York, Springer-Verlag, **1987,** 138-156.
29. Patapoff, T. W., Thomas, G. A., Wang, Y., Peticolas, W. L., *Biopolymers* **1988,** 27, 493.
30. Small, E. W. and W. L. Peticolas *Biopolymers* **1971,** 10, 1377.
31. Erfurth, S. C., Kiser, E.J. and Peticolas, W.L. *Proc. Natl. Acad. Sci.*.**1972,** 69, 938-941; Erfurth, S. C. and W. L. Peticolas *Biopolymers* **1975,** 14, 247-264.

32. Taillandier, E., Liquier, J., and Ghomi, M. *J. Mol. Struct.* **1989,** 214, 185-211; Dohy, D. Ghomi, M., and Taillandier, E. *J. Biomol. Struct. & Dyn.*,**1989,** 6, 741-754.

33. Wartell, R.M. and Harrell, J.T. , *Biochemistry* **1986,** 25, 2664.

34. Peticolas, W.L., Wang, Y., and Thomas, G.A., *Proc. Natl. Acad. Sci. USA*, **1988,** 85, 2579-2583.

35 Benevides, J.M., Wang, A. H.-J., van der Marel, G. A., van Boom, J.H., and Thomas Jr., G. J. *Biochemistry* **1988,** 27, 931-938.

36. Taillandier, E., Ridoux, J.-P., Liquier, J., Leupin, W., Denny, W.A., Wang, Y., Thomas, G. A., and Peticolas W.L. *Bioochemistry* **1987,** 26, 3361-3368.

37. Mayo, S.L., Olafson, B.D., Goddard III, W.A., *J. Phys. Chem.* **1990,** 94, 8897-8909.

38. Holbrook, S.R., Dickerson, R.E. and S.-H. Kim *Acta Crystallographica,* Sect. **1985,** B41, 255.

39. Kintanar, A., Klevit, R.E. and Reid, B.R. (1987) *Nucleic Acids Research*, 15, 5845-62.

40. Nadeau, J.G. and Crothers, D.M. (1989) *Proc. Natl. Acad. Sci. USA*, 86, 2622-26.

41. Koo, H.-S. and Crothers, D.M. *Biochemistry* **1987,** 26, 3745-48.

42. Dai, Z., Dauchez, M., Thomas, G. and Peticolas, W.L. *J. Biomol. Struct. & Dyn.* **1992,** 9, 1155-1183.

43. Sigman, D.S., Spassky, A., Rimsky, S., & Buc, H. *Biopolymers* **1984,** 24, 183-197.

44. Spassky, A. & Sigman, D.S. *Biochemistry* **1985,** 34, 8050-8056.

45. Kuawabara, M., Yoon, D., Goyne, T., Thederahn, T. and Sigman, D. Biochemistry **1986,** 25, 7401-7408.

Chapter 10

Three-Dimensional NOESY–NOESY Hybrid–Hybrid Matrix Refinement of a DNA Three-Way Junction

Varatharasa Thiviyanthan[1], Nishantha Illangasekare[1], Elliott Gozansky[1], Frank Zhu[1], Neocles B. Leontis[2], Bruce A. Luxon[1], and David G. Gorenstein[1]

[1]Sealy Center for Structural Biology and Department of Human Biological Chemistry and Genetics, University of Texas Medical Branch, Galveston, TX 77555–1157

[2]Department of Chemistry, Bowling Green State University, Bowling Green, OH 43403–0213

A computationally efficient hybrid-hybrid relaxation matrix refinement methodology using 3D NOESY-NOESY data is described. The methodology was tested on simulated data, derived from the Dickerson dodecamer, and used to refine the structure of a DNA three way junction from experimentally determined 3D NOESY-NOESY volume data. Our results indicate that the 3D hybrid-hybrid MORASS methodology, by combining the spectral dispersion of 3D NOESY-NOESY spectroscopy and the computational efficiency of 2D refinement programs, provides an accurate and robust means for structure determination of large biomolecules.

For structure determination of large biomolecules by high-resolution proton NMR methods, severe overlap of signal in 2D NOESY spectra is a serious problem. Spectral overlap often prevents adequate measurement of enough NOEs from the 2D spectra severely limiting the precision and accuracy of any refined structure. Spreading the spectral dispersion into a third, or even a fourth, dimension is an appropriate avenue to alleviate the problem of spectral over-crowding as it occurs in the 2D NOESY spectra. While heteronuclear 3D NMR spectroscopy has been used more often to make sequential assignments and obtain distance estimates, homonuclear 3D NOESY-NOESY methods have been found to contain more information for quantitative determination of fine structures once the assignments have been made (1-4). Homonuclear 3D NOESY-NOESY spectroscopy has shown great promise for the structure determination of large biomolecules since the diplor cross-relaxation in biomolecules become more efficient with increasing molecular weight (2).

Several methods have been described for using 3D NOESY-NOESY cross-peak intensities for structural refinement such as the two-spin approximation (4,5), Taylor series expansion of the NOE-rate equation (6), and direct gradient refinement method (7). The two-spin approximation requires that the NOESY derived distances be obtained from vanishingly short experimental mixing times where the build-up of NOE intensity is linear with respect to interproton distance and the effect of spin diffusion (NOE intensity mediated by multiple relaxation pathways) are minimal.

Because most of the structurally important longer range NOEs are not observed at these short mixing times, the use of the two-spin approximation has raised concern over the validity of highly refined NMR structures derived by this methodology. As demonstrated recently (8,9), at realistic mixing times that allow reasonably accurate measurement of 3D NOESY-NOESY volumes, both the two-spin and the Taylor series expansion approximation methods can lead to considerable systematic errors. The direct gradient method provides an accurate and precise means for structural refinement, however it scales with the sixth power of the number of spins. An approximation to the gradient method (10) was proposed that scales with the cube of the number of spins, however, in large systems this will still be computationally intensive.

The deconvolution method proposed earlier by our group, translated the 3D cross peak intensities into 2D NOE intensities, which can be used to derive accurate distance constraints that can be used in structure refinement by an iterative relaxation matrix approach (11). This approach avoids systematic errors while retaining computational efficiency.

In this paper, we report the hybrid-hybrid relaxation matrix refinement tested by simulated refinement calculations on a dodecamer DNA duplex, and the refinement of the three-dimensional structure of a DNA three-way junction (TWJ), using experimental 3D NOESY-NOESY data. The TWJ has previously been studied by 2D NMR spectroscopy (12, 13).

Theory and Method.

In the two spin approximation method the volume of a 3D cross peak,which is derived from interactions between spins i, j and k, is considered to be proportional to the product of the inverse sixth power of the distance between spins i and j and the distance between j and k:

$$A_{ijk}\left(\tau_{m1},\tau_{m2},\right) \propto r_{ij}^{-6}\left(\tau_{m1}\right) \bullet r_{jk}^{-6}\left(\tau_{m2}\right).$$

Where $A_{ijk}(\tau_{m1}, \tau_{m2})$ is the 3D NOESY-NOESY volume and r_{ab} is the inter-proton distance between spins a and b. This model provides a simple description of the 3D NOESY-NOESY interaction, however it does not include the effects of spin diffusion (multiple relaxation pathways). This oversimplification leads to dramatic systematic errors in the form of overestimation of all distances.

A more complete model involves the NOE build-up rate equation. The rate equation, without approximation, takes into account the NOE-type interactions (dipole-dipole interactions) across the entire system and has the following form for a 3D NOESY-NOESY experiment:

$$A_{ijk}\,(\tau_{m1},\tau_{m2}) = \exp(-\tau_{m2}R)_{kj} \bullet \exp(\tau_{m1}R)_{ij}\,A_i\,(0)$$

Where R is the rate matrix that describes the NOE interactions across the system, τ_{m1} and τ_{m2} are the two NOE mixing and A(0) is the initial magnetization. To simplify this equation, a Taylor series expansion of the exponential can be made. Usually, only the first few terms in the expansion are kept for the approximation. The first term approximation is equivalent to the two-spin approximation (5,14). At realistic mixing times (50 ms or more), the Taylor series approximation also yields systematic error in determining the inter-proton distances (9). Figure 1 shows comparison of volumes simulated from the two-term Taylor series approximation and an exact rate-matrix calculation for the Dickerson dodecamer d(CGCGAATTCGCG)$_2$ with a mixing time of 100 ms for both mixing periods. If the error in the approximation were negligibly small, all the data points would lie along the straight line as shown in the figure. However, the overwhelming majority of data, resulting from the two term Taylor series approximation, lie to one side of this line indicating serious systematic error. The sign of the systematic error, whether it be overestimation or underestimation, depends only on the number of terms used in the approximation.

A better approach to structural refinement from 3D NOESY-NOESY data is found by using a hybrid-hybrid matrix method. It avoids the two-spin and Taylor series expansion approximations by employing a matrix eigenvalue/eigenvector solution to the Bloch equations (in the form of a rate matrix). Importantly, this method includes the effects of spin diffusion (3,15-17). For two-dimensional NOE data, accurate distances can be obtained from the complete relaxation method by diagonalizing the 2D volume matrix (8, 18). An eigenvalue/eigenvector solution, however, is very sensitive to the accuracy and completeness of the volume matrix. Due to instrument sensitivity limitations, the experimental NOE data only yields a small fraction of the volumes required for the calculation. The hybrid matrix method was designed to overcome the limited experimental data by combining information from experimental data and calculated volumes based on a model structure (15).

An extension of the hybrid matrix method, known as the 3D hybrid-hybrid matrix method for 3D NOESY-NOESY data analysis, was proposed to generate precise and accurate distances from the 3D data while retaining computational efficiency (11). The method is based on an expression for the 3D NOE peak where the 3D volume is proportional to the product of two 2D volumes (where the effects of spin diffusion have been taken into account),

$$A_{ijk}^{3D} \propto A_{ij}^{2D} \bullet A_{jk}^{2D}$$

where A_{ijk}^{3D} is a 3D NOESY-NOESY volume between spins i, j and k and A_{ab}^{2D} is the 2D NOESY volume between spins a and b, where spin diffusion has been considered. This equation can be rearranged such that a 2D-type volume can be deconvoluted from the 3D volume:

$$A_{ij}^{2D} \propto A_{ijk}^{3D} / A_{jk}^{2D}$$

The problem, of course, is obtaining values for the divisors. In the limits of larger molecular size, there will not be many resolved 2D peaks, which is the reason the 3D experiment is being performed. One solution, similar to that for the 2D hybrid matrix method, is to obtain needed divisors from calculated data based on a model structure. If the devisor volume is available experimentally, then it is taken from experiment.

In simulating the divisors, there is no need to simulate the entire 3D volume matrix. Only volumes needed for the deconvolution and a set of scaling volumes are simulated. The scaling set is actually the same set as those measured experimentally, with respect to the spin indices. This set is used to scale the divisors and then discarded. Once the divisors have been scaled, they are merged with the 3D experimental data and deconvolution is performed. If there is any available 2D NOESY experimental data, it is also scaled and included with the deconvoluted 2D-like data to create a hybrid-hybrid 2D volume matrix. From this point, the method is identical to the standard 2D MORASS hybrid matrix methodology (or any other 2D NOESY refinement protocol can be used) where the experimental data is hybridized with a full 2D volume matrix, which is based upon the model structure. After diagonalization of the matrix, the inter-proton distances are used in restrained molecular dynamics (MD). The output structure from the MD is then used as the new model structure in the next iteration. The entire procedure, including data simulation, is repeated until internal consistency among the output structures is reached. Figure 2 is a flow chart of the 3D hybrid-hybrid matrix method. It is worth noting that only a small fraction of the 3D data needs to be simulated, thus the method only scales with the square of the number of spins.

Simulation Study of 3D NOESY-NOESY Refinement

To test the reliability and computational efficiency of the methodology, refinement was performed on simulated "experimental" data derived from the Dickerson dodecamer. Using a refined structure of the dodecamer, "experimental" data was generated by simulating the volume matrix, creating 3D NOESY-NOESY and 2D NOESY type data and adding integration error and thermal noise. The refined dodecamer also served as the target structure. The 3D hybrid-hybrid refinement process was performed and convergence examined by several refinement indicators; most importantly the Cartesian RMSD (cRMSD) with respect to the target structure. Three different NOESY data sets were simulated in this study. Two of the data sets were of 3D NOESY-NOESY type and a third set was a 2D NOESY type, where the 2D data served as a point of reference for the quality of 3D refinement. A 0.2% random error was added to all three data sets to simulate thermal noise. One 3D data set and the 2D data set had 20% error added in a Gaussian fashion to simulate integration error. The second 3D data set had 50% integration error added. These values for integration error were considered to be reasonable lower and upper limits of integration error for 3D data and 20% is a reasonable value for 2D data. Three different starting models were used in the refinement, including canonical B- and canonical A-DNA. Heating the canonical A-DNA to 1000 K for 2 ps generated a third starting structure, which was called A'-DNA. This third starting model was particularly distinct from target and gave some insight as to the robustness of the methodology. Table I is a summary of the Cartesian RMSD data, which gives the cRMSD of the model and refined structures with respect to the target structure. Table II is a summary of several refinement parameters also examined. All of the simulated 3D refinements produced results comparable to (or better than) the simulated 2D refinements.

Refinement of the structure of a DNA Three-Way Junction Molecule

The Hybrid-Hybrid relaxation matrix refinement method was used to refine a DNA three-way junction (TWJ). The TWJ, made with 3 DNA strands, contains two unpaired bases on one strand at the junction region. Gel electrophoresis and UV melting experiments had shown that two or more unpaired bases stabilize the DNA TWJ (19). This TWJ molecule was chosen as a test molecule because the resonance assignments have been obtained previously (12, 13) and a high quality 2D NOESY data set was available. The G-C rich sequence was designed to include one A-T base pair in each helical arm to serve as spectroscopic marker. The sequence at each A-T base-pair is unique allowing unequivocal assignments to be made. The sequence of the DNA TWJ is shown in Figure 3. The strand is first identified and then the bases are numbered separately for each strand. For example, S1-G1 stands for the first guanine residue on strand 1.

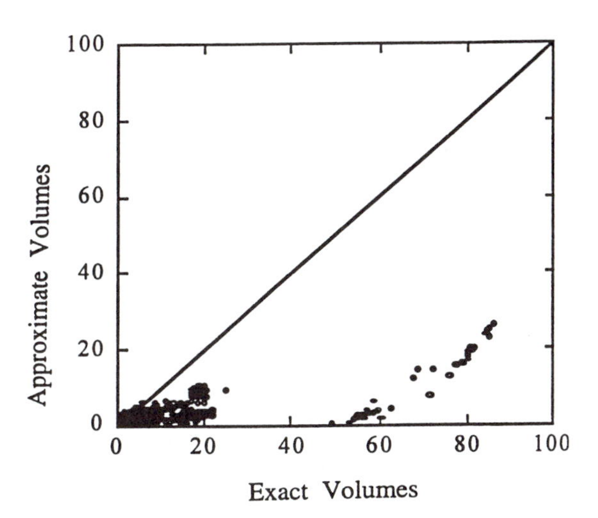

Figure 1. Scatter plot of simulated NOESY-NOESY volumes for a 100 ms mixing time. The vertical axis contains approximate volumes calculated from the Taylor series expansion using two terms and the horizontal axis represents the "exact" volumes calculated from the relaxation matrix. The calculation was based on the dodecamer d(CGCGAATTCGCG)2 with a τ_c= 3.2 ns, assuming a 3D volume, at τ_m= 0, of 1000 where only volumes greater than 1.0 were included in the plots.

Table I Cartesian RMSD of the refined structures to the target duplex

Starting Model		Data set[a] 3D20	3D50	2D20
A-DNA	(3.72)[b]	1.44[c]	1.37	1.23
B-DNA	(2.95)	1.54	1.51	1.10
A•-DNA	(4.21)	1.45	1.60	1.87
Average	(3.63)	1.48	1.49	1.40

[a] Data set nomenclature: 3D20 represents simulated 3D data with 20% integration error added. The others are similar. In the text, each structure is named based on the starting model and the 3D NOE data set used. For example, A•-3D20 is the final structure refined using data set 3D20 from the A•-DNA starting model.
[b] Model Cartesian RMSD error relative to the target structure in Å.
[c] Final structure RMSD error relative to the target structure in Å.
SOURCE: Adapted from ref. 24

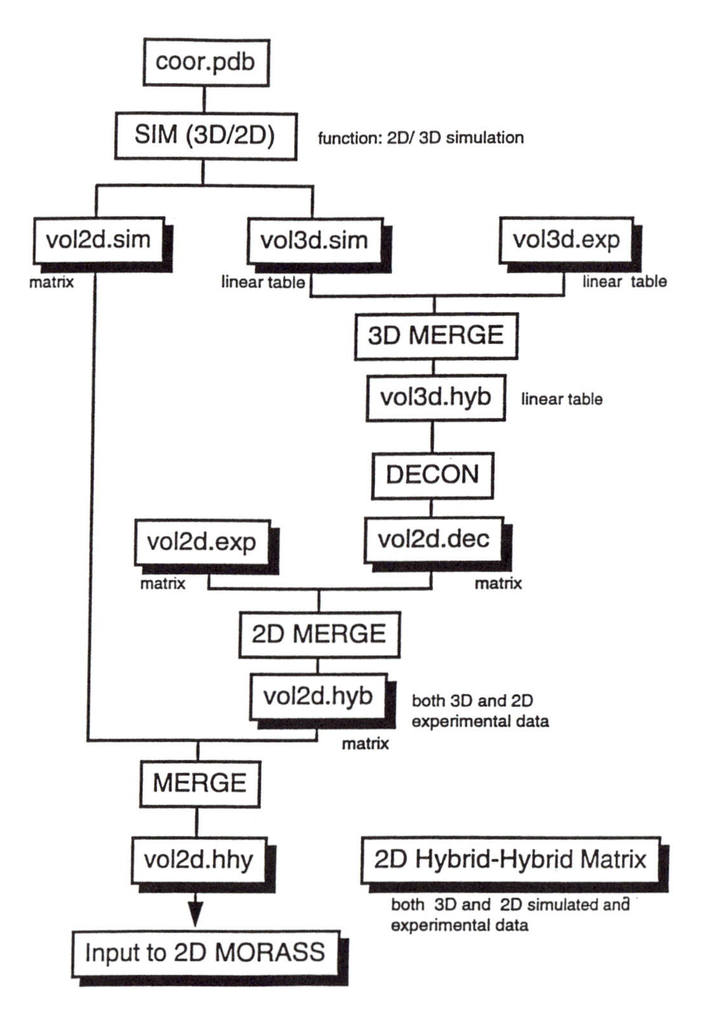

Figure 2. The hybrid-hybrid relaxation matrix refinement procedure for 3D NOESY-NOESY data. A starting model is used to simulate NOE data for the first iteration. Experimental 3D data is scaled and merged with simulated 3D data to produce a linear table of 3D experimental volumes and the simulated volumes needed for deconvolution. Deconvoluted 2D data is merged with any available 2D experimental data and then with a simulated, complete 2D volumes matrix. The structures resulting from the standard 2D MORASS refinement are used in subsequent iterations until convergence is reached. (Reproduced with permission from reference 24. Copyright 1996.)

Table II Quality of the simulated 3D NOE-NOE refinements

Data Set	3D20		3D50		2D20
Number of R_{ij}	481		477		359
A-DNA Results					
%RMS(vol)	72	(41)[a]	133	(32)	32
R-factor	0.31	(0.36)	0.47	(0.18)	0.22
$Q(\frac{1}{6})$	0.095	(0.058)	0.149	(0.035)	0.039
Force (kcal/(mol Å^2))	8		10		8
Error %	12		12		9
No. of Iterations	5		8		5
B-DNA Results					
%RMS(vol)	54	(24)	138	(33)	31
R-factor	0.26	(0.16)	0.46	(0.16)	0.22
$Q(\frac{1}{6})$	0.074	(0.031)	0.014	(0.033)	0.039
Force (kcal/(mol Å^2))	8		10		8
Error %	12		13		9
No. of Iterations	6		8		3
A• -DNA Results					
%RMS(vol)	76	(45)	248	(52)	31
R-factor	0.33	(0.38)	0.52	(0.30)	0.21
$Q(\frac{1}{6})$	0.102	(0.062)	0.183	(0.055)	0.044
Force (kcal/(mol Å^2))	10		10		8
Error %	12		13		9
No. of Iterations	9		7		6

[a] Values in parentheses are calculated from deconvoluted 2D volumes.
SOURCE: Adapted from ref. 24

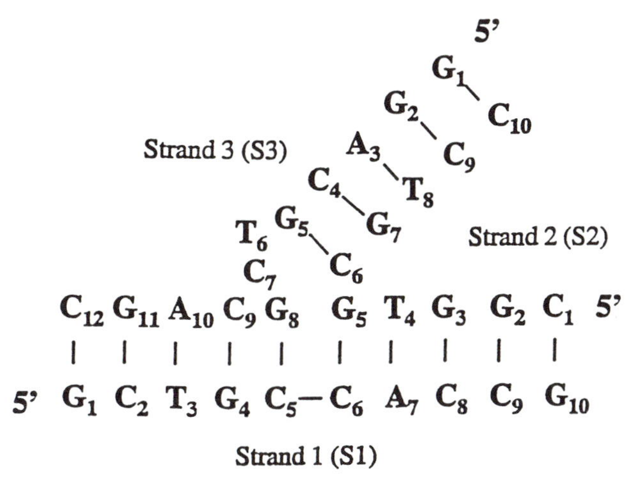

Figure 3. The sequence of the DNA three-way junction molecule with two unpaired bases in the junction region.

The NMR sample was prepared by dissolving stoichiomatric amounts of the oligonucleotide strands in D_2O buffer containing 10 mM sodium phosphate (pH 6.8), 100 mM NaCl, 10 mM $MgCl_2$, and 0.5 mM EDTA to give a final concentration of approximately 2 mM DNA. The 3D NOESY-NOESY data was collected at 28°C at 750 MHz on a Varian UnityPlus instrument. The relaxation delay was set at 0.9 seconds and a sweep width of 7100 Hz was used in all three dimensions. Acquisition time was 72 ms. The two mixing times were both 200 ms. Residual water suppression was achieved by low power saturation at the watefrequency immediately before the first 90° pulse and during the two mixing periods. No attempt was made to suppress zero-quantum interference during the mixing times. In the direct detection dimension 512 complex points were acquired, and for each of the two indirect dimensions 128 complex points were acquired. For each (t_1, t_2) pair, eight scans were acquired. Quadrature detection in F1 and F2 was achieved using the hypercomplex method (20). The total experiment time was 228 hours.

The 3D data set (128*128*512) was processed using Felix software (Biosym, CA) to give a data matrix of 256*256*1024 real data points after zero-filling in all three dimensions. Only the real part of the final spectrum was stored. A 90° shifted sine bell apodization function was used in all three dimensions. The Flat routine available in the Felix program was used for baseline correction.

Assignment of proton chemical shifts was based on previously reported values by Leontis et. al., (12). The cross peaks were picked using the automatic peak-picking option available in the Felix program. The bounding box for each peak of interest was manually adjusted to reflect the actual linewidth of the peak. A total of 6253 peaks were picked by the automatic peak-picking routine. After removing the body diagonals, artifacts and noise peaks, 1635 peaks were retained. A total of 912 3D peaks (ijk and iji type peaks) were used for the deconvolution process. From the 2D NOESY experimental data set, 78 additional volumes not obtained from the deconvoluted 3D spectrum were incorporated with the deconvoluted 2D volumes to form the experimental 2D volume matrix.

Refinement Methods. The 3D NOESY-NOESY volumes were deconvoluted into 2D NOE volumes using the procedures described above (11). The deconvoluted 2D peaks were then used in the standard MORASS/ restrained MD iterative refinement cycles. Starting model coordinates were obtained in X-PLOR(21) from the previous study (22). This model structure was placed in a box of water containing 1719 water molecules, along with 29 counterions (one sodium counterion for each phosphate group) and was equilibrated for 10 ps using the molecular dynamics program AMBER(23). This structure was used as the starting model for all subsequent refinements. To ensure Watson-Crick base-pairing, hydrogen bond constraints were added for each base-pair in the helical arms, except for the base-pairs at the end. Only one hydrogen bond constraint was applied per base-pair to allow propeller twist during the refinement.

In each iteration, experimental 3D volumes were merged with the simulated 3D volumes to create the hybrid data set. Adequate scaling was achieved by using a linear regression procedure between the simulated and experimental 3D volumes. The hybrid data set was deconvoluted to generate 2D volumes. After additional experimental 2D volumes were merged with the deconvoluted volumes, the matrix was merged with the simulated 2D volume matrix to generate a hybrid-hybrid volume matrix (18). This hybrid-hybrid volume matrix was then diagonalized and the relaxation rate constants were calculated. Distance constraints derived from the relaxation rate matrix, were used in the molecular dynamics program AMBER to generate a new structure consistent with theNOE distance constraints. For the MD part of each iteration, the starting structure was first energy minimized with the NOE constraints for 3000 steps. An 8 ps of constrained molecular dynamics protocol with temperature annealing was performed on the energy minimized model. Then the average structure from the last 3 ps of the MD was energy minimized again, and the resulting structure was used as the starting model for the next iteration.

The progress of the iterative refinement process was monitored by several key indicators. The RMS errors in the volumes were used as the first criteria for monitoring the refinements. The %rms (volume) is given by

$$\%\text{rms (volume)} = \sqrt{1/N\sum_{ij}\left(\frac{v_{ij}^{a} - v_{ij}^{b}}{v_{ij}^{a}}\right)^{2}} \times 100\% ,$$

where a or b can be either the experimental/ deconvoluted or theoretical volumes to give the %rms(exp) or %rms(the) respectively.

As can be seen from Table III both these parameters start at relatively higher numbers and gradually settle down to lower values with increased percentage of volume merging between the experimental and theoretical volumes. The energy factors such as the total energy and the constraint energy also were monitored throughout the iterative process. Both the total energy and the constraint energy increased in value as the error bars and force constants on the constraints were tightened in the molecular dynamics refinement. The effect of the force constant was controlled by changing the error bars from a liberal values of 25% to much lower values as the confidence in the intermediate structures was established. The R-factor which is similar to the R-factor used in X-ray crystallography was also used as a refinement criteria. The R-factor is given by,

$$R = \frac{\sum_{ij}\left|v_{ij}^{a} - v_{ij}^{b}\right|}{\sum_{ij} v_{ij}^{a}} .$$

The R factor was found to decrease as the refinement progressed as expected. For 2D matrix methods, we have suggested that the %rms (volume) is a very useful measure of quality of fit to the spectra since it weighs the percentage differences in the theoretical and experimental volumes for both large and small crosspeaks equally. The R-factor is regarded as a poorer measure of the quality of the refined structure since it is often dominated by the largest crosspeaks. Another quality of fit, the Q(1/6) factor, also better reflects the quality of the structure since it more heavily weighs the weak crosspeaks compared to the R-factor. The Q(1/6) factor is defined as,

$$Q(1/6) = \frac{\sum_{ij}\tau_{m}\left|\left(v_{ij}^{a}\right)^{1/6} - \left(v_{ij}^{b}\right)^{1/6}\right|}{\sum_{ij}(1/2)\tau_{m}\left|\left(v_{ij}^{a}\right)^{1/6} + \left(v_{ij}^{b}\right)^{1/6}\right|} .$$

The plot of deconvoluted 2D volumes derived from experimental 3D volumes versus the experimentally determined 2D volumes, shown in Figure 4, yielded a slope of 0.82. Random dispersion of the data points indicates that there is no systematic error from the deconvolution process. Figure 5 shows the number of NOE volumes measured per residue from the 2D and 3D data sets. Except for residues in the junction region, where NOE interactions are weak, the 3D NOESY-NOESY spectra gave higher numbers of measurable NOE peaks than the 2D NOESY. We also observed a tertiary contact between the methyl group of S3-T6 and the H4' of S3-G11. This crucial NOE peak was well resolved in both 2D and 3D spectra (as 2D *iij and ijj* type peak), and was very useful to determine the conformation of the S3-T6 base.

The plot of theoretically calculated 2D volumes for the final structure versus the experimental (deconvoluted 2D plus the experimentally determined 2D) volumes, gave a slope of 0.99 (Figure 6).

Global Structure of the TWJ after refinement. The final structure of the TWJ is shown in Figure 7. As in the preliminary model, the three helical arms form two domains. Two of the helices, helix 1 and helix 2, are stacked on each other forming one continuous helical domain. The other helical domain, formed by helix 3, extends almost perpendicularly from the axis of the first helical domain. The unpaired pyrimidine bases are extra-helical, exposed to solvent and lie along the minor groove of Helix 1. These two unpaired bases are stacked on each other. The helical parameters of all three helical arms exhibit only minor deviations from the typical values for right-handed B-form DNA. Unusual values are observed for the glycosidic angles of S3-T6 and S3-G8. The glycosidic bond of S3-T6 exists in the *syn* conformation, allowing its methyl group to contact the hydrophobic surface of the minor groove of helix 1, at S3-G11.

Advantages of the Hybrid-Hybrid method

This method does not rely on any experimental 2D data. It needs only 3D NOESY-NOESY experimental data and a reasonable starting model. Hence, this method is very suitable to study large biomolecules for which significant numbers of good quality 2D NOESY crosspeaks cannot be resolved. This method provides a simple means to incorporate distance constraints derived from other heteronuclear (e.g., 3D/4D heteronuclear filtered and edited NOESY) experiments. These constraints can be added to the hybrid-hybrid volume matrix along with the deconvoluted volumes. However, one should carefully evaluate the relative accuracy of these constraints which are analyzed at a two-spin approximation level. These less accurate constraints can be added to the distance constraint list with appropriate increase in error bars and decrease in force constants. The method is precise, robust and accurate. The results from these two studies have shown that it has good convergence capability. For biomolecules, both larger molecular size and longer mixing times make spin diffusion a greater problem requiring a complete relaxation matrix analysis. This method uses the more accurate and well established 2D hybrid full relaxation matrix method that takes into account the extensive spin-diffusion occur in large molecules. The method is computationally very efficient as it does not involve the use of any three-dimensional matrix or gradients which scale with the cube of the number of proton spins. Instead, a linear table is used to take advantage of the sparseness of 3D NOESY-NOESY volume matrix.

Table III: Refinement summary of the DNA Three-Way Junction.

Iteration	RMS Difference % vol (exp)	%vol (the)	R-factor	Q (1/6) factor
0	115.8	206.8	0.4459	0.1009
1	109.5	188.9	0.4419	0.0988
2	89.9	163.1	0.4372	0.0969
3	76.7	148.6	0.4035	0.0889
4	83.2	116.6	0.3920	0.0814
5	78.5	117.1	0.3706	0.0774
6	76.8	80.8	0.3749	0.0720
7	76.5	77.0	0.3707	0.0692
Final	79.1	68.6	0.3434	0.0629

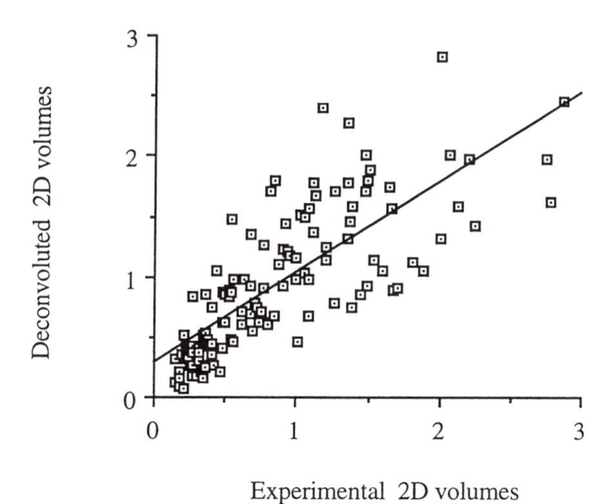

Figure 4. Plot of experimentally determined 2D volumes versus the deconvoluted 2D volumes. Linear regression yields a slope of 0.82

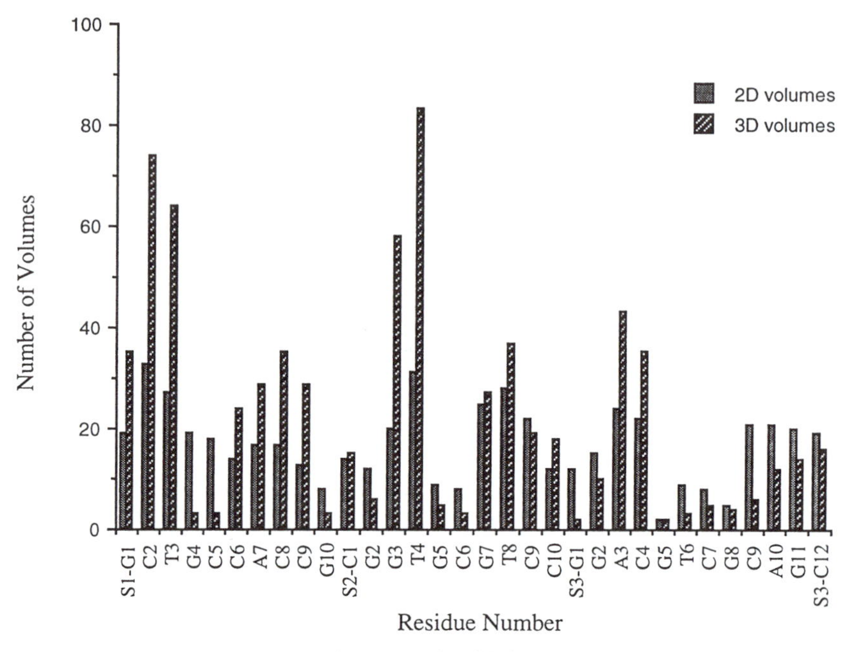

Figure 5. Number of measured NOE integrals per residue.

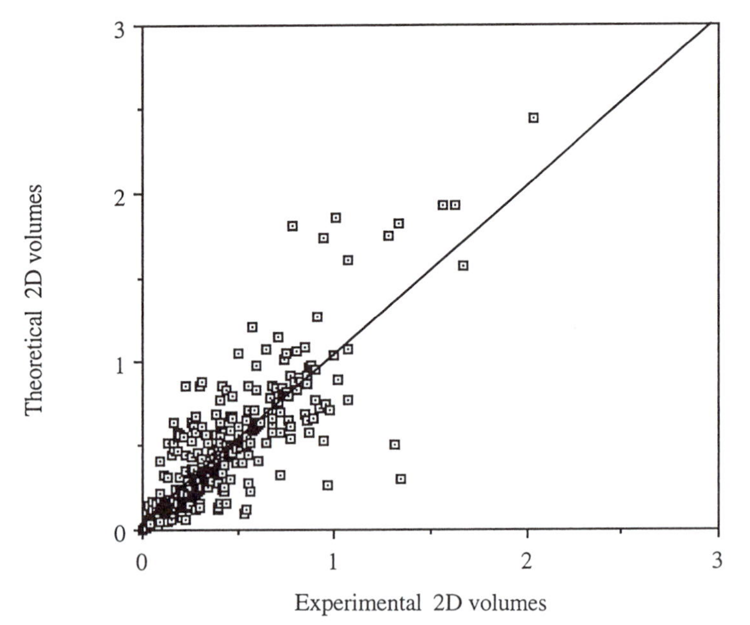

Figure 6. Plot of experimental 2D volumes vs. theoretical 2D volumes calculated from the final structure. The experimental 2D volumes includes deconvoluted 2D volumes (derived from experimental 3D volumes) and experimentally determined 2D volumes. The linear regression yields a slope of 0.99

Figure 7. The Final Structure of the DNA Three-Way Junction molecule. Only bonds between heavy atoms are shown.

Conclusions.

The hybrid-hybrid matrix refinement method yields more accurate structures given the vastly increased number of resolved 3D NOESY-NOESY volumes. This method uses a straightforward deconvolution scheme to obtain 2D NOE volumes. The results from the simulation studies on the DNA dodecamer (24), and the refinement of the TWJ, have clearly shown that our hybrid-hybrid matrix approach can be a very promising method for solving the structures of large biomolecules that requires the use of three-dimensional NMR.

Acknowledgments
Supprted by NIH (AI27744), NIEHS (ES06676), the Welch Foundation (H-1296) and the Sealy and Simith Foundation grants to DGG, and U.S. Public Health Service grants GM41454, and Petroleum Research Fund grant 20871-GB4 to NBL. Building funds were provided by NIH (1CO6CA59098).

References

1. Bonvin, A. M. J. J.; Boelens, R.; Kaptein, R. *J. Magn. Reson.* **1991**, *95*, 626.
2. Berg, J. N.; Boelens, R.; Vuister, G. W.; Kaptein, R. *J. Magn. Reson.* **1990**, *87*, 646.
3. Boelens, R.; Vuister, G. W.; Koning, T. M. G.; Kaptein, R. *J. Am. Chem. Soc.* **1989**, *111*, 8525.
4. Kessler, H.; Seip, S.; Saulitis, J. *J. Biomol. NMR.* **1991**, *1*, 83.
5. Habazettl, J.; Ross, A.; Oschkinat, H.; Holak, T. *J. Magn. Reson.* **1992**, *97*, 511.
6. Habazettl, J.; Schleicher, M.; Otlewski, J.; Holak, T. A. *J. Mol. Biol.* **1992**, *228*, 156.
7. Yip, P.; Case, D. A. *J.Magn. Reson.* **1989**, *83*, 643.
8. Kaluarachchi, K.; Meadows, R. P.; Gorenstein, D. G. *Biochemistry* **1991**, *30*, 8785.
9. Donne, D. G.; Gozansky, E. K.; Gorenstein, D. G. *J. Magn. Reson.* **1995**, *106*, 156.
10. Yip , P. F. *J. Biomol. NMR.* **1993**, *3*, 361.
11. Zhang, Q.; Chen, J.; Gozansky, E. K.; Zhu, F.; Jackson, P. L.; Gorenstein, D.G. *J. Magn. Reson. Ser B.* **1995**, *106*, 164.
12. Leontis, N.B.; Hills, M. T.; Piotto, M.; Ouporov, I. V.; Malhotra, A.; Gorenstein, D. G. *Biophysical J.* **1995**, *68*, 251.
13. Leontis, N. B.; Hills, M. T.; Piotto, M.; Malhotra, A.; Nussbaum, J.; Gorenstein, D. G. *J.Biomolec. Struct. and Dyn.* **1993**, *11*, 215.
14. Boelens, R.; Koning, T. M. G.; van der Marel, G. A.; van Boom, J. H.; Kaptein, R. *J. Magn. Reson.* **1989**, *82*, 290.
15. Nikonowicz, E.; Meadows, R. P.; Gorenstein, D.G. *Biochemistry* **1990**, *29*, 4193.
16. Boelens, R.; Koning, T. M. G.; Kaptein, R. *J. mol. Struct.* **1988**, *173*, 299.
17. Borgias, B. A.; James, T. L. *J. Magn. Reson.* **1990**, *87*, 475.
18. Post, C. B.; Meadows, R. P.; Gorenstein, D. G. *J. Am. Chem. Soc.* **1990**, *112*, 6796.
19. Leontis, N. B.; Kwok, W.; Newman, J.S. *Nucleic. Acids Res.* **1991**, *19*, 759.
20. States, D. J.; Haberkorn, R. A.; Ruben, D. J. *J. Magn.Reson.* **1982**, *48*, 286
21. Brünger, A. T.; **1993**, X-PLOR;version 3.1. A System for X-Ray Crystallography and NMR. Yale University Press. London.
22. Ouporov, I.V.; Leontis, N. B. *Biophysical J.* **1995**, *68*, 266.
23. Weiner, P. K.; Kollman, P. A. *J. Comput. Chem.*, **1981**, *2*, 287
24. Zhu, F. Q.; Donne, D.G.; Gozansky, E.K.; Luxon, B.A.; Gorenstein, D.G. *Mag. Res. Chem.* **1996**, *34*, 125.

Chapter 11

Determination of Structural Ensembles from NMR Data: Conformational Sampling and Probability Assessment

Nikolai B. Ulyanov, Anwer Mujeeb, Alessandro Donati, Patrick Furrer, He Liu, Shauna Farr-Jones, David E. Konerding, Uli Schmitz, and Thomas L. James

Department of Pharmaceutical Chemistry, University of California, San Francisco, CA 94143–0446

A method called PARSE (Probability Assessment via Relaxation rates of a Structural Ensemble) is described for determination of ensembles of structures from NMR data. The problem is approached in two separate steps: (1) generation of a pool of potential conformers, and (2) determination of the conformers' probabilities which best account for the experimental data. The probabilities are calculated by a global constrained optimization of a quadratic objective function measuring the agreement between observed NMR parameters and those calculated for the ensemble. The performance of the method is tested on synthetic data sets simulated for various structural ensembles of the complementary dinucleotide d(CA)·d(TG).

A continuous improvement of high-resolution NMR methodologies and modeling tools during the past decade has made it possible to determine accurate average structures of biomolecules in solution on a routine basis (1-2). At the same time, a strong interest has re-emerged in NMR-based descriptions of flexible molecules in solution, as opposed to the conventional determination of static average structures. It is well-known that many biomolecules in general, and nucleic acids in particular, are flexible (*3-5*). It is hardly news that the flexible nature of a molecule and the consequent time-averaged nature of the NMR-derived structural information pose great challenges for structure determination. Although average NMR structures can be obtained with high precision when a large abundance of nuclear Overhauser effect (NOE) and scalar coupling data are available, structural artifacts can arise due to dynamic averaging; in worst case scenarios the resulting structures may not even be physically meaningful if widely divergent conformations contributed to the observed NMR signal (*6*). It has been shown for mononucleotides (*7*), and subsequently for oligonucleotides that proton-proton scalar coupling constants are not consistent with any single conformation of sugar rings. They can be explained, however, by an equilibrium of northern (N) and southern (S) types of sugar puckers. At present, scalar coupling data are routinely used to assess the dynamic equilibrium of sugars in oligonucleotides (*8-13*). However, this kind of "low-resolution" dynamic description is not easily incorporated into high-resolution structures determined by modern NMR methods. Such an incorporation would require generation of structural ensembles, *i.e.,* sets of structures with explicit or implicit probabilities. In contrast to conventional NMR structure determination, which also typically produces a set of closely related structures individually fitted to all the data, structural ensembles that we are concerned with, may include quite different conformations. Thus, no single member of a structural ensemble should necessarily explain all experimental NMR data, but, the ensemble should.

In recent years, efforts of different research groups yielded a number of significant developments of computational methodologies to determine conformational ensembles from NMR data. Torda *et al.* (*14*) introduced a molecular dynamics (MD)-based method of refinement that requires the experimental restraints to be satisfied only on a time-average basis (henceforth called MDtar), as opposed to the conventional restrained MD (rMD) which enforces all restraints at every time step. MDtar proved to be much more efficient in exploring conformational space of nucleic acids and in generating conformational ensembles that satisfy experimental NMR data better than any single structure (*15-16*, *41*). However, one methodological limitation of the MDtar approach lies in the fact that a relatively small time interval must be used for the averaging, significantly limiting the range of dynamic excursions of the molecule.

Quite recently, Kemmink and Scheek (*17*) and, independently, Bonvin and Brünger (*18*) proposed a method of simultaneous rMD refinement of several copies of the same molecule, assuming that all individual copies contribute equally to the observed NMR parameters. It appears that this method is very useful for situations when just a few conformers with roughly equal populations contribute to the observed NMR parameters. It has been further concluded (*19*) that experimental data typically available are not sufficient to determine reliably the populations of conformers obtained by this method. It is not clear, however, to what extent this conclusion was predetermined by the implicit assumption of equal probabilities of all copies during the refinement. Fennen *et al.* (*20*) proposed a variant of the multiple copy refinement where the contribution of each copy was weighted by a Boltzmann factor calculated using the force field-based conformational energy. While this method certainly allows for uneven distribution of probabilities, the Boltzmann factor weighting might be biased by the force fields used.

Ernst and coworkers (*21*) developed an algorithm called MEDUSA which performs repeated refinements against various internally consistent subsets of experimental restraints, assuming that inconsistencies between groups of restraints is a direct reflection of conformational mobility of the molecule in solution. After generating in this manner a pool of potential conformers, various pairs of conformers are considered, and relative populations of the two conformers in each pair are determined (*22*). The algorithm PARSE developed in our laboratory (*23*) uses a similar approach to produce a pool of potential conformers; at the second stage, the distribution of probabilities is calculated which gives the best agreement with experimental data. In contrast to MEDUSA, PARSE is capable of finding the optimal probability distribution for an arbitrary number of potential conformers with the use of a global constrained optimization method, the quadratic programming algorithm. In a parallel development, Pearlman (*24*) used a genetic algorithm approach to find the optimal probability distribution of potential conformers.

It is noteworthy that in the last class of methods, two distinct problems are separated: (a) generation of potential conformations, and (b) selecting the conformers and their populations which best describe the experimental data. Such a separation has a potential advantage of making it possible to combine the best sampling method (which could depend on a particular system) with the best method to determine the optimal probability distribution. Currently we are working on assessing the PARSE performance on a number of experimental and synthetic systems in combination with different sampling methods. In this chapter we will briefly describe the PARSE algorithm together with other structure determination tools used in this laboratory and also present preliminary results assessing the performance of this method on a series of synthetic NMR data for the complementary dinucleotide d(CA)·d(TG).

Tools for NMR Structure Determination

Determination of solution structure involves three stages: extraction of structural restraints from NMR data; refinement *per se*, *i.e.*, search for a structure satisfying the structural restraints; and comparison of the refined structure with original NMR data (*2*). Restraints typically used for structure determination are interproton distances derived from NOE intensities and torsion angles derived from scalar coupling constants. Although there are presently available a number of programs capable of direct refinement against NOE intensities, such refinement is very time-consuming and, therefore, necessarily limits the extent of conformational sampling. Torsion angle restraints derived from scalar coupling data have been shown to be useful in DNA structure determination (*25-26*); for brevity, we will not consider them in this chapter. In addition, in the case of structural ensemble determination, it is worthwhile to leave scalar coupling constants out of the refinement process for cross-validation purposes (*15,23*).

Calculating NOE Intensities from Structure(s). A matrix of NOE intensities $\mathbf{A}(t_m)$ at mixing time t_m is connected with a matrix of dipole-dipole relaxation rates $\mathbf{R} = \{R_{ij}\}$ by a simple relationship (*27*):

$$\mathbf{A}(t_m) = \exp\{ -\mathbf{R}t_m \} \tag{1}$$

where the off-diagonal relaxation rates R_{ij} are inversely proportional to the sixth power of the interproton distance

$$R_{ij} \sim 1 / r_{ij}^{6} \tag{2}$$

and the proportionality coefficients depend on the motional model. When several conformers in fast exchange with each other contribute to the observed NOE intensities, equation 1 will still be valid, but the effective relaxation rates R_{ij} will be linear averages of rates for individual conformers. For example, for two conformers α and β,

$$R_{ij} = p^{\alpha} R^{\alpha}_{ij} + p^{\beta} R^{\beta}_{ij} \tag{3}$$

where p^{α} and p^{β} are the conformers' probabilities.

Program CORMA (Complete Relaxation Matrix Analysis) calculates relaxation rates and NOE intensities for a given structure (*28-29*) or an ensemble of structures with specified populations (*30*). Typically for short DNA duplexes, a motional model of overall isotropic tumbling is used for all protons except methyl groups which are assumed to undergo fast internal rotation (*31*). If experimental NOE intensities are specified, CORMA also calculates a number of indices of agreement between experimental and simulated data (*2*).

Calculating Interproton Distances from Experimental NOE Intensities. It follows from equation 1 that the relaxation rate matrix **R** could be determined from NOE intensities:

$$\mathbf{R}(t_m) = -\ln\{ \mathbf{A}(t_m) \} / t_m \tag{4}$$

and equation 2 will lead to interproton distances r_{ij}. However, the matrix of experimental NOE intensities **A** is never complete due to finite spectral resolution and signal-to-noise ratio. The iterative algorithm MARDIGRAS solves this problem by combining experimental NOE intensities with intensities calculated for a model structure (*32-33*). Recently, this algorithm was supplemented with an error analysis option (the RANDMARDI option) which determines lower and upper bounds for interproton distances based on analysis of expected errors in experimental intensities (*34*). When multiple conformers contribute to experimental NOE intensities, the dipole-dipole relaxation rates calculated by MARDIGRAS are linear averages of individual conformers' rates (equation 3), and the calculated interproton distances are correspondingly the sixth-root averages.

Structure Refinement. The refinement of a structure involves optimization of a weighted sum of a conformational energy and a penalty function enforcing satisfaction of structural restraints obtained from experimental data. The former is defined by a force field, and the latter has typically a form of a flat-well potential with no penalty when a restrained structural parameter is within the allowed bounds and with a quadratically increasing penalty when the parameter is outside lower or upper bounds. There are many choices for the optimization method; the one most commonly used in this laboratory is an rMD simulated annealing protocol using the AMBER suite of programs (*35*). However, in this chapter we will describe results obtained via restrained Metropolis Monte Carlo (rMC) refinement with the DNAminiCarlo program (*36-38*). This program samples DNA conformational space more effectively and thoroughly by using helical parameters as independent variables. It has been shown that these two approaches, rMD and rMC, result in essentially the same refined structures when the same structural restraints were used (*25,38*).

Probability Assessment of Potential Conformers

Theory. We will assume now that a pool of n potential conformers has been generated, and our goal is to determine the probabilities p^{α}, $\alpha = 1, \ldots, n$, which best describe the observed experimental parameters E_k, $k = 1, \ldots, m$. We will use only those experimental NMR parameters which can be linearly averaged for the ensemble, according to equation 3. In the examples below, we will use for that purpose dipole-dipole relaxation rates calculated from NOE intensities via MARDIGRAS. Theoretical relaxation rates R^{α}_{k} corresponding to the observed parameters E_k can be calculated for each potential conformer α using CORMA (*vide supra*). Then the theoretical relaxation rates for the ensemble defined by the probability distribution $\{p^{\alpha}\}$ are given by

$$T_k = \Sigma p^a R^{\alpha}_{k} \qquad (5)$$

and we can define a quadratic objective function

$$Q^r(\{p^{\alpha}\}) = \Sigma w_k (T_k - E_k)^2 \qquad (6)$$

Here, w_k are weights of individual observables, which are all assumed equal for present purposes. Q^r should be minimized in order to find the optimal probability distribution, subject to constraints

$$\Sigma p^a = 1; p^a \geq 0$$

This problem can be solved using a quadratic programming algorithm, which finds the *global* minimum of a quadratic function subject to a set of linear constraints (*39*); practically it is done with the program PDQPRO (Probability Distribution by Quadratic PROgramming) (*23*).

Model Molecule. We will illustrate the application of this method on a simple model system, the complementary dinucleotide d(CA)·d(TG), because it can be modeled flexible with easy generation of various conformations using DNAminiCarlo. The total number of "observed" NOE intensities, 79, or about 20 per residue, was chosen quite generously; nevertheless, a similar number of restraints can be obtained in practice (*40*), especially for alternating purine-pyrimidine sequences. All intensities correspond to those which can be typically observed for a short DNA duplex; it was assumed that all protons were assigned, except H5', H5", and amino groups.

Model Ensembles. We will consider eleven synthetic structural ensembles which we will attempt to reconstitute using PDQPRO. Each ensemble is a two-state equilibrium of d(CA)·d(TG), composed of an energy-minimized B conformer and an energy-minimized A conformer. The probability of B-form in each two-state ensemble varied from 0 to 1 with a step of 0.1. All NOE intensities for each ensemble were simulated with CORMA, and the 79 intensities (*vide supra*) were selected for further analysis as "observed" NMR data. The "observed" intensities were used as input for the MARDIGRAS program, which calculated the corresponding "experimental" rates (*i.e.*, E_k in equation 6) and lower and upper distance bounds. Typical output distances are illustrated in Figure 1 for an "experimental" ensemble consisting of pure B-form; A-form was used as initial structure for MARDIGRAS calculations. Note that despite significant differences in B-form (target) and A-form (initial structure) distances (Figure 1a), the error analysis option of MARDIGRAS (*34*) produced very accurate distance bounds (Figure 1b). For longer distances the bounds become wider, but importantly, they cover the target distance. In the case of mixed ensembles, the relaxation rates retrieved by MARDIGRAS corresponded very closely to the values expected from equation 3, and the distance bounds covered the corresponding sixth-root average values. In fact, this is a very important result showing that MARDIGRAS is capable of calculating accurate distance bounds from NOE data corresponding not only to a single structure, but from data representing a structural ensemble as well. This constitutes a basis for a future ensemble determination. The calculated rates were subsequently used for PDQPRO calculations, and the distance bounds were used for the structure refinements. Some of the distances calculated for the one-to-one mixture of B- and A-forms are shown in Figure 2. Note that with the sixth-root averaging, the distances derived from the mixture in each case are skewed toward the shorter of the distances in either A- or B-forms and are not arithmetic averages.

Conformational Pool. Potential conformers for the DNA dinucleotide were generated by unrestrained Monte Carlo simulation with DNAminiCarlo at elevated temperature (400 K) in order to exaggerate its flexibility. 400 structures were stored during the simulation (one every 1,000 iterations), which together with B- and A-forms (the two "true" conformers) constituted a pool of 402 potential conformations. The evolution of sugar pseudorotation phase angles during the simulation together with their distributions are shown in Figure 3. The "theoretical" dipole-dipole relaxation rates, $R^{\alpha}{}_{k}$ in equation 5, were calculated for each of them using CORMA.

PDQPRO Results. The PDQPRO program was run separately for the "experimental" rates of each of the eleven two-state B-A equilibria, using a pool of 402 potential conformers. The goal was to pick two correct structures out of 402, and to predict their probabilities. The PDQPRO was successful in all eleven cases, selecting the correct structures (B- and A-forms) and calculating the correct probabilities within the precision of the computer digital representation!

The success of probability assessment in the above example was determined by three main factors: the ability of MARDIGRAS to retrieve correct relaxation rates according to equation 3 from the NOE data; the ability of quadratic programming algorithm to find the global minimum of Q^r (equation 6); and the fact that the correct conformers were present in the conformational pool. However, the latter factor was a rather unrealistic assumption. More typically, one cannot be sure if the "correct" structures have been generated during conformational sampling. To test such a situation, we ran PDQPRO the same way as before, but using a pool of 400 Monte Carlo-generated conformations which lacked the target B- and A-forms. One could

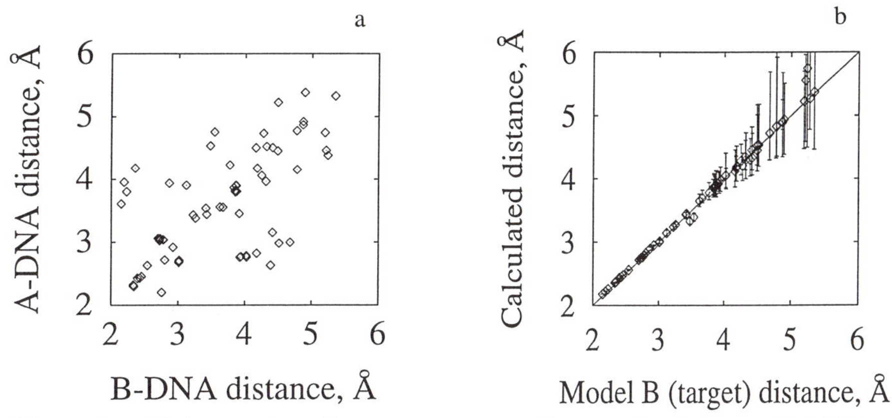

Figure 1. 79 interproton distances corresponding to the "observed" NOE intensities. (a) A-form distances versus corresponding B-form distances. (b) Distances calculated with MARDIGRAS from the B DNA NOE intensities using A DNA as initial structure. Single MARDIGRAS results are shown as diamonds, and results of the error analysis option of MARDIGRAS are shown as error bars.

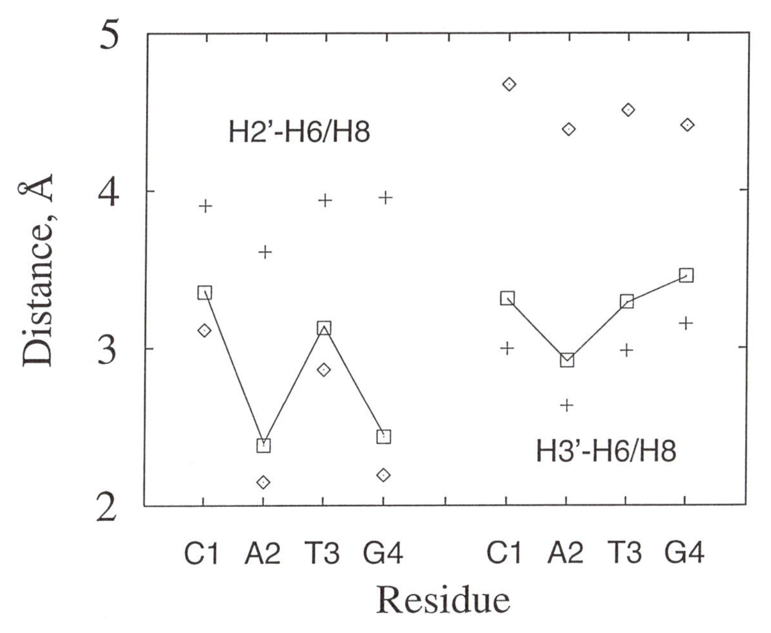

Figure 2. Sequence dependence of selected distances in d(CA)·d(TG). Diamonds: B-form; crosses: A-form; solid line: sixth-root average for one-to-one mixture of B- and A-forms; squares: distances calculated with a single MARDIGRAS run from NOE intensities of the one-to-one mixture of B- and A-forms; error bars: corresponding distances calculated with the error analysis option of MARDIGRAS.

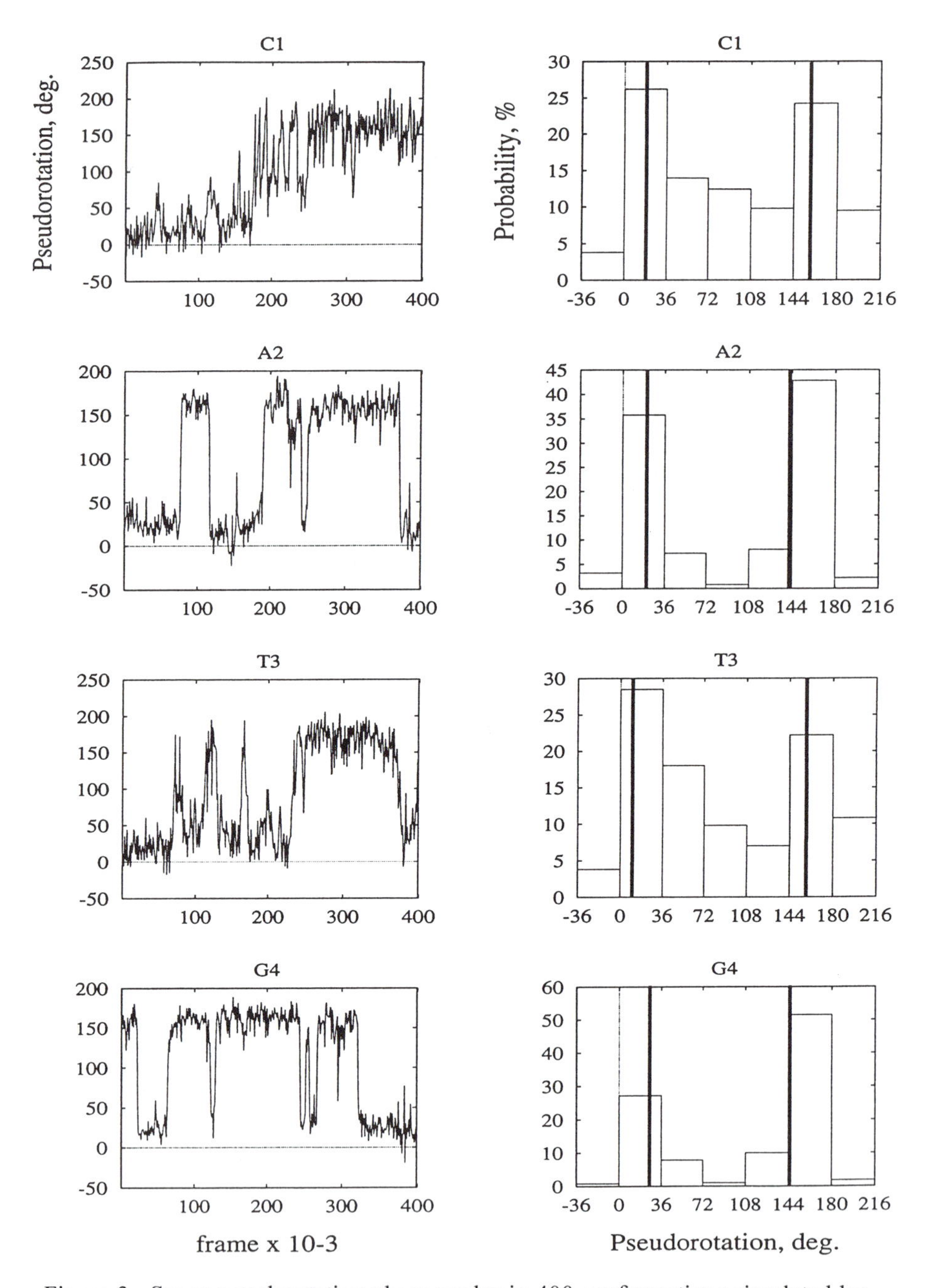

Figure 3. Sugar pseudorotation phase angles in 400 conformations simulated by Monte Carlo. Left panels: evolution of pseudorotation during simulation; right panels: distribution of pseudorotation. Pseudorotation angles of B- and A-forms are shown in thick lines for comparison.

expect that PDQPRO would select a few structures close enough to the B and A conformations. However, the actual ensembles calculated by PDQPRO were rather broad. A typical result is shown in Figure 4 for the target 1:1 mixture of B- and A-forms; pseudorotation phase angles of the target structures are shown with thick lines for comparison. The PDQPRO performance can be considered satisfactory in this case, because it selected 21 out of 400 potential conformations, with the sugar puckers in selected structures clustered around target values with roughly correct populations. Still, this performance was far from perfect, because the sugar conformation distributions were quite wide, and the T3 residue had a significant population of O4'-*endo* pucker.

Conformational Sampling

The above experience emphasizes once again the importance of adequate conformational sampling. Can we generate pools of potential conformations which include the correct structures or, at least, close enough to them? While we are still evaluating different sampling methods, in this section we will present some preliminary results of conformer ensemble refinement using the idea of MEDUSA and PARSE methods (*21-23*). Namely, we will try to partition the set of described above 79 distance restraints into self-consistent subsets, and then carry out separate refinements against each subset. We will illustrate this approach on simulated data for the 1:1 mixture of B- and A-forms of d(CA)·d(TG). However, before we proceed with this, we need to make sure that the distance restraints used indeed can define a structure of a DNA dinucleotide. For that purpose, we will start with refinement of a single structure corresponding to the data simulated for a pure B-form (Figure 1).

The distance restraints with lower and upper bounds (Figure 1) were generated with the error analysis option of MARDIGRAS, as described above, from the set of 79 NOE intensities simulated by CORMA for pure B-form d(CA)·d(TG). The average flat-well width for this restraint set is 0.4 Å, but the maximum is 1.7 Å (Figure 1). The refinement was performed using an rMC simulated annealing protocol with DNAminiCarlo (*38*) starting with A-form DNA. The structure easily converted into the B conformation with atomic root mean square deviation (RMSD) of 0.2 Å from the target B-form (not shown), and residual average distance restraints deviation of 0.02 Å. This shows that the distance restraints used define the structure with very high accuracy.

Now we can proceed with analysis of the data for the one-to-one mixture of B- and A-forms. As noted before, in this case the distances calculated by MARDIGRAS are the sixth-root average of distances in B- and A-forms (see, *e.g.*, Figure 2). It is well known that certain distances, in particular, intraresidue H2'-H6 and H2'-H8 are short in B DNA and they are long in A DNA; while other distances, e.g., intraresidue H3'-H6 and H3'-H8 are short in A DNA and they are long in B DNA. Both types of distances will be short after the sixth-root averaging (Figure 2). This could be used as a basis for the partitioning of the distance restraints into self-consistent subsets (*23*). However, if there existed a stereochemically reasonable structure with both types of distances being short, then we would never be able to distinguish two situations: equilibrium of two structures, or a single "average" structure based on the interproton distances alone. The most straightforward way to answer this question is to attempt

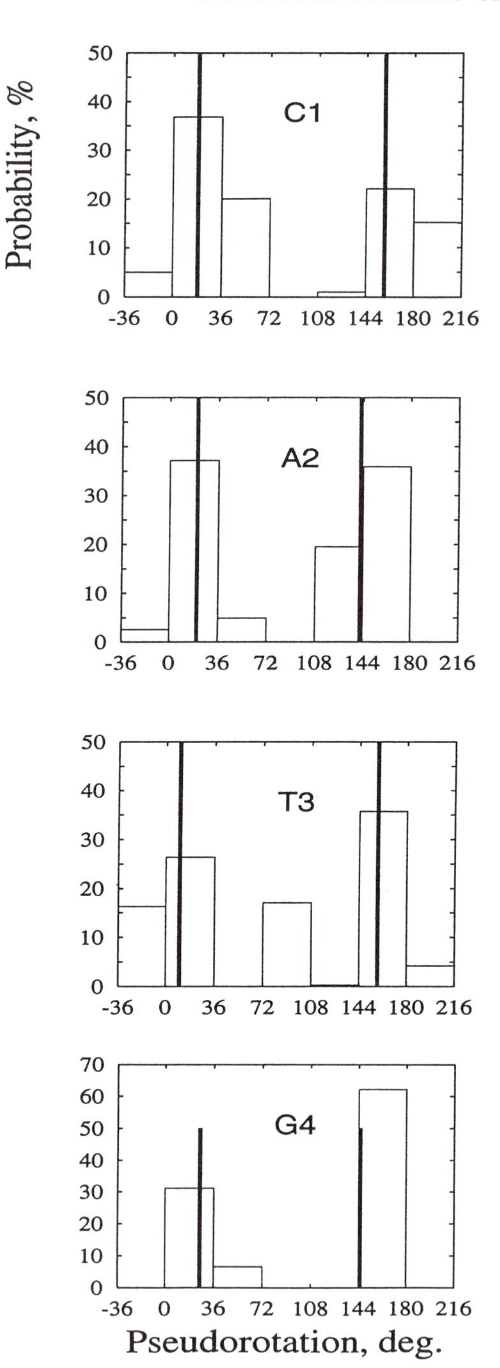

Figure 4. Distribution of pseudorotation phase angles in 21 conformers selected by PDQPRO. Pseudorotation angles of B- and A-forms are shown in thick lines for comparison.

to refine the structure using all existing restraints. The results of such refinement, using the same simulated annealing protocol as above, is shown in Figure 5. Even though the residual average distance restraints deviation was quite low, 0.1 Å, some of individual restraints were violated by more than 0.5 Å (Figure 5). But even more importantly, the resulting sugar puckers were in energetically unfavorable C4'-*exo*/O4'-*endo* conformation (see entry "average" in Table I).

Table I. Results of Conformational Sampling

Conformation	*Sugar Pseudorotation, deg.*				R^{x} [a]
	C1	A2	T3	G4	
1. target B	159	142	158	145	4.84
2. target A	18	19	10	23	5.45
3. "average"	79	49	85	67	4.18
4. refined B	150	60	155	59	3.86
5. refined A	34	37	33	41	4.98
6. ensemble[b]	-	-	-	-	3.03

[a]NOE-based sixth-root weighted R^{x}-factor defined in (*2*).
[b]72% of "refined B" and 28% of "refined A" conformations.

To proceed with the refinement of A-type and B-type conformers, we constructed two subsets of distance restraints. The A-type excluded all distances which are normally short in B conformations: intraresidue H2'-H6/H8 and H2"-H6/H8, and sequential C1H2"-A2H8 and T3H2"-G4H8; altogether 10 distances were removed from the original set of 79. Similarly, to produce the B-type subset of distances, we removed 10 distances which are short in A DNA and long in B DNA: intraresidue H3'-H6/H8 and H4'-H2", and sequential C1H3'-A2H8 and T3H3'-G4H8. The rMC refinement against each of the subsets produced two structures, called "refined B" and "refined A" in Table I. It is seen that while the "refined A" form is relatively close to the target A conformer, the "refined B" resembles more heteronomous than normal B DNA.

In hindsight, this result could be expected. Indeed, despite a strong bias toward short distances, the sixth-root averaged distances used in the restraint subsets are still different from the distances in the target structures (Figure 2). Therefore, the amplitude of variation of sugar puckers in the refined conformers must be smaller than in the target structures, in agreement with the results in Table I. In the original PARSE application for real data for a DNA octamer (*23*), most residues were predominantly in the S-conformation. Therefore, refining B-type conformers using B-type subsets of restraints was relatively straightforward. However, for the minor N-conformers, the difference between the sixth-root averaged distances used for refinement and the real (unknown) distances in the N-conformers was probably even bigger than in our model 50-50% distribution. As a result, the A-type conformers refined using A-type subsets of restraints typically had pseudorotation phase angles in the range of 40-60° (*23*). In order to achieve good agreement with experimental data, we had to supplement the conformations refined against self-consistent subsets of distance restraints with a large number of structures generated by unrestrained calculations, when constructing the conformational pool (*23*). This circumstance should be considered in the future when constructing subsets of restraints for the conformers' refinement.

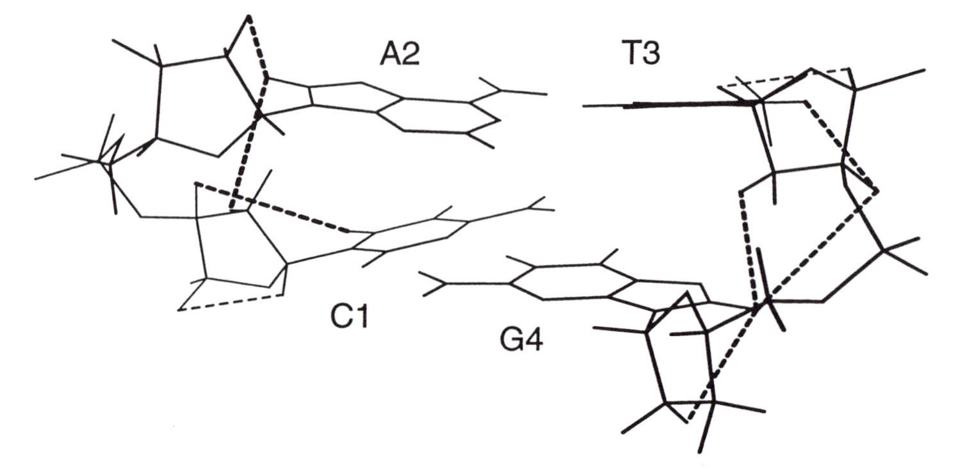

Figure 5. "Average" structure calculated using all distance restraints for a one-to-one mixture of B- and A-forms. Restraints violated by more than 0.5 Å are shown in dashed lines. The image was prepared with the NOESHOW delegate of the MIDASPlus program (*42*).

The PDQPRO program run for the pool of two conformations, "refined B" and "refined A", calculated a 72-28% equilibrium. Even though both conformers and their populations are significantly different from the target 50-50% ensemble of B- and A-forms, the NOE-based R^x-factor is lower for this equilibrium than for any single structure (Table I).

In conclusion, we have shown that correct conformers and their probabilities can be retrieved with the PDQPRO program once they are present in the pool of potential conformations. However, the problem of adequate conformational sampling still remains not completely solved.

Acknowledgments

This work was supported by NIH grants GM 39247 and RR 01081.

References

1. *Methods in Enzymology;* James T. L., Ed.; NMR of Nucleic Acids; Academic Press: New York, NY, 1995, Vol. 261.
2. James, T. L. *Curr. Opin. Struct. Biol.* **1991**, *1*, 1042-1053.
3. Srinivasan, A. R.; Torres, R.; Clark, W.; Olson, W. K. *J. Biomol. Struct. Dyn.* **1987**, *5*, 459-496.
4. Zhurkin, V. B. *Mol. Biol.* **1983**, *17*, 622-638.
5. Hagerman, P. J. *Annu. Rev. Biophys. Biophys. Chem.* **1988**, *17*, 265-286.
6. Jardetsky, O. *Biochim. Biophys. Acta* **1980**, *621*, 227-232.
7. Davies, D. B.; Danyluk, S. S. *Biochemistry* **1975**, *14*, 543-554.
8. Rinkel, L. J.; Altona, C. *J. Biomol. Struct. Dyn.* **1987**, *4*, 621-649.
9. Celda, B.; Widmer, H.; Leupin, W.; Chazin, W. J.; Denny, W. A.; Wüthrich, K. *Biochemistry* **1989**, *28*, 1462-1471.
10. Schmitz, U.; Zon, G.; James, T. L. *Biochemistry* **1990**, *29*, 2357-2368.
11. Mujeeb, A.; Kerwin, S. M.; Egan, W.; Kenyon, G. L.; James, T. L. *Biochemistry* **1992**, *31*, 9325-9338.
12. González, C.; Stec, W.; Kobylyanska, A.; Hogrefe, R. I.; Reynolds, M.; James, T. L. *Biochemistry* **1994**, *33*, 11062-11072.
13. Conte, M. R.; Bauer, C. J.; Lane, A. N. *J. Biomol. NMR* **1996**, *7*, 190-206.
14. Torda, A. E.; Scheek, R. M.; van Gunsteren, W. F. *J. Mol. Biol.* **1990,** *214,* 223-235.
15. Schmitz, U.; Ulyanov, N. B.; Kumar, A., James, T. L. *J. Mol. Biol.* **1993,** *234,* 373-389.
16. González, C.; Stec, W.; Reynolds, M.; James, T. L. *Biochemistry* **1995**, *34*, 4969-4982.
17. Kemmink, J.; Scheek, R. M. *J. Biomol. NMR* **1995**, *6*, 33-40.
18. Bonvin, A. M.; Brünger, A. T. *J. Mol. Biol.* **1995,** *250*, 80-93.
19. Bonvin, A. M.; Brünger, A.T. *J. Biomol. NMR* **1996**, *7*, 72-76.
20. Fennen, J.; Torda, A. E.; van Gunsteren, W. F. *J. Biomol. NMR* **1995**, *6*, 163-170.
21. Brüschweiler, R.; Blackledge, M.; Ernst, R. R. *J. Biomol. NMR* **1991**, *1*, 3-11.
22. Blackledge, M. J.; Brüschweiler, R.; Griesinger, C.; Schmidt, J. M.; Xu, P.; Ernst, R. R. *Biochemistry* **1993**, *32*, 10960-10974.
23. Ulyanov, N. B.; Schmitz, U.; Kumar, A.; James, T. L. *Biophys. J.* **1995**, *68*, 13-24.
24. Pearlman, D. A. *J. Biomol. NMR* **1996**, *8*, 49-66.

25. Schmitz, U.; Sethson, I.; Egan, W. M.; James, T. L. *J. Mol. Biol.* **1992**, *227*, 510-531.
26. Mujeeb, A.; Kerwin, S. M.; Egan, W. M.; Kenyon, G. L.; James, T. L. *Biochemistry* **1993**, *32*, 13419-13431.
27. Macura, S.; Ernst, R. R. *Mol. Phys.* **1980**, *41*, 95-117.
28. Keepers, J. W.; James, T. L. *J. Magn. Reson.* **1984**, *57*, 404-426.
29. Borgias, B. A.; James, T. L. *J. Magn. Reson.* **1988**, *79*, 493-512.
30. Schmitz, U.; Kumar, A.; James, T. L. *J. Amer. Chem. Soc.* **1992**, *114*, 10654-10656.
31. Liu, H.; Thomas, P. D.; James, T. L. *J. Magn. Reson.* **1992**, *98*, 163-175.
32. Borgias, B. A.; James, T. L. *Meth. Enzymol.* **1989**, *176*, 169-183.
33. Borgias, B. A.; James, T. L. *J. Magn. Reson.* **1990**, *87*, 475-487.
34. Liu, H.; Spielmann, H. P.; Ulyanov, N. B.; Wemmer, D. E.; James, T. L. *J. Biomol. NMR* **1995**, *6*, 390-402.
35. Pearlman, D. A.; Case, D. A.; Caldwell, J. C.; Seibel, G. L.; Singh, U. C.; Weiner, P.; Kollman, P. A. *AMBER, version 4.0*; University of California: San Francisco, CA, 1990.
36. Ulyanov, N. B.; Gorin, A. A.; Zhurkin, V. B. In *Proceedings Supercomputing'89: Supercomputer Applications*; Kartashev, L. P.; Kartashev, S. I., Eds.; Int. Supercomputing Inst., Inc.: St. Petersburg, FL, 1989, pp 368-370.
37. Zhurkin, V. B.; Ulyanov, N. B.; Gorin, A. A.; Jernigan, R. L. *Proc. Natl. Acad. Sci. U.S.A.* **1991**, *88*, 7046-7050.
38. Ulyanov, N. B.; Schmitz, U.; James, T. L. *J. Biomol. NMR* **1993**, *3*, 547-568.
39. Fletcher, R. *Practical Methods of Optimization*. Vol. 2, Constrained Optimization; Wiley & Sons: New York, NY, 1981; pp. 1-220.
40. Weisz, K.; Shafer, R. H.; Egan, W. M.; James, T. L. *Biochemistry* **1994**, *33*, 354-366.
41. Yao, L.; Kealey, J. T.; Santi, D. V., James, T. L.; Schmitz, U. *J. Biomol. NMR* **1997**, *9*, 229-244.
42. Ferrin, T. E.; Huang, C. C.; Jarvis, L. E.; Langridge, R. *J. Mol. Graphics* **1988**, *6*, 13-27.

Chapter 12

NMR Studies of the Binding of an SPXX-Containing Peptide from High-Molecular-Weight Basic Nuclear Proteins to an A-T Rich DNA Hairpin

Ning Zhou and Hans J. Vogel

Department of Biological Sciences, Universityof Calgary, Calgary, Alberta T2N 1N4, Canada

The amino acid sequence SPXX in nuclear proteins has been identified as a motif that recognizes the minor groove of A-T rich DNA. Phosphorylation of the serine residue by proline-directed kinases diminishes its DNA binding capacity. To learn more about this abundant protein-DNA interaction motif, we have studied the synthetic 14-mer peptide MRSRSPSRSKSPMR (derived from sperm chromatin of winter flounder), and its binding to a d(T6C4A6) DNA hairpin by one dimensional NMR titration experiments and two dimensional NOESYand TOCSY NMR experiments. Our results show that the first equivalent of the peptide binds to the hairpin loop, while the second equivalent interacts with the minor groove of the hairpin stem with an extended structure. These observations suggest that binding of SPXX motifs to DNA loops or single-sranded regions may be significant.

In studies of protein-DNA interactions, it has long been recognized that contact of the protein with the DNA bases is important in providing sequence discrimination. Bases located in the major groove show a greater degree of sequence dependence in terms of their hydrogen-bonding and van der Waals interaction patterns than those in the minor groove. Therefore it is not surprising that most sequence specific DNA-binding proteins characterized thus far make contact with their target DNA in the major groove[1]. Minor groove binding proteins identified earlier, such as the nucleolins[2] and the HU family proteins[3], appear to lack a high degree of sequence specificity in their binding to DNA and show only some preference for A.T rich or G.C rich sites. The protein motifs used for binding to the minor groove are still not well characterized. Putative α-helical motifs and proline-rich motifs have been proposed (for a review, see 4).

Several recent developments have, however, expanded and changed our notions about minor groove binding. For example, the TATA-box sequences in promoter regions are recognized by the TATA-box binding protein (TBP); its binding is the first step in transcription initiation. The structure of TBP[5,6] and its complexes with the 8-basepair TATA-box sequences[7,8] were solved by X-ray crystallography. TBP utilizes an antiparallel β-sheet structural element to contact the minor groove of the TATA sequence, distorting the B-DNA structure considerably. Specific hydrogen bonding, as well as hydrophilic and hydrophobic interactions between the TBP protein and bases in the minor groove were identified. TBP represents the first example of an antiparallel β-sheet motif, which binds to the minor groove to confer sequence specificity. More recently the structures of the DNA-binding HMG (high-mobility-group) domain of mouse LEF-1 and human SRY, complexed with their respective cognate DNA, were determined by NMR spectroscopy[9,10]. The HMG domain of these two proteins utilizes a concave surface formed mainly by three α-helices to bind in the DNA minor groove; a large number of specific interactions between hydrophobic amino acid sidechains on the surface and the DNA bases were identified. The binding also causes large scale kinking of DNA. Along a separate line of research, Dervan, Wemmer and their co-workers have succeeded in designing minor groove binding peptides that could not only distinguish G.C or C.G from A.T basepairs but also were able to discriminate a G.C basepair from a C.G basepair[11,12]. These peptides contain imidazole and pyrrole rings. Hydrogen bonding between the imidazole nitrogens of the peptides and guanine amino groups in the DNA minor groove were shown to be critical in determining this specificity. This work has set the stage for the design of other minor groove binding molecules to target specific base sequences. Another interesting group of minor groove binding ligands is comprised of carbohydrate-based compounds[13]. These compounds display a different kind of specificity for oligopyrimidine sequences. The progress in these unrelated projects pointed to the diverse ways in which DNA-protein binding specificity could be achieved. It also highlighted the importance of sequence dependent DNA structural features, such as groove width and backbone flexibility, in conferring binding specificity, especially for proteins recognizing the minor groove of B-DNA.

The amino acid sequence SPXX has been proposed by Suzuki and others[14-16] as a minor groove binding motif that recognizes A.T rich DNA segments. In this sequence X often represents a basic amino acid K or R. The SPXX sequence was identified primarily through footprinting experiments and competition experiments with the minor groove binding drug Hoechst 33258. The high frequency with which this sequence is found in chromosomal proteins and other regulatory DNA binding proteins seems to support a function in DNA binding. It was proposed on the basis of NMR experiments that two joint SPXX motifs formed a crescent-shaped structure that binds into the narrow minor groove of A.T rich DNA in a way similar to the drug netropsin[17]. Phosphorylation of the serine residue in SPXX was shown to diminish its DNA binding[18], therefore serine phosphorylation by proline-directed protein kinases in this motif can serve as a modulation of DNA binding. A study on an octamer $(SPRK)_2$, however, found that it did not bind specifically to A.T rich DNA[19]. In order to further investigate the interaction of the proposed SPXX motif with DNA, we have studied the interaction of a proline-rich, serine-rich 14-mer peptide (MRSRSPSRSKSPMR), with an A-T rich DNA hairpin. This peptide corresponds to a repeated sequence found in the high molecular weight basic nuclear proteins that are present in the sperm chromatin of winter flounder[20]; it contains two SPXX sequences. The oligo-DNA used in this study is d(T6C4A6). A DNA hairpin was chosen, because of its greater thermodynamic stability for the relatively short basepaired oligonucleotide. The interaction between the peptide and the DNA was studied by ^1H,

^{31}P and ^{13}C NMR spectroscopy. Our work led to the surprising conclusion that the peptide binds with a higher affinity to the hairpin loop than to the minor groove in the A.T stem.

Experimental.

The peptide was synthesized by the Peptide Synthesis Facility at Queens University. The DNA was provided by the Nucleic Acid Synthesis Facility at the University of Calgary. The peptide MRSRSPSRSKSPMR was dissolved in aqueous (90%/10% H_2O/D_2O) solution to a final concentration of 5 mM, the pH of the solution was adjusted to 6.5. The oligo-nucleotide d(T6C4A6) was dissolved in aqueous (90%/10% H_2O/D_2O) solution with 100mM phosphate buffer (pH 6.5) and 150mM KCl to a concentration of 2.5mM. Aliquots of the peptide stock solution were added to the d(T6C4A6) solution for the titration experiments.

NMR experiments were carried out at 15°C on a Bruker AMX-500 spectrometer equipped with a 5 mm inverse detection probe and an X-32 computer. All 1D spectra were recorded with a 5500 Hz spectral width and 8k data points. The water resonance was suppressed either by a pre-saturation irradiation or by using a tailored jump-return excitation pulse[21]. Phase-sensitive detection in the t1 dimension of 2D experiments was achieved using the time-proportional phase incremental scheme[22].

For natural abundance ^{1}H, ^{13}C correlation experiments, the 2D HMQC pulse sequence as described by Bax et al.[23] was used. A proton spectral width of 5000 Hz and a carbon-13 spectral width of 25000 Hz were used, with 2K X 512 data points in the t2 and t1 dimensions respectively. ^{1}H, ^{31}P correlation experiments were obtained using a hetero-TOCSY sequence[24], with an isotropic mixing time of 67 ms. The spectral width was set to 4000 Hz for protons and 607 Hz for phosphorus, with typically 2K X 256 data points in the two dimensions respectively. The proton 2D DQF-COSY and TOCSY experiments were recorded with standard pulse sequences; the data size was 2K X 512, with spectral width of 5500 Hz in both dimensions. The NOESY sequence with a jump-return excitation pulse was used for optimal imino proton detection[21]. Mixing times from 100 to 300 ms were used. The spectral widths were 11000 Hz in both dimensions.

2D data processing consisted of zero-filling once and apodization by a 90°-shifted sine-square-bell function in both dimensions before Fourier transformation. Proton and carbon-13 chemical shifts are referenced to TSP as an internal standard. Phosphorus-31 chemical shifts are referenced to 85% phosphoric acid as an external standard.

Results and Discussion.

Our first task was to characterize the NMR spectra of the oligo-nucleotide d(T6C4A6); this was accomplished by recording NOESY and TOCSY experiments. Below 30°C in aqueous solution (pH 6.5), it forms a hairpin structure with a stem consisting of T6.A6 and a loop formed by the four C residues. The NOEs observed between neighboring A H2 protons (Fig. 1A), between the T imino protons and the A H2 protons (Fig. 1B), and between neighboring T imino protons (Fig. 1C) indicate that Watson-Crick A.T basepairs were formed holding the T6.A6 stem together. However, in addition, cross-strand NOEs between the T H1' protons and the A H2 protons were detected (see Fig.1, Scheme). These cross-strand NOEs are characteristic of oligo-(dA).oligo-(dT) sequences; they are particularly strong when the A.T basepairs have a high propeller twist[25,26], and have been linked to DNA bending and bendability[25,27]. The NOE patterns observed for the d(T6C4A6) molecule resemble those reported for a longer homologous hairpin [d(T8C4A8)] which we have characterized in a previous study[27]. Further, the first C residue in the loop had an abnormal proton chemical shift profile

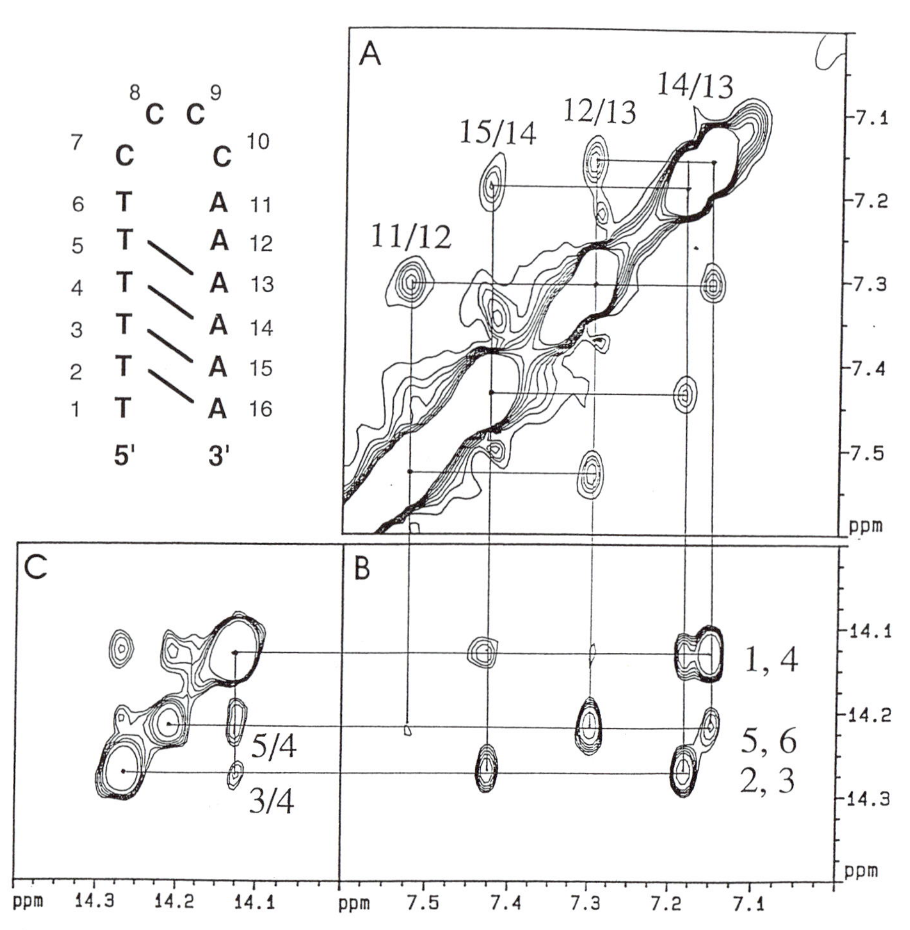

Figure 1. Contour plots of a NOESY experiment for the d(T6C4A6) hairpin in 90%/10% H_2O/D_2O solution. The d(T6C4A6) hairpin sequence and the numbering of the residues are shown in the scheme. The bars connecting the T and the A residues in the scheme represent observed cross-strand NOEs between the T H1'and the A H2 protons. (A) The region showing the sequential NOE crosspeaks between the H2 protons of neighboring A residues (labeled with the two residue numbers involved). (B) The region for the cross peaks between the A H2 and the T imino protons. The crosspeaks are labeled with the T residue numbers involved. (C) The region for the T imino protons. The crosspeaks between neighboring T imino protons are labeled with the two corresponding residue numbers.

during thermal melting (data not shown), which is again similar to that of the corresponding C residue in the longer hairpin. We have previously modeled the structure of the d(T8C4A8) hairpin using our 2D NMR data and restrained molecular dynamics simulation[27]. The stem of this hairpin is bent with a slightly narrower minor-groove compared to a standard B-DNA conformation, and the first C residue in the loop is not stacked but exposed to the solvent. The present measurements on the shorter d(T6C4A6) hairpin demonstrate that it has a similar structure as the longer d(T8C4A8) hairpin.

No regular secondary structure can be detected through 2D NMR experiments for the 14-mer peptide. The NOESY experiments showed few crosspeaks (data not shown) which is characteristic of a flexible peptide backbone. Upon addition of the peptide to the d(T6C4A6) sample, the peptide amide signals appear at shifted resonance positions and with increased linewidths, indicating that binding takes place (Fig. 2). Addition of aliquots of the peptide sample up to one equivalent to the DNA sample causes no significant changes of the imino proton signals; surprisingly only further additions of peptide causes a noticeable change in the chemical shifts of these T imino protons (Fig. 3). The hairpin stem T-imino protons are involved in the base-pairing of the stem; their NMR signals are sensitive to ligand binding in the DNA grooves[28].

To further characterize the changes that occurred in the 1:1 peptide-DNA sample, the chemical shifts of the DNA base aromatic protons, aromatic and deoxyribose carbons and phosphorus were monitored and compared with that of the hairpin. The ^{31}P chemical shift changes (Fig. 3A) for the 1:1 peptide:DNA sample are only observed for the loop region and for the two A residues closest to the loop. The changes in the chemical shifts of the aromatic ^{13}C give a similar profile (Fig. 3A), here only the aromatic C6 of the four loop C residues experienced chemical shift changes. In addition, the deoxyribose C1' of the two A residues close to the loop experienced small chemical shift changes as well, also the C1' resonances of the two A residues at the end of the hairpin changed; but the loop C C1' resonances did not (Fig. 3A). The chemical shift changes for the aromatic proton are plotted in Figure 3B. The loop C H6 as well as the A H8 protons in the 1:1 peptide-DNA sample undergo some changes; while the T H6 and the A H2 proton chemical shifts remained almost identical. These results indicate that up to a 1:1 ratio, the peptide interacts primarily with the loop region of the hairpin.

At peptide to DNA ratios higher than 1, both A H2 and T imino protons showed significant chemical shift changes, while changes in the loop region were minimal (Fig. 3B and 2C). Since all of the A H2 protons are situated in the minor groove of the stem, these changes suggest that at higher than 1:1 peptide to DNA ratios, binding of the peptide to the minor groove of the stem takes place.

Except for all the amide proton signals of the peptide, which show chemical shift changes in the peptide/DNA samples, most of the non-solvent-exchangeable proton signals do not display significant chemical shift changes in these samples. The few that had been identified include the N-terminal Met α-proton and several Arg δ–protons. Since it is likely that the peptide undergoes exchange among different bound and free forms, its proton chemical shift do not give a direct indication of the conformation of the bound peptide. However, in NOESY experiments recorded for the 2:1 peptide-hairpin sample, a significantly increased number of crosspeaks involving the peptide protons were detected (in comparison with a corresponding experiment for the sample of the isolated peptide in aqueous solution). This resulted from an increase of the correlation time of the peptide in the bound form(s). An analysis of the NOESY patterns indicated that the peptide backbone was extended. Therefore an extended peptide which interacts with the convex minor groove of the bent stem through some of its amide and arginine sidechain groups is consistent with these NMR results.

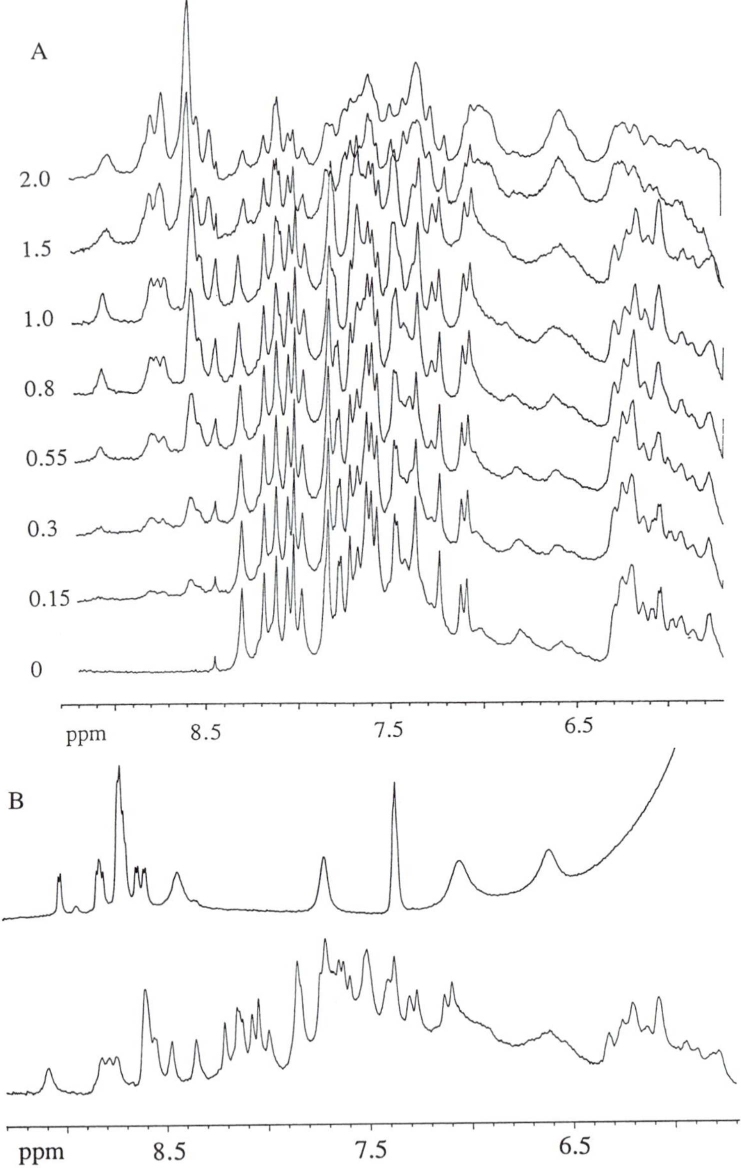

Figure 2. (A) The aromatic proton region of the NMR titration experiments of d(T6C4A6) with the peptide MRSRSPSRSKSPMR. The equivalent amount of the peptide present in the sample is indicated by the number beside each spectrum. (B) Upper trace, the amide and aromatic region of the peptide MRSRSPSRSKSPMR proton spectrum. Lower trace, the same region of the spectrum of the 1:1 peptide-DNA hairpin sample. (C) The imino proton region of the d(T6C4A6) with varying equivalent amounts of the 14-mer peptide present as indicated by the number beside each spectrum. The assignments of the imino protons are indicated by residue numbers.

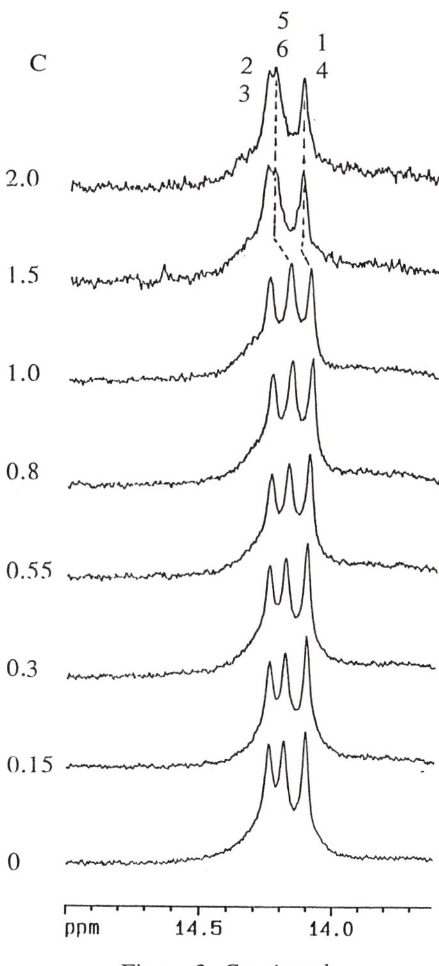

Figure 2. *Continued.*

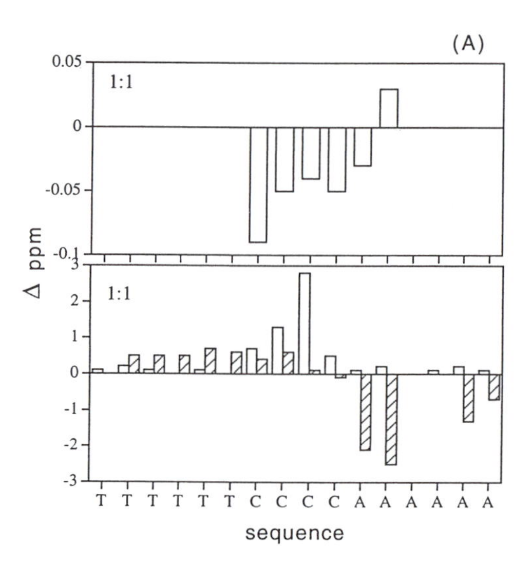

Figure 3. Diagrams showing the chemical shift changes observed in the peptide-DNA hairpin samples relative to the isolated DNA hairpin sample. (A) (upper panel) Phosphorus-31 chemical shift changes of the 5'- P for each residue of the 1:1 peptide-hairpin sample; and (lower panel) carbon-13 chemical shift changes of the aromatic C6/C8 (open bar) and the deoxyribose C1' (thatched bar) signals in the 1:1 peptide-hairpin sample. (B) Aromatic proton chemical shift changes for the 1:1 (upper panel) and the 2:1 (lower panel) peptide-hairpin samples. The changes for the H6/H8 protons (open bar), the A H2 and the C H5 protons (thatched bar) are plotted against the hairpin sequence.

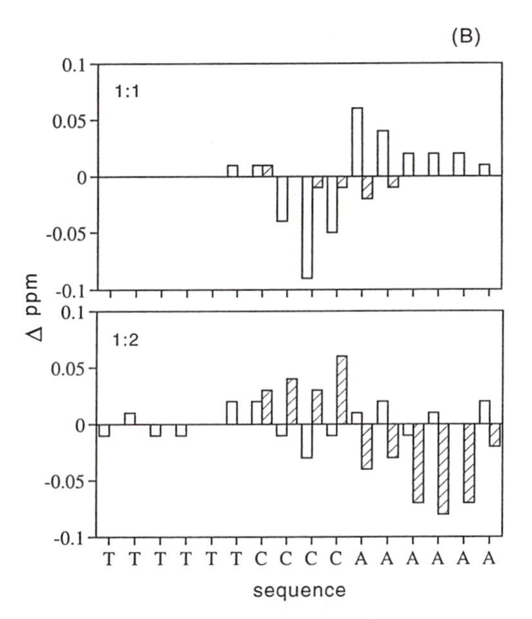

Figure 3. *Continued.*

Taken together, our results indicate first of all that the SPXX-containing peptide interacts preferentially with the single-stranded hairpin C loop region. Secondly we find that at higher than 1:1 peptide-DNA ratios, the peptide binds to the minor groove of the T.A stem in an extended conformation. These results indicate that binding of SPXX motifs to loop regions might be more prevalent than previously considered.

Acknowledgments.

We are indebted to Dr. Peter Davies (Queens University) for providing the peptide and to Dr. Richard Pon (University of Calgary) for synthesizing the DNA hairpin. Financial support by the Alberta Heart and Stroke Foundation is gratefully acknowledged. HJV is a Scientist of the Alberta Heritage Foundation for Medical Research.

References

1. Steitz, T. A. *Q. Rev. Biophys.* **1990**, 23, 205..
2. Lapeyre, B., Bourlion, H. and Amalric, F. *Proc. Natl. Acad. Sci. USA* **1987**, 84, 1472.
3. White, S. W., Appelt, K., Wilson, K. S. and Tanaka, I *Prot. Struct. Funct. Genet.* **1989**, 5, 281.
4. Churchill, M. E. A. and Travers, A. A. *Tr. Biochem. Sc.* **1991**, 16, 92.
5. Nikolov, D. B., Hu, S., Lin, J., Gasch, A., Hoffmann, A., Horikoshi, M., Chua, N., Roeder, R. G. and Burley, S. K. *Nature* **1992**, 360, 40.
6. Chasman, D. I., Flaherty, K. M. Sharp, P. A. and Kornberg, R. D. *Proc. Natl. Acad. Sci. USA* **1993**, 90, 8174.
7. Kim, Y., Geiger, J., Hahn S. and Sigler, P. *Nature* **1993**, 365, 512.
8. Kim, J. L., Nikolov, D. B. and Burley S. K. *Nature* **1993**, 365, 520.
9. Love, J. J., Li, X., Case, D. A., Giese, K., Grosschedl, R. and Wright, P. E. *Nature* **1995**, 376, 791.
10. Werner, M. H., Huth, J. R., Gronenborn, A. M. and Clore, G. M. *Cell* **1995**, 81, 705.
11. Geierstanger, B. H., Mrksich, M. Dervan, P. B. and Wemmer, D. E. *Science* **1994**, 266, 646.
12. Trauger, J. W., Baird, E. E. and Dervan, P. B. *Nature* **1996**, 382, 559.
13. Kahne, D. *Chem. & Biol.* **1995**, 2, 7.
14. Suzuki, M. *J. Mol. Biol.* **1989**, 207, 61.
15. Churchill, M. E. A. and Suzuki, M. *EMBO J.* **1989**, 8, 4189.
16. Hill, C. S., Packman, L. C. and Thomas, J. O. *EMBO J.* **1990**, 9, 805.
17. Suzuki, M., Gerstein, M. and Johnson, T. *Protein Eng.* **1993**, 6, 565.
18. Green, G. R., Lee, H. and Poccia, D. L. *J. Biol. Chem.* **1993**, 268, 11247.
19. Geierstanger, B. H., Volkman, B. F., Kremer, W. and Wemmer, D. E. *Biochemistry* **1994**, 33, 5347.
20. Kennedy, B. P. and Davies, P. L. *J. Biol. Chem.* **1984**, 260, 4338.
21. Plateau P. and Gueron, M. *J. Am. Chem. Soc.* **1982**, 104, 7310.
22. Marion, D. and Wuthrich, K. *Biochem. Biophys. Res. Commun.* **1983**, 113, 967.
23. Bax, A., Griffey, R. H. and Hawkins, B. L. *J. Magn. Reson.* **1983**, 55, 301.
24. Kellogg, G. W. *J. Magn. Reson.* **1992**, 98, 176.
25. Kintanar, A., Klevit, R. E. and Reid, B. R. *Nucleic Acids Res.* **1987**, 15, 5845.
26. Aymami, J., Coll, M., Frederick, C. A., Wang, A. H.-J. and Rich, A. *Nucleic Acids Res.* **1989**, 17, 3229.
27. Zhou N. and Vogel, H. J. *Biochemistry* **1993**, 32, 637.
28. Feigon, J. Denny, W. A. , Leupin, W. and Kearns, D. R. *J. Med. Chem.* **1984**, 27, 450.

Secondary Structure Prediction

Chapter 13

Thermodynamics of Duplex Formation and Mismatch Discrimination on Photolithographically Synthesized Oligonucleotide Arrays

Jonathan E. Forman, Ian D. Walton, David Stern, Richard P. Rava, and Mark O. Trulson[1]

Affymetrix, 3380 Central Expressway, Santa Clara, CA 95051

Oligonucleotide probes immobilized on solid supports are finding increasingly widespread application in genetic analysis. We seek to provide a better understanding of the fundamental thermodynamic aspects of duplex formation in these systems. Equilibrium melting curves for the adsorption of 10-, 20-, 30-, and 40-base oligonucleotides to 10-, 12-, 14-, 16-, 18-, and 20-base probe sites on an Affymetrix GeneChip® array have been investigated.[1] The melt curves are depressed and broadened relative to corresponding solution phase species. Melt curves show multistep behavior above a threshold target concentration that decreases for longer target DNA. Melting temperatures (T_m) are weakly dependent on probe length. Mismatches introduce T_m depressions that are nearly normal at 10-mer probe sites but decrease with increasing probe length. Saturating adsorption densities at perfect match probe sites are less than 10% of the probe density and exceed the corresponding mismatch densities in all cases. The anomalies in the equilibrium data cannot be fully attributed to the presence of truncated probes, and implicate target DNA binding to multiple probes ("probe bridging") and probe-probe interactions that compete with 1:1 probe:target binding.

Recognition of DNA or RNA from solution using arrays of immobilized nucleic acid probes (Figure 1) is finding increased application in a variety of areas, including polymorphism detection, gene expression, and sequence analysis.[2,3] A number of embodiments have been described in which the DNA probes are immobilized onto appropriate support materials (usually glass or polymer surfaces); these arrays provide a rapid means by which nucleic acid sequences can be analyzed using a minimum amount of sample.[2,3] Our photolithographic synthesis methodology allows probe arrays to be prepared with as many as 132,000 synthesis sites (35 μm × 35 μm feature size), capable of interrogating every position on a target of up to 33 kb (16.5 kb if both sense and antisense strands are being screened).[2b]

Duplex formation in solution has been extensively studied, and the kinetics and thermodynamics of such interactions for short oligonucleotides can be described by a model involving nucleation followed by helix zipping (Figure 2a).[4a] In immobilized

[1]Corresponding author

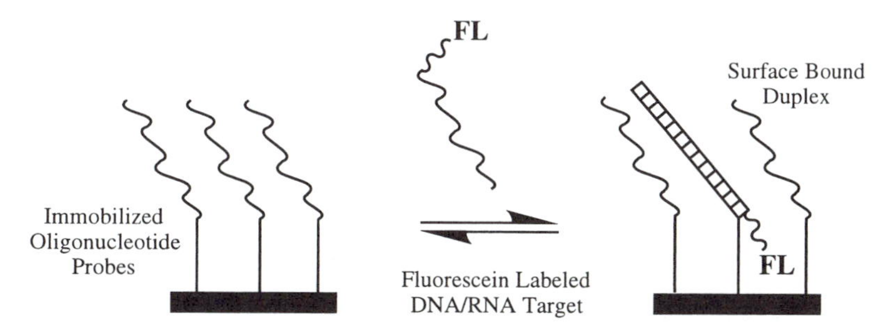

(a) Nucleic acid recognition by immobilized probes

(b) Fluorescence microscope image of probe array with adsorbed target

Figure 1. (a) Recognition of a solution-borne nucleic acid (target) by a nucleic acid probe immobilized on a surface. (b) Fluorescence microscope image of a 1.28 cm × 1.28 cm probe array with 200 μm × 200 μm synthesis sites on which labeled target oligonucleotides have adsorbed.

probe systems, the behavior may be more complex due to the combined influences of bulk diffusion, surface diffusion, and small intermolecular distances typically encountered in these systems (Figure 2b).[5] The small intermolecular distances on the probe array raise the potential for interactions of the target with multiple probes, requiring an annealing step to occur in order to form the optimized duplex structure. Indeed, the equilibrium state may not be made up entirely of ideal duplexes. Additionally, probe-probe interactions may limit the number of sites available for target binding and/or reduce the stability of the probe:target duplexes relative to solution. The stability of the surface bound duplexes may be further affected by differences in dielectric environment, ionic strength, or pH relative to the solution from which the target adsorbs. We are interested in physically characterizing the factors that govern the adsorption of complex solution-borne nucleic acid targets to surface bound oligonucleotide probes. This paper represents our first efforts toward understanding the thermodynamics of oligonucleotide recognition by immobilized probe arrays.

Materials and Methods

Probe arrays and DNA target. The studies described here have been carried out with arrays of 10-, 12-, 14-, 16-, 18-, and 20-base oligonucleotide probe synthesis sites 200 μm × 200 μm in size. The probes are covalently attached to a siloxane derivatized 15 mm × 15 mm × 0.7 mm borosilicate glass substrate and were synthesized using photolithographic techniques that have been described elsewhere.[2ij]

Target oligonucleotides were labeled at the 5' end with fluorescein; samples were obtained from GENSET (HPLC purified) and are estimated to have 99% purity. Table I lists the sequences of the probes and the 20 base oligonucleotide target used in these studies. This particular target sequence was chosen to minimize the chance of intra- and intermolecular secondary structure and has a GC content of 50%, yielding a fairly average stability for a duplex of this length. All studies were carried out using 6X SSPE buffer (1 M NaCl, 0.07 M Na_2HPO_4, 0.07 M EDTA) at pH 7.8, stabilized with 0.005% Triton-X. The probe arrays were pretreated with this buffer for 30-60 minutes before target adsorption studies were undertaken.

Table I. Sequences of oligonucleotides employed in these studies. fl = Fluorescein, su = surface.

	Position	1	2	3	4	5	6	7	8	9	10	11	12	13	14	15	16	17	18	19	20	
Target	5'-fl-	C	T	G	A	A	C	G	G	T	A	G	C	A	T	C	T	T	G	A	C	-3'
10-mer probe	su-3'-						G	C	C	A	T	C	G	T	A	G						-5'
12-mer probe	su-3'-					T	G	C	C	A	T	C	G	T	A	G	A					-5'
14-mer probe	su-3'-				T	T	G	C	C	A	T	C	G	T	A	G	A	A				-5'
16-mer probe	su-3'-			C	T	T	G	C	C	A	T	C	G	T	A	G	A	A	C			-5'
18-mer probe	su-3'-		A	C	T	T	G	C	C	A	T	C	G	T	A	G	A	A	C	T		-5'
20-mer probe	su-3'-	G	A	C	T	T	G	C	C	A	T	C	G	T	A	G	A	A	C	T	G	-5'

Figure 1 shows a fluorescent microscope image of the probe array with adsorbed fluorescein labeled target. Each of the bright cells is a 200 μm × 200 μm synthesis site containing probes of a given sequence. The layout of the probe array includes nine repeats of a six-block set of probe synthesis sites (Figures 1 and 3); each

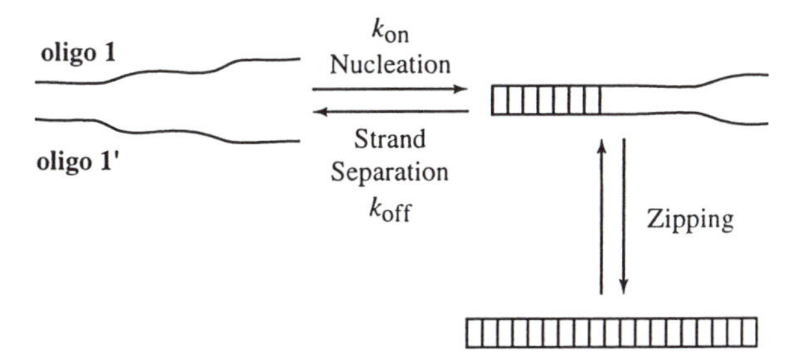

(a) Duplex formation in solution

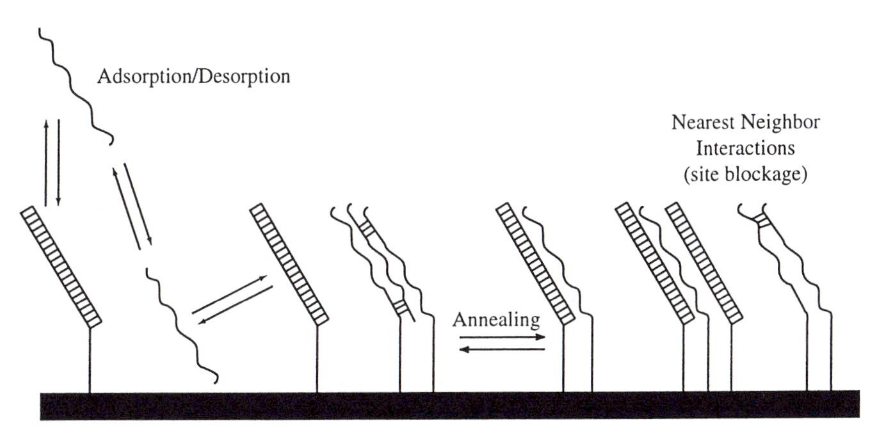

(b) Adsorption to probes at an interface

Figure 2. Oligonucleotide interactions in solution (a) and at an interface with surface bound probes (b). While the mechanism of the solution phase process is well understood, duplex formation at an interface is complicated by a number of processes resulting from adsorption mechanisms and probe distribution.

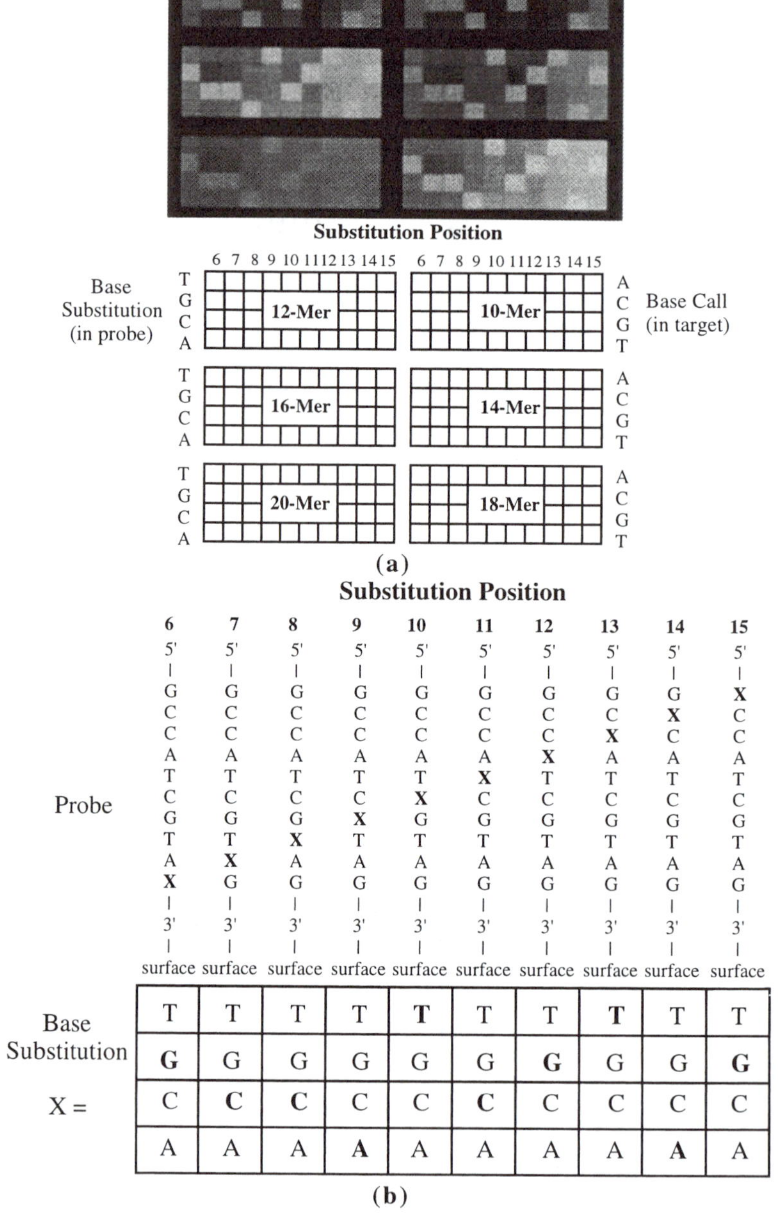

Figure 3. Probe array features examined in this study. (a) Fluorescence microscope image and map of probe types and base substitution sequences. (b) Illustration of the tiling for the 10 sequence positions within a given set of probes. We have used the 10-mer probes for illustration; matched positions are in bold-face type.

of these six blocks represent the six 10-20 base probes with substitution at the 6-15th positions of each of the four possible naturally occurring bases (Table I). As shown in Figure 3, each block of probe sites of a given length contains 10 matched sequence cells (the brightest within each column of four cells) and 30 single base mismatches, making it possible to read the sequence of the target by simply identifying the brightest cells across the 10×4 block of probe sites. The blank lanes between the 10×4 blocks were used as background for the signals measured on the probe array surface.

Instrumentation and procedures. The adsorption measurements were performed with an in-house built confocal laser scanning fluorescence microscope employing a 50 mW Ar^+ laser (488 nm excitation). Probe arrays were mounted to an anodized aluminum flow cell equipped with water jacket for controlling temperature by means of a VWR 1167 Forma bath. Target solution was continuously pumped through the system using a peristaltic pump (VWR Variable Flow Tubing Pump, using 3/16 OD, flow rate ~1.5 ml/minute). The temperature of the target solution was controlled through a sample reservoir equilibrated with a water bath whose temperature was maintained by the same Forma bath used to control the flow cell temperature. Additionally, the fluidic lines from the target reservoir were routed through the water flow lines of the Forma bath to provide better pre-equilibration of sample temperature.

Solution phase melting studies were performed by measuring absorption hypochromism[4c,6] on a HP8453 UV/Visible spectrometer at 260 nm using a minimum of three concentrations of unlabeled oligonucleotides with total oligonucleotide concentration ranging from 5×10^{-7} to 3×10^{-6} M (1:1 ratio of probe:target sequences) in 6X SSPE. The solution measurements were performed for the 20-mer target/20-mer probe and the 20-mer target/10-mer probe perfect match duplexes, as well as single base substitution at the central point of the duplex. van't Hoff plots were used to determine thermodynamic parameters to allow extrapolation of T_m's for solution phase duplexes to sub μM concentrations.[4ac,6]

All experiments were performed with the probe array continually exposed to a constant concentration of target solution. The flow cell sample chamber volume is 250 μl; however total volume of target solution in the flow cell and target reservoir was maintained at ~6 ml, to ensure there was no depletion of target. Kinetic experiments were carried out under constant temperature conditions, with initial focusing of the scanner, and fluorescence images taken at appropriate time intervals. For variable temperature equilibrium experiments, a 30 minute equilibration time was allowed at each temperature point, the scanner was focused and four scans taken. The fluorescence intensities reported at a given feature site for a given temperature point are the average of the four scans. Equilibrium experiments were run in both the high-low and low-high temperature directions to check for hysteresis.

The experimentally accessible target concentration range for equilibrium measurements was 1 nM to 500 nM. The low concentration limit was dictated by the adsorption kinetics, which have been observed to be proportional to target concentration. The high concentration limit was set by the ability of the confocal imaging system to reject the luminescence of the bulk fluorophore.

The conversion from fluorescence intensity (counts/pixel) to adsorbed target density (molecules/μm^2 or pmol/cm^2) was accomplished by measuring the system gain of the optical scanner. The system gain was determined from the derivative of the axial dependence of the signal from a solution of fluorophore labeled oligonucleotide of known quantum yield and concentration. The measured adsorbed target densities are estimated to be accurate to $\pm$ 0.2 pmol/cm^2.

Adsorption Kinetics

Initial inspection of the adsorption time course suggests that rates of adsorption for all probe lengths level off within an hour of exposure for a 10 nM target concentration, with longer probe lengths adsorbing a greater amount of target (Figure 4a). Observed

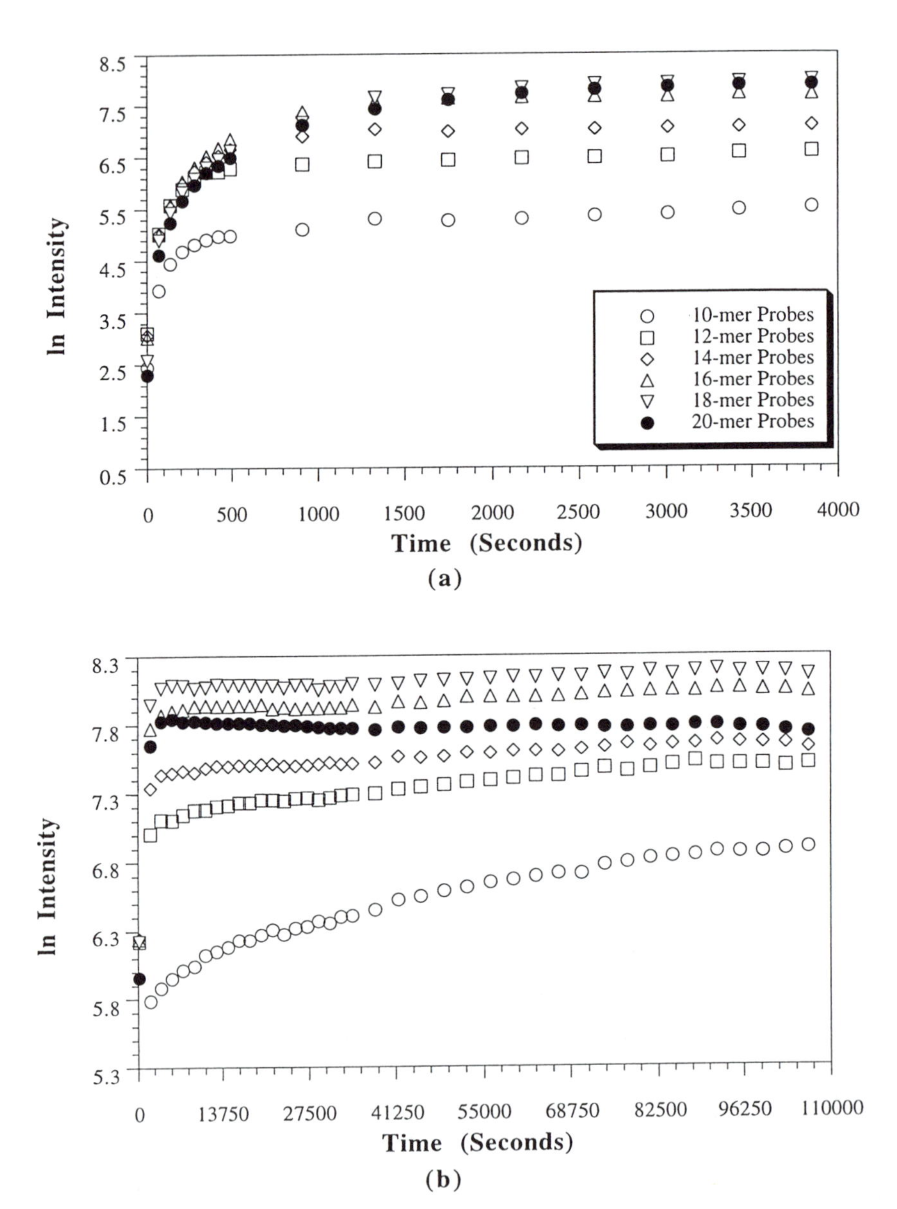

Figure 4. Change in adsorption with time for matched probes at 25°C with 10 nM target. (a) One hour time course. (b) 30 hour time course.

adsorption rate constants are comparable to duplex formation rate constants in solution ($\sim 10^4$ - 10^5 M^{-1} sec^{-1})[4c,7]; however, variations in adsorption rates have been observed with different mixing schemes, and our measured rate constants therefore should be regarded as a lower limit. The initial adsorption rates for mismatched probe sites were within 5% of those to the matched sites, but at all times the densities of adsorbed target were lower (Figure 5).

Experiments carried out beyond the 1 hour time course showed a continued change in adsorbed density. At 25°C, the adsorbed density for longer probe sites leveled off after several hours, but within short probe sites adsorption was still increasing after 30 hours! (Figure 4b). The increase in adsorbed material from the initial hour to 30 hours is greatest for the shorter probe sites, with the 10-mer sites showing 175% and the 18-mer sites showing 108% of the density at the 1 hour time point. The 20-mer probe sites, while following the trend of 10-mer > 12-mer > 14-mer > 16-mer > 18-mer > 20-mer adsorbate change over time, differ from the other probe sites by *decreasing* in adsorbate density relative to the 1 hour time point (90% of one-hour signal at 30 hours). After equilibration, the 20-mer probe sites show the lowest adsorption levels of all the probes (Figure 6). These long term changes in adsorbed density are strongly indicative of some type of structural reorganization within the probe array. This apparent annealing process has been found to be driven by the adsorption of target and is not a consequence of structural rearrangements following hydration of the array: The association kinetics are essentially unchanged by 72 hours of soaking in buffer prior to the start of the kinetic experiment.[8]

Adsorption Equilibrium

Probe length dependence. The isotherms shown in Figure 6 represent the concentration range from 10-500 nM and it can be seen that at all target concentrations (except 150 nM) the adsorption densities follow the order: 18-mer > 16-mer > 14-mer > 12-mer > 10-mer > 20-mer. An unusual aspect of these studies is the shape of the isotherms. At concentrations as high as 100 nM, the adsorption isotherms appear to be made up of of two components with binding constants of <10 nM and 80 nM, respectively. The decrease in apparent binding densities above 100 nM is puzzling. An obvious candidate culprit is quenching of the fluorescein label. We do not believe this to be occurring. At all probe sites the decrease in fluorescence intensity occurs at the same bulk concentration and not at the same adsorbate density, hence intermolecular distance on the surface.[4b]

Melting curves obtained at lower target concentrations (10 nM - 100 nM) are more gradual than solution melt curves (Figure 7a). The melting transition temperature ranges, expressed as the temperature range over which the adsorbed target density changes from 10% to 90% of maximum, are in the neighborhood of 40 °C, compared to observed transition temperature ranges of 5-10 °C in solution. Furthermore, the melting curves do not show the simple sigmoidal shape that is characteristic of solution melting curves of oligonucleotide duplexes, suggesting the presence of multiple thermodynamic components in the melting transition.

At concentrations above 100 nM, coincident with the appearance of the second adsorption component in the isotherms, the melting curves show dips and step-like structure (Figure 7b). The structure persists in melting curves obtained at still higher concentrations (Figure 7c), despite the apparent disappearance of the second adsorption component in the isotherms. It is important to note that the structure is essentially reproducible when measuring melting curves in both high-low and low-high directions and is therefore not a kinetic artifact (Figure 8).

There is little apparent dependence of melting temperatures on probe length or target concentration. At all target concentrations, the melt temperatures for 10 - 18-mer sites were in the range of 34 to 37 °C and the 20-mer sites in the range of 43 to 44 °C. It should be pointed out that, although the adsorbed densities appear to be approaching an

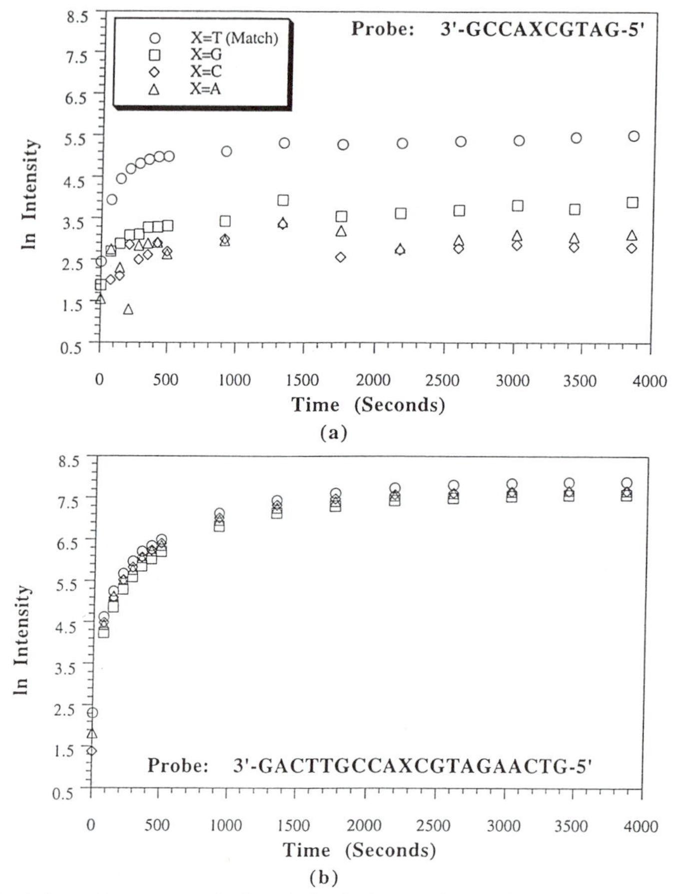

Figure 5. Adsorption to matched and single-base mismatched probes (position 10) at 25°C with 10 nM target; illustrated for (a) 10-mer and (b) 20-mer probes.

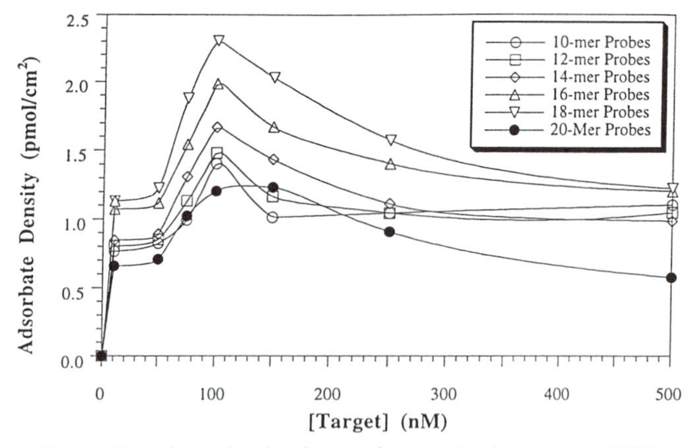

Figure 6. Adsorption isotherms for matched probes at 25°C.

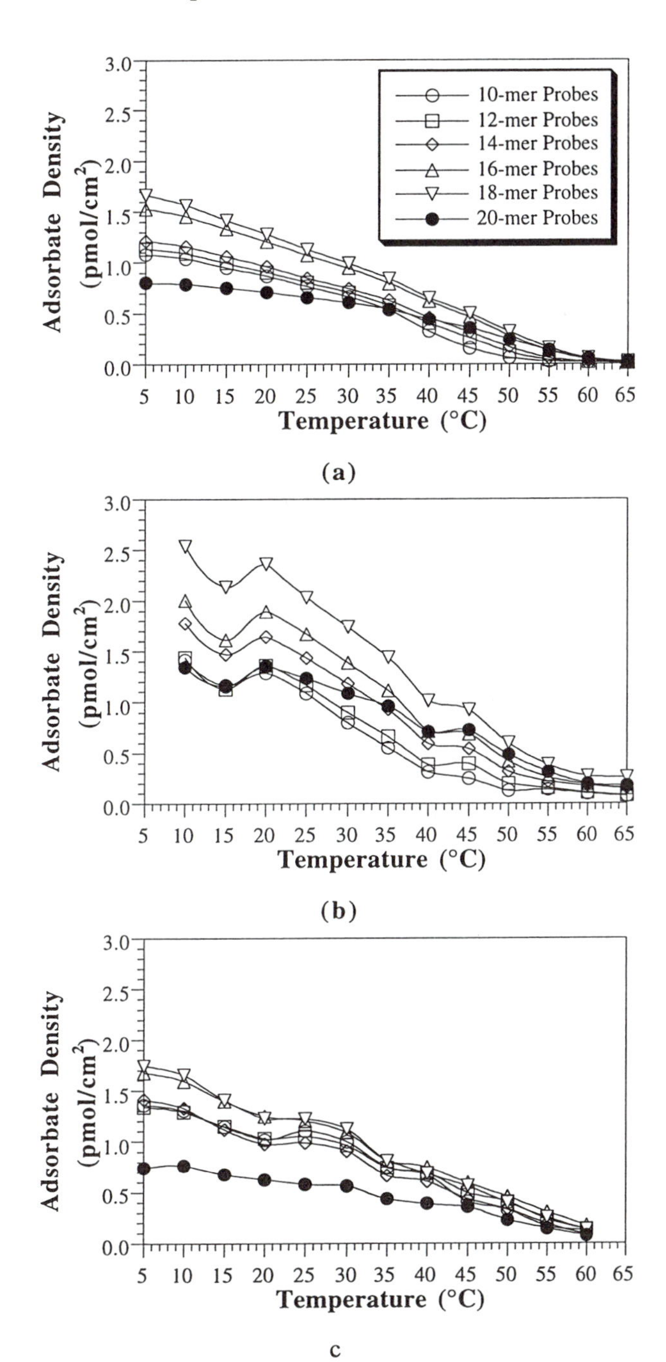

Figure 7. Melting curves for matched probes obtained at target concentrations of (a) 10 nM, (b) 150 nM, and (c) 500 nM.

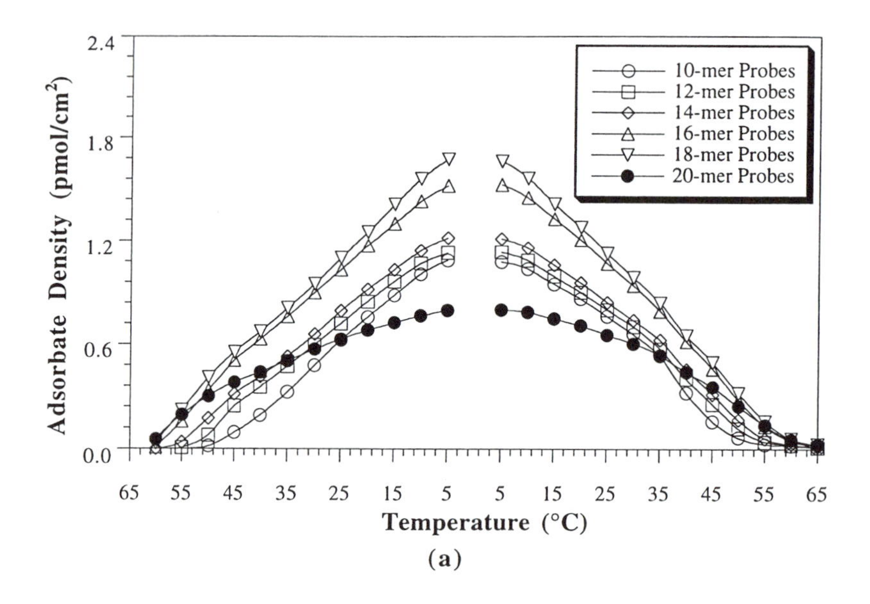

(a)

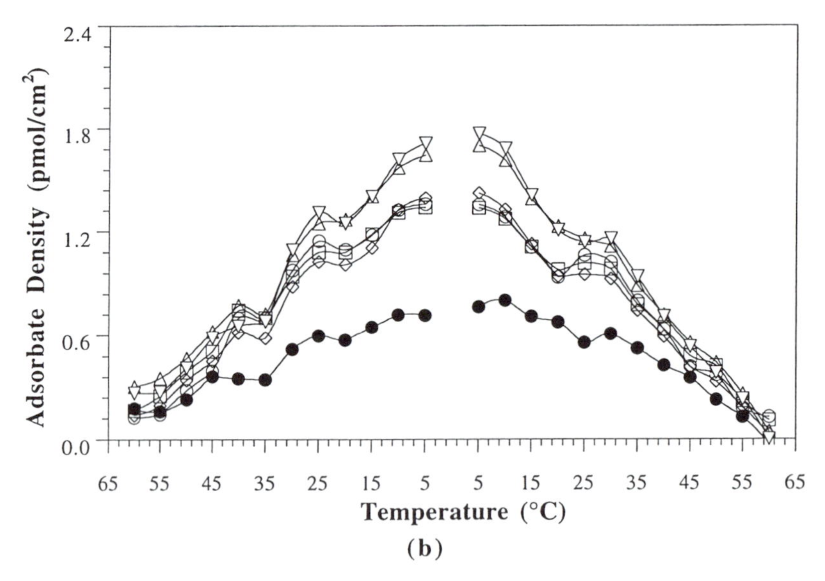

(b)

Figure 8. Melting curves obtained in high-low and low-high temperature directions on the same set of matched probe synthesis sites at (a) 10 nM and (b) 500 nM.

asymptote at the low temperature limit of 5 °C, only for the 20-mer probe sites does the asymptote appear to have been reached. The densities at shorter probe sites appear to fall short of an asymptote by a few percent. Although this limitation introduces a small uncertainty into the estimated melting temperatures, it is clear that the melting profiles at different probe sites are remarkably similar.

Mismatch dependence. The melting curves for perfect match and single base substitutions at the central T position (position 10, Figure 3) for 10-mer and 20-mer probe sites are shown in Figure 9. The melting temperatures of the 10-mer mismatch sites range from 7 to 12 degrees below those of the 10-mer match site (Figure 9a), while the melting point depressions for the 20-mer mismatch sites averaged 6 degrees (Figure 9b). These results are in contrast with the 14 degree melting temperature depression induced by mismatches for the same species in solution.

Figure 10 depicts the dependence of the average mismatch discrimination ratios on temperature, concentration, and probe length. The discrimination ratios are defined here as the average of the density at the 10 perfect match probe sites to the average of the densities of the thirty single base mismatch sites. In all cases, the best mismatch discrimination was obtained at or slightly above the apparent match T_m.[9] This behavior is similar to solution behavior, although the peak discrimination ratios are lower on the array surface.

The discrimination ratios decrease to a limiting value as the temperature is lowered below the T_m, but unlike solution behavior the limiting value is greater than unity. This is a remarkable observation when it is recognized that the densities and distributions of probe lengths at match and corresponding mismatch sites are nearly identical to one another.[10] This result is contrary to the prediction of Langmuir adsorption theory that, given identical densities of binding sites, equal adsorption densities should be obtained under conditions of sufficiently low stringency, i.e. low temperature or high concentration. The limiting discrimination ratios at low temperature were highest (roughly 2:1) for the 10-mer probe sites at 10 nM concentration (Figure 9a). For the 20-mer sites, the limiting ratio was much smaller, of the order of 1.1 (Figure 9b).

Target length dependence. The small interprobe distances on the array surface (< 50 Å) open up the possibility of target binding to two or more probes simultaneously. To address the issue of such "bridged" structures, melting experiments were performed with 10-base, 30-base, and 40-base targets (Table II). The 10-mer target covers positions 6-15 of the 20-mer target and can form duplexes with all probe lengths. The 30-mer and 40-mer targets are the 20-base sequence with runs of 5 or 10 adenines appended to the 5' and 3' ends. The appended bases are unable to form specific duplexes with any of the probes. The melting curves obtained at 10 nM target concentration are shown in Figure 11.

Two aspects of the melt profiles of the 30 and 40 base targets are notable. First, at low temperature the adsorption densities are greater than those of the 20-mer perfect complement. Second, the 30-mer and 40-mer melting curves show the structure seen with only higher concentrations of 20-mer target (Figures 7bc and 11bc), while the 10-mer target shows the structureless melt curve of the 20-mer target (Figures 7a and 11a). These observations support the notion that noncomplementary sequence appended to the 20-mer complementary sequence leads to enhanced adsorption via binding to multiple probes, and that a signature for such "bridging" interactions is structure in the melt curve.

Discussion

The probe array is known to differ from the bulk aqueous reference system in three principal ways. First, each synthesis site contains a substantial number of truncated probes in addition to the desired full length probes. Second, the probe array resides at an abrupt dielectric discontinuity. Third, local concentrations of probes are known to be

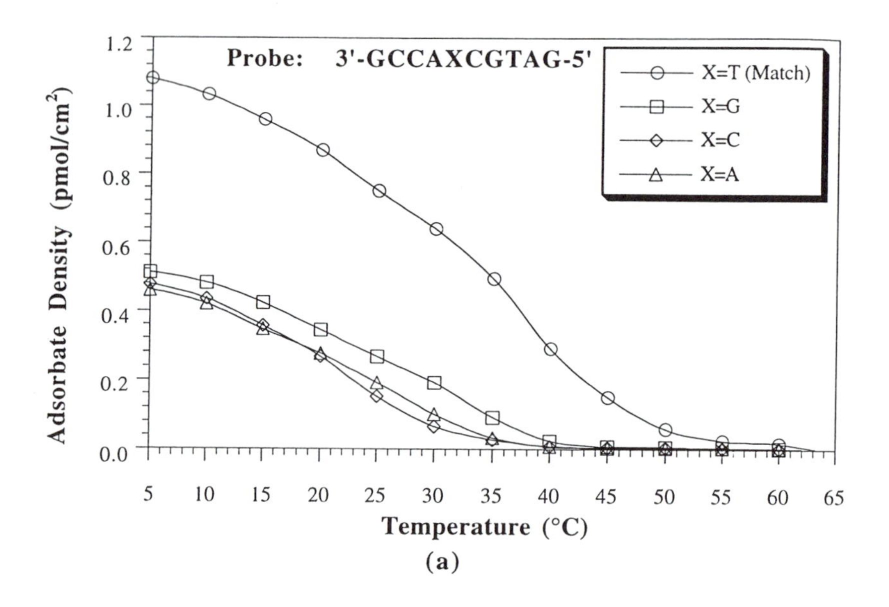

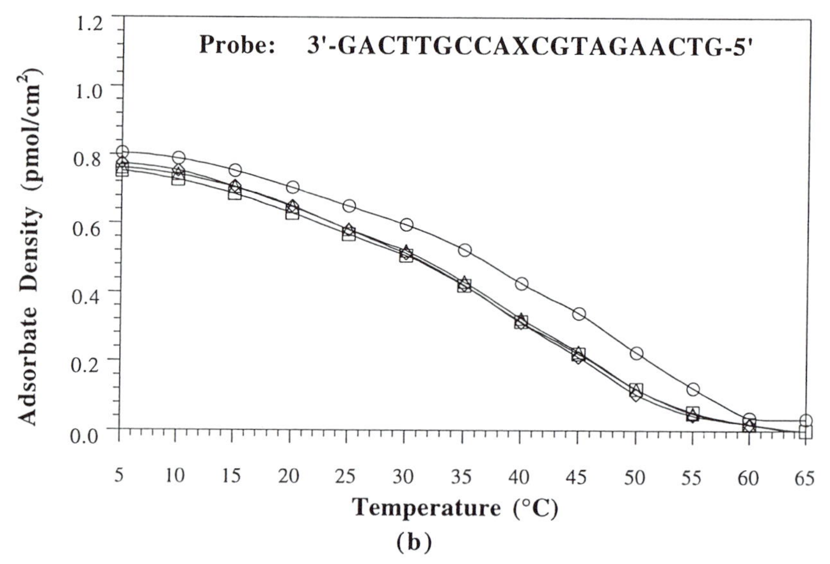

Figure 9. Melting curves for matched and single-base mismatched probes (position 10) obtained at a target concentration of 10 nM; illustrated for (a) 10-mer and (b) 20-mer probes.

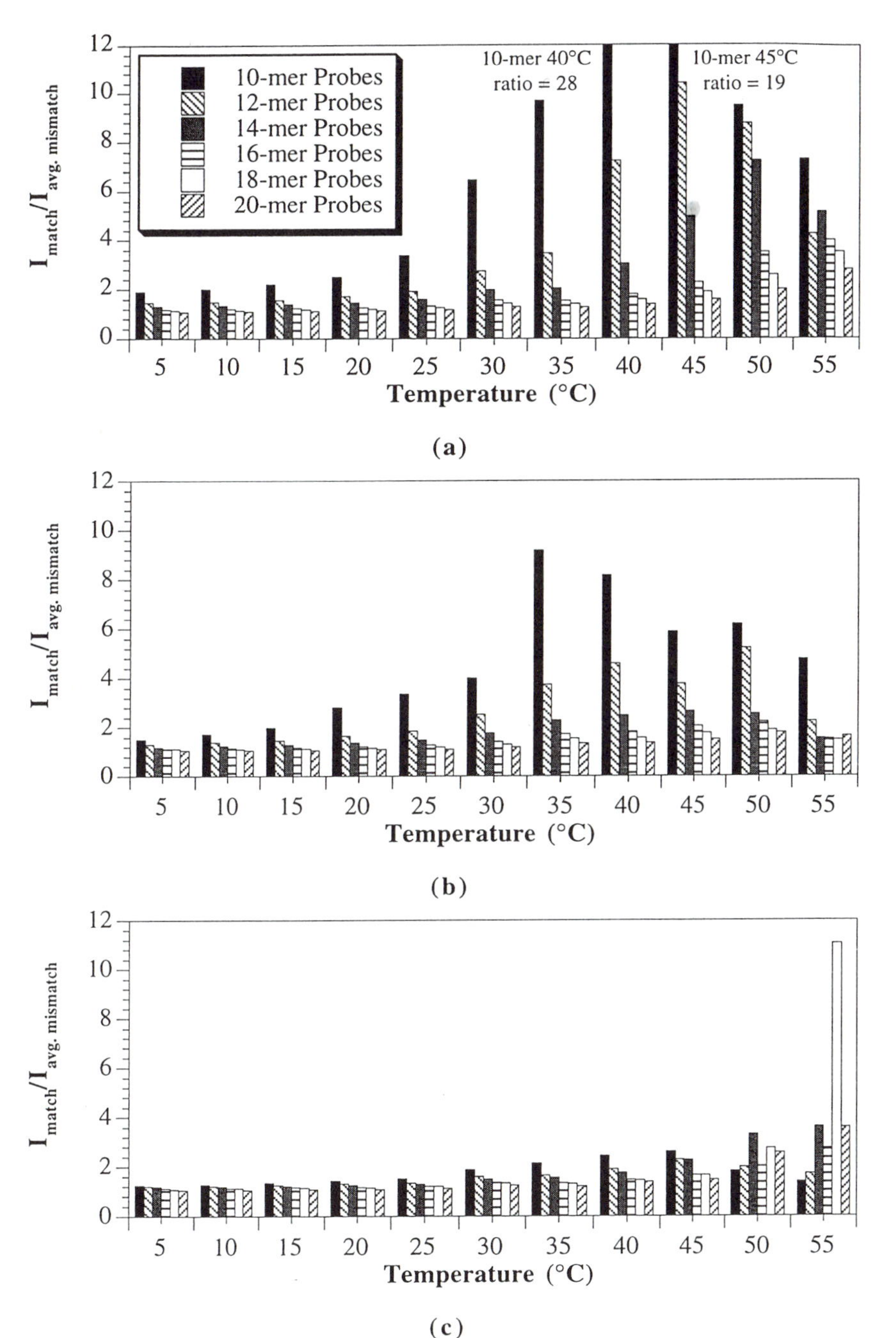

Figure 10. Average ratio of match signal to mismatch signal obtained at target concentrations of (a) 10 nM, (b) 150 nM, and (c) 500 nM.

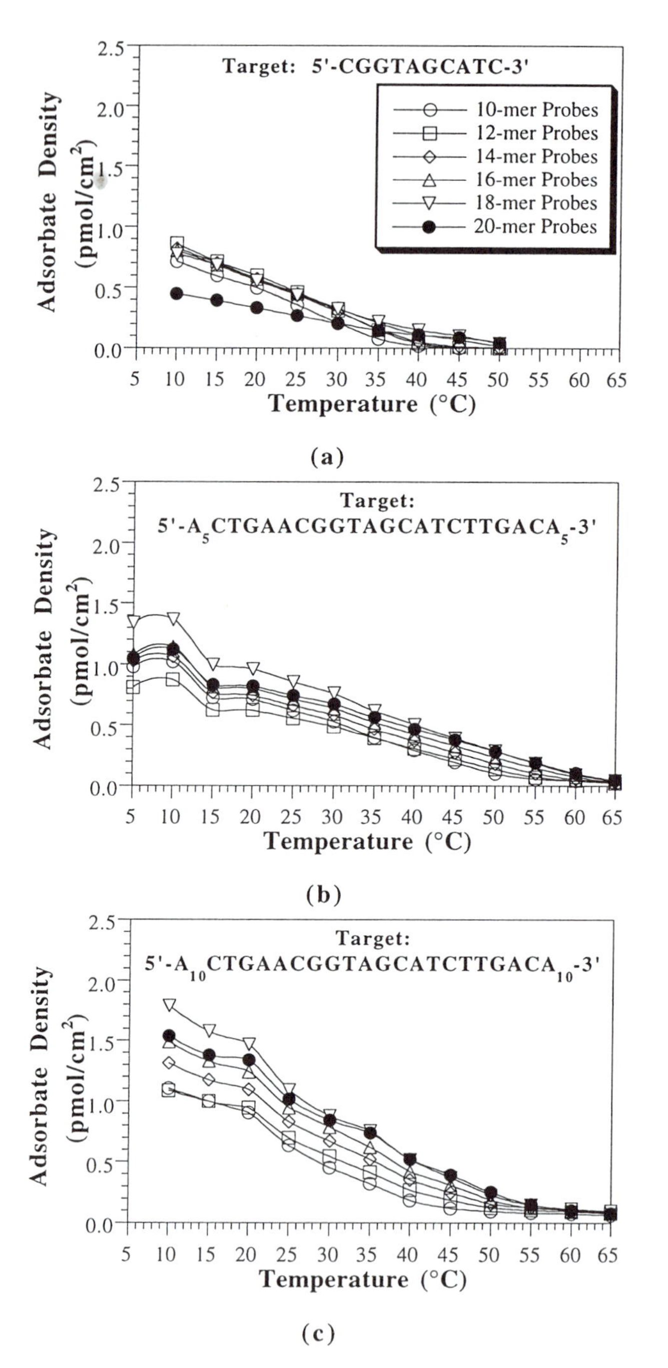

Figure 11. Melting curves for matched probes at 10 nM target concentration obtained with (a) 10-mer target, (b) 30-mer target, and (c) 40-mer target.

quite high. The discussion that follows attempts to address the ways in which these factors may lead to the observed differences between adsorption behavior on the probe array surface and duplex formation in solution.

Table II. Sequences of alternate length target oligonucleotides. fl = Fluorescein.

N-mer	Position	1	2	3	4	5	6	7	8	9	10	11	12	13	14	15	16	17	18	19	20	
10	5'-fl-						C	G	G	T	A	G	C	A	T	C						-3'
20	5'-fl-	C	T	G	A	A	C	G	G	T	A	G	C	A	T	C	T	T	G	A	C	-3'
30	5'-fl-A_5-	C	T	G	A	A	C	G	G	T	A	G	C	A	T	C	T	T	G	A	C	-A_5-3'
40	5'-fl-A_{10}-	C	T	G	A	A	C	G	G	T	A	G	C	A	T	C	T	T	G	A	C	-A_{10}-3'

Probe Length Heterogeneity. The step-wise yields for synthesis of probes are approximately 90% in this system.[11] Figure 12 depicts the oligonucleotide length distributions within synthesis sites of the indicated nominal probe length. With an initial density of reactive hydroxyl groups of 27 pmol/cm^2, only 14 pmol/cm^2 of the probes are of length $\geq$ 6 bases, the minimum length expected to have a T_m above room temperature with our 20-mer target. The proportion of full length probes decreases with nominal length for a given probe site; for example, in 10-mer and 20-mer probe sites only 40% and 15% of the probes are full length, respectively. The impurity sequences are primarily 5'-truncated species.[12]

To investigate the role of probe length heterogeneity in the broadening and depression of the melting curves, simulations of the melting curves for probe sites containing the chemical make-up shown in Figure 12 were carried out. The simulated melt curves for each probe site are population-weighted superpositions of melting curves for each probe length in the distributions of Figure 12. Each probe length subpopulation is assumed to behave according to Langmuir adsorption equilibrium theory,[13] whose principal underlying assumptions include a fixed density of non-interacting adsorption sites with identical adsorption thermodynamics. The standard enthalpies and entropies for each probe length subpopulation were taken from the measured solution phase energetics for our sequences,[14] and 10 nM target concentration was assumed. The results are shown in Figure 13 for comparison to Figure 7a.

The simulated melting curves differ from the experimental curves most visibly in three ways. First, the simulated melting curves show a well defined "step" at high temperature corresponding to the melting of the full length probes that is not seen in the experimental curves. Second, the observed melting curves diverge widely at low temperatures, contrary to the simulated behavior. Finally, the observed saturating densities of adsorbed target are at most 10% of the predicted densities. The assumption of Langmuir adsorption theory that no nearest neighbor interactions are taking place leads to the prediction that 100% of the adsorption sites will be occupied under conditions of sufficiently low stringency.[15] By contrast, under all conditions explored in this study the majority of probes are apparently unavailable to target binding.

Interfacial Environment. The energetics of DNA duplex formation are strongly dependent on the dielectric constant and ionic strength of the immediate environment due to the increase in linear charge density accompanying double strand formation. The probes are covalently attached to a siloxane layer through a hexaethylene glycol (HEG) linker.[1ij,11] The siloxane layer is itself attached to the glass substrate *via* a solution dip process, which is thought to form a film that contains a non-negligible proportion of polymerized siloxanes, is thicker than a monolayer, and of uncertain topography.[15] The dielectric constant within the HEG/siloxane layer is 10 or lower,[16] much lower than that of the aqueous environment from which the target adsorbs. The local concentration of target and counterions may be altered thereby, and the stability of duplexes may be reduced to the extent that any intermingling of the DNA probes with the siloxane/HEG layers occurs. Heterogeneity in the local dielectric

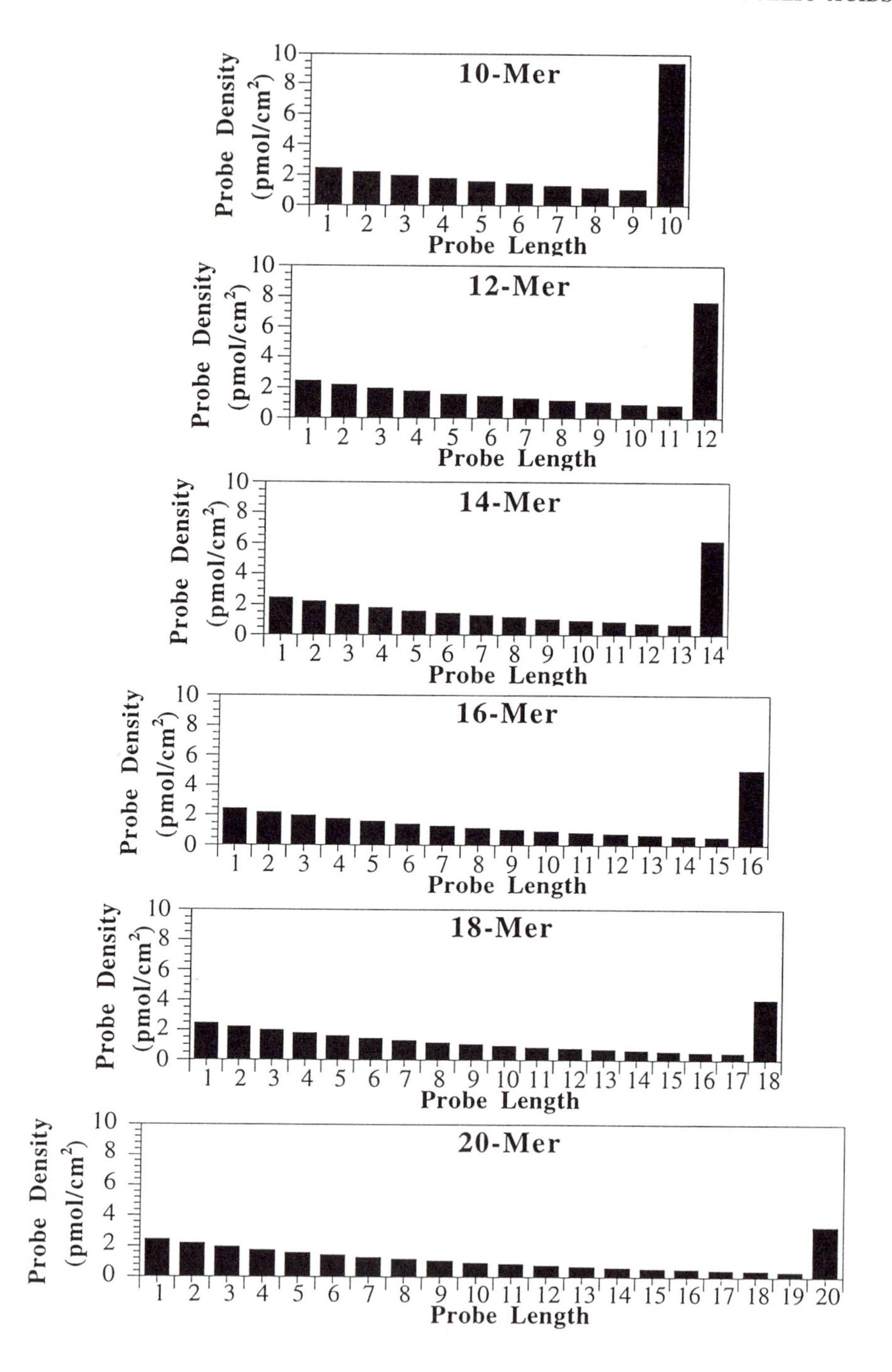

Figure 12. Probe distributions within synthesis sites designed to contain 10, 12, 14, 16, 18, and 20-mer probe sequences.

environment may generate a distribution of melting temperature depressions, and may be responsible in part for the observed depression and broadening of the experimental melting curves. The dependence of melting curves on the nature and deposition method of the siloxane and the length and chemical nature of the linker are expected to elucidate the importance of these factors.

High probe density. Despite the low yield of full length probes, the high initial density of synthesis sites leads to crowded conditions on the array surface. Under the assumption of a square lattice, in all probe sites the average nearest neighbor distance for probes of length 6 or greater is 35Å, and the average distance between full length probes ranges from 42 Å in 10-mer sites to 70 Å in 20-mer sites. Indeed, the distances estimated here are likely an upper limit.[17] The lengths of the probes and HEG linkers therefore allow probes to come into contact. Such high local DNA probe concentrations generate high local charge densities, encourage the association of probes that may interfere with target binding, may permit the simultaneous binding of target molecules to multiple probes, and may sterically prohibit access of target to the probes.

Electrostatic factors. The probe array carries a high charge density due to the phosphate backbones of the nucleic acids. The projected surface density of phosphodiester groups ranges from 160 to 210 pmol/cm^2 within 10-mer and 20-mer probe sites, respectively. Estimated probe array layer depths of 30 to 60 Å lead to volume concentrations in the range of 0.3 to 0.5 M, which may generate an energetic penalty to adsorption of the polyanionic target. Long range hydration forces may be operative in the presence of high densities of polyanions.[19] Additionally, the proximity of probes, in conjunction with an altered dielectric environment, may perturb counterion uptake and release processes thereby altering duplex stability.

Despite the concerns raised about high charge density on the array surface, there is some evidence to indicate that it is of little consequence to the melting behavior. First, the studies described here have been carried out in aqueous buffer with bulk NaCl concentrations of 1M, which are thought to effectively shield charge over distances shorter than the estimated average nearest neighbor distances. Second, adsorbed 40-mer target densities have been observed to exceed the 20-mer target densities by a small margin (Figure 11), despite the doubling of charge and the absence of specific duplex forming capability introduced by the appended $(dA)_{10}$ segments. The dependence of adsorption behavior on electrolyte valency and concentration, as well as the behavior of uncharged target and/or probe species such as peptide nucleic acids[20] should provide more insight into the importance of electrostatic factors in this system.

Competing Equilibria. We propose that interactions between probes that compete with target binding can lead to some of the anomalies in the equilibrium data, and are indeed necessary to explain in particular the differences in saturating densities at match and mismatch sites (Figure 9a). The essential features of the model are discussed in the context of probe:probe (P:P) interactions that may render the participating probes unavailable to target binding (P:T interactions). It should be recognized, however, that a variety of competing interactions with different stoichiometries may be at work. For example, a probe may be rendered only partially inaccessible to target binding, yielding a formal P:P:T species. A target molecule may "bridge" two probes, yielding a formal P:T:P structure.

It seems reasonable to postulate that the probes interact with one another through the same types of hydrogen bonding and stacking interactions that drive duplex formation. The average number of stabilizing interactions between noncomplementary probes is necessarily small, implying that the enthalpy changes that drive the P:P equilibria are substantially smaller than typical duplex formation enthalpies. In order to compete with duplex formation, the entropic penalty for the P:P equilibria must be proportionally smaller. This does not appear to be unreasonable in view of the immobilized environment of the probes. Hydrogen bonding and stacking of bases from two different strands that are fixed in near contact is likely to incur a smaller loss of conformational entropy and a smaller entropy loss due to counterion release than the same process in solution. Additionally, the free energy for association of surface bound species lacks the -RT ln[target] term that applies to the T:P equilibrium,[21] which amounts to a +11 kcal/mol penalty at 10 nM and room temperature. For these reasons, it is plausible that P:P associations that involve relatively few residues may compete with the formation of P:T duplexes.

Because of the distribution of discrete probe lengths within each synthesis site, the free energies for P:T binding are in turn discretely distributed according to duplex length. The distribution of interprobe nearest neighbor distances within each synthesis site is expected to generate a distribution of P:P formation free energies. In this heterogeneous surface environment, the discrete spectrum of P:T duplex formation equilibria takes place in the presence of the background of competing P:P interactions that cover an essentially continuous spectrum of free energies spanning the free energy of the P:T equilibria. These relationships are illustrated in Figure 14. Any P:P structures whose free energy fall at lower ΔG than a particular P:T structure prohibit that mode of target binding; the fraction of probes available for target binding is proportional to the area of the P:P free energy distribution that falls above the P:T duplex formation free energy.

The model described above predicts some important effects upon the equilibrium behavior that are consistent with some of the observed differences between surface and solution behavior. First, at temperatures well below the T_m of the P:T duplexes, some fraction of P:P complexes may fall at free energies below the P:T free energy. Thus, at all temperatures below the T_m, *the saturating adsorbed target density is predicted to be lower than the probe density*. Second, the free energy distribution for P:T complexes in mismatch sites are shifted to higher free energy than those of the match sites, whereas the free energy spectra of the competing P:P equilibria are expected to be far less dependent upon such single base substitutions. At all temperatures, a larger fraction of P:P equilibria is therefore competitive with P:T mismatch binding than with P:T match binding: *Saturating mismatch adsorption densities are predicted to be lower than match densities*. Finally, it is expected that because of the smaller average number of residues participating in the formation of P:P structures, the free energies of the P:P equilibria should be more weakly dependent on probe length than the ΔG of the P:T equilibria: *The divergence in saturating densities should be most pronounced at 10-mer probe sites*, again consistent with our observations. In support of this prediction is the observation that the 10-mer sites show the greatest percent change in adsorbate density during the long-term annealing observed in the adsorption kinetics, likewise pointing to a greater relative importance of competing P:P equilibria within shorter probe sites.

Concluding Remarks

We have presented preliminary investigations of the thermodynamics of DNA oligonucleotide adsorption to oligonucleotide probes at a solid-liquid interface, which represents a necessary first step toward understanding the behavior of the biological assays to which DNA probe arrays are applied. Much work lies ahead. The dependence of the melting behavior on such factors as base composition and sequence must be characterized. The nature of the nucleic acid target (DNA vs. RNA, degree of fragmentation, propensity for secondary structure formation, etc.) should be investigated. The quantitation of the effects of alternate bases and alteration of the chemical nature and charge of the polynucleotide backbone is another goal. Finally, the influence of the solution factors such as the nature and composition of the electrolyte and the use of agents such as tetramethylammonium chloride (TMAC), formamide, and cetyltrimethylammonium bromide (CTAB) on the adsorption kinetics and equilibria will be investigated.

Some of the broadening and depression of the melting curves has been shown to be a consequence of the presence of truncated probes. The melting behavior of probes synthesised by alternate methodoligies with higher stepwise yields should permit the

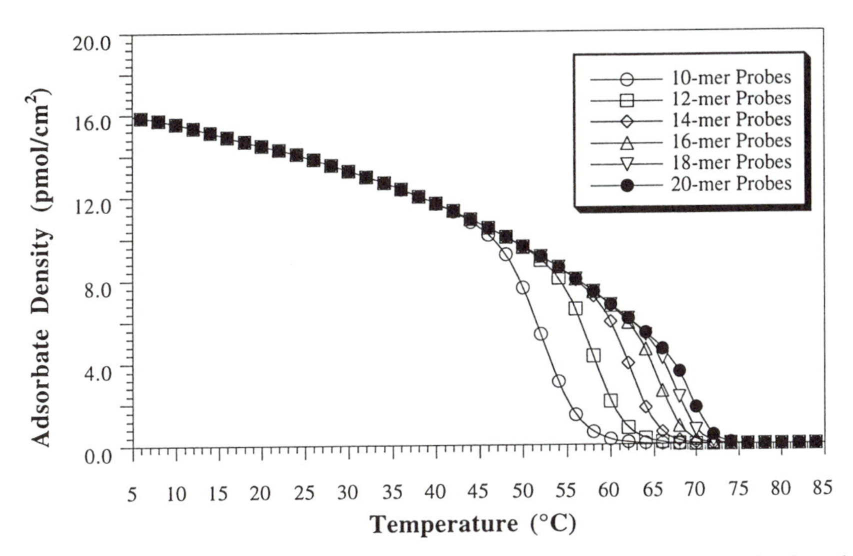

Figure 13. Melting curves expected for synthesis sites with the probe length distributions of Figure 12 assuming solution phase energetics, Langmuir adsorption, and 10 nM target concentration.

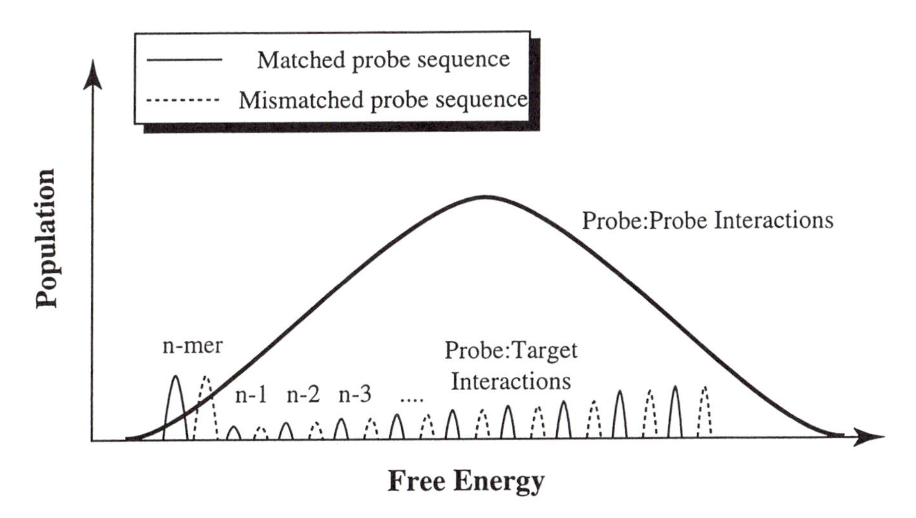

Figure 14. Free energy distributions for probe:target binding and probe:probe associations within a given synthesis site. The details of the distributions are heuristic value only and not intended to be quantitative.

effects of probe length heterogeneity to be better elucidated. The data suggest the importance of bridging interactions that become more pronounced at higher target concentrations, and associations between probes that compete with target binding. Although the data appear consistent with these phenomena, only through systematic and controlled reductions in probe density may this prediction be confirmed.

The peculiar maximum near 100 nM in the measured adsorption isotherms remains a puzzle. Although fluorescence quenching appears to be an unlikely explanation, detailed measurements of the fluorescence quantum yield as a function of adsorbed density have not been carried out and are warranted to address this issue. In any case, the anomalous behavior of the isotherms occurs at concentrations far higher than typically employed in biological applications of the probe arrays (1 pM to 10 nM) and is therefore not expected to be relevant to them.

It should be noted here that the divergence in saturating match and mismatch adsorption densities provides an unanticipated benefit to the execution and performance of assays based on probe arrays. For reasons of simplicity and economy it is desirable that the assay be performed under a single set of of stringency conditions, despite the wide range of duplex stabilities that are likely to be encountered over many kilobases of sequence. In contrast to solution behavior, mismatch discrimination is possible at any temperature below the T_m, dramatically relaxing the stringency conditions necessary for base calling within any particular segment of sequence. If the divergence in saturating match and mismatch densities is indeed a consequence of competing P:P equilibria, it may not be desirable from the standpoint of assay simplicity to eliminate such interactions. The test of this hypothesis awaits the results of probe density reduction experiments.

Acknowledgments

We thank Dale Barone, Jody Beecher, Mark Chee, Erik Gentalen, David Lockhart, Glenn McGall, and Audrey Suseno for general help and discussion, Paul Berg and Lubert Stryer for comments on our variable temperature results, and Tom Ryder for reviewing our manuscript.

References and Footnotes

1. GeneChip® is a U.S. registered trademark of Affymetrix, Inc.

2. (a) Lockhart, D. J.; Dong, H.; Byrne, M. C.; Follettie, T.; Gallo, M. V.; Chee, M. S.; Mittmann, M.; Wang, C.; Kobayashi, M.; Horton, H.; Brown, E. L.; *Nat. Biotechnol.* **1996**, *14*, 1675-1680. (b) Chee, M.; Yang, R.; Hubbell, E.; Berno, A.; Huang, X. C.; Stern, D.; Winkler, J.; Lockhart, D. J.; Morris, M. S.; Fodor, S. P. A.; *Science* **1996**, *274*, 610-614. (c) Kozal, M. J.; Shah, N.; Yang, R.; Fucini, R.; Merigan, T. C.; Richman, D. D.; Morris, D.; Hubbell, E.; Chee, M.; Gingeras, T. R.; *Nat. Med.* **1996**, *2*, 753-759. (d) Hacia, J. G.; Brody, L. C.; Chee, M. S.; Fodor, S. P. A.; *Nat. Genet.* **1996**, *14*, 441-447. (e) Shoemaker, D. D.; Lashkari, D. A.; Morris, D.; Mittmann, M.; Davis, R. W.; *Nat. Genet.* **1996**, *14*, 450-456. (f) Cronin, M. T.; Fucini, R. V.; Kim, S. M.; Masino, R. S.; Wespi, R. M.; Miyada, C. G.; *Hum. Mutat.* **1996**, *7*, 244-255. (g) Sapolsky, R. J.; Lipshutz, R. J.; *Genomics* **1996**, *33*, 445-456. (h) Lipshutz, R. J.; Morris, D.; Chee, M.; Hubbell, E.; Kozal, M. J.; Shah, N.; Yang, R.; Fodor, S. P. A.; *Biotechniques* **1995**, *19*, 442-447. (i) Pease, A. C.; Solas, D.; Sullivan, E. J.; Cronin, M. T.; Holmes, C. P.; Fodor, S. P. A.; *Proc. Natl. Acad. Sci. USA* **1994**, *91*, 5022-5026. (j) Fodor, S. P. A.; Read, J. L.; Pirrung, M. C.; Stryer, L.; Lu, A. T.; Solas, D.; *Science* **1991**, *251*, 767-773.

3. (a) Heller, R. A.; Schena, M.; Chai, A.; Shalon, D.; Bedilion, T.; Gilmore, J.; Woolley, D. E.; Davis, R. W, *Proc. Natl. Acad. Sci. USA* **1997**, *94*, 2150-2155. (b) Sosnowski, R. G.; Tu, E.; Butler, W. F.; O'Connell, J. P.; Heller, M. J.; *Proc. Natl. Acad. Sci. USA* **1997**, *94*, 1119-1123. (c) Shchepinov, M. S.; Case-Green, S. C.; Southern, E. M.; *Nucleic Acids Res.* **1997**, *25*, 1155-1161. (d) Drobyshev, A.; Mologina, N.; Shik, V.; Pobedimskaya, D.; Yershov, G.; Mirzabekov. A; *Gene* **1997**, *188*, 45-52. (e) Southern, E. M.; *Trends Genet.* **1996**, *12*, 110-115. (f) Yershov, G.; Barsky, V.; Belgovskiy, A.; Kirillov, E.; Kreindlin, E.; Ivanov, I.; Parinov, S.; Guschin, D.; Drobishev, A.; Dubiley, S.; Mirzabekov, A.; *Proc. Natl. Acad. Sci. USA* **1996**, *93*, 4913-4918. (g) Stimpson, D. I.; Hoijer, J. V.; Hsieh, W. T.; Jou, C; Gordon, J.; Theriault, T.; Gamble, R.; Baldeschwieler, J.; *Proc. Natl. Acad. Sci. USA* **1995**, *92*, 6379-6383. (h) Lamture, J. B.; Beattie, K. L.; Burke, B. E.; Eggers, M. D.; Ehrkich, D. J.; Fowler, R.; Hollis, M. A.; Kosicki, B. B.; Reich, R. K.; Smith, S. R.; Varma, R. S.; Hogan, M. E.; *Nucleic Acids Res.* **1994**, *22*, 2121-2125.

4. Cantor, C. R.; Schimmel, P. R.; *Biophysical Chemistry-Parts I, II, and III*; W. H. Freeman and Company, New York, 1980. (a) *Part I* (b) *Part II* (c) *Part III.*

5. (a) Chan, V.; Graves, D. J.; Fortina, P.; McKenzie, S. E.; *Langmuir* **1997**, *13*, 320-329. (b) Chan, V.; Graves, D. J.; McKenzie, S. E.; *Biophys. J.* **1995**, *69*, 2243-2255.

6. (a) Doktycz, M. J.; Morris, M. D.; Dormady, S. J.; Beattie, K. L.; Jacobson, K. B.; *J. Biological Chem.* **1995**, *270*, 8439-8445. (b) Aboul-ela, F.; Koh, D.; Tinoco, I., Jr.; *Nucleic Acids Res.* **1985**, *13*, 4811-4824. (c) Nelson, J. W.; Martin, F. H.; Tinoco, I., Jr.; *Biopolymers* **1981**, *20*, 2509-2531.

7. Williams, A. P.; Longfellow, C. E.; Freier, S. M.; Kierzek, R.; Turner, D. H.; *Biochemistry* **1989**, *28*, 4283-4291.

8. Lyon, W. A.; Forman, J. E.; *Unpublished Results.*

9. The apparently high discrimination ratio for the 18-mer site at 55°C and 500 nM target concentration is an artefact of the high bulk fluorescent background and low surface signals encountered under these conditions.

10. The sequence of photodeprotection and chemical coupling at match probe sites and corresponding mismatch probe sites of length L differ at only two steps out of a total of 4L synthesis steps. The probe densities and length distributions are thus expected to be quite similar.

11. McGall, G. H.; Barone, A. D.; Diggelmann, M.; Fodor, S. P. A.; Gentalen, E.; Ngo, N.; *J. Am. Chem. Soc.* **1997**, *119*, 5081-5090.

12. Probes that are truncated at the 3' end, formed if hydroxyl groups become exposed as the synthesis proceeds, cannot be detected with the HPLC analysis described in Figure 7. The proportion of truncated species may therefore be underestimated.

13. (a) Somorjai, G. A.; *Introduction to Surface Chemistry and Catalysis*, John Wiley and Sons, New York, 1994. (b) Parfitt, G. D.; Rochester, C. H. (eds); *Adsorption from Solution at the Solid/Liquid Interface*, Academic Press, London, 1983.

14. The energetics of the intermediate length duplexes were estimated by linear interpolation between our measured values for the 10-mer:20-mer and 20-mer:20-mer duplexes.

15. (a) *Gelest, Inc. Chemical Catalog*; Arkles, B. A. (ed); Gelest Inc., 1995. (b) Arkles, B. A.; *Personal communication.*

16. Weast, R. C.; Astle, M. J.; Beyer, W. H. (eds); *CRC Handbook of Chemistry and Physics*, CRC Press; Boca Raton, Fl.; 1987.

17 Ideally, the probes are distributed according to Poisson statistics,[18] introducing some degree of clustering with consequent reduction of average nearest neighbor distances. In practice, some additional clustering may take place due to chemical factors such as formation of domains. Average nearest neighbor distances computed under the assumption of a square lattice must therefore be regarded as an upper limit.

18. Cox, D. R.; Lewis, P. A. W.; *The Statistical Analysis of Series of Events*; Chapman and Hall, London 1963; page 213.

19. (a) Rau, D. C.; Parsegian, V. A.; *Biophys. J.* **1992**, *61*, 246-259. (b) Leikin, S.; Rau, D. C.; Parsegian, V. A.; *Physical Review A* **1991**, *44*, 5272-5278. (c) Podgornik, R.; Rau, D. C.; Parsegian, V. A.; *Macromolecules* **1989**, *22*, 1780-1786. (d) Rau, D. C.; Lee, B.; Parsegian, V. A.; *Proc. Natl. Acad. Sci. USA* **1984**, *81*, 2621-2625.

20. (a) Nielsen, P. E.; Haaima, G.; *Chem. Soc. Rev.* **1997**, *26*, 73-78. (b) Dueholm, K. L.; Nielsen, P. E.; *New J. Chem.* **1997**, *21*, 19-31.

21. Alberty, R. A.; *Physical Chemistry*, John Wiley and Sons, New York, 1987.

Chapter 14

RNA Folding Dynamics: Computer Simulations by a Genetic Algorithm

A. P. Gultyaev[1,2], F. H. D. van Batenburg[2], and C. W. A. Pleij[1]

[1]Leiden Institute of Chemistry, Department of Biochemistry, Leiden University, P.O. Box 9502, 2300 RA Leiden, Netherlands
[2]Theoretical Biology Section, Institute of Evolutionary and Ecological Sciences, Leiden University, Kaiserstraat 63, 2311 GP Leiden, Netherlands

Genetic algorithms exploit principles of natural evolution for optimization procedures. It is shown that RNA folding pathways may be simulated by a genetic algorithm. In addition to improvements of RNA structure predictions, an implementation of genetic algorithm allows one to simulate essential features of RNA dynamics, like e.g. the folding of a molecule during its synthesis, pseudoknot formation, functional metastable structures and conformational transitions. Thus it is shown that such an approach can also serve as a tool to study RNA dynamics.

Physical-Chemical Features of RNA Folding Process

RNA Folding Dynamics. For more than 30 years, it is known that the nucleotide sequence is not the only factor to determine RNA structure formation. In 1966, alternative conformations were revealed for a tRNA molecule, yeast $tRNA^{Leu}$ (*1*). Furthermore, it was shown that the alternative non-functional structure was not in thermodynamic equilibrium with the more stable functional folding, i.e. the RNA was captured in a metastable state during the folding process. Thus at physiological conditions the energy barrier between the metastable structure and the stable one resulted in a very long time (2 days) of transition to the energy minimum. Changes in temperature and ionic strength lowered the barrier so that the transition occurred faster (2 minutes). The conclusion was made that such a transition was not a simple superposition of tertiary structure on an already formed secondary structure, but rather some refolding that required disruption of a misfolded conformation (*1*).

Actually this means that the process of the functional structure formation may depend on a sequence of folding events and on their kinetics. In other words, some folding pathway seems to guide a molecule to the native state through a conformational space which may have an astronomic number of structures. This problem is extensively studied for protein folding, but similarities in folding dynamics of RNA and proteins are only starting to be recognized (*2-4*). An interest in RNA folding kinetics has been revived recently, mostly because of data on kinetics of tertiary structure formation in catalytic RNAs (*4-10*). The functional activity of ribozymes turned out to be dependent on specific structures which are formed stepwise with some steps relatively slow (on the timescale of minutes), and the rates of intermediate transitions can be measured by various methods.

Kinetics of RNA Secondary Structure Formation. At the level of secondary structure, RNA folding also occurs through intermediate conformations. Even in such a short molecule as tRNA (about 75 nucleotides), intermediate stages of the folding pathway can be detected and kinetic parameters for some separate steps were measured by NMR experiments with *E.coli* $tRNA^{fMet}$ (*11*). In this case folding is rather fast, with relaxation times in the range between a few microseconds and ten milliseconds. Therefore it can be concluded that the folding and thermal unfolding are reversible processes so that the final structure is the lowest energy state which is reached by a relatively fast pathway. In general, there is an hierarchy in the RNA folding process with initial formation of secondary structure followed by the folding of the less stable tertiary structure (*11,12*). However, in relatively long RNAs the folding may be more complicated. For example, a pseudoknotted mRNA fragment of 127 nucleotides from the regulatory region in a ribosomal protein operon was shown to fold *via* a pathway where a simple model of sequential formation of secondary and tertiary structures does not hold (*13*).

Furthermore, RNA folding is not necessarily a reversible process even at the step of secondary structure formation and dynamic factors are very important, especially for large molecules. Functional secondary structures of 16S-like ribosomal RNAs (more than 700 nucleotides) deviate from the lowest energy states, predicted by current methods for minimizing the free energy of secondary structure. In many cases the functional folding and the minimum energy structure share less than 50% of their helices (*14*). Although a correlation between such a deviation and RNA length is not straightforward, a contribution of kinetic factors seems to increase with the length, whereas relatively small domains still fold into structures of minimal free energy (*15*).

There is some hierarchy in the formation of RNA secondary structure. A rate of single helix (stem) formation depends mostly on the energy barrier of a loop which is formed while the reverse reaction of stem melting is determined by its stacking energy (*16-18*):

$$k_{folding} \sim \exp(-\Delta G_{loop} / RT)$$
$$k_{melting} \sim \exp(\Delta G_{stack} / RT).$$

This approximation assumes that the rate-limiting step for stem folding is the formation of the first basepair, followed by rather fast "zippering" of the stem by formation of adjacent pairs. The destabilizing loop contribution has mainly entropic character and increases with loop size. Thus, short-range hairpins are formed much faster than long-range interactions and the formation of some of the long-range pairings would require the disruption of the previously folded structures. Even if these structures are not the most stable ones, the disruption may be rather slow, being mostly determined by stacking energies, often with rather high values. Furthermore, if structural rearrangement requires disruption of several helices, the effective barrier is the sum of stacking values and may result in a very long time of disruption of all stems. This effect is seen in deviations from the lowest energy states in large domains of rRNAs, in contrast to small domains in the same molecules, which have rather low free energies (*15*). The analysis of the favourable structures of long mRNA transcripts also indicates that they seem to fold sequentially, keeping the stable hairpins and rearranging the long-range stems (*19*). Therefore initially formed hairpins guide a subsequent folding so that the final structure may strongly depend on the first nucleation steps. A change in conditions of the RNA folding process, e.g. in renaturation experiments, may capture many RNA molecules in misfolded conformations ("alternative conformer hell") for a long time (*20*).

A very important factor in RNA folding is the transcription rate. The folding process starts during the synthesis, when RNA transcription is not yet completed. Therefore some structures may be folded that are favourable only at an intermediate RNA length. In some cases they are disrupted when more stable structures become possible upon subsequent RNA elongation. However, many RNAs are trapped in such conformations for a considerable time. Furthermore, these metastable structures may be functional and sometimes they represent the more active conformation in contrast to the less active final folding. One of the best studied examples is represented by viroid RNAs which require metastable hairpins for efficient replication. Multi-branched hairpin-containing conformations of a viroid RNA are formed during transcription and can be detected right after RNA synthesis (*21*). Although at physiological conditions viroid RNAs are folded into very stable rod-like structures, the alternative metastable hairpins were shown to be very essential for replication intermediates (*22*). A similar situation was observed for SV11 (*23*), a small RNA species (115 nucleotides), efficiently replicated into a metastable hairpin-rich conformation by a phage replicase while the final folding into a single stable stem-loop structure was inactive (see also below). Monte Carlo simulations of the SV11 folding kinetics indicate that the metastable structure formation depends on the folding during the RNA synthesis, because this structure is predicted for the growing chain, but the stable conformation can be yielded when folding is simulated for the entire molecule (*24*).

It seems that an accurate formation of many functional RNA structures require a coordination between the rates of synthesis and folding. This is well illustrated by experiments with molecules synthesized by T7 RNA polymerase, which transcribes RNA at a high speed: about 250 nucleotides per second in vitro compared with 50 for *E.coli* RNA polymerase (*25*). Such a high transcription rate might capture an RNA in a metastable structure if the molecule did not have enough time to refold during transcription. This is observed for various RNAs. For example, misfolded conformations were suggested for T7 transcripts of *Tetrahymena* pre-rRNA, which were relatively slowly processed (*6*). The rate of splicing could be raised by a thermal denaturation/renaturation treatment, indicating the presence of non-equilibrated foldings after transcription. Similarly, inactive ribosomal particles were obtained when T7 RNA polymerase was used for rRNA synthesis (*26*). An elevated transcription rate was shown to be responsible for the effect, because an almost normal pattern was observed upon slower transcription at a lowered temperature by the same enzyme. Another example is overproduction of non-functional molecules of plasmid RNA primer, when T7 RNA polymerase is used (*27*).

In some cases the rate of RNA folding may play a functional role and be exploited by natural systems. Such kinetic models were proposed for the control of translation by RNA secondary structure (*28*, *29*). In these cases the translation is inhibited by a structure in the mRNA that sequesters the translation initiation region. Therefore the level of protein synthesis is very sensitive to the time of folding which actually regulates translation. For example, a long-distance interaction (LDI) was suggested to repress the synthesis of one of the proteins in phage MS2 (*29*). However, changing the thermodynamic stability of the LDI did not alter the repression. On the other hand, the mutations thought to change the folding rate did result in different levels of protein, indicating that the main regulatory factor is the time required to form the final structure rather than the stability of this folding at equilibrium (*29*). Such effects are not unique for translational control: a metastable structure of ColE1 plasmid RNA primer was suggested to regulate the plasmid copy number by a delay in the formation of folding that is sensitive to the antisense RNA inhibition (*30*).

Predictions of RNA Structure by Simulation of Folding Pathways

Optimal and Suboptimal RNA Structures. Upon implementation of recursive, or dynamic programming, algorithms, the calculation of the secondary structure with the minimal free energy has become possible (*31,32*). Such algorithms minimize free energy, using known or approximated energy parameters for elementary structural elements (helical stacks and loops). Thanks to the improvements in these parameters (*33, 34*) and to the progress in hardware development, prediction of the lowest free energy structure is performed in reasonable time even for relatively long RNA sequences and is widely used in laboratory practice. However, it turns out that for a given sequence numerous foldings exist with energy values which are very close to the optimal one. Such suboptimal solutions can also be computed and finding structures within some vicinity of the energy minimum can partly compensate for uncertainties in the thermodynamic data (*35*). Nevertheless, kinetic effects in RNA folding may trap a molecule in a metastable structure with an energy that is essentially higher than the global minimum. This suggests that a proper simulation of the folding process may predict the functional structure when it is not in the lowest energy state. Recently, it has been shown that predictions of suboptimal structures can be improved considerably if the hierarchy of RNA folding is mimicked interactively using folding constraints known from experiments or theoretical analysis, e.g. phylogeny (*36*).

Simulations of Folding Pathways. The algorithms for RNA folding simulation usually consider the folding as a stepwise process with elementary steps of formation and disruption of base pairs. The nucleation of a double helical region (stem) is much slower than its extension, therefore folding can be approximated with stems as elementary units. The simplest way is to start simulation from the most stable stem and to continue adding at every step the most stable stem from those compatible with the previously formed structure (*37-39*). Such an approach simulates some folding pathway, starting mostly from the most favourable short-range pairings followed by long-range interactions. A caveat of these algorithms is that sometimes they trap RNA in a wrong structure because of preference for single, relatively strong, stems instead of more stable combinations of stems that have smaller individual energy contributions. The favourable stem combinations can be considered by using a Monte Carlo procedure, which is less deterministic and allows to evaluate local structures prior to stem addition (*38, 40*).

However, the main advantage of stochastic approaches is the opportunity to implement kinetic features of elementary folding steps (*17, 24*). As mentioned above, a rate-limiting step for single stem folding is probably the formation of the first basepair, therefore the energy barrier may be approximated by the energy of the loop. In this approximation, the rate of the reverse reaction, i.e. stem melting, is determined by the stacking energy. If elementary folding steps are defined as reactions of formation or disruption of single stems, RNA folding may be simulated by a Monte

Carlo routine with probabilities of single events proportional to the calculated rates. Such a procedure can predict structures of relatively short RNAs and estimate kinetic ensembles of molecules (*17, 24*).

One of the essential problems in Monte Carlo simulations is a very broad range of rate constants in RNA folding, which vary by several orders of magnitude. This results in a serious technical difficulty of repeated iterations that follow the fastest transitions back and forwards without essential improvement of free energy that may be achieved by a slower reaction. Therefore an accurate simulation would require an enormous computation time. One of the ways to solve the problem is to define groups of structures that are in local equilibrium due to very fast transitions between them, and to represent the folding process as a network of slower reactions between these clusters (*41*). It remains to be seen how to implement this idea for successful predictions in relatively large RNAs. A promising approach to circumvent a tendency of Monte Carlo simulations to halt in local minima is the procedure of "simulated annealing" (*42*).

Simulations of RNA folding allow one to implement sequential folding during transcription, because a routine usually has some iterations of folding calculations which may be repeated for successively longer transcripts (see e.g. *24, 40, 42, 43*). Another important advantage of folding pathway simulations, compared to the search for the lowest energy, is an opportunity to include the formation of RNA pseudoknots, the simplest elements of tertiary structure (*39, 40*).

We have recently proposed to simulate RNA folding pathways by using a so-called genetic algorithm, which is also based on a stochastic procedure (*44, 45*). Below we describe possible ways of implementation and show how the main features of RNA dynamics can be simulated by the algorithm. It can be suggested that the principles of the genetic algorithm make it one of the most powerful approaches for RNA structure prediction and a very flexible routine to implement the data and ideas about the RNA folding process.

Implementation of the Principles of Genetic Algorithm for RNA Folding Simulation

Principles of Genetic Algorithms (GA). Genetic algorithms are a class of algorithms that have got their name because they exploit the basic principles of natural genetic evolution for optimization procedures (see e.g. *46, 47*). GA's are a very powerful technique, mostly in combinatorial problems that require a lot of computation for an exhaustive search. Details of the algorithms may vary, but basically all of them mimic Darwinian evolution of problem solutions in the computer, dealing with a population of these solutions that are mutated and selected according to some criterion. An individual solution ("chromosome") is stored as a combination of some elements ("genes"). Chromosomes are reproduced by random changes of genes ("mutations") and recombinations ("crossover") of subsets of genes from parent chromosomes. One of the most simple examples is the population of bit strings with random changes of ones to zeros and vice versa. If a combination of elements in every solution can be evaluated according to some "fitness", mutational operations can be followed by selection of the fittest solutions, and the next iteration of GA starts with a new population. There is a lot of freedom for modifications of the algorithm to improve its effectiveness, e.g. in deviations from an uniform random process, maintaining the diversity of the population etc. In an ideal case, repeated iteration of GA converges all solutions to the optimal combination, but the convergence can also result in a suboptimal solution that is relatively good in terms of problem definition.

GA's have received much attention in recent years. In chemistry, they can be used for a search in conformational space which very often involves combinations of many parameters. In particular, there were several attempts to use them for protein structure prediction (reviewed in *48*). Recently, GA's were suggested to use in three rather different aspects of RNA structure:conformational search for stem-loop structures (*49*), prediction of optimal and suboptimal secondary structures (*50*) and simulation of RNA folding pathways (*44,45*). In the case of RNA folding simulation, a GA is also very attractive because it allows to simulate the process, in addition to obtaining a final solution.

GA for RNA Folding. The main principles of GA for simulation of RNA folding (*44, 45*) are rather simple (Fig. 1). Every solution (structure) can be described as a set of stems. The operation of mutation is then disruption of some of these stems followed by formation of new ones. Crossover is the construction of a new structure that contains stems from parent solutions. The free energy of a given structure is the measure of fitness. Apparently the folding during transcription can be easily included, if structures are initially built for the 5'-proximal region with a gradual increase of chain length in subsequent GA iterations.

Apparently, the core of the procedure is the mutation operation, because this routine folds new structures, derived from those previously formed. Also, it is clear that this is the part of the algorithm, which allows to implement kinetic effects into the simulation. As mentioned above, disruption and formation of stems may be considered as elementary reactions in multistep RNA folding. Similar to the Monte Carlo procedure, at any given step of simulation a choice between different possible reactions can be made using their probabilities that depend on rate constants. Therefore addition and removal of stems with non-uniform probability distributions can mimic dynamic RNA folding, and different approximations may increase the efficiency of the algorithm. We consider separately some of these approximations, because they represent essential features of RNA dynamics.

Formation of Stems. Many algorithms for RNA folding simulation contain a routine that adds a new stem to the previously formed structure. Essentially this is done by calculating a list of stems that are compatible with the structure, together with free energies that could be gained (or lost) if particular stems are formed. The next step, to choose a particular stem, actually determines the differences between algorithms. For example, in the "greedy" algorithm (*39*) the most energetic stem is formed, in Monte Carlo procedure the selection is done using probabilities that depend on energy gains and/or energy barriers of loops (*17, 24, 38, 40*). Repeated additions of new stems represent folding steps.

We tested different ways to calculate probabilities of such steps. When stem selection followed a uniform probability distribution, that is, at every step the chances were equal for all stems from a list, the free energies of the structures did not improve very much from iteration to iteration. This is explained by the low probability of encountering an improving stem among others. However, a mere preference for the strongest (at a given step) stems did not produce correct structures either: a procedure to add stems with probabilities proportional to individual energy gains usually yielded incorrect long-range interactions that evolved just after the formation of the most stable hairpins.

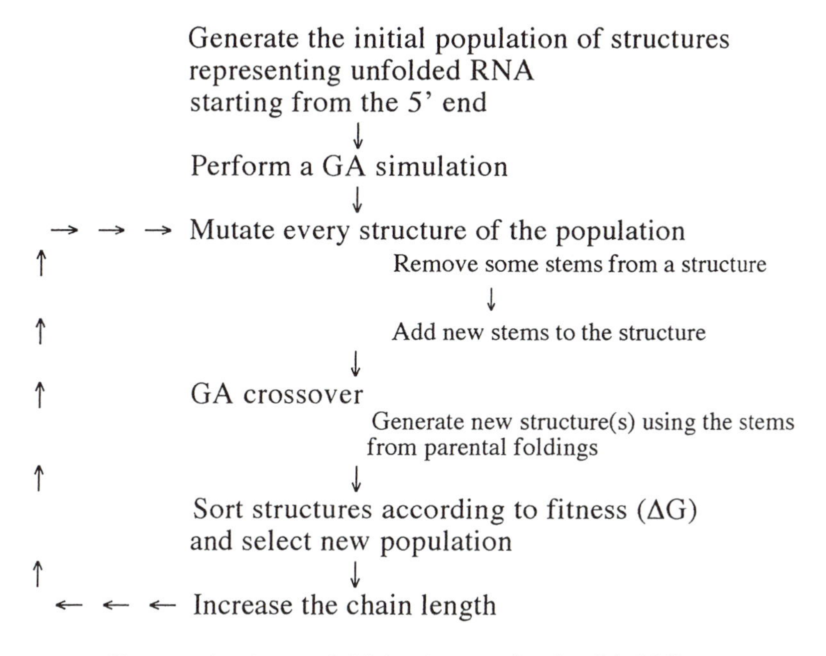

Figure 1. A scheme of GA implementation for GA folding

This suggests that the algorithm has to follow a more realistic pathway with short-range interactions prevailing over long-range pairings. Therefore a preference should be given to the stems which would result in relatively small loops. Such a preference was introduced into the algorithm (*45*) by making the probability of incorporating a given stem into the structure proportional to the free energy gain of the stem addition divided by the destabilizing energy of the new loop to be formed:

$$f \sim \Delta G(gain) / \Delta G(loop).$$

Although this approximation does not follow real exponential dependence of rate constants on a loop barrier (see above), the modification did improve the results considerably, both in terms of free energies of generated structures and in terms of consistency with phylogenetic data. More "strong" dependences on the loop energies resulted in a strong preference for short-range weak stems that captured simulation in unfavourable local minima, like in Monte Carlo routines. It seems that due to unavoidable hardware limitations some compromise has to be found between prediction of correct structures and using very close approximations for kinetic steps . One may also speculate that an account for free energy gain indirectly takes a reverse elementary reaction into account, that is, qualitatively "integrates" a reversible elementary step of RNA folding that includes both folding and melting of a stem. Of course, future studies of RNA kinetics and/or improvements of computer facilities will probably suggest better approximations, which are very easy to implement and to test in a genetic algorithm.

Disruption of Stems. In our implementation of GA (*45*) the formation of stems is preceded by a disruption of some stems in the previously folded structure (except the first steps starting with unfolded RNA chain). Again, we did not use an explicit formula for exponential dependence of disruption probability of a given stem on its stacking value. Instead, stems are disrupted in groups, located in some RNA region, with a preference for regions of relatively "weak" local structures. The probabilities of disruption of stems in particular regions were calculated by ascribing to every position in the sequence a value equal to some constant minus the sum of stacking energy values of stems embracing this position (*45*). Stems located in a vicinity of the selected point are disrupted. Such a strategy attempts to mimic sequential RNA folding (see *e.g. 19*) that mostly involves rearrangements of long-range pairings rather than interiors of local structures. In order to raise a chance for new long-range interactions, stems were always disrupted in two sequence regions, selected using described approximation for probabilities.

Energy Barriers of Refolding. Thus every mutation operation of the GA started with a disruption of some stems in a given structure (an individual of the genetic algorithm population). Then after calculation of a list of stems compatible with the remaining structure, new stems were added by stepwise selection from the list, using a probability distribution for stem formation. The procedure continued until no stem could be added.

It is clear that such a manipulation of an RNA structure in GA is equivalent to climbing some energy barrier due to stem disruption, followed by descending to a (probably new) local minimum of free energy determined by stem formation. Repeated GA mutations represent a walk in a free energy landscape for a given sequence, that is, RNA folding pathway (*44*). As mentioned in the first section, the time to overcome some of the barriers is very long, therefore the kinetics of the pathway may retain a molecule in some valley of the landscape, even in the case when another valley has a deeper minimum. This important kinetic feature was also implemented in the algorithm: relatively big barriers required for stem disruption were discouraged by introducing relatively low probabilities to overcome such barriers (*45*). Thus, when the removal of stems resulted in a loss of more than 10 kcal/mol, this barrier could be passed by with a probability inversely proportional to the squared value of the total loss in ΔG:

$$f = 100 / (\Delta G)^2.$$

Therefore some stem could fold back with a probability $1 - f$. The choice of such stems was made similar to the operation of stem addition described above, the loss in energy being recalculated after each addition of a stem. It should be noted that introduction of such a smooth (compared to exponential) barrier dependence did improve the predictions (see also below).

Selection of Structures. Every iteration of the GA (*45*) included a mutation (stem disruption/formation) of every structure in the population, thus doubling this population. In addition, a single operation of GA crossover was implemented, using all structures as parental solutions. Compared to more typical crossover operations between the most fittest parents, such a modification was better because it did not result in very fast convergence in some local minimum. Thus all stems from all parental solutions were mixed in one list and for a new solution stems were selected by the same procedure that was used for adding mutations, that is, with probabilities proportional to free energy gain divided by loop energy.

Mutations and crossover are followed by selection of a new generation of structures, creating a new population of initial size. Free energy is the main fitness criterion, therefore the best structure was always selected for a new population. Also, the diversity of the population was maintained by taking structural differences between solutions into account in order to favour those that deviated from the best one and to avoid very fast convergence. So for each structure the fitness parameter was calculated as the difference between its energy and that of the best structure, divided by the number of stems that were present only in one of these solutions. This modification raises the chance that structures of moderate fitness could be improved in subsequent iterations of the algorithm.

Upon the selection of a new population of structures, a GA iteration is accomplished and the next iteration starts with structures that are generally better in terms of free energy (at least not worse). However, this does not necessarily mean that the algorithm eventually finds the lowest energy state, both because of the enormous number of possible solutions and the implementation of kinetic features of RNA folding.

Chain Growing. The folding during transcription is easily introduced by restricting the folding at GA iterations to intermediate transcript sizes (with the same 5' end), with gradual increase of the length from iteration to iteration until reaching the full-length value. The described implementation of GA does not have an explicit timescale. However, for an approximation of the apparent relationship between arbitrary units of folding and elongation rates in the computer simulation, we made the chain growth increment dependent on the rate of energy improvement produced by the current iteration as compared with the previous ones: the greater the improvement, the smaller the increase of chain length. Such an approximation follows the qualitative assumption that stable local structures are folded relatively quickly. In our implementation (*45*) an elongation at any given step never exceeded 10 nucleotides.

Parameters of GA. An implementation of GA for RNA folding is very flexible for various modifications which improve the calculations. Some of such modifications may be relevant to the natural dynamics of RNA, others are just technical improvements of the algorithm. For example, in our implementation (*45*) stems with positive free energies (probably requiring some intermediate to be formed) were excluded from lists used for stem addition at a given mutation cycle. Such a stem could appear at the next step, if a loop contribution had changed due to formation of other helices; this could reflect a stepwise dynamics of RNA folding. On the other hand, many parameters of simulation are just adjustments of the procedure. One of the essential parameters is the population size. For RNAs without important metastable structures, the optimum found was about five solutions (*45*). Smaller numbers of structures were not sufficient for good results, but increasing this value increased the time of calculations without significant improvement. On the other hand, for molecules with high barriers between metastable and stable structures an increase in the population size may be useful (Gultyaev, A.P. *et al.*, unpublished data). Of course, formulas given above for barrier estimates are only approximations; they could be improved or implemented in some other way.

Simulation of RNA Folding as a Tool to Study RNA

Quality of Predictions. The capability of the algorithm to predict the final RNA structure was tested using phylogenetically proven structures. The GA predictions were compared with supported foldings and with the predictions produced by the search for the lowest free energy. The GA simulations proved to predict RNA structures reasonably well, usually yielding about 80% of correct structure (*45*). In addition, some of the predictions turned out to be better than those given by the global minimum solution, despite the use of the same energy parameters. This is explained by the specific folding pathways, generated by GA. In case of rRNA folding, the effects of such pathways are probably one of the reasons for the large deviations of structures from the lowest energy structure (*14*). Although a GA simulation does not take other important effects (like e.g. ionic environment or protein binding) into account, it simulates a folding pathway that avoids prediction of structures that cannot be reached due to very high energy barriers.

Predictions of Pseudoknots. As could be expected, the biggest difference between predictions using the GA simulations and those given by the search for minimal free energy is observed for foldings of pseudoknot-containing RNAs (*45*). This emphasizes the importance of including pseudoknots into RNA structure predictions and strongly advocates using GA simulations for pseudoknot predictions, because current implementations of free energy minimization exclude pseudoknot formation (*32*). In particular, the advantages of taking pseudoknots into account are pronounced in predictions for molecules with many pseudoknots like e.g. some plant viral RNAs (Figure 2). In principle, a (sub)optimal secondary structure for a pseudoknot-containing RNA may be predicted by energy minimization, if calculations take into account some experimental data for the conditions (e.g. elevated temperatures) that disrupt tertiary structure while secondary structure is still retained (*12*). However, this

a TMV

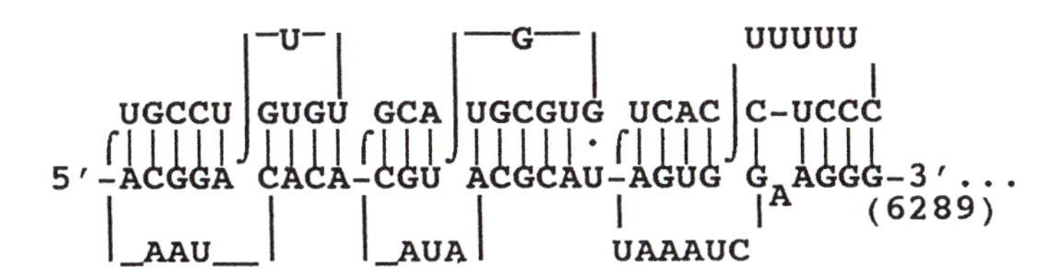

Figure 2. Examples of multiple pseudoknots predicted in the plant viral RNAs of (a) tobacco mosaic virus (*39*) and (b) satellite tobacco mosaic virus (*52*).

Continued on next page

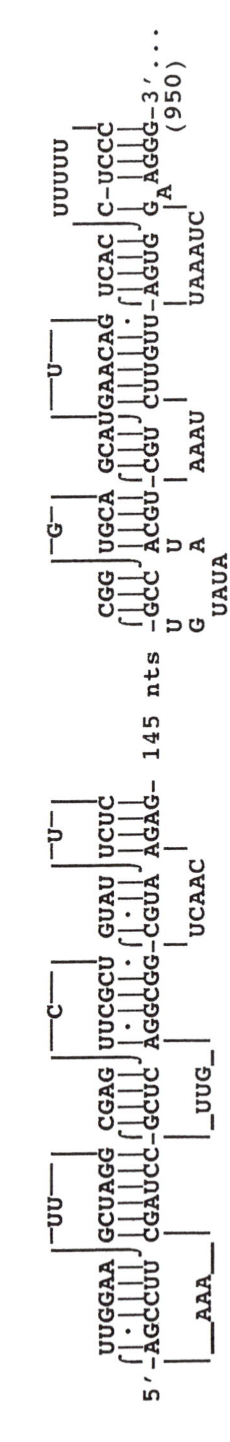

Figure 2. *Continued.*

demands extensive experimental study and a preliminary knowledge that pseudoknots do fold in the RNA in question. Also, in some cases tertiary structure elements are probably formed together with a secondary structure rather than as a separate step (*13*). Although usually pseudoknots are less stable compared to the secondary structure, in the case of RNA with pseudoknots a program neglecting them will predict some, probably weak, secondary structure elements that could be misleading. Therefore simulations of RNA folding including pseudoknots, even with approximated loop energies, are very useful for pseudoknot predictions, especially when pseudoknot formation is not known in advance (see e.g. *39, 51, 52*).

Relevance of GA Simulations to the Real Folding Pathways. At every GA iteration, the simulation can display the best structure found so far, and the sequence of these intermediate structures may be considered as the simulated folding pathway. It is likely that at least some of the predicted intermediates do constitute essential steps of the natural pathway. This can be tested using predictions for molecules that are known to have a relatively long lifetime of metastable structures.

It turned out that GA-simulated folding pathways can be very close to the natural ones, despite the fact that GA simulations do not have an explicit timescale of folding. The most clear illustration is the folding of SV11, a molecule of 115 nucleotides, selected in experiments on replication by phage Qβ replicase (*23*). SV11 is an active template in a metastable conformation while the final structure is inactive. The simulated folding pathway is summarized in Figure 3 (*45*). The simulation for the folding of the growing chain predicted the consecutive formation of stems that eventually yielded the metastable structure (Figure 3a). After this initial folding, the GA simulation did not produce any change for a considerable number of iterations, because newly formed structures were less stable compared with that already predicted. However, after a substantial number of GA cycles (varying in different runs) some new stems were found that could rearrange in the much more stable final structure (Figure 3b). Thus it is seen that the whole spectrum of processes, known for the folding of this molecule (*23*), may be simulated by the GA procedure: the folding during replication, the capture of a molecule in the metastable state for a considerable time and the final transition to the structure of free energy minimum (*45*).

Thus, in addition to the prediction of RNA structures, GA simulation predicts their folding pathways. Presumably this is due to qualitative mimicking of multistep dynamics of RNA folding. It is noteworthy that the algorithm did not produce reliable final predictions when the options modelling kinetic steps were not included. Only implementation of important kinetic features (like e.g. barrier restrictions, discrimination between fast and slow processes, folding during transcription) resulted in succesful predictions of phylogenetically proven structures. This clearly demonstrates the potential to use GA for RNA structure predictions, and as a tool for the study of the folding process itself. Of course, the GA approach may be further improved by a knowledge about RNA folding, but this knowledge may also benefit from GA simulations, because different ideas about RNA dynamics can be easily tested by "computer experiments".

Revealing Important Metastable Structures by GA. Metastable RNA structures are transient conformations and therefore are rather difficult targets for an investigation. Simulations of RNA folding by GA can reveal such structures. For example, GA simulations predicted that the RNA primer for replication of ColE1 plasmid is folded through an intermediate metastable structure that has a different activity compared to

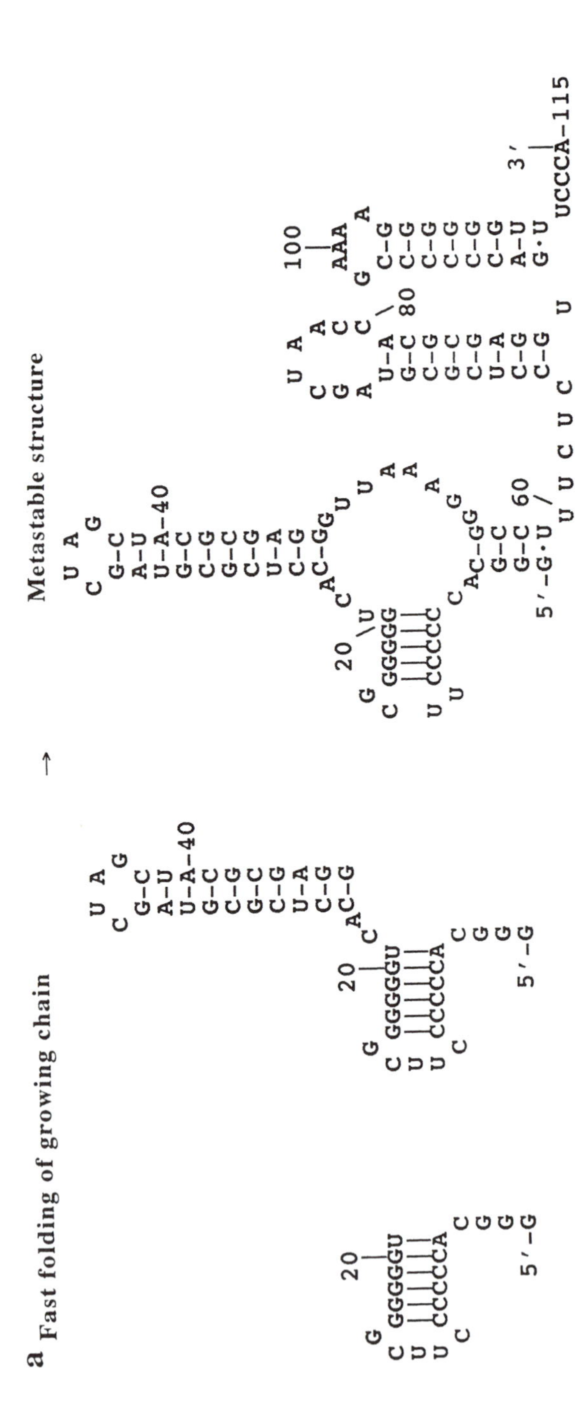

Figure 3. The folding pathway simulated for SV11 RNA. (Adapted with permission from reference 45. Copyright 1995 Academic.)

Continued on next page

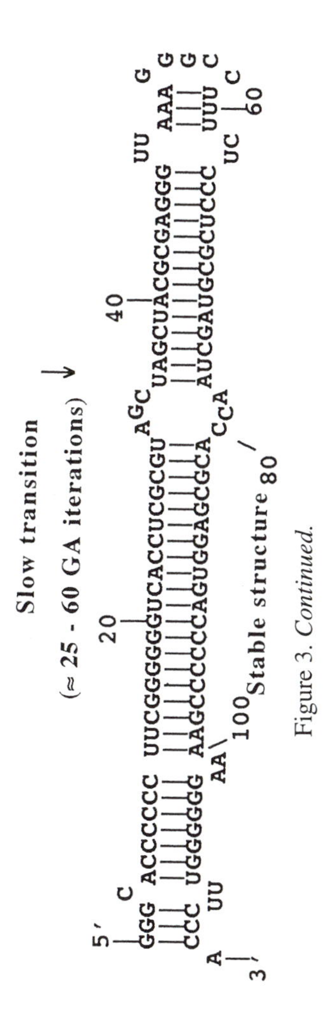

Figure 3. *Continued.*

the final stable conformation (*30*). Therefore, the lifetime of the metastable folding modulates the function of the RNA primer and thereby the efficiency of plasmid replication. In this case the folding pathway, suggested by GA predictions, can be used for estimates of kinetic curves, using exponential dependences for rate constants of elementary steps in the model of multistep RNA refolding (*18, 30*). It is shown that the effect of some plasmid copy number mutations is to stabilize the metastable structure, which then causes a delay in the final structure formation, while the properties of the final folding itself seem to remain unchanged.

Recently we have revealed a conserved metastable hairpin in a group of viroid RNAs (Gultyaev, A.P. *et al.*, in preparation). This hairpin, which may be formed in replicative minus-strand intermediates, is suggested to guide the folding of active RNA templates that escape from the folding into the stable rod-like structure and form functional metastable structures (*21, 22*). Presumably, detection of functional transient RNA foldings is one of the most interesting applications of GA.

Concluding Remarks

Simulations of RNA folding using a GA demonstrate that this approach is a very useful one in predicting RNA secondary structures, especially when they are not in the global minima of free energy or contain RNA pseudoknots. However, it seems that the main advantage is the ability of simulations to mimic the RNA folding process. This gives an opportunity to predict metastable foldings that may be very important for RNA functioning, despite the fact that such foldings could have a transient nature. Even simplified kinetic models, incorporated in simulations, can give reliable predictions. On the other hand, the results of using various dynamic models in the algorithm, as seen in comparisons of predictions with known data, may be used for conclusions about the validity of these ideas. This suggests an exciting field of studies on RNA dynamics by "computer experiments".

Acknowledgment

The investigations were supported by The Netherlands Foundation for Chemical Research (SON) and Foundation for Life Sciences (SLW) with financial aid from The Netherlands Organization for Scientific Research (NWO).

Literature Cited

1. Fresco, J.R.; Adams, A.; Ascione, R.; Henley, D.; Lindahl, T. *Cold Spring Harbor Symp. Quant Biol.* **1966,** *31*, 527-537.
2. Thirumalai, D.; Woodson, S.A. *Acc. Chem. Res.* **1996,** *29*, 433-439.
3. Draper, D.E. *Nature Struct. Biol.* **1996,** *3*, 397-400.
4. Zarrinkar, P.P.; Williamson, J.R. *Nature Struct. Biol.* **1996,** *3*, 432-438.
5. Bevilacqua, P.C.; Kierzek, R.; Johnson, K.A.; Turner, D.H. *Science* **1992,** *258*, 1355-1358.
6. Emerick, V.L.; Woodson, S.A. *Biochemistry* **1993,** *32*, 14062-14067.
7. Zarrinkar, P.P.; Williamson, J.R. *Science* **1994,** *265*, 918-924.
8. Banerjee, A.R.; Turner, D.H. *Biochemistry* **1995,** *34*, 6504-6512.
9. Zarrinkar, P.P; Wang, J.; Williamson, J.R. *RNA* **1996,** *2*, 564-573.
10. Downs, W.D.; Cech, T.R. *RNA* **1996,** *2*, 718-732.
11. Crothers, D.M.; Cole, P.E.; Hilbers, C.W.; Shulman, R.G. *J. Mol. Biol.* **1974,** *87*, 63-88.
12. Banerjee, A.R.; Jaeger, J.A.; Turner, D.H. *Biochemistry* **1993**, *32*, 153-163.

13. Gluick, T.C.; Draper, D.E. *J. Mol. Biol.* **1994,** *241*, 246-262.
14. Konings, D.A.M.; Gutell, R.R. *RNA* **1995,** *1*, 559-574.
15. Morgan, S.R.; Higgs, P.G. *J. Chem. Phys.* **1996,** *105*, 7152-7157.
16. Porschke, D. *Biophys. Chem.* **1974,** *1*, 381-386.
17. Mironov, A.A.; Dyakonova, L.P.; Kister, A.E. *J. Biomol. Str. Dyn.* **1985,** *2*, 953-962.
18. Tacker, M.; Fontana, W.; Stadler, P.F.; Schuster, P. *Eur. Biophys. J.* **1994,** *23*, 29-38.
19. Nussinov, R.; Tinoco, I. *J. Mol. Biol.* **1981,** *151*, 519-533.
20. Uhlenbeck, O.C. *RNA* **1995,** *1*, 4-6.
21. Hecker, R.; Wang, Z.; Steger, G.; Riesner, D. *Gene* **1988,** *72*, 59-74.
22. Loss, P.; Schmitz, M.; Steger, G.; Riesner, D. *EMBO J.* **1991,** *10*, 719-727.
23. Biebricher, C.K.; Luce, R. *EMBO J.* **1992,** *11*, 5129-5135.
24. Higgs, P.; Morgan, S.R. In *Advances in Artificial Life;* Moran, F; Moreno, A.; Merelo, J.J; Chacon, P., Eds.; Lecture Notes in Artificial Intelligence; Springer: Berlin, 1995, Vol. 929; pp 852-861.
25. Golomb, M.; Chamberlin, M. *J. Biol. Chem.* **1974,** *9*, 2858-2863.
26. Lewicki, B.T.U.; Margus, T.; Remme, J.; Nierhaus, K.H. *J. Mol. Biol.* **1993,** *231*, 581-593.
27. Chao, M.Y.; Kan, M.-C.; Lin-Chao, S. *Nucleic Acids Res.* **1995,** *23*, 1691-1695.
28. Ma, C.K.; Kolesnikow, T.; Rayner, J.C.; Simons, E.L.; Yim, H.; Simons, R.W. *Molec. Microbiol.* **1994,** *14*, 1033-1047.
29. Groeneveld, H.; Thimon, K.; van Duin, J. *RNA* **1995**, *1*, 79-88.
30. Gultyaev, A.P.; Batenburg, F.H.D. van; Pleij, C.W.A. *Nucleic Acids Res.* **1995,** *23*, 3718-3725.
31. Nussinov, R.; Jacobson, A.B. *Proc. Natl. Acad. Sci. USA* **1980,** *77*, 6309-6313.
32. Zuker, M.; Stiegler, P. *Nucleic Acids Res.* **1981,** *9*, 133-148.
33. Freier, S.M.; Kierzek, R.; Jaeger, J.A.; Sugimoto, N.; Caruthers, M.H.; Neilson, T.; Turner, D.H. *Proc. Natl. Acad. Sci. USA* **1986,** *83*, 9373-9377.
34. Jaeger, J.A.; Turner, D.H.; Zuker, M. *Proc. Natl. Acad. Sci. USA* **1989,** *86*, 7706-7710.
35. Zuker, M. *Science* **1989,** *244*, 48-52.
36. Gaspin, C.; Westhof, E. *J. Mol. Biol.* **1995,** *254*, 163-174.
37. Nussinov, R.; Pieczenik, G. *J. Theor. Biol.* **1984,** *106*, 245-259.
38. Martinez, H.M. *Nucleic Acids Res.* **1984,** *12*, 323-334.
39. Abrahams, J.P.; van den Berg, M., Batenburg, E. van.; Pleij, C. *Nucleic Acids Res.* **1990,** *18*, 3035-3044.
40. Gultyaev, A.P. *Nucleic Acids Res.* **1991,** *19*, 2489-2494.
41. Mironov, A.A.; Lebedev, V.F. *BioSystems* **1993,** *30*, 49-56.
42. Schmitz, M.; Steger, G. *J. Mol. Biol.* **1996,** *255*, 254-266.
43. Mironov, A.; Kister, A. *J. Biomol. Str. Dyn.* **1986,** *4*, 1-9.
44. Batenburg, F.H.D. van; Gultyaev, A.P.; Pleij, C.W.A. *J. Theor. Biol.* **1995,** *174*, 269-280.
45. Gultyaev, A.P.; Batenburg, F.H.D. van; Pleij, C.W.A. *J. Mol. Biol.* **1995,** *250*, 37-51.
46. Holland, J.H. *Adaptation in natural and artificial systems;* University of Michigan Press: Ann Arbor, 1975.
47. Forrest, S. *Science* **1993,** *261*, 872-878.
48. Pedersen, J.T.; Moult, J. *Curr. Opinion Struct. Biol.* **1996,** *6*, 227-231.
49. Ogata, H.; Akiyama, Y.; Kanehisa, M. *Nucleic Acids Res.* **1995,** *23*, 419-426.
50. Benedetti, G.; Morosetti, S. *Biophys. Chem.* **1995,** *55*, 253-259.
51. ten Dam, E.; Pleij, C.W.A.; Draper, D. *Biochemistry* **1992,** *31*, 11665-11676.
52. Gultyaev, A.P.; Batenburg, F.H.D. van; Pleij, C.W.A. *J. Gen. Virol.* **1994,** *75*, 2851-2856.

Chapter 15

An Updated Recursive Algorithm for RNA Secondary Structure Prediction with Improved Thermodynamic Parameters

David H. Mathews[1], Troy C. Andre[1], James Kim[1], Douglas H. Turner[1,3], and Michael Zuker[2]

[1]Department of Chemistry, University of Rochester, Rochester, NY 14627–0216
[2]Institute for Biomedical Computing, Washington University, St. Louis, MO 63110

An updated recursive algorithm that minimizes free energy predicts 82.5% of phylogenetically determined base pairs from sequence in four small subunit rRNAs, four group I introns, three group II introns, and 41 tRNAs. The rRNAs and group II introns were folded in phylogenetically determined domains of no more than 500 nucleotides. The algorithm incorporates recently determined thermodynamic parameters for the free energies of internal loops of 2 by 1 and 2 by 2 nucleotides. New free energy bonuses for tetraloops and triloops have been developed by consideration of the database of phylogenetically determined structures. Finally, new rules for coaxial stacking have been applied. This new version will be available in FORTRAN for Unix machines and a C++ version is now available for use on Personal Computers with Windows 95 or Windows NT. The program was used to explore structures predicted to have a free energy near the minimum. On average, a structure with 92% of phylogentically determined base pairs is found within 2% of the minimum free energy. For a roughly 400 nucleotide RNA, this is typically 2.3 kcal/mol above the minimum free energy. Implications for determining RNA secondary structure from sequence are discussed.

Sequence information is being collected at a rate faster than a million nucleotides per day. These data are having a large impact on our understanding of biology and medicine, but to extract the maximum amount of information, additional computational analysis of the database will be necessary. One area of analysis is the prediction of RNA secondary structure from sequence.

When a large number of phylogenetically related sequences are known, sequence comparison is the standard technique for determining RNA secondary structure (*1*). Commonly, however, there are no related sequences or too few sequences for rigorous comparative methods. Thermodynamics can be used to predict

[3]Corresponding author

secondary structure when only one or a few sequences are available and to facilitate sequence comparison when many sequences are available (*2-6*).

Several approaches have been investigated for the prediction of RNA secondary structure from a single sequence. The recursive algorithm used in this study predicts a lowest free energy structure and a set of structures, called sub-optimal structures, within a given increment of free energy (*4,7*). It can also be used to generate an energy dot plot, a graphical representation of all base pairs involved in suboptimal structures. A dot plot can identify base pairs that are well determined in the predicted structure (*8*). Another recursive algorithm, by McCaskill (*9*), calculates a partition function, thus providing the probability of pairing of individual bases and the probability of individual base pairs. Another approach for RNA structure prediction is the genetic algorithm (*10,11*). The genetic algorithm has the advantage of being able to simulate possible kinetic contributions to folding. It can also predict pseudoknots, which recursive algorithms cannot. The genetic algorithm, however, takes much more time to predict a structure than the recursive algorithms.

All three approaches to predicting secondary structure from a single sequence rely upon nearest neighbor free energy parameters. Parameters for Watson-Crick helical regions are well determined experimentally (*12,13*), but the parameters for unpaired regions are an area of active research (*14-18*). It is apparent that many of the free energy models are over-simplified because they neglect the sequence of unpaired regions. Presently, it is impossible to study every possible variation for even small structural motifs. For example, a symmetric internal loop of four with two unpaired nucleotides per strand has 4,704 distinct sequences when the closing base pairs are considered. Therefore, the free energies of these motifs must be deduced by other methods. Development of new experimental approaches, however, may eventually allow measurement of most sequence variations (*19,20*).

Several methods are available for approximating the free energy parameters of structural motifs. Often, an empirical model is developed that is based on free energies determined by optical melting experiments for representative sequences (*12-18*). Another method for assigning stability to unpaired regions is to base it on the frequency of occurrence of different sequences for motifs in a database of phylogenetic structures (*3*). Lastly, parameters can be varied so as to optimize the accuracy of computer folded sequences against their phylogenetically determined structures (*2,3,21*). The last two approaches may partially simulate the effect of tertiary contacts on the stability of secondary structure. A combination of all three methods were used to update the parameters in this study. Eventually, it may be possible to calculate these parameters explicitly using methods such as free energy perturbation.

This preliminary report presents recent progress in secondary structure prediction based on free energy minimization. The following changes have been implemented: The method for forcing base pairs has been improved. A filter that removes isolated Watson-Crick or G-U base pairs (those that cannot stack on any other Watson-Crick or G-U pair) has been incorporated. Recently measured free energies for 2 by 2 internal loops (Xia, T.; McDowell, J. A.; Turner, D. H. In preparation.), 2 by 1 internal loops (*15*), and hairpin loops (*18*) have also been incorporated. Finally, a new model for coaxial stacking of helixes has been developed.

Methods

Folding Algorithm. The folding algorithm is freely available in several formats. A C++ coded version, called RNAstructure version 2.0, for personal computers running Windows 95 or Windows NT is available on the Turner lab homepage at http://rna.chem.rochester.edu. This Windows version has a graphical user interface. The computational time for a typical sequence, 433 nucleotides of the group I LSU intron in *Tetrahymena thermophila*, is 135 seconds on a Pentium 120 with 48 MB of RAM running Windows 95. The structures for this paper were predicted with the C++ code compiled for a Silicon Graphics work station.

A version of the algorithm in Fortran, called Mfold, for use on Unix machines will be at Michael Zuker's ftp directory, ftp://snark.wustl.edu/pub. This version will also be available for online folding through a World-Wide-Web interface at the Zuker homepage, http://www.ibc.wustl.edu/~zuker/rna/form1.cgi.

Thermodynamic Parameters. The thermodynamic parameters are taken from previous studies of RNA folding (*2,3*) with the exception of recently studied motifs. Changes have been made in the stabilities of internal loops of 1 by 2 nucleotides (one unpaired nucleotide opposite two unpaired nucleotides) and 2 by 2 nucleotides (two unpaired nucleotides opposite two unpaired nucleotides), hairpin loops, and multibranch loops. The complete tables of parameters are available on the Turner Lab Homepage at http://rna.chem.rochester.edu.

Internal Loops. The free energy parameters of 2 by 2 internal loops (also called tandem mismatches) have been studied for the cases of symmetric mismatches (*14*) and non-symmetric mismatches (Xia, T.; McDowell, J. A.; Turner, D. H. In preparation.). A preliminary model by Xia et al. for approximating the stabilities of 2 by 2 loops that have not been measured was used to fill a table of thermodynamic parameters for all possible 2 by 2 mismatches and closing base pairs for a total of 4,704 parameters. The algorithm now consults this table when determining the free energy of a 2 by 2 internal loop.

Adjustments to the table of 2 by 2 internal loops were made based on comparisons of predicted and known secondary structures. The stability of $\begin{matrix}\text{GACU}\\\text{CACA}\end{matrix}$ was increased from 3.7 to 2.2 kcal/mol. The parameters for $\begin{matrix}\text{GGXU}\\\text{CAYA}\end{matrix}$ with XY as AA, AC, AG, CA, CC, CG, CU, GA, GG, UC, and UU were made 0.7, 0.4, -0.7, 0.4, 0.4, 1.2, 0.4, -0.7, 0.3, 0.6, and 0.5 kcal/mol, respectively. These are within 0.5 kcal/mol of the preliminary model, with the exception of CG, which is improved in stability by 0.8 kcal/mol.

The revised free energy rules for the 2,304 possible 2 by 1 internal loops are based on the work of Schroeder et al. (*15*). A table containing a free energy for each possible 2 by 1 loop and closing base pair was added to the algorithm. The parameters for C-G closures were those measured by Schroeder et al. (*15*). The parameters for other closing pairs were estimated by adding 0.7 kcal/mol for each A-U or G-U closure.

Hairpin Loops. The free energy of hairpins is based on the model of Serra and co-workers (*16-18*). In this model, hairpin stability for loops larger than three nucleotides is independent of sequence with the exception of the first mismatch and closing pair. The free energy for a hairpin loop, ΔG°_{HL}, is:

$$\Delta G_{HL}^\circ = \Delta G^\circ_i + \Delta G^\circ_{stack} \tag{1}$$

where ΔG°_i is the free energy penalty for the closure of a loop of length i and ΔG°_{stack} is the free energy for stacking of the first mismatch on the helix and is approximated by:

ΔG°_{stack} (kcal/mol) = ΔG°_{mm} + 0.6 (if closed by an A-U or U-A pair) - 0.7 (if first mismatch is GA or UU) (2)

ΔG°_{mm} is the free energy for a mismatch at the end of a helix (*12*). For the algorithm, the parameters for ΔG°_{stack} were varied within 1.0 kcal/mol from the model to maximize the number of correctly predicted basepairs in the database of structures presented below. This was done by repeatedly making changes based on comparisons of predicted structures to phylogenetic structures. The final free energies are shown in Table I. The parameter ΔG°_i is 4.9, 5.0, 5.0, 5.0, 4.9, and 5.5 kcal/mol for loops of length 4-9, respectively (*18*). This model is known to predict free energy changes for the formation of several naturally occurring hairpin loops, but underestimates the stability reported for the hexanucleotide hairpin loop ACAGUGCU (*22*).

For loops of three nucleotides, the free energy is entirely independent of loop sequence:

ΔG°_{37} (kcal/mol) = 4.8 + 0.6 (if closed by an A-U or U-A pair) (3)

Closure of hairpins of less than three nucleotides are not considered in the folding algorithm.

Specific tetraloops and triloops, i.e. hairpin loops of four and three nucleotides, respectively, are given an enhanced stability. Some of these hairpins are known to be more stable than predicted by the model above (*23-27*) and it is known that these motifs are an important component of tertiary structure (*28-30*) and are therefore stabilized by tertiary contacts. Previous studies have given stability bonuses to tetraloops according to the sequence of the unpaired nucleotides alone (*2,3*). A search of the database of phylogenetic structures of small and large subunit ribosomal RNA and Group I introns (*31,32*) shows that the occurrence of tetraloops also depends on the closing base pair (*33*). Thus we tried assigning bonuses to specific tetraloops and triloops according to the sequence of the unpaired nucleotides and the closing base pair. Table II shows the bonuses given to tetraloops and triloops. The bonuses are based on phylogenetic occurrence and on optimizing the accuracy of folding. There are 50 tetraloops with varying bonus stability, whereas the study of Walter et al. (*2*) gave a flat -2.0 kcal/mol to 78 tetraloops when all the closing base pair possibilities are considered.

Table I. Stability of Closing Base pair and First Mismatch in Hairpin Loops.

Base pair	X↓	Y→ A	ΔG°_{37} (kcal/mol) C	G	U
AX UY	A	-0.0	-0.4	+0.3	-0.3
	C	-0.0	-0.1	-1.5	+0.5
	G	-0.5	-1.1	-0.2	+0.1
	U	-0.3	-0.2	-0.0	-0.5
CX GY	A	-1.5	-1.1	-1.4	-1.8
	C	-1.0	-0.9	-2.9	-0.6
	G	-1.9	-2.0	-1.6	-1.2
	U	-1.7	-1.4	-1.9	-1.5
GX CY	A	-0.7	-1.8	-1.5	-2.1
	C	-1.1	-0.5	-3.0	-0.5
	G	-2.1	-2.9	-1.4	-1.4
	U	-1.9	-1.0	-2.1	-1.4
GX UY	A	-0.2	-0.4	-0.0	-0.3
	C	-0.3	-0.1	-1.5	-0.2
	G	-0.9	-1.1	-0.2	+0.1
	U	+0.1	-0.2	-0.4	-0.9
UX AY	A	-0.2	-0.2	-0.1	-0.5
	C	-0.1	-0.0	-1.2	-0.2
	G	-1.2	-1.2	-0.6	-0.0
	U	-0.3	-0.0	-0.5	-0.9
UX GY	A	-0.4	-0.0	-0.5	-0.5
	C	-0.1	-0.0	-1.7	+0.1
	G	-1.2	-1.2	-0.6	-0.7
	U	-0.6	-0.0	-0.5	-0.6

Multibranch Loops. Coaxial stacking of helixes and a Jacobson-Stockmayer function (*34*) for the free energy of multibranch loops (also called junctions) cannot be incorporated into a computationally efficient recursive algorithm for secondary structure prediction. Models for both are included in a program called efn2 (for second energy function) that re-calculates the free energy of each sub-optimal structure. These free energies are used to re-order the structures by over-all stability and the lowest free energy structure after the efn2 calculation is the predicted structure.

Table II. Tetraloop and Triloop Bonuses in kcal/mol.

Tetraloop:		Tetraloop:	
AAUCAU	-1.0	GUGAAC	-1.5
AGAAAU	-2.0	GUUCGC	-1.5
AGAGAU	-1.5	UCAGGG	-1.5
AGCAAU	-2.0	UGAAAA	-2.0
AGUAAU	-2.0	UGAGAG	-2.5
AGUGAU	-1.5	UGCAAA	-1.5
CCAAGG	-1.5	UGCAAG	-1.5
CCUUGG	-1.5	UGCCAA	-1.5
CGAAAG	-3.5	UGGAAA	-2.0
CGAGAG	-1.0	UGGAAG	-1.5
CGAGAG	-3.0	UGUAAA	-1.5
CGCAAG	-3.5	UGUGAA	-1.5
CGCAUG	-1.5	UUAGGG	-1.5
CGCCAG	-2.0	UUCCAA	-1.5
CGCGAG	-2.0	UUCCCA	-3.0
CGGAAG	-3.0	UUCCGG	-1.5
CGUAAG	-3.0	UUGAGG	-1.5
CGUGAG	-1.5	UUUAGG	-1.5
CGUGAG	-3.0	UUUCGG	-1.5
CUAAGG	-1.5		
CUACGG	-2.5		
CUUCGG	-3.5	Triloop:	
GCUUGC	-1.5	AAAAU	-1.5
GGAAAC	-3.0	AGACU	-1.5
GGAGAC	-1.0	CAAAG	-2.0
GGCAAC	-2.5	CGACG	-2.0
GGCGAC	-1.5	GAAAC	-2.5
GGGAAC	-2.0	GAUUU	-0.5
GGUAAC	-1.5	GGACC	-2.5
GGUGAC	-1.5	UAAAA	-1.0
GUGAAC	-1.5	UGACA	-1.0

In the recursive algorithm, the free energy penalty for closing a multibranch loop is given by the linear approximation:

$$\Delta G^{\circ}\ (\text{kcal/mol}) = a + bn + ch \qquad (4)$$

where n is the number of unpaired nucleotides in the loop, h is the number of helixes that branch from the loop, and a = 4.6, b = 0.2, and c = 0.1 (*3,10*).

The Jacobson-Stockmayer function is non-linear and can therefore not be used in a recursive algorithm. In efn2, however, the Jacobson-Stockmayer function is used for multibranch loops larger than six nucleotides and the free energy penalty is given by:

$$\Delta G^\circ \text{ (kcal/mol)} = a + 6b + (1.1) \ln (n/6) + ½ + ch. \quad (5)$$

The linear approximation of equation 1 is used for loops of six and fewer nucleotides.

In the recursive algorithm, a free energy bonus for base stacking is assigned for each unpaired nucleotide 3' or 5' to helixes exiting the loop. These parameters are dangling end free energies determined experimentally (*12*). In efn2, the base stacking bonuses are only assigned to helixes not involved in coaxial stacking.

Efn2 gives an enhanced stability for coaxial stacking of adjacent helixes with one intervening nucleotide at most. This stability is based on the work of Walter et al. (*2*) and Kim et al. (*35*). When helixes have no intervening nucleotides, a table is consulted for the stacking bonus. This bonus is the free energy parameter for stacking of base pairs in a helix made more favorable by 1.0 kcal/mol. The addition of the -1.0 kcal/mol was found to improve the accuracy of structure prediction. With one intervening nucleotide, coaxial stacking is allowed when there is a nucleotide (5' to the 5' helix or 3' to the 3' helix) that can make an intervening mismatch. There are two distinct stacks and parameters for each are contained in separate tables (Figure 1). The first stack is on the side of the continuous backbone which, in Figure 1, is a 5' A-G 3' on a U-A. The parameters for this stack are set equal to the parameter of a terminal mismatch in a hairpin loop (Table I). The second stack is on the side where the backbone opens for the entering and exiting strands. In Figure 1, this is an A-G on a C-G. This is mostly sequence independent, with a value of -1.5 kcal/mol. The exceptions to this are that A-A mismatches have a stack of -1.0 kcal/mol; A-U and C-G intervening "mismatches" of either orientation stack with -2.5 kcal/mol; and G-U wobbles are given -2.3 kcal/mol. To find the lowest free energy possible for a multibranch loop, efn2 uses a recursive algorithm to search for the most favorable combination of interactions. This is necessary because helixes involved in coaxial stacking cannot have dangling ends and cannot coaxially stack on more than one other helix.

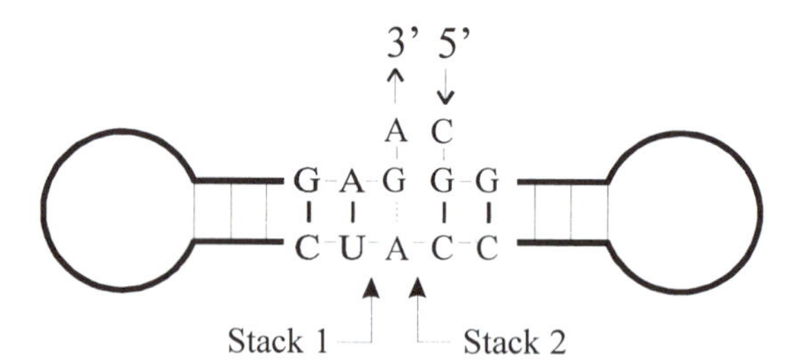

Figure 1. Coaxial Stacking of two helixes with an intervening mismatch. Stack 1 is the stack of the mismatch with a continuous backbone. Stack 2 is the stack of the mismatch with an open backbone.

Exterior Loops. Exterior loops are open loops that contain the ends of a sequence. This version of the algorithm gives bonuses for dangling ends and coaxial stacking in exterior loops using the same model as multibranch loops.

Removal of Isolated Base Pairs. To improve the accuracy of folding, isolated base pairs (those not adjacent to a possible Watson-Crick or G-U pair) are not allowed because they are rare in the database of known secondary structures. This is accomplished during the fill routine of the dynamic programming algorithm. Before calculating $V_{i,j}$, the lowest free energy for the sequence fragment from nucleotides i to j, with i and j base paired, the algorithm checks whether an adjacent canonical pair (Watson-Crick or G-U) is possible (either i+1 with j-1 or i-1 with j+1). If not, $V_{i,j}$ is set to a large integer used to represent an infinite free energy (1600 kcal/mol). This filters out isolated base pairs.

Constraining a Nucleotide to a Specific Base Pair. The folding algorithm allows the user to specify base pairs that are to occur in the final structure. In prior versions of the program, a large free energy bonus was assigned to base pairs specified by the user. This had the disadvantage of distorting the energy dot plot and making it difficult to compute a representative sample of foldings. Furthermore, for large structures, the base pair might not occur because the sum of other interactions might be more favorable than the arbitrary bonus.

The folding algorithm now forces a base pair by not allowing the two constituent nucleotides to be in the structure in any form other than the specified base pair. In other words, a large free energy (1600 kcal/mol) is given to an occurrence of the nucleotide not in the specified base pair, i.e. either single stranded or in any other base pair. This has the advantage of not complicating the determination of the free energy of a structure. The folding algorithm also uses a similar technique to force individual nucleotides to be double-stranded.

Scoring. Structures were scored against phylogenetic structures by:

score = (# base pairs correctly predicted)/(total bp in phylogenetic structure) (6)

A phylogenetic base pair was considered correctly predicted if it was identical to a base pair in the structure or if the structure contained a base pair in which one of the two bases was slipped by one nucleotide. That is, given a phylogenetic base pair between nucleotides i and j, then a pair of i—j, (i-1)—j, (i+1)—j, i—(j-1), or i—(j+1) would be counted as a correct base pair. This allows for the common occurrence of helixes that have similar stabilities when slipped by one nucleotide.

Results

The accuracy of the folding algorithm was tested on a database of structures that contains four small subunit ribosomal RNAs (rRNA), four Group I introns, three Group II introns, and 41 transfer RNAs (tRNA). The rRNAs and Group II introns were folded in domains of less than 500 nucleotides (*3*). Table III summarizes the percent of correctly predicted base pairs for this version of the algorithm and the last

Table III. Percent of Correctly Predicted Base Pairs.

		% of phylogenetic base pairs predicted		
Structure*:	total bp	Walter et al. (2)	Current optimal after efn2	Current best suboptimal
Small Subunit rRNAs:				
E. coli	443	75.6	77.0	94.4
Rat mitochondria	216	56.0	65.3	79.2
H. volcanii	433	85.9	86.1	92.4
C. r. chloroplast	413	64.2	86.1	92.4
Group I Introns:				
LSU (I)	128	63.3	76.6	89.8
Yeast OX5α (I)	96	89.6	87.5	91.7
ND1 (I)	149	43.6	70.5	78.5
T4D (I)	75	74.7	73.3	89.3
Group II Introns:				
Yeast A1 (II)	188	75.5	86.2	94.7
Yeast A5 (II)	206	90.3	90.8	93.2
Yeast B1 (II)	219	89.5	89.5	93.2
tRNAs:	860	83.4	88.8	95.1
total†:	3426	76.6	82.5	91.6

*Structures, except tRNAs, are those used by Jaeger et al. (*3*). For 16s rRNA: *E. coli* (*32,36,37*), Rat mitochondria (*32,37*), *H.* (*Halobacterium*) *volcannii* (*32,37*), and *C. r.* (*Chlamydomonas reinhardtii*) chloroplast (*32,37*), structures are divided into four domains for folding as described by Jaeger et al. (*3*). The group I introns are LSU (from *Tetrahymena thermophila*), Yeast OX5α (from *Saccharomyces cerevisiae*), ND1 (from *Podospora anserina*), and T4D (*31,38*). Large regions of undetermined structure were replaced with four unpairing nucleotides as described in Jaeger et al. (*3*). Group II introns (*39,40*), all from *Saccharomyces cerevisiae*, are split into two domains as described by Jaeger et al. (*3*). The tRNAs (*41*) are a250, a590, c250, d250, d590, e250, e590, f250, f590, g235, g251, h250, h780, i203, i250, k590, k780, l235, l250, m235, m250, n110, n250, p235, p255, q250, q530, r235, r250, s250, s590, t250, t590, v250, v590, w250, w570, x250, x530, y250, and y590. Modified nucleotides that cannot pair or stack are forced to be single stranded except for modified nucleotides in the yeast phenylalanine tRNA, f590, which is known to adopt its secondary structure without modified nucleotides (*42*). All basepairs were correctly predicted for tRNA f590.

†Column total percentages are found by dividing the sum of correctly predicted base pairs in that column by the sum of phylogenetic base pairs.

version (*2*). Suboptimal structures were generated with a window size of zero, 10% sort in energy, and a maximum of 1000 structures for rRNAs and introns. For tRNAs, a maximum of 100 structures were generated. The window size of zero was chosen to maximize the number of subtle structural variations for efn2 to sort.

For the database of structures, 82.5% of base pairs were correctly predicted in the lowest free energy or optimal structures. When the suboptimal structures were sorted for accuracy, the most accurate structures were found to contain 91.6% of phylogenetic base pairs. For the rRNA and introns, the most accurate structure is on average 2.1% or 2.3 kcal/mol higher in free energy than the lowest free energy structure. In tRNAs, the average increase in free energy between the lowest free energy and the most accurate structures is 1.5% or 0.5 kcal/mol. In both cases, the most accurate structure is always within 6% of the free energy of the lowest free energy structure.
Another method of scoring is to examine whether a phylogenetic base pair occurs in any suboptimal structure. For this database, 98.0% of phylogenetic base pairs are found in at least one suboptimal structure.

Conclusions

Recent experimental studies show that the thermodynamic stabilities of non-Watson-Crick regions in RNA can be very sequence dependent (*14,15*). We have revised a recursive algorithm for prediction of RNA secondary structure (*2,4,7*) to allow inclusion of some of this sequence dependence. Since a limited amount of experimental data are available, the sequence dependence was initially approximated by crude models based on experimental results and frequencies of natural occurrence. These parameters were then adjusted to optimize the prediction of 3,426 base pairs contained in four small subunit rRNAs, four group I introns, three group II introns, and 41 tRNAs. A set of parameters were found that places 82.5% of the phylogenetically determined base pairs in the predicted lowest free energy structure. When a set of up to 1000 higher free energy structures is generated for each RNA and the structure most consistent with the phylogenetic structure is chosen for each RNA, 92% of known base pairs are found. This suggests that secondary structure is largely determined by free energy minimization. The small number of experimental free energy parameters and the small size of the secondary structure database used means the free energy parameters are underdetermined. Thus more experimental data and phylogenetically derived structures are required to further test this hypothesis.

For domains of about 400 nucleotides, hundreds of potential secondary structures are found within 2 kcal/mol of the lowest free energy structure. In contrast, biochemical experiments indicate that RNAs often form a single secondary structure in solution. This suggests there is much to learn about the factors determining RNA structure. For example, a single active structure may be stabilized by favorable tertiary interactions between non-Watson-Crick regions, and these interactions are not yet included in the folding algorithm. Experimental data on the unfolding of tRNA (*43,44*), group I introns (*45,46*), and a domain from 23S rRNA (*47*), however, indicate that single secondary structures form in the absence of tertiary interactions. Another possibility is that the kinetics of folding selects one of many energetically similar secondary structures. Many RNAs, however, are known to fold into their active structure both during transcription, when the molecule begins to fold before it is completely synthesized, and during renaturation from a denatured state, when the entire molecule is available for folding. There is no fundamental reason to expect the same kinetics in both cases. Possibly, the stability of the non-Watson-Crick regions, which are the subject of current study, can selectively stabilize the native structure compared to other structures. Presumably, understanding the reason for the prevalence of a single structure will allow better prediction of secondary structure from sequence. Given that an RNA of 400 nucleotides has more than 10^{100} possible secondary structures (*48*), it is encouraging that free energy minimization can provide a filter that narrows consideration to only about 10^3 structures. Selection of the native structure is then often possible from experimental data, such as chemical mapping and site directed mutagenesis, or by comparisons of sequences with similar functions (*5,6*).

Acknowledgments

The authors thank T. Xia, S. Schroeder, M. J. Serra, and T. W. Barnes for helpful discussions concerning the new thermodynamic parameters and M. E. Burkard and J. A. McDowell for discussions about computer programming. Thanks to R. R. Gutell for making phylogenetic secondary structures available. This work was supported by National Institutes of Health Grants GM22939 to D.H.T. and GM54250 to M.Z.. D.H.M. received support from National Institutes of Health Grant 5T32 GM07356.

Literature Cited

1. James, B. D.; Olsen G. J.; Pace N. R. *Methods Enzymol.* **1989**, *180*, 227-239.
2. Walter, A. E.; Turner, D. H.; Kim, J.; Lyttle, M. H.; Müller, P.; Mathews, D. H.; Zuker, M. *Proc. Natl. Acad. Sci.* **1994**, *91*, 9218-9222.
3. Jaeger, J. A,; Turner, D. H.; Zuker, M. *Proc. Natl. Acad. Sci. U.S.A.* **1989**, *86*, 7706-7710.
4. Zuker, M. *Science.* **1989**, *244*, 48-52.,
5. Mathews, D. H.; Banerjee, A. R.; Luan, D. D.; Eickbush, T. H.; Turner, D. H. *RNA.* **1997**, *3*, 1-16.
6. Lück, R; Steger, G.; Riesner D. *J. Mol. Biol.* **1996**, *258*, 813-826.
7. Zuker, M.; Stieger, P. *Nucleic Acids Res.* **1991**, *9*, 133-148.
8. Zuker, M.; Jacobson, A. B. *Nucleic Acids Res.* **1995**, *23*, 2791-2798.
9. McCaskill, J. S. *Biopolymers.* **1990**, *29*, 1105-1119.
10. Gultyaev, A. P.; Batenburg, F. H. D. van; Pleij, C. W. A. *J. Mol. Biol.* **1995**, *250*, 37-51.
11. Batenburg, F. H. D. van; Gultyaev, A. P.; Pleij, C. W. A. *J. Theor. Biol.* **1995**, *174*, 269-280.
12. Serra, M. J.; Turner, D. H. *Methods Enzymol.* **1995**, *259*, 242-261.
13. Freier, S. M.; Kierzek, R.; Jaeger, J. A.; Sugimoto, N.; Caruthers, M. H.; Neilson, T.; Turner D. H. *Proc. Natl. Acad. Sci. U.S.A.* **1986**, *83*, 9373-9377.
14. Wu, M.; McDowell, J. A.; Turner, D. H. *Biochemistry.* **1995**, *34*, 3204-3211.
15. Schroeder, S.; Kim, J.; Turner, D. H. *Biochemistry.* **1996**, *35*, 16105-16109.

16. Serra, M. J.; Lyttle, M. H.; Axenson, T. J.; Schadt, C. A.; Turner, D. H. *Nucleic Acids Res.* **1993**, *21*, 3845-3849.
17. Serra, M. J.; Axenson, T. J.; Turner, D. H. *Biochemistry.* **1994**, *33*, 14289-14296.
18. Serra, M. J.; Barnes, T. W.; Betschart, K.; Gutierrez, M. J.; Sprouse, K. J.; Riley, C. K.; Stewart, L.; Temel, R. E. *Biochemistry.* **1997**, *36*, 4844-4851.
19. Fodor, S. P. A.;Read, J. L.; Pirrung, M. C.; Stryer, L.; Lu, A. T.; Solas, D. *Science.* **1991**, *270*, 467-470.
20. O'Donnell-Maloney, M. J.; Smith, C. L.; Cantor, C. R. *Trends Biotechnol.* **1996**, *14*, 410-407.
21. Papanicolaou, C.; Gouy, M.; Ninio, J. *Nucleic Acids Res.* **1984**, *12*, 31-44.
22. Laing, L. G.; Hall, K. B. *Biochemistry.* **1996**, *35*, 13586-13596.
23. Jucker, F. M.; Pardi, A. *Biochemistry.* **1995**, *34*, 14416-14427.
24. SantaLucia, J., Jr.; Kierzek, R.; Turner, D. H. *Science.* **1992**, *256*, 217-219.
25. Antao, V. P.; Tinoco, I., Jr. *Nucleic Acids Res.* **1992**, *20*, 819-824.
26. Antao, V. P.; Lai, S. Y.; Tinoco, I., Jr. *Nucleic Acids Res.* **1991**, *19*, 5901-5905.
27. Varani, G.; Cheong, C.; Tinoco, I., Jr. *Biochemistry.* **1991**, *30*, 3280-3289.
28. Lehnert, V.; Jaeger, L.; Michel, F.; Westhof, E. *Chemistry & Biology.* **1996**, *3*, 993-1009.
29. Cate, J. H.; Gooding, A. R.; Podell, E.; Zhou, K.; Golden, B. L.; Kundrot, C. E.; Cech, T. R.; Doudna, J. A. *Science.* **1996**, *273*, 1678-1685.
30. Westhof, E.; Michel, F. *J. Mol. Biol.* **1990**, *216*, 585-610.
31. Damberger, S. H.; Gutell, R. R. *Nucleic Acids Res.* **1994**, *22*, 3508-3510.
32. Gutell, R. R. *Nucleic Acids Res.* **1994**, *22*, 3502-3507.
33. Woese, C. R.; Winker, S.; Gutell, R. R. *Proc. Natl. Acad. Sci. U.S.A.* **1990**, 87, 8467-8471.
34. Jacobson, H.; Stockmayer, W. H. *J. Chem. Phys.* **1950**, *18*, 1600-1606.
35. Kim, J.; Walter, A. E.; Turner, D. H. *Biochemistry.* **1996**, *35*, 13753-13761.
36. Moazed, D.; Stern, S.; Noller, H. F. *J. Mol. Biol.* **1986**, *187*, 399-416.
37. Gutell, R. R.; Weiser, B.; Woese, C. R.; Noller, H. F. *Prog. Nucleic Acid Res. Mol. Biol.* **1985**, *32*, 155-216.
38. Waring, R. B.; Davies, R. W. *Gene.* **1984**, *28*, 277-291.
39. Michel, F.; Umesono, K.; Ozeki, H. *Gene.* **1989**, *82*, 5-30.
40. Michel, F.; Jacquier, A.; Dujon, B. *Biochimie.* **1982**, *64*, 867-881.
41. Sprinzl, M.; Hartmann, T.; Weber, J.; Blank, J.; Zeidler, R. *Nucleic Acids Res.* **1989**, *17*, supplement, 1-189.
42. Hall, K. B.; Sampson, J. R.; Uhlenbeck, O. C.; Redfield, A. G. *Biochemistry.* **1989**, *28*, 5794-5801.
43. Hilbers, C. W.; Robillard, G. T.; Shulman, R. G.; Blake, R. D.; Webb, P. K.; Fresco, R.; Riesner, D. *Biochemistry.* **1976**, *15*, 1874-1882.
44. Crothers, D. M.; Cole, P. E.; Hilbers, C. W.; Shulman, R. G. *J. Mol. Biol.* **1974**, *87*, 63-88.
45. Banerjee, A. R.; Jaeger, J. A.; Turner, D. H. *Biochemistry.* **1993**, *32*, 153-163.
46. Jaeger, L.; Westhof, E.; Michel, F. *J. Mol. Biol.* **1993**, *234*, 331-346.
47. Laing, L. G.; Draper, D. E. *J. Mol. Biol.* **1994**, *237*, 560-576.
48. Zuker, M; Sankoff, D. *Bull. Math. Biol.* **1984**, *46*, 591-621.

Molecular Dynamics Simulation

Chapter 16

Modeling of DNA via Molecular Dynamics Simulation: Structure, Bending, and Conformational Tansitions

D .L. Beveridge, M. A. Young, and D. Sprous

Chemistry Department and Molecular Biophysics Program, Wesleyan University, Middletown, CT 06459

Molecular Dynamics (MD) computer simulation is a powerful computational approach to the study of macromolecular structure and fast motions. For DNA sequences, an accurate dynamical model obtained from MD can, in principle, provide a general theoretical basis for understanding the intrinsic nature of the dynamical structure and variations on the theme such as sequence dependent structural irregularities, axis bending and helical flexibility and deformation. With the latest developments in supercomputer power and the adaptation of MD code for parallel processing, MD trajectories extending into the nanosecond regime on DNA oligonucleotides including water, counterions, coions and organic additives have been performed for the first time. Analyses of the results have revealed interesting new details about DNA dynamics and the structure and motions of the various components of the solvent. This article provides a concise overview of MD methodology and analysis, and a review of recent nanosecond MD trajectories on DNA oligonucleotide sequences carried out in this Laboratory. The accompanying figures illustrate both the results from our recent work and the rich variety of computer graphics techniques available for the display and analysis of MD simulations on DNA. This article is not a comprehensive review; the results and perspectives of our colleagues in the field are presented in the other articles of this symposium volume. In particular, a complementary focus on RNA sequences is provided by Westhof (*1*).

Methods

MD simulation (*2*) is a computer "experiment" in which the atoms of a stipulated system execute Newtonian dynamics on an assumed potential energy surface. The model system chosen for study, the assumed energy surface, and the simulation protocol are all operational variables. The simulation *per se* begins with the choice of an initial configuration, typically the crystal structure of the macromolecule, and an arbitrary arrangement of solvent. In DNA problems, the canonical forms obtained from fiber diffraction studies are sometimes chosen to serve as unbiased macromolecular starting points. The initial configuration of the system is first subjected to energy minimization (EM) to relieve any major stresses. The first stage of MD is a brief period in which the velocities on the particles are increased to the temperature of interest (heating). Then the simulation seeks out a thermally bounded state (equilibration) and proceeds to sample it (production), reporting back dynamical characteristics of the region as reflected in the time series development of structural parameters. A realistic goal of a macromolecular MD simulation is thus to characterize a thermally bound state in the vicinity of the assumed initial structure and to determine the extent to which the calculated properties agree with corresponding experimentally observed quantities. When sufficient agreement is established, the wealth of molecular detail obtained as a function of time in the simulation can form the basis for interpreting and understanding diverse experimental results. MD procedures specific for biological macromolecules are described in more detail in several monographs (*3, 4*) and a recent review article emphasizing methodology (*5*).

The analysis of an MD trajectory involves following the manifold structural changes that occur in the molecule as a function of time, and the average behavior of physical properties over time. The ensemble of MD structures obtained in an MD trajectory is a representation of the statistical state of the system. The MD structures can be presented in panels of "snapshots" or superimposed to convey an idea of the dynamic range of structure covered by the simulation. Analysis of the conformational dynamics of the system can be carried out based on well-defined conventions for the definitions of DNA backbone torsions (*6*). There is no unique approach to analysis of the helix (helicoidal) parameters. One approach, exemplified by the program *Newhelix* (*7*) is based on a local axis convention for base pair steps (*8*). The approach taken in the *Curves* program (*9*) determines helicoidical parameters with respect to a global helix axis computed by means of a spline fit to the atomic coordinates. A computer graphics utility, *Dials & Windows* (*10*), allows one to monitor conformational transitions and to report the conformational and helicoidal parameters of the dynamical model over the entire time course of the simulation. Plots of root mean square deviations (RMSD) vs. time for the MD structures compared with the initial structure, reference canonical forms, or multiple other structures in the trajectory convey additional information about the stability of the dynamical structure and the location of the thermally bounded states in configuration space. Superimposition of the helical axes from *Curves* for the various MD snapshots provides an indication of the dynamic range of axis bending. A definition of a persistence length index based on the statistical theory of chain molecules has been developed to ascertain quantitatively the degree of linearity, the lateral direction of bending and the flexibility or the lack thereof in local regions of a DNA sequence (*11*). A graphics presentation, *Bending Dials*, has been created to analyze the bending in structures from crystallography, NMR or MD and to compare the results (*12*). The various methods essential for the analysis of MD on DNA (and proteins) have been collected in a UNIX-based suite of programs and computer graphics routines called the *Molecular Dynamics Tool Chest* (*13*).

A description of the molecular force field underlies a particular dynamical model obtained from MD. The force field is computed from a relatively simple analytical energy function consisting of contributions from bond stretching, angle bending, torsional displacements and non-bonded interactions such as electrostatic, dispersion and van der Waals repulsions. The energy function is an empirical approximation to the corresponding quantum mechanical Born-Oppenheimer potential energy surface, a procedure necessitated by the need for rapid machine calculation of the atomic forces. The energy functions incorporated in the suites of programs readily available for MD simulation such as *AMBER* (*14*), *CHARMM* (*15*) and *GROMOS* (*16*) each have a full set of parameters for nucleic acids, developed based on extensions of polypeptide and protein force fields and supplemented with additional parameters developed from quantum chemical calculations and experimental data on prototype cases. Two new versions of these force fields have been described in the last year, one based on the *AMBER* format (*17*) and the other based on *CHARMM* (*18*), each with proposed improvements in the description of nucleic acids in simulations including solvent. Application of these parameter sets to duplex DNA has led to a series of new MD simulations. Our studies, as reviewed herein, were based initially on the GROMOS force field, and more recently on the force field of Cornell et al. as incorporated in *AMBER* 4.1 (*19*).

The treatment of long-range interactions in the computation of forces in MD is a serious operational issue in MD simulation (*20, 21*). The idea is to avoid the repetitive evaluation of negligible contributions to the forces while not eliminating anything of significance. One approach is to shift the origin of the potential arbitrarily to zero at the cutoff distance potential. Here an error is introduced into the energy but not necessarily the forces. Another popular method is truncation followed by application of a switching function around the cutoff to feather the potentials smoothly off to zero, so that the calculation of the forces remains well-conditioned. The truncation error depends on the extent to which the various kinds of interactions fall off with distance, r. Dispersion energies diminish with distance dependence of r^{-6}, dipole- dipole interactions as r^{-3}, and simple Coulombic electrostatic effects as r^{-1}. The longer- ranged Coulomb forces may still be relatively large at the cutoff, and are thus most affected by truncation effects. Coulomb forces are especially important in MD on DNA, due to the polyionic character of the DNA backbone and the important interactions of the DNA with mobile counterions. If the range of the switching function is too narrow in potentials used for MD on DNA, artifacts may be introduced as a consequence of charged groups tending to cluster at the cut-off limit (*21, 22*). This can be remedied by extending the range of the switching function and thus modulating the effects more gradually.

An alternative to truncation of any kind is the use of Ewald Sums, as originally introduced to calculate the lattice energy of ionic crystals. A fast algorithm, "Particle Mesh Ewald (PME)" (*23*), has recently been devised which makes this approach viable in full-scale MD on macromolecules in solvent. The initial round of PME simulations on DNA demonstrated a higher degree of stability and less deviation from the starting structure (*24, 25*). The accuracy of the overall dynamics with PME is a matter of concern (*26*), but recent results (*vide infra*) of MD on DNA with PME in which considerable axis deformations are involved lead to a cautious optimisim on this issue.

MD Studies of a Prototype B-DNA Oligonucleotide: d(CGCGAATTCGCG)

Recent reviews from this Laboratory provide an overview of the literature of MD simulations on DNA oligomers through 1993 (*27*) and theoretical and computational aspects of DNA hydration (*28*) and counterion atmosphere (*29*). References to the most recent literature can be found in (*30*). Experimental data for comparison with MD results are available for crystal structures in the Nucleic acids Data Bank (NDB) (*31*), and for NMR structure in a review by Ulyanov and James (*32*). The research described in this article is directed towards understanding the dynamical structure of the various right-handed helical forms of DNA, their deformations and interconversions. The canonical A and B structures of DNA are shown for reference in Figure 1. The A and B forms are distinguishable in three major ways: the displacement of nucleotide base pairs from the helix axis, the inclination of base pairs with respect to the helix axis, and sugar puckers. Details on these and other structural features of DNA relevant to MD analysis is readily available (*33*).

The d(CGCGAATTCGCG) duplex assumes the B form in the crystalline state and in solution, and serves as a prototype for MD characterization studies. Some 20 crystal structures have been reported for this sequence, both in the free state under various conditions of temperature and in complexation (*34*). The crystal structures are all essentially in accord on several structural observations: a) a narrowing of the minor groove in the central AATT region of the structure, which in uncomplexed form supports an ordered network of solvent peaks in the electron density known as the "spine of hydration," b) sequence dependent variations in helicoidal parameters, propeller twist and base pair buckling, and c) bending in the helix axis to the extent of ~19°, accomplished mainly by alterations in base pair roll parameters with respect to values observed in canonical B DNA. The axis bending is localized at two points, at or near the junction between the CGCG and AATT tracts. The extent to which these effects are intrinsic to the DNA or, alternatively, a consequence of crystal packing effects (*35*) has yet to be unequivocally determined, and the influence of solvent effects in the crystal is a matter of current debate (*36, 37*). The structure of d(CGCGAATTCGCG) in complex with the restriction enzyme EcoRI endonuclease has been reported (*38*). The axis bending is more pronounced at the roll points in the protein bound form of the sequence, and involves local unwinding. In addition, a quite extreme local deformation is found near the center of the duplex.

NMR structures of d(CGCGAATTCGCG) have been reported by Nerdal et al. (*39*) and Lane et al. (*40*) based on 2D-NOESY data. Nerdal et al. proposed a highly underwound structure with kinks at the primary and secondary roll points and at the ApT step. Lane et al. propose an NMR structure much closer to that of conventional B DNA. The limit of the nuclear Overhauser effect to interproton distances ~4.5 Å makes unequivocal determination of the helix parameters in an

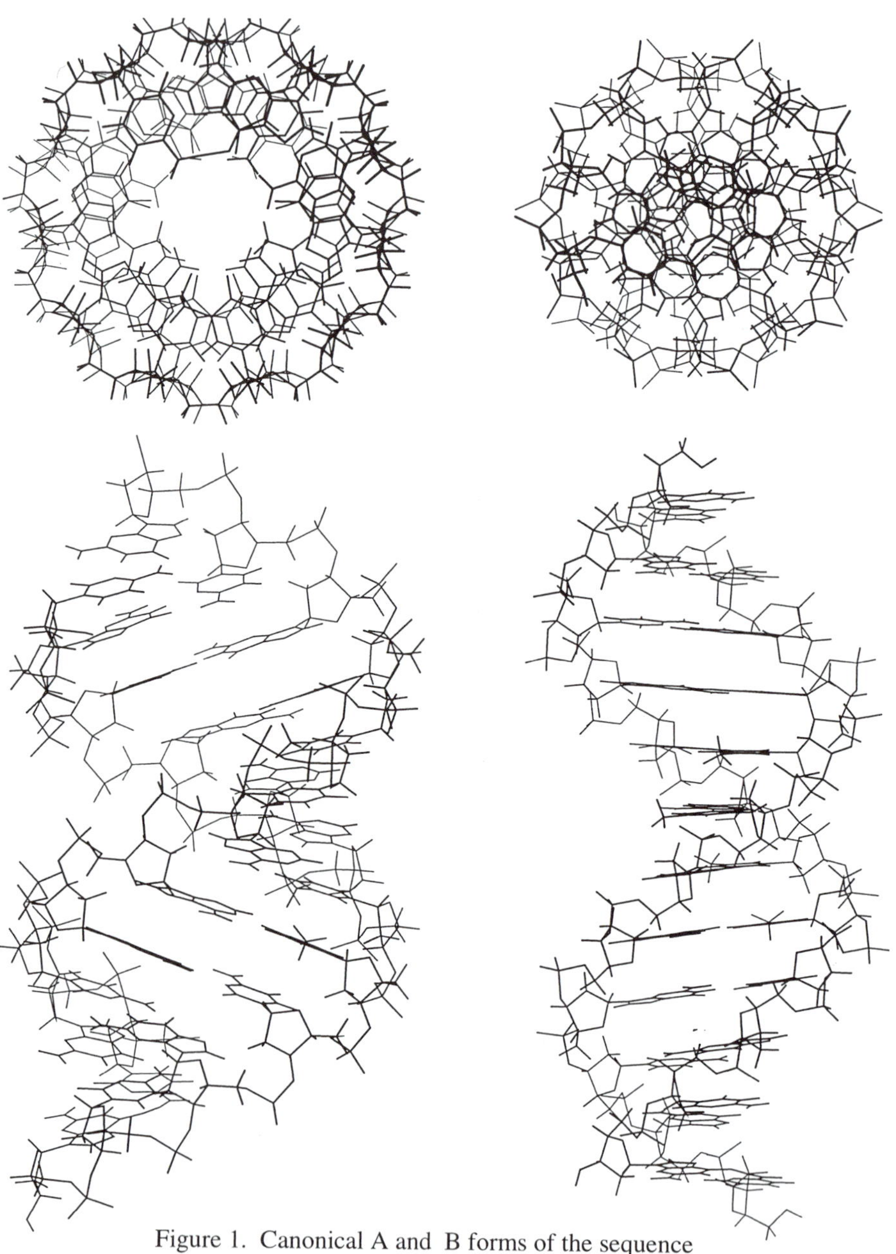

Figure 1. Canonical A and B forms of the sequence d[CGCGAATTCGCG].

oligonucleotide duplex difficult, and some aspects of sequence dependent fine structure are indeterminable (*38*). A most recent detailed study of this issue is given by Ulyanov et al. (*41*).

Results based on the GROMOS Force Field

An MD simulation on the d(CGCGAATTCGCG) duplex together with 1,927 water molecules and 22 Na^+ counterions was reported from this Laboratory based on the GROMOS force field (*42*). Extensive Monte Carlo equilibration of the solvent was used to prepare the system in a suitable state to perform a stable dynamical trajectory. Results based on GROMOS *per se* showed base pair opening. Subsequently, we supplemented the force field with a restraint potential for maintaining Watson-Crick (WC) base pairing. The initial structure in the simulation was a canonical B-form and the MD trajectory was extended for 140 ps. At the termination of the trajectory, the structure resided clearly in the B DNA family, ~2.3 Å RMS deviation from the corresponding canonical form and ~2.5 Å RMS deviation from the corresponding crystal structure.

The analysis of the simulation revealed agreement with some but not all features seen in the x-ray crystal structure of the dodecamer. The local axis deformation around the GC/AT interfaces in the sequence and large propeller twist in the base pairs were reproduced in the MD model. The DNA base pairs showed a consistent inclination in the simulation at variance with canonical B, but in accord with an interpretation of results obtained from flow dichroism studies (*43*). The negative propeller twist found in the crystal structure was clearly observed in the MD results, but was not that different for AT and GC, in spite of the expectation that it would be larger for the AT base pairs. Narrowing of the minor groove in the AT region of the crystal structure was not observed over the time course of the GROMOS MD simulation. On the other hand, DNA footprinting studies provide quite strong evidence for sequence dependent fine structure, which Tullius and coworkers (*44*) suggest is linked to minor groove narrowing. Theoretical studies of the mechanics of groove deformations have recently been discussed in detail by Zakrzewska and coworkers (*45*). The MD results do not support a strict two-state model for sugar puckering, but a distribution of values between canonical B and A form puckers similar to that anticipated by Olson (*46, 47*). Some transitions from B to A sugar puckering are seen, particularly at points of axis bending.

The trajectories obtained for the GROMOS simulation in solvated d(CGCGAATTCGCG) were subsequently used to calculate sugar puckers (*48*) and *de novo* theoretical nuclear Overhauser effect buildup curves for all the observable cases (*49-51*). NMR measurements of the buildup curves of NOE cross-peak volumes for the dodecamer were carried out, and the overall agreement between theoretically calculated and experimentally measured intensities was determined for each residue of each of the different models. The comparison between the theoretical and experimental results including motional considerations demonstrated that both the anisotropic tumbling and local motions of the DNA need to be explicitly included for accurate modeling of DNA in solution. The NOE build-up curves calculated from the ensemble of thermally accessable structures generated in the MD was reported to be improved over that of single structures, indicating that the dynamical models may be a better description of DNA structure in solution than the crystallographic or canonical forms alone.

Three subsequent *in aquo* MD simulations were performed on the d(CGCGAATTCGCG) duplex based on the GROMOS86 force field, augmented with a hydrogen-bond function applied to Watson-Crick nucleotide base pairs (*52*). Reduced charges on the phosphates were used in lieu of explicit counterions. The three simulations were configured identically except for starting structures and run lengths: the canonical B80 fiber diffraction form (500 ps), the Drew-Dickerson crystal form (500 ps), and the protein-bound form of the dodecamer from the crystal structure of the EcoRI endonuclease complex (1 ns). The stability and convergence behavior of the simulations were monitored via two-dimensional root-mean-square deviation (2D-RMSD) maps, which confirmed that the three simulations converged on a common B-form dynamical structure. Analyzed in more detail, MD results suggest the presence of distinguishable substates in the B DNA family, with lifetimes of the order of hundreds of picoseconds. Poncin et al. (*53*) have recently considered the issue of substates in DNA from an alternative point of view. MD simulation carried out with the EcoRI complexed form of the d(CGCGAATTCGCG) duplex as the initial structure, which bears initially an extreme kink at the A6 step, indicated that the distortion relaxed within 20 picoseconds of MD, indicating it to be a strained form induced by the protein rather than a distinctly separate intermediate structure on the B-DNA potential surface consistent with the interpretation advanced earlier by Kumar et al. (*54-56*).

Results based on the *AMBER* force fields

We previously reported results on DNA based on the Weiner et al. force field (*57*) using AMBER 3.0 (*58*), and extensive references to AMBER-based MDs on DNA have been provided (*27*). Young et al. (*59-60*) have carried out MD simulations well into the nanosecond regime on the DNA duplex of sequence d(CGCGAATTCGCG), including explicit consideration of water molecules and 22 Na^+ counterions and based on the Cornell et al. force field (*17*) with PME used in the treatment of long-range interactions. Simulations were performed both on the crystallographic unit cell (crystal structure model) and for a solution structure model, with the DNA surrounded by ~10 Å of water molecules. After 2 ns of simulation, the crystal structure MD (*60*) showed an RMSD of only .65 Å from experiment (Figure 2), and a close agreement on the structure of the minor groove width (Figure 3). The solution structure MD results (*59, 61*), shown in Figure 4, now extended to the unprecedented run length of 13 ns, support a dynamical model of B-DNA closer to the B form than any of our previously reported MDs, and the results were shown to be similar for three different choices of initial configuration for the arrangement of counterions. The MD results were compared with corresponding crystallographic and NMR studies on the d(CGCGAATTCGCG) duplex, and placed in the context of observed behavior of B DNA by comparisons with the complete crystallographic data base of B form structures. The narrowing of the minor groove in the AATT region persists in solution, a distinct

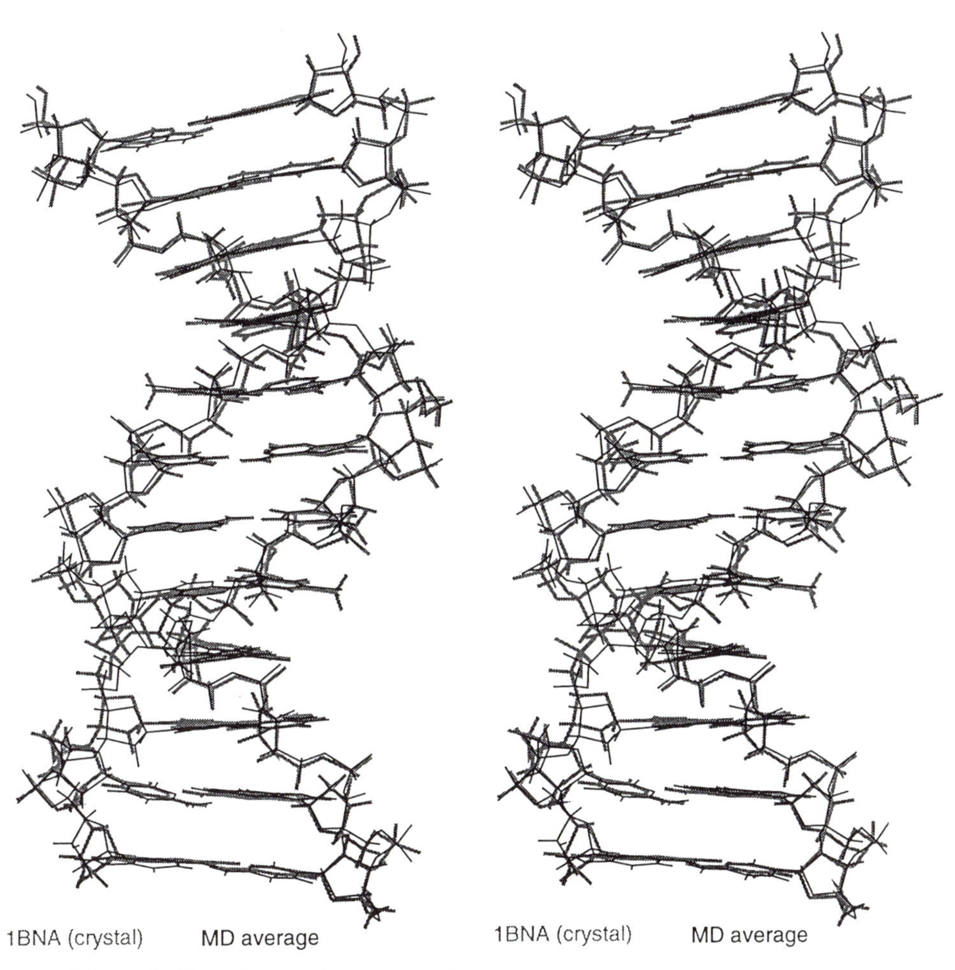

Figure 2. Superimposed structures of duplex d(CGCGAATTCGCG) from crystallography (*90*) and MD simulation on the crystallographic unit cell (*60*).

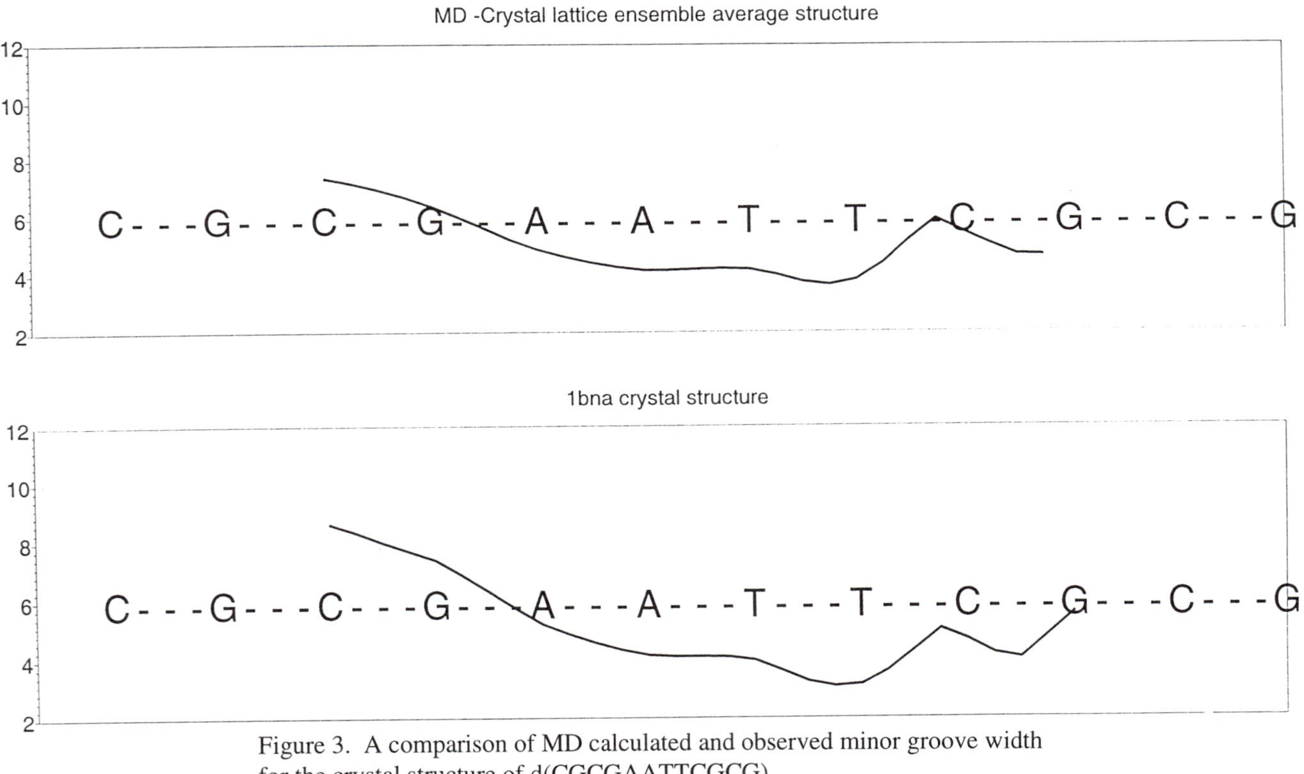

Figure 3. A comparison of MD calculated and observed minor groove width for the crystal structure of d(CGCGAATTCGCG).

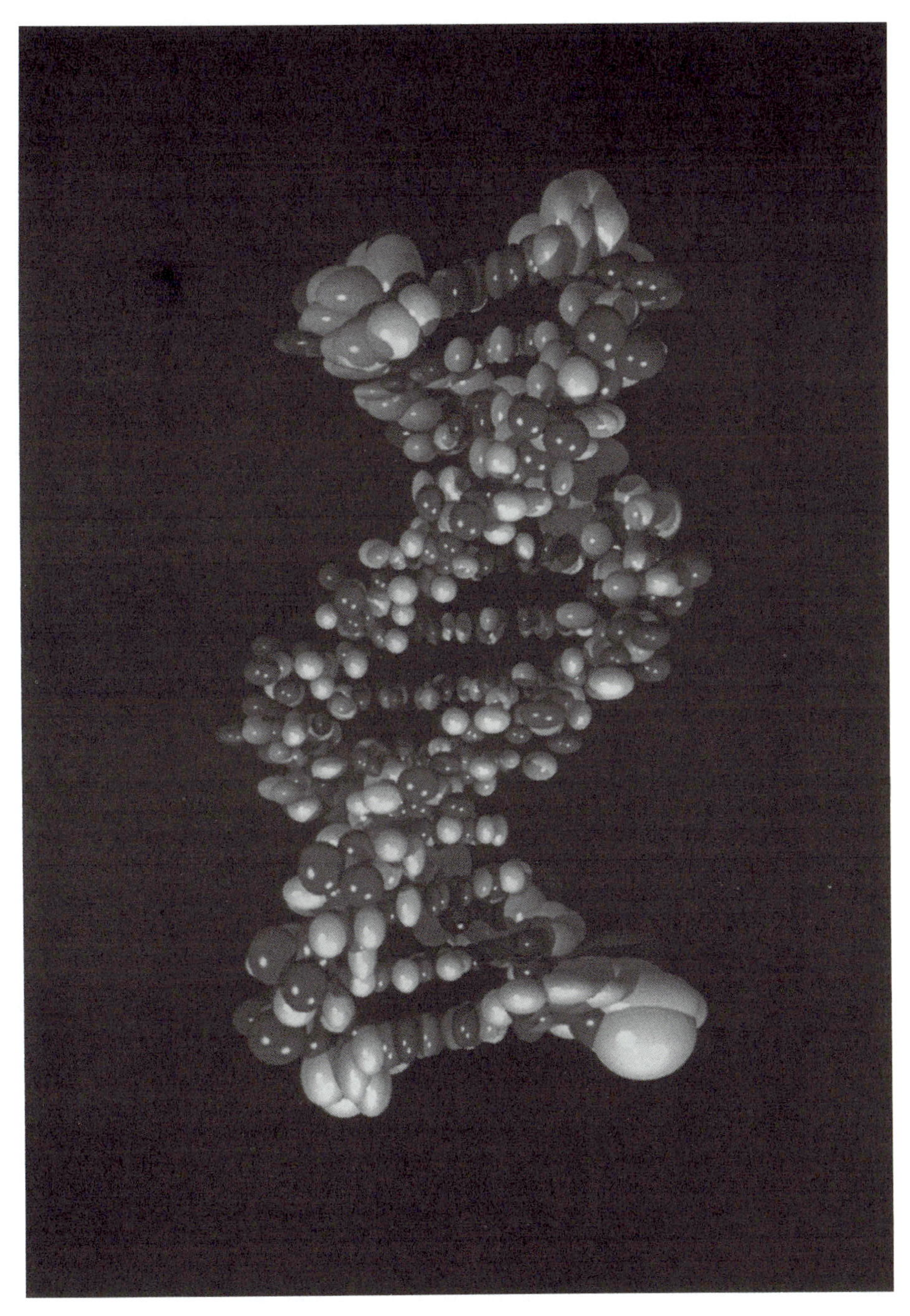

Figure 4. MD calculated solution structure for duplex d(CGCGAATTCGCG) in solution (*59*).

improvement over prior dynamical models. An *a priori* description of ion atmosphere was found to be well-developed in this trajectory (Figure 5) (*64*) which supports the core concept of "Counterion Condensation" (*63*).

Two further issues are of interest in comparing the MD results with the predictions of counterion condensation theory: concentration dependence of the ion atmosphere and specific interactions of ions with the DNA. The former question (*64*) awaits a more extensive series of simulations to provide MD perspectives. On the latter issue, details of the calculated ion distribution in the solution structure MD (*61*) suggest that expanded consideration of specific interactions of counterions with the DNA may be warranted. While much of the MD shows the spine of hydration in the minor groove as shown in the crystal structure, significant fractional occupancy is indicated for Na^+ ions in the minor groove localized at the ApT step (Figure 6). This position, termed herein the "ApT pocket", was noted previously (*65*) to be of uniquely low negative electrostatic potential relative to other positions of the groove, a result supported by the location of a Na^+ ion in the crystal structure of the rApU mini-duplex (*66*) and continuum electrostatics (*61*). The Na^+ ion in the ApT pocket interacts favorably with thymine 02 atoms on opposite strands of the duplex, and is well-articulated with the water molecules which constitute the remainder of the minor groove spine. The minor groove in the junction of the AATT and CGCG steps also shows significant fractional occupancy of counterions. Thus, MD predicts that counterions may intrude on the minor groove spine of hydration on B form DNA. NMR support for the fractional occupation of (divalent) ions in the minor groove of B DNA has recently been obtained by Feigon and Hud (*67*). The 'pocket' concept (Figure 7) is now being pursued further for other sequences. The idea of localized complexation of otherwise mobile counterions in electronegative pockets in the grooves of DNA helices introduces a heretofore underappreciated source of sequence dependent effects on local conformational, helicoidal and morphological structure, and may have important implications in understanding the functional energetics and specificity of the interactions of DNA and RNA with regulatory proteins, pharmaceutical agents, and other ligands. Novel properties of DNA grooves have been noted by Lamm and Pack (*68*).

MD Studies of DNA Curvature

The deformation of the DNA double helix plays an important role in structural biology, and the molecular nature of sequence-dependent axis bending, both intrinsic and induced, is a problem of considerable current interest and import. Sequences of DNA containing stretches of adenine bases (A-tracts) positioned in phase with the 10.5 base pair helical repeat of B-form DNA have been known for some time to exhibit a variety of unique structural features, including an enhanced tendency for axis bending. A series of models have been proposed to account for the data (*69, 70*), but there is still not an unequivocal interpretation of the results. To explore this issue with MD, nanosecond length trajectories based on the Cornell et al. force field supplement with the parameters of Aqvist (*71*) for ions were carried out for a DNA oligonucleotide duplex featuring phased A-tracts,

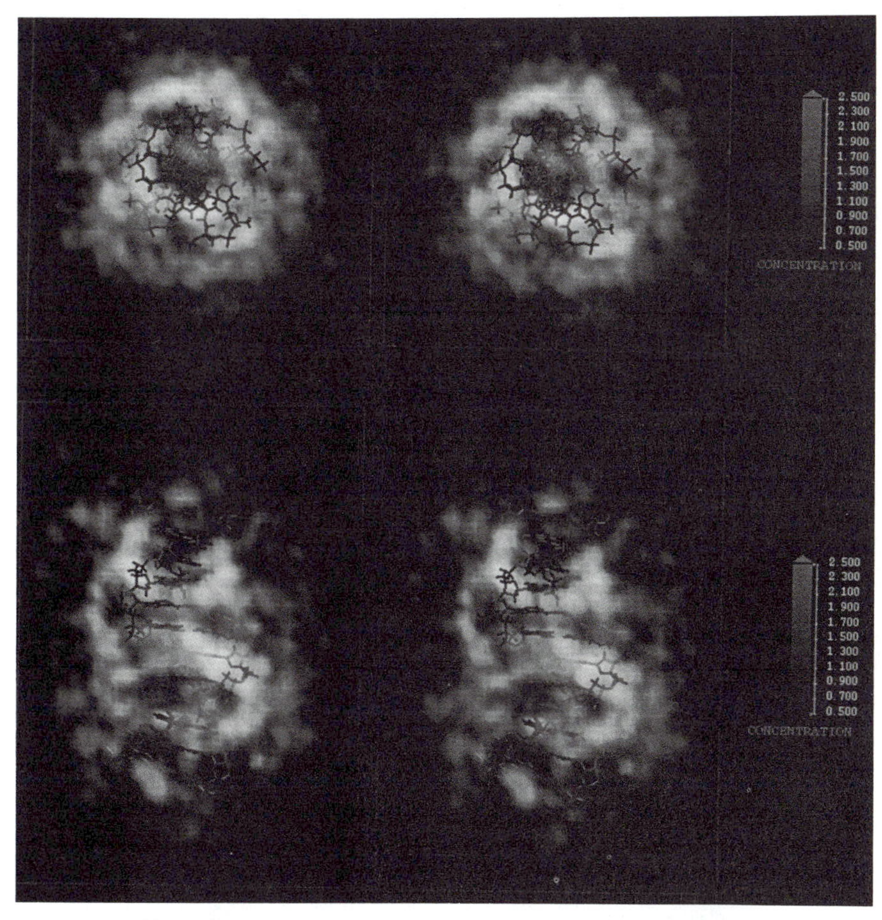

Figure 5. MD calculated counterion density around duplex d(CGCGAATTCGCG) in solution (*59*).

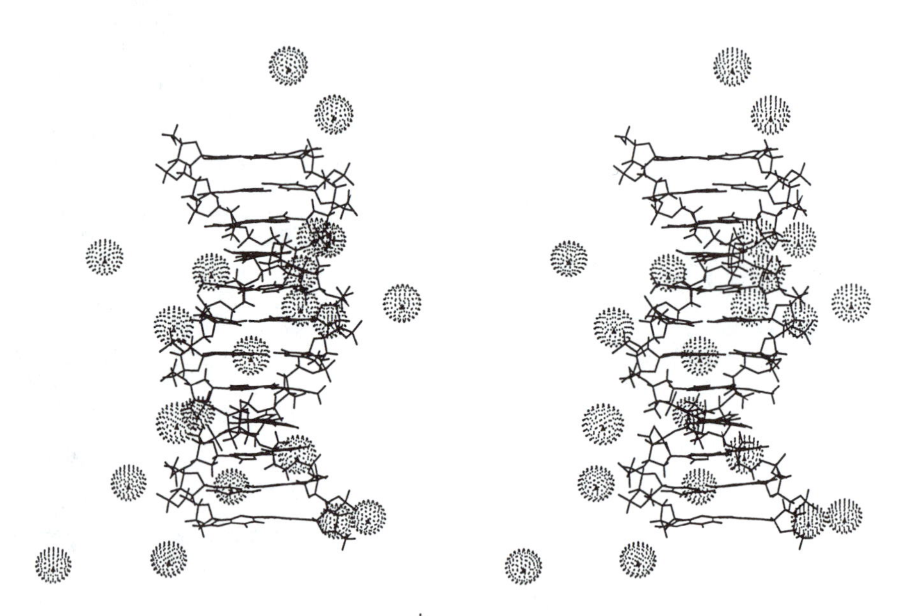

Figure 6. Structural detail of an Na^+ ion at the ApT step in the minor groove of duplex d(CGCGAATTCGCG) as predicted by MD simulation (*62*).

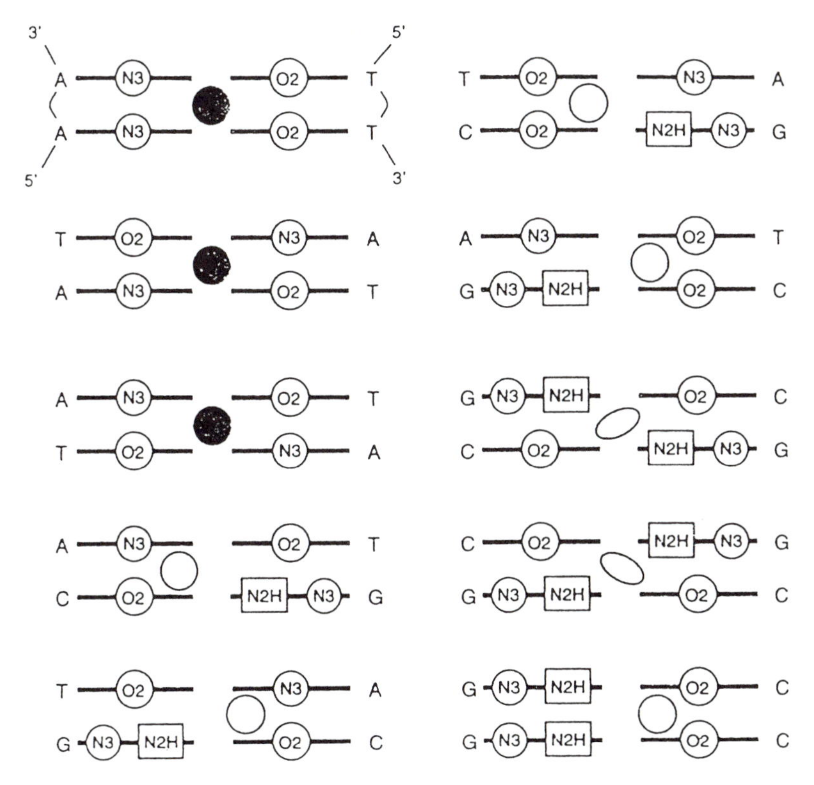

Figure 7. Possible locations of electronegative pockets in the minor groove of B-DNA (schematic); shading proportional to putative efficacy of the site (*62*).

d(ATAGGCAAAAAATAGCICAAAAATGG) in water with various concentrations of ionic species (K^+, Na^+, Mg^{2+}, and Cl^-) (*72*). Oligonucleotides of comparable lengths composed from DNA sequences complexed with the proteins CAP and BAMH1 were studied in parallel as controls. The MD results presented in Figure 8 show that in the presence of divalent cations the MD model of the phased A-tract 25-mer exhibits spontaneous axis bending in a concerted direction, with an average magnitude of -16.5° per A-tract (~33° overall). This result compares qualitatively with the axis bending anticipated for A-tracts phased by a full helix turn, and the bending per turn of approximately 17°-21° inferred from cyclization experiments. The model exhibits a distinct, progressive 5' -> 3' narrowing of the minor groove (Figure 9), a feature inferred from extensive results from DNA footprinting studies by Tullius and coworkers (*73*) of phased A-tract sequences. The MD structures of the control sequence lacking phased A-tracks did not show this effect. Analysis of the MD results supports a bending model with relatively straight A-tracts and pronounced deformations at the junctions between A-tracts and flanking sequences. There is some larger amplitude axis curvature within the flanking sequences as well, and the MD thus far supports a modified junction model.

In order to explore further the ability of molecular dynamics to accurately reproduce DNA bending phenomena, Sprous et al. (*74*) performed analogous MDs on two 30mer DNA segments: $d[G_5\text{-}(GA_4T_4C)_2\text{-}C_5]$ and $d[G_5\text{-}(GT_4A_4C)_2\text{-}C_5]$. Gel migration studies of the sequence motifs A_4T_4 and T_4A_4 by Hagerman (*75, 76*) show the former to be retarded (curved) and the latter normal, a consequence of the differential apposition of the 5' and 3' ends of the individual A-tract strands in the two motifs. The MD simulations were configured in water and 10 mM Mg^{++}, 50 mM K^+ and 70 mM Cl^- above simple neutralization of the DNA backbone, solution conditions targeted to match a ligase buffer system of a particular series of phased A-tract experiments (*77, 78*). The results were analyzed both by visual inspection of the structures (Figure 10) and by following $(L - \ell)/L$, where ℓ is the end-to-end distance and L is the contour length, the angle formed by the first and last dyad axes in the 30-mer as a function of time, and by examining minor groove width in 100 ps windows. The $d[G_5\text{-}(GA_4T_4C)_2\text{-}C_5]$ MD shows stronger signatures of curvature than the $d[G_5\text{-}(GT_4A_4C)_2\text{-}C_5]$ trajectory. The $d[G_5\text{-}(GA_4T_4C)_2\text{-}C_5]$ trajectory shows only a single pattern in which a minor groove width minima is centered in each $d[A_4T_4]$ motif. Analysis with *Curves* reveals that the curvature is towards the minor groove near the center of each $d[A_4T_4]$ motif and away from the minor groove near the central d[CG] dinucleotide step, in general accord with the direction of curvature assigned to A-tracts by Crothers and coworkers (*79*) and by Trifonov and coworkers (*80, 81*). The $d[G_5\text{-}(GT_4A_4C)_2\text{-}C_5]$ trajectory (400-2700 ps) exhibits no gross curvature and a variable minor groove width throughout the sequence. These results agree specifically with the work of Hagerman (*75*) and Tullius and coworkers (*73, 82*).

Thus, dynamical models of oligonucletide sequences based on the Cornell et al. force field and *AMBER 4.1* MD give a reasonable account at the molecular level of many of the salient features observed for the sequence dependent helix morphology of sequences with A-tracts phased by a full helical turn. A series of

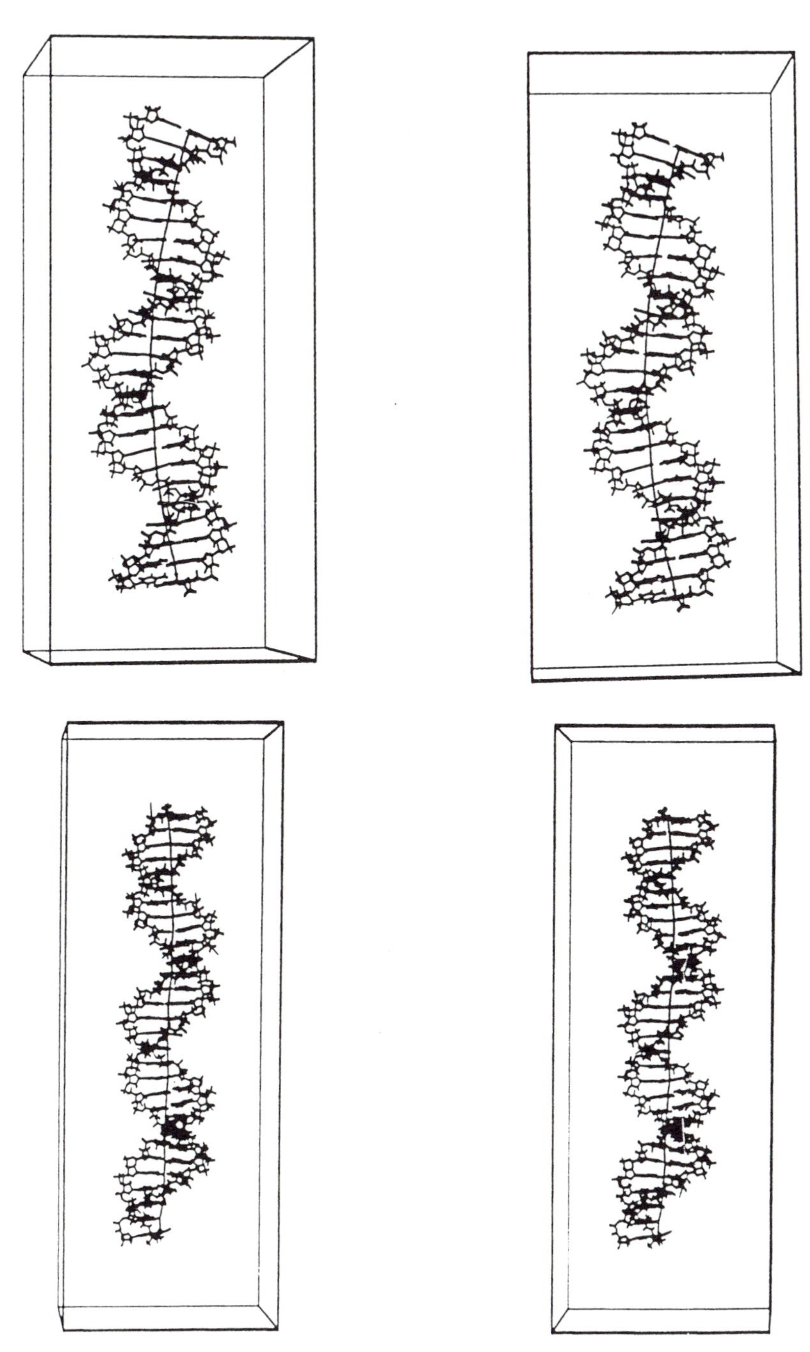

Figure 8. Comparison of MD calculated average structures for a) a DNA oligonucleotide with A-tracts phased by 10 base pairs; and b) that of a control sequence without phased A-tracts (*72*).

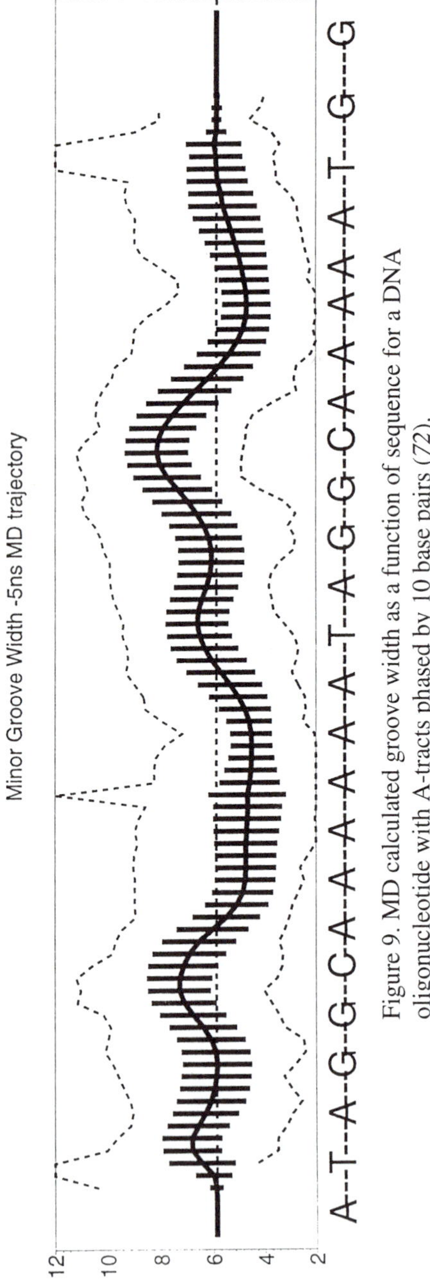

Figure 9. MD calculated groove width as a function of sequence for a DNA oligonucleotide with A-tracts phased by 10 base pairs (72).

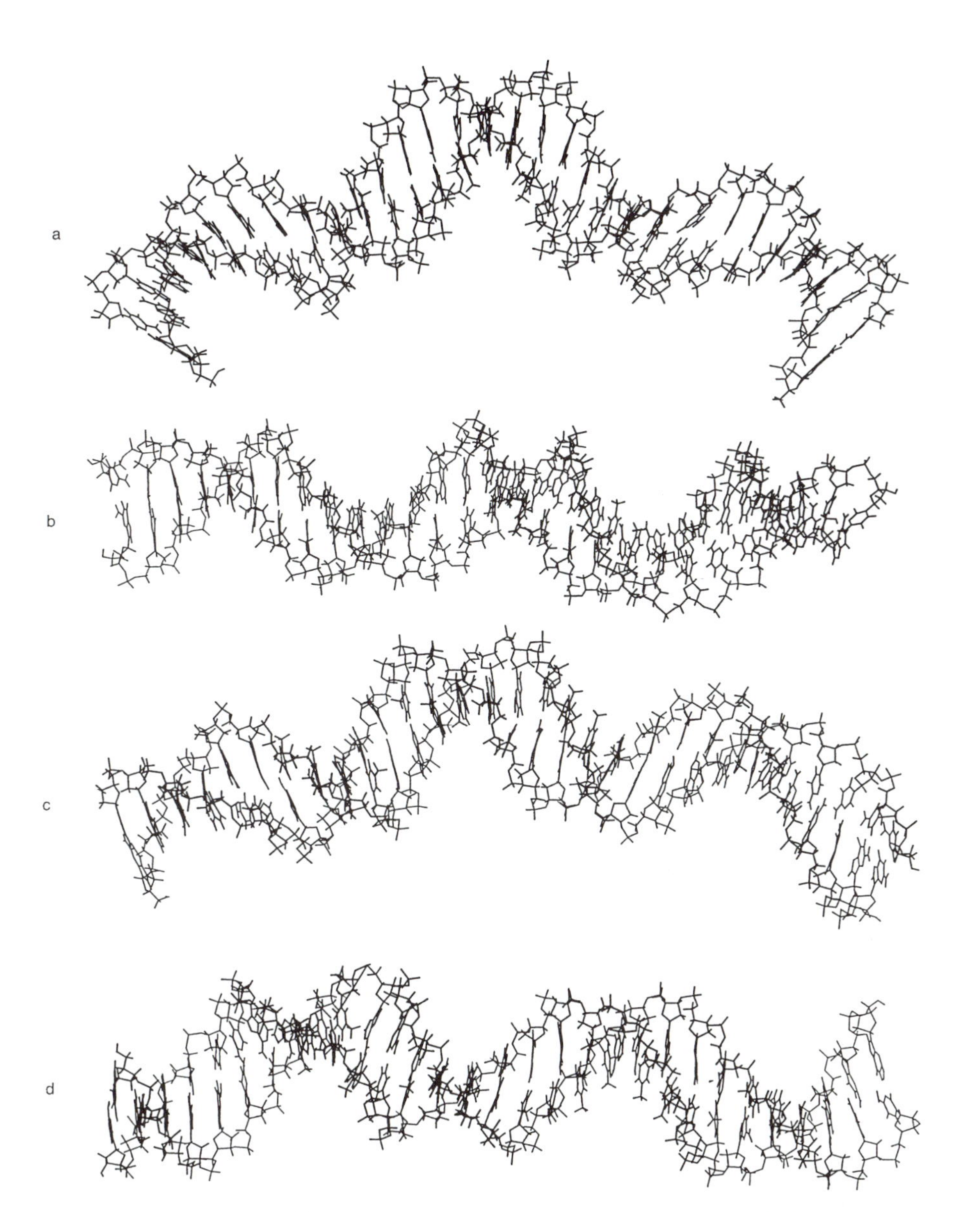

Figure 10. Maximally and minimally curved structures from the d[G_5-{GA_4T_4C}$_2$-C_5] trajectories. a) for d[G_5-{GA_4T_4C}$_2$-C_5] at 950 ps, shortening = 20%; b) for d[G_5-{GA_4T_4C}$_2$-C_5] at 2450 ps, shortening = 3%; c) for d[G_5-{GT_4A_4C}$_2$-C_5] at 250 ps, shortening = 10% and d) for d[G_5-{GT_4A_4C}$_2$-C_5] at 1350 ps, shortening = 3%.

further MD studies is directed at elucidating axis deformations of DNA sequences involved in protein-DNA complexes in order to understand the extent to which structural changes in the DNA observed in complexes are induced by the protein or are, to some extent at least, an intrinsic propensity of the cognate sequence. Studies on the sequences cognate to EcoRI Endonuclease (*59-60*), CAP sequence (*72*), and TATA box sequences (*83*) all provide some indication that a cognate DNA sequence may have an intrinsic propensity to distort in the direction required for complex formation. A comprehensive study of this requires a larger database of results, but the idea introduces a uniquely dynamical dimension to the induced fit problem and the understanding of the nature of DNA protein specificity at the sub-molecular level.

MD Studies of DNA Conformational Transitions

The A to B transition in DNA, based on evidence from CD results on DNA in ethanol and other organic alcohol mixtures as a function of composition (*84 , 85*), is known to be driven by changes in water activity. For sequences conducive to the transition (*86*), the B form is favored at high water activity and the transition to the A form is engendered by lowering the water activity. There are salt effects observed for the transition as well, but control experiments indicate that the transition can be driven purely by hydration. There have been a number of hypotheses proposed (*28*) such as our own favorite (unsubstantiated) account of this transition as based on an "oil drop model" with the hydrophobic pressure on A form sugars increasing with increased water activity effectively squeezing the structure into B form, spine of hydration, the "economics of phosphate hydration", and groove volume effects. One can now apply full-scale molecular dynamic (MD) simulation including counterions, water, and apolar solvent to an appropriate sequence and analyze the results in detail (*87, 88*). Sprous et al. (*89*) have studied the A/B conformational preferences of d(CGCGAATTCGCG) in water and 85% ethanol and other mixed solvent systems using AMBER 4.1 MD and the Cornell et al. force field with PME boundary conditions. The question addressed is simply characterization: whether or not the basic conformational preferences expected for this sequence at high and low water activity are produced in MD based on the Cornell et al. force field. At the nanosecond level of sampling and beginning MDs from both A and B form, the results so far (Figure 11) indicate that A form remains A in 85% ethanol and transitions to B in water as expected, over a time course of 1 ns. The B form starting structure for this sequence remains firmly B in water, as demonstrated by Young et al. (*59*). The MD of a B form starting structure in ethanol has been carried out for 2 ns, but the anticipated transition to the A form had not yet occurred. For the model system, this implies that either a) there is a barrier to the B⇒A transition which cannot be overcome by thermal effects, or b) the transition state on the surface has not yet been encountered in the MD trajectory. The possibility that the Cornell et al. force field slightly overestimates the stability of B form, resulting in an elevated barrier to the A transition cannot be excluded; see also Langley's contribution to this symposium (*91*). MD studies in progess in the DNA hexamer d(GCCGGC) duplex show B form in both water and

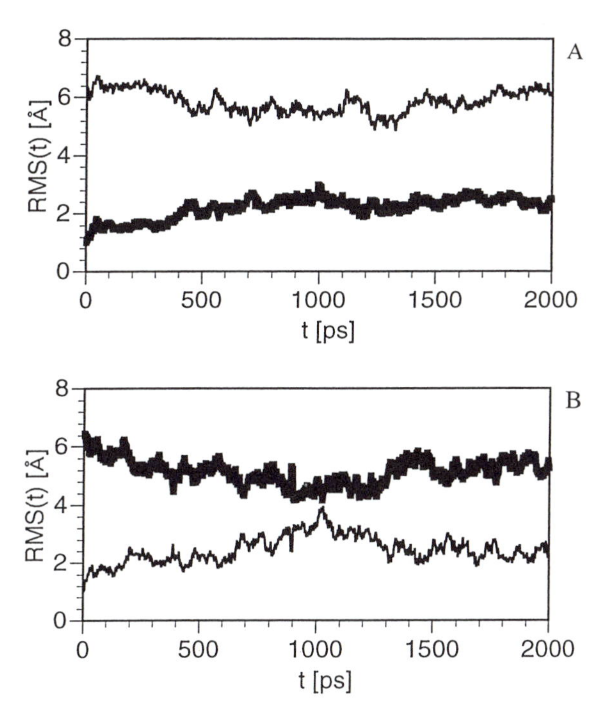

Figure 11. RMS(t) taken against canonical A-DNA (heavy line) and canonical B-DNA (light line) for two d[CGCGAATTCGCG] simulations in 85% OPLS-ethanol:TIP3P-water. For A the starting structure was canonical A-DNA while for B it was canonical B-DNA.

85% ethanol. Describing accurately the conformational preferences of helical forms of DNA in water as a function of added organic solvent and salts and developing a detailed understanding of the dynamical nature of the A to B transition is an important benchmark in defining the role of MD simulation to the field.

Acknowledgments

This research in this area has been supported by NIH Grant #GM 37909 from the National Institutes of General Medical Sciences, and #RR 07885 from the Division of Research Resources. We acknowledge support for M.A. Young from a Traineeship for Molecular Biophysics Training Grant #GM 08271 and Dennis Sprous from NIH postdoctoral fellowship #GM18117. Supercomputer time was generously provided for these projects by the Pittsburgh Supercomputer Center and the Frederic Biomedical Supercomputer Center of the National Cancer Institute, NIH. Access to AMBER code was generously provided by Prof. Peter Kollman. Conversations on this subject with Dr. D.R. Langley and Profs. S.C. Harvey, H.M. Berman, R. Lavery, P. Kollman and T.E. Cheatham III are gratefully acknowledged.

Literature Cited

1. Westhof, E.; Auffinger, P. in *Molecular Modeling and Structure Determination of Nucleic Acids*; Leontis, N.; Santa Lucia Jr., J., Eds. American Chemical Society: Washington, D.C., **1997**.
2. Allen, M.P.; Tildesly, D.J. *Computer Simulation of Liquids*, Clarenden Press: Oxford, **1987**.
3. McCammon, J.A.; Harvey, S.C. *Dynamics of Proteins and Nucleic Acids*, Cambridge University Press: Cambridge, **1987**.
4. Brooks III, C.L.; Karplus, M.; Pettitt, B.M. *Proteins: A Theoretical Perspective of Dynamics, Structure and Thermodynamics*, Advances in Chemical Physics, John Wiley and Sons: New York, NY, **1988**.
5. van Gunsteren, W.F.; Berendsen, H.J.C. *Agnew. Chem. Int. Ed. Engl.*, **1990**, *29*, 992-1023.
6. Dickerson, R.E.; Bansal, M.; Calladine, C.R.; Diekmann, S.; Hunter, W.N.; Kennard, O.; von Kitzing, E.; Lavery, R.; Nelson, H.C.M.; Olson, W.K.; Saenger, W.; Shakked, Z.; Sklenar, H.; Soumpasis, D.M.; Tung, C.S.; Wang, A.H.J.; Zhurkin, V.B. *EMBO J.*, **1989**, *8*, 1-4.
7. Dickerson, R.E. *NEWHELIX93*. University of California: Los Angeles, CA, **1993**.
8. Babcock, M.S.; Pednault, E.P.D.; Olson, W.K. *J. Mol. Biol.*, **1994**, *237*, 125-156.
9. Lavery, R.; Sklenar, H. *J. Biomol. Struct. Dyn.*, **1988**, *6*, 63-91.
10. Ravishanker, G.; Swaminathan, S.; Beveridge, D.L.; Lavery, R.; Sklenar, H. *J. Biomol. Struct. Dyn.*, **1989**, *6*, 669-699.

11. Prevost, C.; Louise May, S., Ravishanker, G.; Lavery, R.; Beveridge, D.L. *Biopolymers*, **1993**, *33*, 335-350.
12. Young, M.A.; Ravishanker, G.; Beveridge, D.L., Berman, H.M. *Biophysical. J.*, **1995**, *68*, 2454-68.
13. Ravishanker, G. *Molecular Dynamics Toolchest*, 1.0 Ed., Wesleyan University: Middletown, CT, **1991**.
14. Weiner, S.J.; Kollman, P.; Case, D.A.; Singh, C.U.; Ghio, C.; Alagona, G.; Profeta, S.; Weiner, P. *J. Am. Chem. Soc.*, **1984**, *106*, 765-784.
15. Brooks, B.R.; Bruccoleri, R.E.; Olafson, B.D.; States, D.J.; Swaminathan, S.; Karplus, M. *J. Comput. Chem.*, **1983**, *4*, 187-195.
16. van Gunsteren, W.F.; Berendsen, H.J.C. *GROMOS86: Groningen Molecular Simulation System*; van Gunsteren, W.F.; Berendsen, H.J.C, Eds.; University of Groningen, Groningen, **1986**.
17. Cornell, W.D.; Cieplak, P.; Bayly, C.I.; Gould, I.R.; Merz Jr., K.M.; Ferguson, D.M.; Spellmeyer, D.C.; Fox, T.; Caldwell, J.W.; Kollman, P.A. *J. Am. Chem. Soc.*, **1995**, *117*, 5179.
18. Mackerell Jr., A.D.; Wiorkiewicz-Kuizera, T.; Karplus, M. *J. Am. Chem. Soc.*, **1995**, *117*, 1194.
19. Pearlman, D.A.; Case, D.A.; Caldwell, J.W.; Ross, W.S.; Cheatham III, T.E.; Fergusen, D.M.; Seibel, G.L.; Singh, U.C.; Weiner, P.; Kollman, P. *AMBER 4.1*; 4.1 Ed.; Pearlman, D.A.; Case, D.A.; Caldwell, J.W.; Ross, W.S.; Cheatham III, T.E.; Fergusen, D.M.; Seibel, G.L.; Singh, U.C.; Weiner, P.; Kollman, P., Eds. UCSF: San Francisco, CA, **1995**.
20. Loncharich, R.J.; Brooks, B.R. *Proteins*, **1989**, *6*, 32-45.
21. Auffinger, P.; Beveridge, D.L. *Chem. Phys. Lett.*, **1995**, *234*, 413.
22. Ravishanker, G.; Auffinger, P.; Langley, D.R.; Jayaram, B.; Young, M.A.; Beveridge, D.L. *Rev. Comp. Chem.*, **1997**, *11*, 317-372.
23. York, D.M.; Darden, T.A.; Pedersen, L.G. *J. Chem. Phys.*, **1993**, *99*, 8345.
24. York, D.M.; Yang, W.; Lee, H.; Darden, T.; Pedersen, L.G. *J. Am. Chem. Soc.*, **1995**, *117*, 5001.
25. Cheatham III, T.E.; Miller, J.L.; Fox, T.; Darden, T.A.; Kollman, P.A. *J. Am. Chem. Soc.*, **1995**, *117*, 4193.
26. Smith, P.E.; Pettitt, B.M. *J. Chem. Phys.*, **1996**, *105*, 4289.
27. Beveridge, D.L.; Swaminathan, S.; Ravishanker, G.; Withka, J.M.; Srinivasan, J.; Prevost, C.; Louise-May, S.; Langley, D.R.; DiCapua, F.M.; Bolton, P.H. in *Water and Biological Molecules;* Westhof, E., Ed.; The Macmillan Press, Ltd.: London, **1993**; 165-225.
28. Westhof, E.; Beveridge, D.L. in *Water Science Reviews: The Molecules of Life*; Ed. F. Franks; Cambridge University Press: Cambridge, **1989**, *5*; 24-136.
29. Jayaram, B.; Beveridge, D.L. *Ann. Rev. Biophys. Biomol. Struct.*, **1996**, *25*, 367-394.
30. Beveridge, D.L. in *Encyclopedia of Computational Chemistry;* von Rague Schleyer, P., Ed.; John Wiley and Sons: New York, NY, **1997**, *in press*.

31. Berman, H.M.; Olson, W.K.; Beveridge, D.L.; Westbrook, J.; Gelbin, A.; Demeny, T.; Hsieh, S.-H.; Srinivasan, A.R.; Schneider, B. *Biophys. J.*, **1992**, *63,* 751-759.
32. Ulyanov, N.B.; James, T.L. *Appl. Magn. Reson.*, **1994**, *7,* 21-42.
33. Swaminathan, S.; Ravishanker, G.; Beveridge, D.L.; Lavery, R.; Etchebest, C.; Sklenar, H. *Proteins.*, **1990**, *8,* 179-93.
34. Dickerson, R.E. *Methods Enzymol.*, **1992**, *211,* 67-111.
35. Dickerson, R.E.; Goodsell, D.S.; Kopka, M.L.; Pjura, P.E. *J. Biomol. Struct. Dyn.*, **1987,** *5,* 557-579.
36. Sprous, D.; Zacharias, W.; Wood, Z.A.; Harvey, S.C. *Nucleic Acids Res.*, **1995**, *23,* 1816-1821.
37. Harvey, S.C.; Dlakic, M.; Griffith, J.; Harrington, R.; Park, K.; Sprous, D. *J. Biomol. Struct. Dyn.*, **1995**, *13,* 301-307.
38. McClarin, J.A.; Frederick, C.A.; Wang, B.-C.; Greene, P.; Boyer, H.W.; Grabel, J.; Rosenberg, J.M. *Science*, **1986**, *234,* 1526-1540.
39. Nerdal, W.; Hare, D.R.; Reid, B.R. *Biochemistry*, **1989**, *28,* 10008-10021.
40. Lane, A; Jenkins, T.C.; Brown, T.; Neidle, S. *Biochemistry*, **1991**, *30,* 1372-1385.
41. Ulyanov, N.B.; Gorin, A.A.; Zhurkin, V.B.; Chen, B.-C.; Sarma, M.H.; Sarma, R.H. *Biochemistry*, **1992**, *31,* 3918-3930.
42. Swaminathan, S.; Ravishanker, G.; Beveridge, D.L. *J. Am. Chem. Soc.*, **1991**, *111,* 5027-5040.
43. Dougherty, A.M.; Causley, G.C.; Johnson, W.C.J. *Proc. Natl. Acad. Sci.(USA)*, **1983**, *80,* 2193-2195.
44. Burkhoff, A.M.; Tullius, T.D. *Cell*, **1987**, *48,* 935-943.
45. Boutonnet, N.; Hui, X.; Zakrzewska, K. *Biopolymers*, **1993**, *33,* 479-490.
46. Olson, W.K.; Sussman, J.L. *J. Am. Chem. Soc.*, **1982**, *104,* 270-278.
47. Olson, W.K. *J. Am. Chem. Soc.*, **1982**, *104,* 278-286.
48. Srinivasan, J. *Characterization of the Solution Structure of DNA Dodecamers using Molecular Dynamics Simulations and Nuclear Magnetic Resonance Methods*. Wesleyan University, 1993.
49. Withka, J.M. *Structural Studies of Intact and Damaged DNA by NMR. Structure Determination of DNA by NMR and Computational Methods*. Wesleyan University, 1991.
50. Withka, J.M.; Swaminathan, S.; Beveridge, D.L.; Bolton, P.H. *J. Am. Chem. Soc.*, **1991**, *113,* 5041-5049.
51. Withka, J.M.; Swaminathan, S.; Srinivasan, J.; Beveridge, D.L.; Bolton, P.H. *Science*, **1992**, *55,* 597-599.
52. McConnell, K.J.; Nirmala, R.; Young, M.A.; Ravishanker, G.; Beveridge, D.L. **1994**, *J. Am. Chem. Soc.*
53. Poncin, M.; Hartmann, B.; Lavery, R. *J. Mol. Biol.*, **1992,** *226,* 775-794.
54. Kumar, S. *Dynamical Behavior of DNA*. Kumar, S., Ed.; University of Pittsburgh: **1990**.
55. Kumar, S.; Kollman, P.A.; Rosenberg, J.M. *J. Biomol. Struct. Dyn.*, **1991**, *8,* a114.

56. Kumar, S.; Duan, Y.; Kollman, P.A.; Rosenberg, J.M. *J. Biomol. Struct. Dyn.*, **1994**, *12,* 487-525.
57. Weiner, S.J.; Kollman, P.A.; Nguyen, D.T.; Case, D.A. *J. Comput. Chem.*, **1986**, *7,* 230-238.
58. Srinivasan, J.; Withka, J.M.; Beveridge, D.L. *Biophysical J.*, **1990**, *58,* 533-47.
59. Young, M.A.; Ravishanker, G.; Beveridge, D.L. *Biophysical J.*, **1997**, *in press.*
60. Young, M.A.; Beveridge, D.L., *in preparation.*
61. Young, M.A.; Jayaram, B.; Beveridge, D.L. *J. Am. Chem. Soc.*, **1997**, *119,* 59-69.
62. Young, M.A.; Jayaram, B.; Beveridge, D.L., *J. Am. Chem. Soc.*, **1997**, *119,* 59-69.
63. Manning, G.S. *Quart. Rev. Biophys.*, **1978**, *11,* 179-246.
64. Sharp, K.A.; Honig, B. *Curr. Opin. Struct. Biol.*, **1995,** *5,* 323-328.
65. R. Lavery and B. Pullman, *J. Biomol. Struct. Dyn.*, **1985**, *5,* 1021-32.
66. Seeman, N.C.; Rosenberg, J.M.; Suddath, F.L.; Kim, J.J.P.; Rich, A. *J. Mol. Biol.*, **1976**, *104,* 109-144.
67. Feigon, J.; Hud, N.V. *J. Am. Chem. Soc.*, **1997**, 119, 5756-5757.
68. Lamm, G.; Pack, G.R. *Proc. Natl. Acad. Sci. USA*, **1990,** *87*, 9033.
69. Sundaralingam, M.; Sekharudu, Y.C. in *Structure and Expression Vol 3: DNA Bending and Curvature*; Olson, W.K.; Sarma, M.H.; Sarma, R.H.; Sundaralingam, M., Eds.; Adenine Press: New York, **1988**; 9-23.
70. Olson, W.K.; Zhurkin, V.B. in *Biological Structure and Dynamics*; Sarma, R.H.; Sarma, M.H., Eds.; Adenine Press: Schenectady, **1996**.
71. Aqvist, J. *J. Phys. Chem.*, **1990**, *94,* 8021-8024.
72. Young, M.A.; Beveridge, D.L. *in preparation.*
73. Burkhoff, A.M.; Tullius, T.D. *Nature*, **1988**, *331,* 455-457.
74. Sprous, D.; Young, M.A.; Beveridge, D.L. *in preparation.*
75. Hagerman, P.J. *Nature*, **1986**, *321,* 449-450.
76. Hagerman, P.J. *Annu. Rev. Biochem.*, **1990**, *59,* 755-781.
77. Gartenberg, M.R.; Crothers, D.M. *J. Mol. Biol.*, **1991**, *219,* 217-30.
78. Koo, H.S.; Wu, H.M.; Crothers, D.M. *Nature*, **1986**, *320,* 501.
79. Zinkel, S.S.; Crothers, D.M. *Biopolymers*, **1990**, *29,* 29-38.
80. Ulanovsky, L.E.; Trifonov, E.N. *Nature*, **1987**, *326,* 720-722.
81. Bolshoy, A.; McNamara, P.; Harrington, R.E.; Trifonov, E.N. *Proc. Natl. Acad. Sci. U. S. A.*, **1991**, *88,* 2312-6.
82. Ganunis, R.M.; Hong, G.; Tullius, T.D. *Biochemistry*, **1996**, *35,* 13729-13732.
83. Flatters, D.; Young, M.A.; Beveridge, D.L.; Lavery, R. *J. Biomol. Struct. Dyn.*, **1997**, *14,* 1-8.
84. Saenger, W. *Principles of Nucleic Acid Structure*, Springer Verlag, New York, **1983**.
85. Ivanov, V.I.; Krylov, D. *Methods Enzymol.*, **1992**, *211,* 111-27.

86. Basham, B.; Scroth, G.P.; Ho, P.S. *Proc. Natl. Acad. Sci. USA*, **1995**, *92*, 6464-6468.
87. Yang, Y.; Pettitt, B.M. *J. Phys. Chem.*, **1996**, *100,* 2564-2566.
88. Cheatham III, T.E.; Kollman, P.A. *J. Mol. Biol.*, **1996**, *259,* 434-444.
89. Sprous, D.; Beveridge, D.L. **1997**, *in preparation.*
90. Drew, H.R.; Wing, R.M.; Takano, T.; Broka, C.; Tanaka, S.; Itikura, K.; Dickerson, R.E. *Proc. Natl. Acad. Sci. (USA)*, **1981**, *78,* 2179-2983.
91. Langley, D.R. in Molecular Modeling and Structure Determination of Nucleic Acids; Leontis, N.; Santa Lucia Jr., J., Eds. American Chemical Society: Washington, D.C., **1997**.

Chapter 17

Molecular Dynamics Simulations on Nucleic Acid Systems Using the Cornell et al. Force Field and Particle Mesh Ewald Electrostatics

T. E. Cheatham, III[1,2], J. L. Miller[1,3], T. I. Spector[1,4], P. Cieplak[1,5], and P. A. Kollman[1]

[1]Department of Pharmaceutical Chemistry, University of California, San Francisco, CA 94143–0446
[2]CombiChem, Inc., 9050 Camino Sante Fe, San Diego, CA 92121
[3]Laboratory of Structural Biology, National Institutes of Health, Bethesda, MD 20892–5626
[4]Department of Chemistry, University of San Francisco, San Francisco, CA 94117–1080
[5]Department of Chemistry, Universityof Warsaw, Pastera 1, 02–093 Warsaw, Poland

We present a review of our recent applications of molecular dynamics simulations with explicit inclusion of solvent and neutralizing ions on DNA and RNA systems, using the Cornell *et al.* molecular mechanics force field and the particle mesh Ewald approach to describe long range electrostatics. We show that this model does a good job of describing the sequence dependent structural properties of DNA duplexes in aqueous solution, in ethanol and in mixed water-ethanol environments, in the presence of $Co(NH_3)_6^{3+}$, in describing the structural properties of phosphoramidate and photochemically damaged DNA duplexes and in describing the properties of RNA duplexes and an RNA hairpin loop.

Molecular dynamics simulations of DNA duplexes go back nearly 15 years and useful insights into the structure, dynamics and hydration of nucleic acids has emerged from these studies. Nonetheless, the increase in computer power, allowing one to simulate into the nanosecond time range, with the inclusion of explicit solvent and counterions and improvements in the simulation protocols (force fields and efficient ways to include long range electrostatic effects) has made the last few years particularly fruitful.

Levitt has noted the importance of smooth truncation of atomic forces in leading to stable trajectories of proteins (*1*) and methods for the smooth truncation of atomic forces are discussed in detail by Steinbach and Brooks (*2*) (see also Allen & Tildesley (*3*)). Although there are other approaches to more accurately represent all of the electrostatic interactions, the Ewald (*4*) approach is one of the most accurate ways to include all of the electrostatic energies and forces for periodic crystals. However, for simulations of DNA in solution, the standard Ewald approach is computationally expensive and moreover one is concerned that the periodicity imposed on the system may lead to computational artifacts. The use of fast Fourier transforms in the particle mesh Ewald (PME) (*5*) approach makes the computations tractable and Smith and Pettitt have shown that periodicity artifacts should be small if a sufficient box size is chosen and a reasonably high dielectric solvent is used (*6, 7*). Thus, the PME model appears to be a useful approach to apply in periodic box simulations of highly charged systems like nucleic acids.

In recent years, one has also realized the importance of employing force fields with balanced solute-solvent and solvent-solvent interactions. OPLS (*8*), CHARMM23 (*9*) and Cornell *et al.* (*10*) are examples of force fields that have been developed with the goal of such balance in mind. The combination of such force fields and accurate inclusion of long range electrostatic interactions has been shown to lead to much improved behavior in the molecular dynamics of nucleic acid systems (*11-13*). In this symposium we have had a number of other exciting examples of the application of such simulation approaches, including those by Beveridge and Young, Langley, MacKerrell, Pastor and Weinstein. Below we briefly describe the highlights of our work in this area.

Reasonable representation of nucleic acid structure: Simulations of DNA and RNA.

The first applications of the particle mesh Ewald method to the simulation of nucleic acid structure involved the study of crystalline DNA (*14-17*). These simulations demonstrated that a simulation protocol employing an Ewald approach could lead to stable simulation of nucleic acid crystals. A drawback of simulations of this type is that the tight packing in the crystal, constant volume conditions and possibly high pressures can inhibit movement of the DNA away from the crystal structure. This suggests that the close agreement between the crystal structure and those structures visited in nanosecond length molecular dynamics simulations may not necessarily validate the force field but instead be fortuitous and imply that conformational sampling is inhibited. An additional drawback of simulations of this type is that it is well known that crystal packing is a strong determinant of structure in nucleic acid crystals. Very similar sequences crystallize into different structures depending on the packing environment or crystal space group (*18-20*). In fact, the commonly quoted helical twist value of 36° for DNA, representing exactly 10 base pairs per repeat, could be an artifact of end-on crystal packing seen in crystal studies of decamers. In solution, both NMR (*21*) and other solution experiments (*22, 23*) suggest values closer to ~34°. However, despite issues regarding crystal packing and perhaps inhibited conformational sampling, the ability to reasonably represent the structure with a simulation protocol employing Ewald methods was a major advance. Prior to this, simulations of nucleic acids were plagued by instability likely due to neglect of the long ranged electrostatic interactions, abrupt truncation of the electrostatic forces, and the lack of a sufficient balance in the force field (for review see (*24, 25*)). We were interested in the structure and dynamics of nucleic acids in solution, an environment arguably more representative of physiological conditions. In collaboration with Tom Darden (National Institute of Environmental Health Sciences) who developed the particle mesh Ewald (PME) code and Michael Crowley and other scientists from the Pittsburgh Supercomputing Center who parallelized the AMBER/PME code to run on the Cray T3D, T3E and other parallel machines, we embarked upon the journey described herein.

The first simulations by our group with the particle mesh Ewald method (*5, 26*) and the Cornell *et al.* (*10*) force field involved the test and development of a simulation protocol applying the new methods and force field. The initial test cases were the small protein BPTI, the DNA decamer $d(CCAACGTTGG)_2$ and an RNA hairpin loop, each in explicit water with counterions with the goal of mimicking *solution* rather than crystal conditions. To do this, the molecule of interest is placed into a box of water with roughly 10 Å of water surrounding the molecule in each direction. Although the Ewald method does impose true periodicity on the molecule, or in other words the simulation cell is replicated in each direction and therefore the molecule may "feel" the influence of its images in neighboring cells in the periodic lattice, the simulations did not impose crystal packing. It was hoped that no undue effects from the periodic images, perhaps as reduced atomic positional fluctuations or inhibited conformational sampling, would be manifest in the simulation. Therefore, in these early simulations of ours with the new protocol, particular care was taken to analyze the dynamics. The results showed that in nanosecond length simulations, the structures were stable and moreover reasonable atomic positional fluctuations were observed (*13*). This suggested that the artificial periodicity imposed by the Ewald treatment did not inhibit the motion of the molecules which was later confirmed in explicit tests by Smith & Pettitt (*6, 7*). Around the same time, a number of other papers came out which showed excellent behavior in the simulations of a DNA triplex (*11*) and RNA hairpin loop (*12*) in solution when an Ewald approach was used.

Further characterization of the DNA decamer in molecular dynamics simulations with the new protocol suggested that sequence specific bending patterns (such as TpG and CpG bends into the major groove) were reasonably well represented and also suggested that the DNA undergoes constant motion, as evident by sugar repuckering throughout the sequence and correlated transitions in the backbone angles. Moreover, spontaneous A-DNA to B-DNA transitions were observed in nanosecond length simulations of $d(CCAACGTTGG)_2$ which further suggested that the conformational dynamics are not inhibited (*27*). In addition to demonstrating that the dynamics are not inhibited, these simulations also validated the force field by demonstrating stabilization of B-DNA over A-DNA in solution, as is expected. As discussed in the introduction, the balance in the force field is critical; unexpected B-DNA to A-DNA transitions have been seen with other force fields (*28, 29*).

The reasonable representation of DNA structure and dynamics in nanosecond length simulations is made possible not only by the simulation protocol but the fact that the structure is very dynamic on a nanosecond time scale. The spontaneous observation of a A-DNA to B-DNA transition further suggests that the barriers to interconversion are not too high and can be surmounted in nanosecond length simulations. Given the tremendous motion and surmountable barriers to conformational transition, this suggests that the representation of the subtle equilibrium between A-DNA and B-DNA is within reach. In the following sections, discussion is presented which suggests that the B-DNA to A-DNA transition is not limited to the d(CCAACGTTGG)$_2$ sequence, that simulation can reasonably represent the preference for "A-DNA" seen with chemical modifications to the backbone and structural differences observed from photodamage, and also that some representation of the environment can be represented in simulations of DNA in mixed water and ethanol or in the presence of hexaamminecobalt(III) which shifts the equilibrium to favor A-DNA structures.

However, before leaving this section it should be noted that there are still many issues to be resolved with respect to the force field and conformational sampling. In the simulations of B-DNA, the sequence dependence of the helical twist and average twist values are not well reproduced by the current generation of the Cornell *et al.* (*10*) force field. Uniformly a lower than expected helical twist is encountered as is a lower than expected average sugar pucker and χ angle. Despite this, long simulations do show reasonable agreement with the averages obtained from crystallography (*30*). Additionally, the Cornell *et al.* (*10*) force field can reasonably represent the structure of A-RNA and the intermediate A-form and B-form structure found in DNA/RNA hybrid duplexes. Simulations of r(CCAACGUUGG)$_2$ and r(CCAACGUUGG)-d(CCAACGTTGG) show that A-RNA is stable and that the RNA/DNA hybrid has a structure that has A-like helicoidal parameters, groove widths intermediate between A-DNA and B-DNA and predominantly C3'-endo sugar puckers in the RNA strand (*31*). A major issue that still needs to be addressed is conformational sampling in RNA systems. The RNA duplex simulations, in addition to showing stable A-form geometries, showed stable "B-RNA" (*31*). However, "B-RNA" has never been seen experimentally. Very likely, this is an artifact of the larger barrier to conformational transition due to the higher barrier to sugar repuckering (*32*) and the extra hydrogen bond interactions possible with the O2' hydroxyl which can interact with neighboring hydrogen bonding donors or acceptors to stabilize a given sugar pucker. Nanosecond length simulations may not be sufficient to overcome the barriers and in fact when the "B-RNA" simulation reported in (*31*) is continued out to 10 ns, the structure does not interconvert to A-RNA. It is not clear whether the force field is overstabilizing B-RNA, the force field creates an artificially larger than expected barrier to conformational transition, or whether the barrier is inherently too large to be surmounted in nanosecond length simulations. The metastability of various RNA conformations was studied in depth in the simulation of RNA hairpin loops as is presented in a later section of this review.

A test case for modeling chemical changes to DNA: Phosphoramidate modified DNA antisense oligonucleotides.

Moving beyond standard DNA and RNA duplexes, simulations of chemically modified DNA provide a further test of the simulation methods. To this end, we performed unrestrained molecular dynamics simulations on a standard $d(CGCGAATTCGCG)_2$ dodecamer duplex in aqueous solution and its phosphoramidate (N3'-P) analog using the particle mesh Ewald summation technique (*5*) and recent AMBER force field (*10*). In the modified dodecamer each phosphodiester has been replaced by a phosphoramidate unit with a N3'-P5' internucleoside linkage.

The synthetic oligonucleotides and their analogs have become an object of intensive study since they may be used as highly specific therapeutic agents and as powerful diagnostic tools (*33*). The chemically modified analogs are usually more stable and degraded more slowly by cellular nucleases than unmodified oligonucleotides.

Recently, Gryaznov *et al.* (*34-39*) have initiated studies on very promising modified oligonucleotides, where the O3'-P bonds are replaced by N3'-P linkages. Such phosphoramidate analogs form very stable duplexes with single-stranded DNA, RNA and with themselves (*i.e.* enhanced melting temperatures of the duplexes by 2.2-2.6° C per modified linkage compared with normal phosphodiesters (*36*)). The phosphoramidate analogs are able also to form stable triplexes with double stranded DNA and RNA (*36, 38*). They have been tested as antisense agents (*39*) and have good water solubility (*35, 36*).

Phosphoramidate oligonucleotide analogs demonstrate very interesting conformational properties. Based on CD and NMR spectroscopy it has been shown that complexes formed by phosphoramidates correspond to an A form of DNA (*36, 37*), whereas standard deoxy-oligonucleotide duplexes adopt B-form only (*40*). According to Thibaudeau *et al.* (*41*) the N/S sugar puckering equilibrium observed in the A and B forms of DNA oligonucleotides depends strongly on the electronic nature of the C3' substituent. For the 2'-deoxyribonucleosides, introducing the less electronegative substituent (*i.e.* $3'NH_3$) relative to the OH group, leads to a predominance of the N (or C3'-endo) conformation of the furanose ring. In the case of 3'OH substituent the S (or C2'-endo) conformation predominates.

The conformational behavior of the normal and N3'-P modified oligonucleotides in aqueous solution described above has been studied using the AMBER 4.1 (*42*) program, the Cornell *et al.* force field (*10*) and the particle mesh Ewald method (*5*) to properly take into account long range electrostatic effects. Molecular dynamics simulations have been performed for the sequence $d(CGCGAATTCGCG)_2$, for which an abundance of experimental data are available (*43-45*) and for which theoretical analysis has also been done (*25*). Experimental data (NMR, CD) on phosphoramidate analogs of this sequence are also available (*37*). These calculations represent another stringent test of the generality of the force field and simulation protocol.

We have carried out unconstrained molecular dynamics simulations in aqueous solution on both normal and modified duplexes starting in both A and B conformational forms. Two "control" simulations have been performed for standard dodecamers $d(CGCGAATTCGCG)_2$, starting in the canonical B form (1000ps long) and starting in the canonical A form (1500ps). Two other simulations have also been performed on the modified dodecamer containing N3'-P5' linkages between all nucleosides. Since an accurate structure of such a modified dodecamer is not yet available from crystallography or NMR measurements we created our initial structures from standard canonical A and B forms by replacing the O3' atoms by an N-H group. For those modified oligonucleotides 1500ps and 1000ps of molecular dynamics simulations were performed for the A-like and B-like structures, respectively.

The SHAKE (*46*) procedure for constraining all hydrogen atoms, 2 fsec time step and the 9 Å cutoff has been applied in all simulations. Calculations have been performed at the temperature of 300K. Results have been analyzed based on the trajectories saved after each 1 psec of simulations. The technical details concerning molecular dynamics simulations protocol and the results are described in Cieplak *et al.* (*47*).

The main results can be summarized as follows. Molecular dynamics simulations for the normal dodecamer starting in canonical A and B structures reveal that both of the structures converge to a single well defined structure within the RMSd value of 1.2 Å (or 1.0 Å for the central eight base pairs). These structures are 2.7-3.1 Å (all atoms) and 1.9-2.2 Å (central 8 residues) from the X-ray structure (9bna) (*43, 44*), which is of the B-type. As discussed previously, similar behavior was also found by Cheatham and Kollman for another shorter duplex (*27*). It was also found that the α, β, γ, ε and δ angles are all very similar to the crystal (9bna) and a canonical B-form. The χ angle is calculated to be somewhat smaller for simulated structures than for the crystal (9bna) and the sugar pucker parameters lead to a ratio of C2'-endo to C3'-endo structures of approximately 80% to 20%. Other helical parameters are also close to what is found in the 9bna structure, but the twist values are closer to canonical A form. Observation of the time evolution of structural parameters during the A to B conformational transition revealed that it takes 350-500 psec for the inclination, pucker, x-displacement and rise parameters to adopt the B like values. During the A to B transition from 490 to 1100 ps we also observed spontaneous, water assisted, breaking and reforming of the terminal GC base pair hydrogen-bonding. Such a phenomena has never been observed before in an unrestrained molecular dynamics simulations.

Molecular dynamics simulations performed on the NP-modified sequence starting in A and B form also lead to structures which are rather close to each other (within 1.9 Å for all atom, or 1.5 Å for the central 8 residues) and belong to the "A" family. These structures are also 1.7-2.3 Å (or 1.4-2.0 Å for the 8 core residues) from canonical A DNA and are 6.4-7.0 Å (3.0-5.0 Å) from canonical B DNA. Examining the structural parameters we could also demonstrate that the χ angle, sugar pucker, x-displacement and twist are much closer to those characteristic of A than B structures, although the standard deviations for those parameters are higher than for the normal dodecamer simulations. Similar to what was observed in the normal DNA simulations, it takes 350-500 psec for the sugar repuckering and B to A form transition in the molecular dynamics simulations (depending on the starting model as is discussed in more detail in (*47*)).

These simulations confirm the generality of the Cornell *et al.* (*10*) force field by demonstrating the ability to reasonably simulate the structure of DNA decamers and chemical modified DNA. As was seen in the earlier simulations, the DNA is found to be very flexible and the structure is very dynamic. In fact, in the middle of the simulation of $d(CGCGAATTCGCG)_2$, the terminal GC base pair opens at ~400 ps and reforms at ~1100 ps. This fraying of the terminal base pair and its spontaneous reforming is discussed in much more detail in Cieplak *et al.* (*47*); however its observation deserves some comment here. In early simulations, terminal base pair fraying was commonly observed and led the community to believe that terminal base pair fraying (on a ps to ns time scale) was a common event. Extending this observation, perhaps "end effects" may explain differences in structure seen in the simulation of short duplexes compared to crystals and fibers. However, base pair opening is known to be a slow process (*48*) and extensive base pair opening should not be seen in nanosecond length simulations. This is supported by the observation of little base pair opening in the extensive set of the simulations of DNA run by our group using the recent simulation protocol.

Understanding DNA damage: Unrestrained molecular dynamics of photodamaged DNA in aqueous solution:

Another chemical modification studied in molecular dynamics simulations (employing the same simulation protocol) involves photodamaged DNA; this provides a further test of the methods and force field and the results can be directly correlated with experiment. Spectroscopic studies of the pyrimidine cis-syn cyclobutane dimer and 6-4 adduct formed by UV irradiation of DNA have begun to reveal the relationship between the structure of photodamaged DNA oligomers with their potential carcinogenicity (*49, 50*). Recently, using a simulation protocol similar to that applied by our group, Miaskiewicz *et al.* have reported the first accurate simulation of a photodamaged DNA dodecamer for structure activity analysis (*51*).

In a related study, using these same computational methods, we have performed 800 ps simulations on the photodamaged cis-syn dimer, 6-4 adduct and undamaged (native) B-DNA duplex decamer of $d(CGCATTACGC)_2$. The advantage of using this sequence is that both its cis-syn and 6-4 thymine dimer modified duplexes have been the subject of high resolution NMR studies (*50*). Thus, starting from the canonical B forms of the photolesion containing decamers and the native duplex, minimization, equilibration and 800 ps unrestrained MD simulations were performed to yield reasonably converged structures (*52*). Analysis of the pertinent backbone and base parameters for the 580-800 ps simulated structures reveals that for both photodamaged decamers the greatest differences in the torsional and helicoidal parameters relative to the native duplex occur primarily at the lesion sites. As has been reported by other researchers, the most notable distortions in the torsion angles of the cis-syn dimer are the A-DNA-like δ at T6 and syn χ at T5 (*51, 53-54*) and the large positive buckle and tilt, and negative propeller and roll, at the first TpA base step.

As anticipated, the 6-4 lesion creates a larger disturbance in the structure of the decamer than the cis-syn dimer. Most notably, there are substantial deviations in the sugar and backbone parameters α, ε, ζ and χ at the lesion site. More significant however, are the observed changes in the base parameters relative to the native duplex which reflect the experimentally observed break in the standard sequential NOEs to the 3' side of the 6-4 lesion (*50*). For the 6-4 adduct there is a large opening between the oxidized base T6 and A15. A moderate opening is also found for the T5-A16 base pair of the cis-syn dimer. These openings are consistent with the "hole" at the photodimer site reported by Vassylyev for the crystal structure of a DNA duplex-T4 endonuclease V complex (*55*).

A comparison of the torsion angles suggested from the NMR refined structure (*50*) of the 6-4 lesion with our simulated values finds considerable coincidence in the sugar and backbone torsions β, δ, ε and χ. The backbone torsions, α and ζ, which are further from the lesion site, are not the same as in the NMR structure. Specifically, while we observe an α of g- and a ζ of g+, Kim *et al.* reports a g+ torsion for α and a g- torsion for ζ. Unfortunately, because it is not possible to experimentally determine the difference between these torsion angles without applying phosphorus NMR we can not determine the significance of this difference between the simulated and experimentally derived structures.

The experimental NOEs (*50*) and the interproton distances for the simulated structures were compared with the assumption that any distance whose value (including standard deviations) is < 5Å should yield an experimentally observable NOE. Based on this assumption our interproton data suggests a CB+ conformation and standard Watson-Crick bonding at the cis-syn lesion correspondent with the NMR structure (*50*). As reported previously, the hydrogen bond distance between the lesion base pairs of the cis-syn dimer shows that there is a slight lengthening of the hydrogen bond between T5(H3)-A16(N1) (*51*). It is worthy of note that, consistent with our simulation results, the X-ray structure of the T4 endonuclease/cis-syn dimer duplex complex (*55*) finds the 5'T adenine-thymine hydrogen bond broken; this suggests that this hydrogen bond is weaker than that of the 3'T of the cis-syn dimer.

As anticipated for the 6-4 adduct our interproton distances reflect the R stereochemistry of the linkage between the adjacent thymines found experimentally (*50*). More interesting from both an experimental and computational perspective is the unusual A4(H1')-T6(CH_3) NOE observed for the 6-4 adduct (*50*). This NOE was also reflected in the interproton distance determined for these same protons in the simulated structure. In fact, analysis of the simulated structure finds the T6 methyl significantly closer to the A4 deoxyribose than it is in the refined NMR structure. Thus we believe the simulated structure better accounts for the experimental observation of this unusual NOE. On further inspection it is also apparent that the relatively close proximity of the T6 methyl to the A4 sugar in the simulated structure is a result of the geometric constraints imposed by hydrogen bonding/dipole-dipole interactions between the carbonyl of the T6 pyrimidone and the A15 NH2 not found in the NMR derived structure.

Studies on the relative repair rates of the cis-syn dimer and 6-4 adduct postulate that the larger deformations in a 6-4 adduct containing DNA duplex lead to greater recognition by repair enzymes (*56*). We also found that the 6-4 decamer deviates most, and the control deviates least, from canonical B-DNA helicity. Most notably, the total angle of curvature between the first and last helical axis segments (*i.e.*, the overall helical bend into the major groove) of the averaged simulated structures of the cis-syn dimer, 6-4 adduct and native duplex are 22.3, 13.6 and 8.2 degrees respectively.

Based on our calculations and in contrast with conclusions derived from experiment, these overall bending angles are presumably not the determinant of difference in the repair of photodamaged DNA. Likely, either variances in the local helicoidal parameters from that of native B-DNA, or the weakness of the hydrogen bonds at the lesion site are the determinants of repair enzyme action. Specifically we found relatively large local distortions in the axis curvature of each base step of the 6-4 adduct relative to the cis-syn dimer and native duplex. In addition, as detailed earlier for the torsion, helicoidal and hydrogen bonding parameters, substantial differences were found in the local distortions from helicity for the three decamers. Thus, the native duplex has only slight local deviations from helicity relative to canonical B-DNA while the cis-syn dimer and 6-4 adduct show significant distortions at the T5-A16 basepair.

In conclusion, unrestrained PME molecular dynamics calculations have provided the first reasonably realistic simulated structure for a 6-4 adduct containing DNA duplex, as well as a cis-syn dimer containing DNA decamer and the native decamer of the same sequence. The results of these simulations yielded structures in very good agreement with experiment (*50*). Specifically, there was good correspondence between our interproton distances, torsional and helicoidal parameters and the experimentally determined NOEs and torsion angles of the photodamaged structures. Overall, based on the results of this study (*52*) we can now confidently simulate other non-canonical or damaged DNA structures.

Molecular dynamics simulations of an RNA tetraloop: Can we get there from here?

Simulations of an RNA hairpin loop showed that its structure was tremendously stable on a nanosecond time scale (*13*) and simulations of a RNA duplex showed that in addition to A-RNA, "B-RNA" is also stable on a nanosecond time scale (*31*). This brings up the issue of conformational sampling in simulations of RNA with the Cornell *et al.* force field which has been extensively studied in our lab. At the meeting, we presented the results of five unrestrained molecular dynamics simulations of an RNA tetraloop, describing our efforts to observe the conversion between an incorrect and a correct loop conformation. We have utilized two separately determined and published NMR structures of this tetraloop (*57, 58*) as the starting structures in our simulations. The second NMR determination, which was refined with many more restraints than the first one, resulted in a different conformation for the loop portion of the structure. Because the differences in the two loop conformations were mainly in the geometry of the U:G base pair of the tetraloop, our hope was that we could observe the conformational transition from the old, incorrect loop structure to the new, correct one. Nanosecond length simulations of both of these NMR structures using the Cornell *et al.* (*10*) force field and a particle mesh Ewald treatment (*5*) were performed and these provide a direct comparison between our calculations and both sets of NMR data.

Our simulations starting from each of the NMR structures not only stay very close to the initial models and preserve the non-bonded interactions in the loops, but they also maintain most of the NMR-derived distances used in their respective structure refinements. These results indicate that our simulations sampled conformational states only in the vicinity of the starting structures, suggesting that the barriers to conformational change are very high in RNA systems. We did not observe the transition between the incorrect and correct loop structures even though one of the simulations was run for 2+ ns. We have also run a separate 1 ns simulation of the first NMR structure and found that it, too, stayed very close to the starting structure.

In an attempt to observe the conformational change, we ran two subsequent simulations of this tetraloop but with the loop riboses replaced with deoxyribose. These simulations were undertaken under the assumption that by removing the 2′OH moieties from the loop domain we would lower the transitional barriers and observe the conformational change. Again, we used each of the NMR structures as starting points for these simulations. Not only were we were able to increase the flexibility of the loops in both simulations, but the simulation which started in the incorrect structure underwent a transition to the correct conformation. The simulation which started in the correct conformation, although it demonstrated increased sampling, remained in the correct conformation. To our knowledge, this is the first example of an observation of a transition in a non-helical nucleic acid system using MD. There have been a few reports of transitions observed in DNA helices, including the expected A-DNA to B-DNA transitions seen in our lab and discussed previously (*27, 47*) and B-DNA to A-DNA transitions seen when using the CHARMM23 all-hydrogen parameter set and an Ewald (*28*) or atom-based force shift (*29*) treatment of the electrostatics. However, similar transitions in RNA seem to be inhibited by the large barriers to structural interconversion. Based on our results with this small model system, we feel that the method of mutating the RNA into DNA for part of the simulation is worthy of consideration during the refinement of nucleic acid structures. We feel that it could be particularly useful when applied to nonhelical regions where the structure is often not well defined by the experimental data.

In the meeting we also reported on a fifth simulation which was run starting from the average structure calculated from the RNA-DNA chimera after it underwent the conformational change. In this simulation, we added the 2′ hydroxyls back to the average structure prior to the dynamics. This structure was the most stable of the five presented here. The effect of the hydroxyl atoms was to lock the loop into a single conformation. Perhaps the most remarkable result presented here is that, after re-introducing the riboses into the loop, the RMSd between the average structure from this simulation and that from the simulation of the correct NMR structure is only 0.5 Å.

These recent studies suggest that the multiple molecular dynamics (MMD) method of Westhof and coworkers (*59-61*) may not be as successful as hoped as a way to increase the sampling in RNA simulations. The MMD method uses different initial velocities for the same starting geometry in order to obtain independent trajectories which sample more of phase space around an "equilibrium" geometry. Our two simulations of the old NMR structure were trapped in the incorrect conformation even though the simulations differed in the timestep used and the frequency of the non-bonded pairlist update. Similar to the MMD method, these differences in the simulation conditions generated independent trajectories which should have increased the overall sampling. However, because of the intrinsic rigidity of RNA, we observed increased sampling only when we changed the starting geometry of the system from RNA to DNA.

We have also examined the behavior of the U5:2′OH group in the simulation starting from the correct NMR structure and in the simulation starting from the converged structure. Allain and Varani, in their in vacuo refinement of the new loop conformation, had found that this moiety hydrogen bonds to the O6 atom of the loop closing guanosine residue (G8). They postulated that this hydrogen bond was responsible for the remarkable thermodynamic stability of the UNCG class of RNA tetraloops. However, our results show that once the system is solvated, the preferred hydrogen bond acceptor for the U5:2′OH is not the G8:O6 atom, but rather the backbone O5′ atom of the following residue (U6). This hydrogen bond is very stable during the simulation which is consistent with the reduced rate of exchange observed in both of the NMR experiments. Also, it does not violate any of the five distance restraints derived from the NOEs. The interactions of this particular hydroxyl may indeed provide the stability to this system, but our results from solvated simulations indicate a different interaction than the in vacuo results do.

We are continuing our MD investigations into the atomic-level interactions of the loop hydroxyls of this tetraloop in an attempt to understand the remarkable stability of this system. Presently, these investigations are restricted to MD simulations and it may be necessary to perform free energy calculations in order to fully understand which hydroxyls provide the stability. However, we feel that the results presented here are very encouraging to not only those engaged in theoretical studies of nucleic acids but also to those researchers interested in the unusual structural characteristics found in many RNA sequences.

Representing the effect of the environment: Simulations of DNA under conditions expected to stabilize A-DNA.

The work presented to this point shows that the simulation protocol and force field applied are able to reasonably represent B-DNA structures. Structures that in the native environment are A-form are also found to maintain a stable A-form geometry, such as RNA and the phosphoramidate modified DNA. Stabilization of an A-form geometry in these models is not surprising. In the case of RNA, it is expected that the electronegativity of the O2' hydroxyl will shift the equilibrium sugar pucker average towards C3'-endo. When the sugars are all held in a C3'-endo conformation, A-form structures are clearly stabilized. Within the molecular mechanical model, this shift towards stable C3'-endo sugar pucker is due to the additional O-C-C-O torsion value (*32*), the charge on the O2' atom and the O2' hydroxyl groups ability to hydrogen bond to nearby groups. With the phosphoramidate DNA, the stabilization of an A-form geometry is likely due to the absence of a V_2 term in the torsional potential of backbone to sugar N-C-C-O dihedral angles; this term's absence is justified based on the observation that 3'-NH_2 has the same sugar pucker ratio as 3'-H (which is very different from 3'-OH). With the Cornell *et al.* force field, a small V_2 term on the O-C-C-O torsions is present to represent the gauche tendency of such torsions. When this term is absent, the sugar pucker distribution clearly shifts towards C3'-endo. It is clear from these examples that a subtle modification in the force field (due to a chemical change) is the operative influence in the shift in the conformational preference. To move to the next level in terms of force field validation, the more stringent test of the force fields is its ability to represent a shift in the conformational preferences by a change in environment alone. In other words, without modifying the force field at all, can changes to the environment such as those occurring due to ligand binding or changes in the solvent be correctly represented?

Clearly this test of environmental influence requires a well chosen test case. The limited conformational sampling seen in RNA simulations rules this system out as a candidate. In a similar manner, any conformational change which requires overcoming a significant conformational barrier, such as a B-DNA to Z-DNA transition, cannot reasonably be represented in nanosecond length simulations without methods applied to artificially boost the conformational sampling. Since we were able to see spontaneous B-DNA to A-DNA transitions within a nanosecond time period, since the DNA is very flexible and dynamic, and since the equilibrium between A-DNA to B-DNA is very strongly influenced by the environment, this seemed like a reasonable test case.

The equilibrium between A-DNA and B-DNA is very strongly and subtly influenced by the environment. From the early fiber studies, where conformational transitions to A-DNA were seen at low relative humidity, the prevailing dogma emerged that it is changes in the water activity (specifically lowering the water activity) that lead to stabilization of A-DNA. Although the effective water concentration is clearly a factor in the stabilization of A-DNA, the effect is more subtle than simply “lowering” the effective water concentration. Based on the experimental observations, two basic mechanisms can be distinguished. These are a general mechanism which likely involves dehydration, such as is seen when the relative humidity is lowered, and a specific mechanism, such as is seen when ligands or proteins bind to DNA to induce a conformational change from B-DNA to A-DNA. Examples of the general mechanism include the B-DNA to A-DNA transition observed in 76%, 80%, or 84% ethanol (v/v) solutions of DNA fibers in the presence of Na^+, K^+, or Cs^+ respectively (*62*). The transition is subtle since in the presence of Li^+ or Mg^{2+} a transition from B-DNA to C-DNA (*63*) or P-DNA (*64*) is observed instead and in the presence of methanol and Na^+, transitions from B-DNA to A-DNA are not seen (*65*). Examples of the specific mechanism include the binding of small acid-soluble spore proteins to DNA in gram positive bacteria which leads to a B-DNA to A-DNA transition (which increases the bacteria’s resistance to UV photodamage) (*66*) or the binding of poly-cationic ions, such as hexaammine cobalt(III), neomycin and spermine, which lead to B-DNA to A-DNA transitions in solution for DNA with GpG sequences (*67*). The specific mechanism is also subtle since the polycationic ligand $Pt(NH_3)_4^{2+}$ does not lead to transitions to A-DNA.

In our studies, we investigated both the general and specific mechanism for inducing a transition from B-DNA to A-DNA. The initial goal was to determine if the force fields and methods were able to represent these environmentally induced structural transitions. If this is possible, then it is hoped that the simulations can give insight into the stabilization of A-DNA in solution and ultimately suggest a unification of the specific and general mechanism. The results were somewhat surprising in that the simple pairwise molecular mechanical force field (albeit a state of the art force field) combined with a reasonably stable simulation protocol could reasonably represent this environmental influence despite the force fields fairly clear bias towards B-DNA and C2’-endo sugar puckers. This was shown by demonstrating the stability of A-DNA in ~85% ethanol and water mixtures (*68*) and spontaneous B-DNA to A-DNA transitions seen with 4:1 $Co(NH_3)_6^{3+}$ (*69*).

The simulation of an A-DNA model of $d(CCAACGTTGG)_2$ was performed with an all atom ethanol model in both pure ethanol and ~85% ethanol-water as described by Cheatham *et al.* (*68*). In pure ethanol, both A-DNA and B-DNA structures diverged significantly from the starting structures during nanosecond length simulations. The structures were characterized by local distortions in the helicoidal parameters. In constrast, when solvated A-DNA was placed into a box of ethanol (leading to ~85% ethanol) the structure remained in a canonical A-DNA geometry for more than three nanoseconds. The stabilization of the A-DNA structure occurred despite repuckering away from C3’-endo and significant C2’-endo populations. When the average ion, water and ethanol density around the DNA is investigated, significantly ion association and hydration is found in the major groove. The ethanol does not disrupt this hydration of the major groove and instead preferentially interacts with the backbone and in the minor groove.

These simulations suggest that the current force fields and simulation protocols can represent some effect of the environment, however in the example discussed the simulation does not completely represent the subtle equilibrium between B-DNA and A-DNA. This is true since no spontaneous B-DNA to A-DNA transitions were observed in multi-nanosecond simulations without subtle modifications to the O-C-C-O torsion terms. As introduced previously, lowering the V_2 term of the O-C-C-O torsion from the 1.0 kcal/mol-deg value in the Cornell *et al.* force field shifts the equilibrium sugar pucker distribution towards more C3'-endo sugar puckers. While this simple fix to shift the pucker distribution to favor more A-like sugar puckers (*i.e.* V_2 from 1.0 to 0.3 kcal/mol for the O-C-C-O torsions) allows a spontaneous transition from B-DNA to A-DNA in mixed water and ethanol, the transition from A-DNA to B-DNA in water leads to an average structure that is less B-DNA like. Therefore, reduction of this torsion potential is not recommended as the general "fix".

An additional issue with these simulations relates to the omnipresent conformational sampling problem. Under the assumption that DNA is never completely dehydrated (even under extremely dehydrating conditions, DNA still has tightly associated water), the simulation of DNA in mixed water and ethanol was started by immersing pre-solvated DNA into a box of pure ethanol rather than immersing *in vacuo* DNA into an equilibrated box of water and ethanol. The amount of water solvating the DNA represented the hydration of B-DNA with ~20 waters per nucleotide and ~6 waters per counterion. It is possible that after the three nanoseconds of simulation the water and ethanol mixture is still not equilibrated and perhaps the A-DNA is trapped in a metastable state and undergoing a slow transition to a more B-DNA like structure. The less than 500 ps time scale of the A-DNA to B-DNA transitions seen in a variety of systems in water, ranging from the published A-DNA to B-DNA transitions of $d(CCAACGTTGG)_2$ (*69*) and $d(CGCGAATTCGCG)_2$ (*47*) to unpublished A-DNA to B-DNA transitions seen with 10-mer duplexes of poly(A)-poly(T), poly(G)-poly(C) and $d(ATATATATAT)_2$ would suggest otherwise unless the presence of ethanol in the simulation seriously damps the motion. In these simulations no significant damping of the atomic positional fluctuations is seen. Of course it is possible that the lower dielectric constant of ethanol increases the artifacts from the periodicity in the Ewald simulation. To further validate the stabilization of A-DNA and influence of the environment, ideally we would like to be able to show titratable B-DNA to A-DNA transitions and back. Progress on this front was presented in the talk by David Langley during this symposia. Despite the issues of convergence and conformational sampling, the 3 ns stabilization of A-DNA does give insight into the stabilization of A-DNA. In these simulations, hydration is preferentially removed from the backbone and minor groove. The major groove remains extensively hydrated and is also well associated with counterions. Ion association in the major groove and extensive major groove hydration was also seen in the simulations of A-DNA in the presence of hexaamminecobalt(III) or $Co(NH_3)_6^{3+}$.

The simulation of $d(ACCCGCGGGT)_2$ with 4 $Co(NH_3)_6^{3+}$ in both a canonical A-form and B-form geometry was performed as described by Cheatham & Kollman (*69*). In the simulations, one $Co(NH_3)_6^{3+}$ ion was hand docked into each of the GpGpG pockets and the additional two $Co(NH_3)_6^{3+}$ ions were placed near the backbone. Additional Na^+ ions were added to neutralize the system and the whole system was solvated. These simulations showed stable A-DNA, and B-DNA to A-DNA transitions, in a series of ~3 ns simulations. In addition to ions in the GpGpG pockets, $Co(NH_3)_6^{3+}$ ions were found interacting with phosphates from both backbones bridging the bend across the major groove. We speculate that this is an additional element stabilizing the A-DNA which in part explains the need of a 4:1 $Co(NH_3)_6^{3+}$:DNA ratio to complete the transition to A-DNA seen experimentally (*67*). The A-DNA structure found for this sequence is very similar to the A-DNA crystal of the same sequence in the presence of Ba^{2+} (*70*). As in the case of the simulations in ethanol, to better show the influence of the environment in these simulations, the ions should be placed away from the DNA to remove the initial bias. Ideally this would show stable B-DNA, followed by slow ion association by diffusion to interact with the DNA which ideally then will lead to a spontaneous B-DNA to A-DNA transition. However the time scale for the diffusion of the ions to the DNA may make this simulation unfeasible at present.

These simulations, exemplifying both the general and specific mechanisms for the B-DNA to A-DNA transition and stabilization of A-DNA, suggest that A-DNA is in part stabilized by extensive hydration and counterion association in the major groove. This is not unreasonable based on what has been seen experimentally and based on simple arguments about conformational features of A-DNA. A-DNA has a wider and shallower minor groove which increases the separation of the phosphates across the groove and exposes more hydrophobic groups making the A-DNA minor groove more favorable in low dielectric solvents or at low relative humidity. Likewise, an economy of hydration of the phosphates in A-DNA compared to B-DNA suggests less water (*71*) interacting with the backbone stabilizes A-DNA. In contrast, in A-DNA the major groove has the phosphates rotated into it and the bends across the major groove lead to close approach of opposing strands. Extensive hydration and counterion association will tend to stabilize these A-DNA conformational features. This stabilization motif is valid in both the specific mechanism of A-DNA stablization by proteins binding into the minor groove (disrupting the water hydration there) and poly-cationic ligands associating in the major groove (stabilizing A-DNA) and the general mechanism of dehydration (taking place primarily in the minor groove and along the backbone). Crystallography of A-DNA supports this general claim; in all the A-DNA crystals, packing of the terminal residues of each duplex is seen into the minor groove of symmetry related duplexes (disrupting the minor groove hydration) and the major groove is extensively hydrated (*72*). In the one case where a protein is crystallized to DNA with a locally A-DNA structure, specifically the TATA binding protein with the TATA box DNA, the protein is bound into the minor groove (*73, 74*).

Force fields and conformational sampling: Have we reached the limit?

As mentioned in the introduction, thanks to advances in computer power, the development of state of the art "balanced" force fields and the application of methods which remove the discontinuities in the forces, we are now able to reliably simulate the structure of nucleic acids in solution. Caveats are issues related to conformational sampling, the need to tweak the force field to improve the helical twist and average sugar pucker values and to better represent the environmental dependence, and the tremendous cost of these calculations. The last point deserves some elaboration. Most of the simulations discussed in this review represent on the order of 10,000 to 15,000 atoms. Simulation of one of these systems for 1 ns requires on the order of one month on a typical workstation (such as the SGI R8000), 1 week on 16 processors of the Cray T3D or approximately 1 day on 64 processors of the Cray T3E. This represents a tremendous amount of computational power. Therefore, potential users of these methods should be aware of the costs and determine whether less costly *in vacuo* methods, like internal coordinate treatments such as JUMNA (*75*), may be appropriate to answer the question at hand. However it is our belief that some representation of explicit solvent is necessary to accurately represent nucleic acid structure and dynamics in many cases, as the simulations investigating the environmental dependence of DNA structure attest.

To move again to a new level in terms of the simulation of nucleic acids, there is still a tremendous amount of work that needs to be done. This relates to the development of methods to overcome barriers to conformational sampling and methods to estimate free energies, but also the development of more refined force field representations. To those entering the field, it should be noted that simply showing stability in a simulation with respect to a starting model structure is clearly not sufficient evidence of that model structures validity because limits in conformational sampling can lead to the stable simulation of metastable states. Moreover, there is a need to move beyond the phenomenological demonstration of simple A-DNA and B-DNA transitions, which may represent pre-conceived biases in the force field, into more complex conformational transitions and interactions, such as protein-nucleic acid interaction, B-DNA to Z-DNA transitions, or ligand-DNA interaction. Finally, there is a major need to use the data from all these simulations to compare with experiment. Examples include the estimation of B_I to B_{II} correlated backbone transition probabilities (as discussed in this symposia) and specific water hydration lifetimes, both of which have been measured by NMR. Without better correlation with experiment, it will be difficult to further refine the force fields.

Despite the caveats above, the ability to reliably simulate the structure of nucleic acids is a tremendous advance. It opens up the possibility to use the simulation methods to attack a variety of problems, ranging from the few touched on here, such as photodamage of DNA, to larger issues such as protein-nucleic acid recognition, the design of molecules to specifically bind to DNA to block transcription or translation, or even the design of molecules to compact DNA for enhanced gene delivery in gene therapy.

Acknowledgments

P.A.K. is grateful to acknowledge research support from the NIH through grant CA-25644. T.E.C. would like to acknowledge research support as a NIH Biotechnology Training Grant fellow (GM08388) and UCSF Chancellor's Graduate Research fellow.

J.L.M. would like to acknowledge research support as a NIH Pharmaceutical Training Grant (GM07175) and American Foundation for Pharmaceutical Education fellow. T. I. S. would like to acknowledge research support from the Dean's Office of Arts and Sciences at the University of San Francisco. PC was partially supported by The Polish Committee for Scientific Research, grant KBN-CHEM-BST 562/23/97. We would also like to acknowledge significant computational support from the Pittsburgh Supercomputing Center (MCA93S017P) and Silicon Graphics, Inc; Michael Crowley (PSC) for parallelizing the PME code on the Cray T3D and T3E; Tom Darden (NIEHS) for helpful discussions and release of the PME code; and the Computer Graphics Lab at UCSF (RR-1081).

References

1. Levitt, M. *Chem. Scripta* **1989**, vol. 29A, pp. 197-203.
2. Steinbach, P. J.; Brooks, B. R. *J. Comp. Chem.* **1994**, vol. 15, pp. 667-683.
3. Allen, M. P.; Tildesley, D. J., *Computer simulation of liquids* Oxford University Press: Oxford, UK, **1987**.
4. Ewald, P. *Ann. Phys. (Leipzig)* **1921**, vol. 64, pp. 253-264.
5. Essmann, U.; Perera, L.; Berkowitz, M. L.; Darden, T.; Lee, H.;Pedersen, L. G. *J. Chem. Phys.* **1995**, vol. 103, pp. 8577-8593.
6. Smith, P. E.; Pettitt, B. M. *J. Chem. Phys.* **1996**, vol. 105, pp. 4289-4293.
7. Smith, P. E.; Blatt, H. D.; Pettitt, B. M. *J. Phys. Chem.* **1997**, vol. 101B, pp. 3886-3890.
8. Jorgensen, W. L.; Maxwell, D. S.; Tirado-Rives, J. *J. Amer. Chem. Soc.* **1996**, vol. 118, pp. 11225-11236.
9. Mackerell, A. D.; Wiorkiewiczkuczera, J.; Karplus, M. *J. Amer. Chem. Soc.* **1995**, vol. 117, pp. 11946-11975.
10. Cornell, W. D.; Cieplak, P.; Bayly, C. I.; Gould, I. R.; Merz, K. M.; Ferguson, D. M.; Spellmeyer, D. C.; Fox, T.; Caldwell, J. W.; Kollman, P. A. *J. Amer. Chem. Soc.* **1995**, vol. 117, pp. 5179-5197.
11. Weerasinghe, S.; Smith, P. E.; Mohan, V.; Cheng, Y. K.; Pettitt, B. M. *J. Amer. Chem. Soc.* **1995**, vol. 117, pp. 2147-2158.
12. Zichi, D. A. *J. Amer. Chem. Soc.* **1995**, vol. 117, pp. 2957-2969.
13. Cheatham, T. E., III; Miller, J. L.; Fox, T.; Darden, T. A.; Kollman, P. A. *J. Amer. Chem. Soc.* **1995**, vol. 117, pp. 4193-4194.
14. York, D. M.; Darden, T. A.; Pedersen, L. G. *J. Chem. Phys.* **1993**, vol. 99, pp. 8345-8348.
15. York, D. M.; Yang, W.; Lee, H.; Darden, T. A.; Pedersen, L. *J. Amer. Chem. Soc.* **1995**, vol. 117, pp. 5001-5002.
16. Lee, H.; Darden, T. A.; Pedersen, L. G. *J. Chem. Phys.* **1995**, vol. 102, pp. 3830-3834.
17. Lee, H.; Darden, T. A.; Pedersen, L. G. *Chem. Phys. Lett.* **1995**, vol. 243, pp. 229-235.
18. Dickerson, R. E.; Goodsell, D. S.; Neidle, S. *Proc. Natl. Acad. Sci. USA* **1994**, vol. 91, pp. 3579-83.
19. Ramakrishnan, B.; Sundaralingam, M. *J. Biomol. Struct. Dyn.* **1993**, vol. 11, pp. 11-26.
20. Shakked, Z.; Guerstein-Guzikevich, G.; Eisenstein, M.; Frolow, F.; Rabinovich, D. *Nature* **1989**, vol. 342, pp. 456-60.
21. Ulyanov, N. B.; James, T. L. *Meth. Enzym.* **1995**, vol. 261, pp. 90-120.
22. Rhodes, D.; Klug, A. *Nature* **1980**, vol. 286, pp. 573-578.
23. Peck, L. J.; Wang, J. C. *Nature* **1981**, vol. 292, pp. 375-378.

24. Beveridge, D. L.; Swaminathan, S.; Ravishanker, G.; Withka, J. M.; Srinivasan, J.; Prevost, C.; Louise-May, S.; Langley, D. R.; DiCapua, F. M.; Bolton, P. H. In *Water and Biological Molecules*, E. Westhof, Ed. Macmillan Press: New York, NY, **1993**, pp. 165-225.
25. Beveridge, D. L.; Ravishanker, G. *Curr. Opin. Struct. Biol.* **1994**, vol. 4, pp. 246-255.
26. Darden, T. A.; York, D. M.; Pedersen, L. G. *J. Chem. Phys.* **1993**, vol. 98, pp. 10089-10092.
27. Cheatham, T. E., III; Kollman, P. A. *J. Mol. Biol.* **1996**, vol. 259, pp. 434-44.
28. Yang, L. Q.; Pettitt, B. M. *J. Phys. Chem.* **1996**, vol. 100, pp. 2564-2566.
29. Norberg, J.; Nilsson, L. *J. Chem. Phys.* **1996**, vol. 104, pp. 6052-6057.
30. Young, M. A.; Ravishanker, G.; Beveridge, D. L. *Biophys. J.* **1997** [in press].
31. Cheatham, T. E., III.; Kollman, P. A. *J. Amer. Chem. Soc.* **1997**, vol. 119, pp. 4805-4825.
32. Olson, W. K.; Sussman, J. L. *J. Amer. Chem. Soc.* **1982**, vol. 104, pp. 270-278.
33. Crooke, S. T.; Lebleu, B., *Antisense research and applications* CRC Press: Boca Raton, FL, **1993**.
34. Gryaznov, S. M.; Sokolova, N. I. *Tetrahedron Lett.* **1990**, vol. 31, pp. 3205-3208.
35. Gryaznov, S.; Chen, J.-K. *J. Amer. Chem. Soc.* **1994**, vol. 116, pp. 3134-3144.
36. Gryaznov, S. M.; Lloyd, D. H.; Chen, J.-K.; Schulz, R. G.; DeDionisio, L. A.; Ratmeyer, L.; Wilson, W. D. *Proc. Natl. Acad. Sci.* **1995**, vol. 92, pp. 5798-5802.
37. Ding, D.; Gryaznov, S. M.; Lloyd, D. H.; Chandrasekaran, S.; Yao, S.; Ratmeyer, L.; Pan, Y.; Wilson, W. D. *Nuc. Acid Res.* **1996**, vol. 24, pp. 354-360.
38. Escude, C.; Giovannangeli, C.; Sun, J. S.; Lloyd, D. H.; Chen, J.-K.; Gryaznov, S. M.; Garestier, T.; Helene, C. *Proc. Natl. Acad. Sci.* **1996**, vol. 93, pp. 4365-4369.
39. Gryaznov, S.; Skorski, T.; Cucco, C.; Nieborowska-Skorska, M.; Chiu, C. Y.; Lloyd, D.; Chen, J.-K.; Koziolkiewicz, M.; Calabretta, B. *Nuc. Acid Res.* **1996**, vol. 24, pp. 1508-1514.
40. Saenger, W., *Principles of Nucleic Acid Structure* Springer-Verlag: New York, NY, **1984**.
41. Thibaudeau, C.; Plavec, J.; Garg, N.; Papchikin, A.; Chattopadhyaya, J. *J. Amer. Chem. Soc.* **1994**, vol. 116, pp. 4038-4043.
42. Pearlman, D. A.; Case, D. A.; Caldwell, J. W.; Ross, W. S.; Cheatham, T. E.; Debolt, S.; Ferguson, D.; Seibel, G.; Kollman, P. *Comp. Phys. Comm.* **1995**, vol. 91, pp. 1-41.
43. Dickerson, R. E.; Drew, H. R.; Conner, B. N.; Wing, R. M.; Fratini, A. V.; Kopka, M. L. *Science* **1982**, vol. 216, pp. 475-485.
44. Westhof, E. *J. Biomol. Struct. Dyn.* **1987**, vol. 5, pp. 581-600.
45. Chou, S. H.; Flynn, P.; Reid, B. *Biochem.* **1989**, vol. 28, pp. 2422-2435.
46. Ryckaert, J. P.; Ciccotti, G.; Berendsen, H. J. C. *J. Comp. Phys.* **1977**, vol. 23, pp. 327-341.
47. Cieplak, P.; Cheatham, T. E., III; Kollman, P. A. *J. Amer. Chem. Soc.* **1997**, vol 119, 6722-6730.
48. Cantor, C. R.; Schimmel, P. R., *Biophysical chemistry part III: The behavior of biological molecules* W. H. Freeman and Co.: New York, NY, **1980**.
49. Taylor, J.-S.; Garrett, D. S.; Brockie, I. R.; Svoboda, D. L.; Telser, J. *Biochem.* **1990**, vol. 29, pp. 8858-8866.
50. Kim, J. K.; Patel, D.; Choi, B. S. *Photochem. and Photobiol.* **1995**, vol. 62, pp. 44-50.
51. Miaskiewicz, K.; Miller, J.; Cooney, M.; Osman, R. *J. Amer. Chem. Soc.* **1996**, vol. 118, pp. 9156-9163.

52. Spector, T. I.; Cheatham, T. E., III; Kollman, P. A. *J. Amer. Chem. Soc.* **1997**, vol 119, pp. 7095-7104.
53. Cadet, J.; Voituriez, L.; Hruska, F. E.; Grand, A. *Biopoly.* **1985**, vol. 24, pp. 897-903.
54. Raghunathan, G.; Kieber-Emmons, T.; Rein, R.; Alderfer, J. L. *J. Biomol. Struct. Dyn.* **1990**, vol. 7, pp. 899-913.
55. Vassylyev, D. G.; Kashiwagi, T.; Mikami, Y.; Ariyoshi, M.; Iwai, S.; Ohtsuka, E.; Morikawa, K. *Cell* **1995**, vol. 83, pp. 773-782.
56. Svoboda, D. L.; Smith, C. A.; Taylor, J.-S.; Sancar, A. *J. Biol. Chem.* **1993**, vol. 268, pp. 10694-10700.
57. Allain, F. H. T.; Varani, G. *J. Mol. Biol.* **1995**, vol. 250, pp. 333-353.
58. Varani, G.; Cheong, C.; Tinoco, I. J. *Biochem.* **1991**, vol. 30, pp. 3280-3289.
59. Auffinger, P.; Louise-May, S.; Westhof, E. *J. Amer. Chem. Soc.* **1996**, vol. 118, pp. 1181-1189.
60. Auffinger, P.; Westhof, E. *Biophys. J.* **1996**, vol. 71, pp. 940-954.
61. Auffinger, P.; Louise-May, S.; Westhof, E. *J. Amer. Chem. Soc.* **1995**, vol. 117, pp. 6720-6726.
62. Piskur, J.; Rupprecht, A. *FEBS Lett.* **1995**, vol. 375, pp. 174-178.
63. Bokma, J. T.; Johnson, W. C. J.; Blok, J. *Biopolymers* **1987**, vol. 26, pp. 893-909.
64. Zehfus, M. H.; Johnson, W. C. J. *Biopolymers* **1984**, vol. 23.
65. Ivanov, V. I.; Minchenkova, L. E.; Schyolkina, A. K.; Poletayev, A. I. *Biopolymers* **1973**, vol. 12, pp. 89-110.
66. Mohr, S. C.; Sokolov, N. V. H. A.; He, C.; Setlow, P. *Proc. Nat. Acad. Sci.* **1991**, vol. 88, pp. 77-81.
67. Robinson, H.; Wang, A. H.-J. *Nuc. Acids Res.* **1996**, vol. 24, pp. 676-682.
68. Cheatham, T. E., III; Crowley, M. F.; Fox, T.; Kollman, P. A. *Proc. Natl. Acad. Sci.* **1997** [in press].
69. Cheatham, T. E., III;Kollman, P. A. *Structure* **1997** [in press].
70. Gao, Y.-G.; Robinson, H.; van Boom, J. H.; Wang, A. H.-J. *Biophys. J.* **1995**, vol. 69, pp. 559-568.
71. Saenger, W.; Hunter, W. N.; Kennard, O. *Nature* **1986**, vol. 324, pp. 385-388.
72. Wahl, M. C.; Sundaralingam, M. *Biopolymers* **1997**, vol. 44, pp. 45-63.
73. Kim, Y.; Geiger, J. H.; Hahn, S.; Sigler, P. B. *Nature* **1993**, vol. 365, pp. 512-520.
74. Kim, J. L.; Nikolov, D. B.; Burley, S. K. *Nature* **1993**, vol. 365, pp. 520-527.
75. Lavery, R.; Zakrzewska, K.; Sklenar, H. *Comp. Phys. Comm.* **1995**, vol. 91, pp. 135-158.

Chapter 18

Observations on the A versus B Equilibrium in Molecular Dynamics Simulations of Duplex DNA and RNA

Alexander D. MacKerell, Jr.

Department of Pharmaceutical Sciences, School of Pharmacy, University of Maryland at Baltimore, 20 North Pine Street, Baltimore, MD 21201

Molecular dynamics simulations of duplex DNA or RNA in solution now yield stable structures for a duration of well over 1 ns. Different force fields, however, lead to different equilibrium structures. Presented are details of the equilibrium between the A and B forms of DNA by the CHARMM all-hydrogen force field for several sequences, including previously unpublished results on the d[GCGCGCG]$_2$ and d[ATATATA]$_2$ duplexes in solution. Comments are also included on the performance of other force fields. Based on a combination of observations a dominate role of the sugar and phosphodiester backbone moieties in the equilibrium between the A and B forms of DNA is suggested.

Theoretical calculations of biological molecules in solution based on empirical force field models allow for atomic details relating structure to energetics and function to be investigated (1,2). Empirical force field calculations on biological macromolecules were first successfully applied to proteins. Only recently have successful simulations been performed on lipids (3,4) and nucleic acid duplexes (5-10). A large part of the recent successes in simulations of duplex oligonucleotides has been the development of new force fields for nucleic acids. The most widely used are the new CHARMM (11) and AMBER (12) all-hydrogen parameter sets. Another set that has been successful is the BMS parameter set (D.R. Langley, Personal Communication).

Adequate treatment of the electrostatic interactions is another area important for stable duplex oligonucleotide simulations. Some of the first successful DNA simulations (13-15) , including calculations on a DNA triplex (16), were performed using the Ewald method (17), in some cases with the Particle Mesh Ewald (PME) approach (18). Stable DNA and RNA simulations have also been performed using spherical atom truncation methods (9,19,5). These simulations were performed using a relatively long electrostatic cutoff distance of 12 Å. Thus, it appears that either Ewald sums or atom truncation based methods may be used for MD simulations of nucleic acids. Ewald based methods may be considered more rigorous, but are limited to periodic systems and atom truncation schemes should be used with cutoff distances of 12 Å or longer.

The remainder of this chapter will concentrate on the equilibrium between the A and B forms of DNA and RNA observed in MD simulations in low salt environments using the new CHARMM and AMBER parameter sets and present previously unpublished CHARMM based MD simulations of the d[GCGCGCG]$_2$ and d[ATATATA]$_2$ duplexes.

Review of Published Oligonucleotide Simulations

Both AMBER and CHARMM based calculations show spontaneous transitions between the A and B forms of DNA. CHARMM based calculations using both Ewald sums and atom truncation for the treatment of electrostatics show spontaneous transitions from the B form to the A form (9,7,5) . Simulations with AMBER stay in the vicinity of the B form of DNA while those started in the A form shift to the B form (6) . RNA simulations with AMBER if started in the A form stay A, however, if started in the B form stay in the B form (20). A form RNA simulations using CHARMM stay A (5); RNA simulations starting in the B form have yet to be performed with CHARMM.

An interesting study has been performed on the d(CCCCCTTTTT)$_2$ decamer using both the AMBER and CHARMM force fields (M. Feig and B.M. Pettitt, *J. Phys. Chem.* In Press). Experimental studies show this structure to assume the A form for the C region and the B form for the T region. Starting from the A form of DNA a number of properties were monitored. The AMBER simulation had all sugars in the C1'exo/C2'endo conformation typical for B form DNA while CHARMM produced A and G sugars the C3'endo (A type) sugar conformation while the T and C sugars assumed the B type, though relatively large fluctuations in puckering occurred. The overall conformations were primarily B form for AMBER and A form for CHARMM. In summary, published calculations indicate a trend where the AMBER force field has a tendency towards the canonical B form and CHARMM towards the canonical A form of DNA.

Closer examination of CHARMM results are presented in Table 1 for a collection of DNA duplex MD simulations. Results are for a variety of sequences, with lengths varying from 5mers up to 18mers, including several calculations on the EcoRI dodecamer. Several results are of note. Examination of the entire series shows a trend where the rms differences are getting closer to the B form and further from the A form of DNA as the sequence length increases. The rms differences of the EcoRI dodecamer are similar at different salt concentrations or with atom truncation versus Ewald summation for the treatment of the electrostatic interactions. Comparison of the (AT)$_6$ and (GC)$_6$ dodecamers shows (GC)$_6$ to stay closer to the A form although both structures deviate the same amount from the B form of DNA. Results for the AT and GC 7mers in Table 1 appear to contradict this results, however, the 7mer AT structure is closer to the B form of DNA and analysis of the change in RMS difference versus time shows the 7mer GC structure to relax more quickly to the A form of DNA (see below).

Analysis of the nucleic acid database (NDB)(21) allows for some generalizations to be made concerning the relationship of length and base composition to A versus B form structural properties. The majority of the A form DNA structures are 8mers or less (42 out of 58, June 97) while only 4 out of 80 B form structures are 8mers or less, indicating shorter oligonucleotides to favor the A form. In A form experimental crystal structures GC basepairs dominate. In several A DNA octamers the number of AT basepairs equals the number of GC basepairs, but never exceeds it and for longer sequences the highest ratio has 4 AT basepairs with 6 GC basepairs for a decamer and two dodecamers have 4 AT basepairs with 8 GC basepairs. Although preliminary in nature, comparison of this analysis of the NDB with results in Table 1 indicate that the CHARMM parameters produce DNA structures that are consistent with trends observed in experimental DNA crystal structures concerning length and base composition.

Table 1. RMS differences with respect to canonical A and B DNA from the final time frame of MD simulations using the CHARMM parameters.

Sequence	RMS difference, A	
	vs. A-DNA	vs. B-DNA
TCGCG(500ps, SB)[1]	1.4	2.4
CGCGCG(500ps, SB)[1]	1.5	2.5
ATATATA (800ps)[2]	0.9	3.6
GCGCGC (800 ps)[2]	1.4	4.2
CCAACITTGG (850 ps)[3]	2.3	5.6
EcoRI (no salt, 1000 ps)[4]	2.8	4.1
EcoRI (Mg, 1000 ps)[4]	2.2	5.5
EcoRI ($MgCl_2$, 1000 ps)[4]	2.7	5.2
EcoRI (NaCl, 3500ps, Ewald)[5]	2.0	4.3
CTATAAAAGGGC (550 ps)[3]	3.5	4.2
CCATAAAAGGGC (550 ps)[3]	3.2	5.0
CTATATAAGGGC (550 ps)[3]	3.5	4.7
CTATAAGAGGGC (550 ps)[3]	3.5	5.2
CTTTTATAGGGC (550 ps)[3]	3.2	4.7
ATATATATATAT (550 ps)[3]	4.7	4.2
CGCGCGCGCGG (550 ps)[3]	3.6	4.2
CTCAGAGGCCGAGGCGGC (400 ps)[6]	5.5	3.8

Rms differences are from the final time frame of the simulations for non-hydrogen atoms following least squares fitting of the non-hydrogen atoms to the canonical structures. All simulations are performed with periodic boundary conditions unless noted. 1) Stochastic boundary simulations from (19), 2) simulations presented in this chapter, 3) NVT/PBC simulations in the presence of 0.36 M Na^+ (N. Pastor and H. Weinstein, see contribution by these authors in this volume), 4) NVT/PBC (9), 5) NVT/PBC (7), and 6) NPT/PBC simulations in 0.015 M Na^+ and 20 % formamide (A.D. MacKerell, Jr. unpublished data).

Simulations of d[GCGCGCG]$_2$ and d[ATATATA]$_2$

Better understanding of the B to A transition in DNA with CHARMM can be obtained from simulations performed in our laboratory on the ATATATA and GCGCGCG 7mers using the program CHARMM (22) with the CHARMM all-hydrogen nucleic acid parameters (11). System setup was initiated by generating the two duplexes in the canonical B-DNA conformation (23). The DNA was then overlaid with a preequilibrated solvent box of the CHARMM TIP3P water model (24,25) that contained 1 M NaCl and all solvent molecules with a non-hydrogen atom within 1.8 Å of any DNA non-hydrogen atom deleted. Additional sodiums were then added to the system at random locations to yield an electronically neutral system. The final box size was 42x32.3x32.3 Å, yielding Na^+ and Cl^- concentrations of 0.94 M and 0.49 M, respectively. Atom based spherical truncation with 13 Å list and 12 Å nonbond interaction cutoffs, electrostatic shift and vdW switch smoothing functions with the vdW switching function turned on at 10 Å, list updating every 10 minimization or MD steps and periodic boundary conditions were used in all calculations. MD simulations were performed in the NVT ensemble with an integration time step of 0.002 ps and SHAKE of all covalent bonds involving hydrogens (26). Following the overlay the systems were subjected to 200 adopted-basis Newton-Raphson (ABNR) minimization steps with the DNA atomic positions fixed and 2.0 kcal/mol/Å^2 harmonic constraints applied to all solvent non-hydrogen atoms. This was followed by a 50 ps MD simulation with DNA atoms fixed to equilibrate the solvent around the DNA. The entire system was then minimized for 50 ABNR steps with all non-hydrogen atoms subjected to 2.0 kcal/mol/Å^2 harmonic constraints. These structures were then used to initiate the production simulations.

Presented in Figure 1 are the rms differences versus time with respect to the canonical A and B forms of DNA for all non-hydrogen atoms. Both systems initially relax away from the B form and towards the A form of DNA. At about 200 ps the GC 7mer structure continues to relax while the AT 7mer assumes a conformation that is approximately 2.0 and 3.0 Å from the B and A forms of DNA, respectively. At 500 ps of the simulation the AT 7mer undergoes additional relaxation toward the A form. By the end of the simulation the rms difference of both systems are approximately 4 and 1Å from the B and A forms of DNA, respectively. These results are consistent with the previous simulation results on short oligonucleotides using the CHARMM potential (see Table 1). The slower relaxation of the AT 7mer duplex supports the suggestion that the parameters exhibit base composition dependent structural effects such that AT containing sequences favor the B form of DNA more that GC containing structures.

Details of the structural transition from the A to B form of DNA can be obtained by analysis of individual conformational descriptors of the duplex DNA. As a first step, all the standard conformational descriptors of DNA were analyzed using the Dials and Windows program (27). From this analysis the sugar pseudorotation angle, the glycosidac linkage dihedral chi, the backbone dihedral zeta and the basepair inclination were selected for detailed inspection. These terms were selected due to significant differences between themselves in the A and B forms of DNA and they may be considered as descriptors of different portions of the DNA molecule. Presented in Figure 2a through 2d are the zeta, sugar puckering, chi, and inclination angles, respectively, as a function of simulation time.

Comparison of Figures 1 and 2 allows for relationships between the conversion from the B to the A form, as defined by the rms difference, and changes in individual structural parameters to be analyzed. Figure 1 shows the GC 7mer to initially move towards the A form faster than the AT 7mer despite being initially further from the A form. This pattern is similar with that occurring with the zeta backbone dihedral, which defines the torsion around the O3'-P bond. With zeta, the GC 7mer is initially more B like, but rapidly changes to the A form value. After 200 ps the GC and AT 7mer zeta values are similar and track each other until approximately 400 ps, where the GC 7mer zeta values move towards the A form ahead of the AT structure. Both structures have similar zeta values at the end of the simulations. Concerning sugar puckering both structures change in a similar manner during the first 200 ps, with the GC values slightly closer to the A form. After 200 ps the AT 7mer puckering stays closer to the B form as compared to the GC 7mer, similar to the trend in the rms differences. Over the final 200 ps of the simulation the AT 7mer puckering moves more towards the A form with both structures being dominated by A type puckering at the end of the simulations. Analysis of chi shows a monotonic change through the course of both simulations, indicating little correlation with changes in the rms differences. The inclination as a function of time is similar to the behavior of the sugar puckering although inclinations changes more gradually towards the final value, rather than the abrupt changes in sugar puckering. The two systems track each other closely during the initial 200 ps following which the GC 7mer structure tends to be more A like for the next 400 ps. During this time the AT 7mer inclination gradually drifts towards the A form value, with the GC 7mer and AT 7mer values again being similar at the end of the simulations.

Based on the above analysis some preliminary comments concerning the relationship of the changes in conformation of different moieties of the DNA to the overall change in the structure can be made. The glycosidac linkage, chi, appears to have the least influence on the A to B equilibrium. Inclination appears to be more important than chi, but the gradual drift of the AT structure suggests this term is following other structural alterations. Sugar puckering and the backbone conformation,

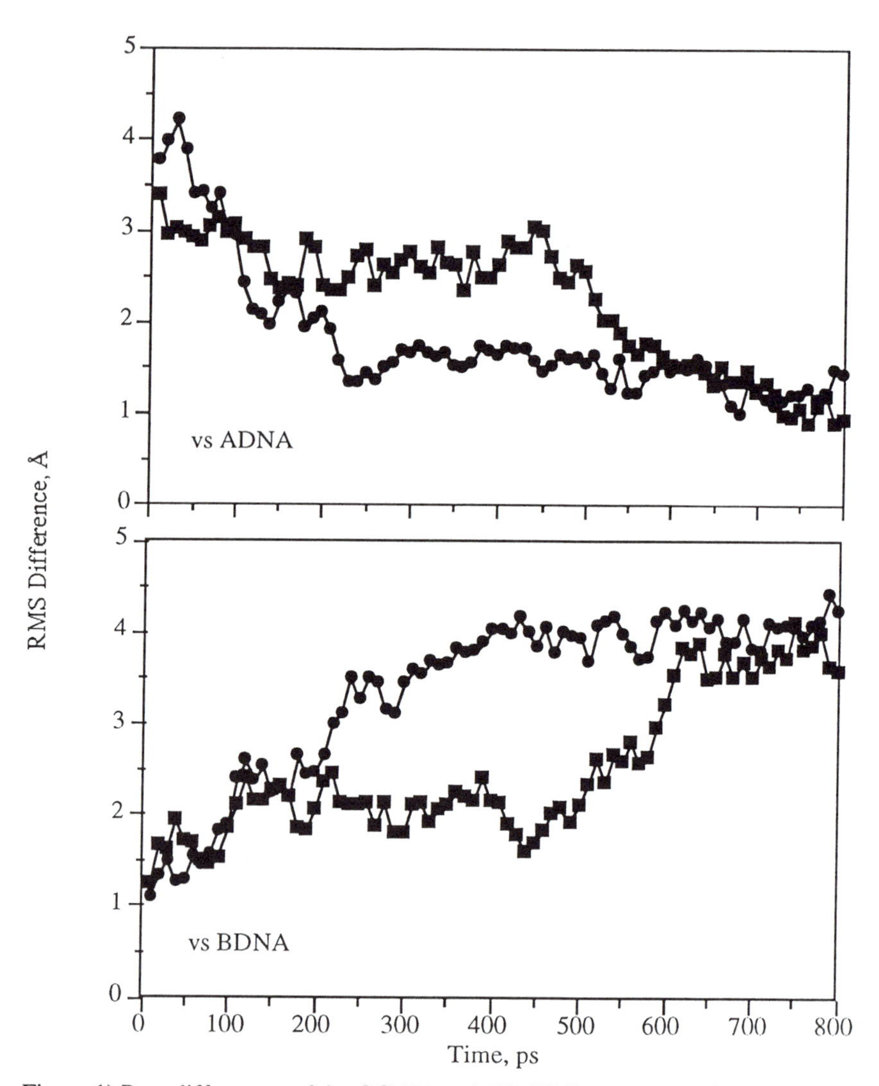

Figure 1) Rms differences of the GC (●) and AT (■) 7mers versus the A (upper) and B (lower) forms of DNA. Rms differences are for all non-hydrogen atoms following least-squares fitting to the canonical forms of DNA. Results are from snapshots taken every 10 ps from the simulations.

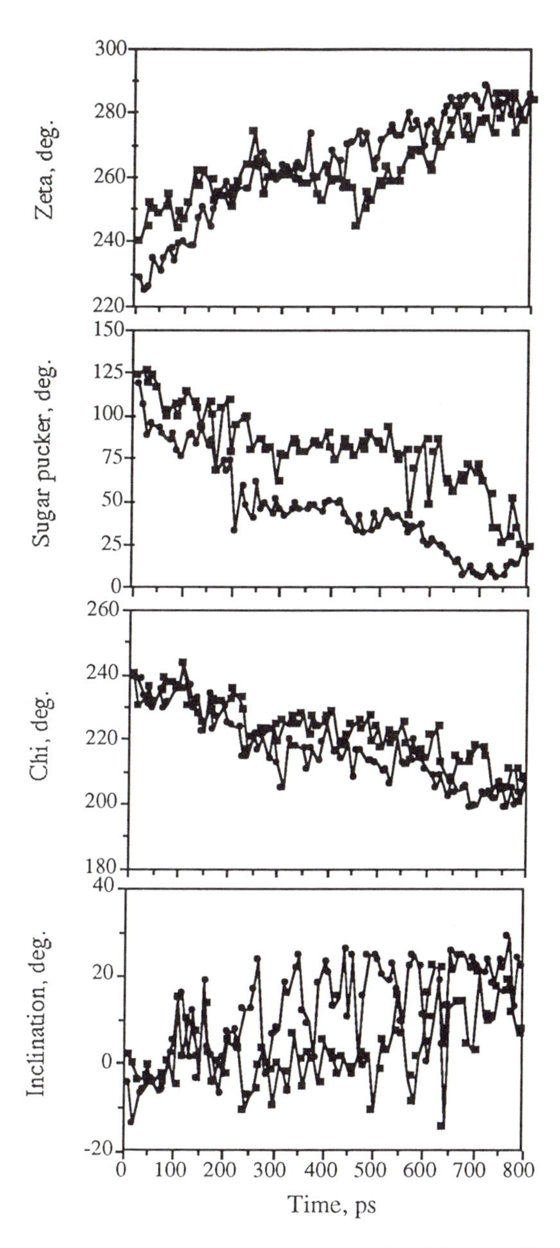

Figure 2) From top to bottom; Zeta, Sugar pucker, chi and base inclination energy 10 ps from the GC (●) and AT (■) 7mer simulations. Each value (e.g. zeta) is averaged over all nucleotides in the duplex 7mers at each timeframe.

based on zeta, may significantly contribute to the equilibrium. Sugar puckering is suggested to be important, since the stable region between 250 and 550 ps is quite similar to that observed in the rms differences. Zeta is suggested to also have a strong influence on the A versus B equilibrium due to the changes in this term being similar to the changes in the rms difference with respect to the A form of DNA during the first 200 ps of the simulations. The present analysis, therefore implicates both sugar puckering and the backbone conformation to have a strong influence on the A versus B equilibrium, although contributions from other degrees of freedom appear to also contribute.

These observations are consistent with previous studies of the equilibrium between the A and B forms of DNA and RNA. Numerous experimental studies, including X-diffraction work in fibers, show high salt concentrations to favor the A form of DNA (28). This role of ions and their potential for interaction with the charged phosphate groups is consistent with the importance of the phosphodiester backbone in the A versus B equilibrium. Calculations with the BMS force field (D.R. Langley, Personal Communication) and AMBER (29) show changes in salt concentrations to influence the A versus B equilibrium, consistent with experiment. Further support for a role of the phosphodiester linkage are theoretical studies on dimethylphosphate and the EcoRI duplex in solution (A.D. MacKerell, Jr., *J. Chim. Phys.*, In press). In the study, QM calculations suggested that interaction of water or sodium with dimethylphosphate alter its conformational properties and that these alterations may effect the equilibrium between the A and B forms of DNA. Inclusion of those alterations in simulations of the EcoRI dodecamer in solution showed the equilibrium to be shifted. Supporting a role of the sugar are calculations on the RNA $(CCAACGUUGG)_2$ duplex using the AMBER 4.1 force field (20). Calculations on RNA starting in the B conformation showed that conformation to be maintained until the sugars were forced to flip from the C2'endo to the C3'endo form, following which the conformation changed to the A form of RNA.

Summary

Current empirical force fields allow for more reliable simulations of oligonucleotides in solution to be performed. Limitations in the force fields are evident, however, from tendencies for biasing structures towards either the A or B forms depending on the force field and starting stucture. Results presented on AT and GC 7mer simulations, along with other work, suggest that the phosphodiester backbone and sugar moieties of DNA dominate the equilibrium between the A and B forms. This information allows for both a better understanding of the role of different chemical moieties in oligonucleotides on their overall structural properties and areas to focus on in the future development of empirical force fields for nucleic acids. Work in progress on a revised version of the CHARMM force field, emphasizing a balance between energetic data from *ab initio* calculations and reproduction of experimental data on duplex DNA and RNA, has produced highly stable structures intermediate to the A and B forms of DNA (N. Foloppe and A.D. MacKerell, Jr., In progress). It is hoped that such an approach will allow for the optimization of force fields that have an accurate balance between the different degrees of freedom in nucleic acids, thereby achieving improvements in the balance between calculated microscopic and macroscopic properties of nucleic acids in solution.

Acknowledgments

Financial support from the National Institutes of Health (GM51501-01) and computational support from the Pittsburgh Supercomputing Center and NCI's Frederick Biomedical Supercomputing Center are appreciated.

References

(1) Brooks, C. L., III; Karplus, M.; Pettitt, B. M. *Proteins, A Theoretical Perspective Dynamics, Structure, and Thermodynamics*; John Wiley and Sons: New York, 1988; Vol. LXXI.

(2) McCammon, J. A.; Harvey, S. *Dynamics of Proteins and Nucleic Acids*; Cambridge University Press: , 1987.

(3) Pastor, R. W. *Curr. Opin. Struct. Biol.* **1994**, *4*, 486-492.

(4) Tobias, D. J.; Tu, K.; Klein, M. L. *Curr. Opin. Coll. & Inter. Sci.* **1997**, *2*, 15-26.

(5) Norberg, J.; Nilsson, L. *J. Chem. Phys.* **1996**, *104*, 6052-6057.

(6) Cheatham, T. E., III; Kollman, P. A. *J. Mol. Biol.* **1996**, *In Press*.

(7) Yang, L.; Pettitt, B. M. *J. Phys. Chem.* **1996**, *100*, 2550-2566.

(8) Young, M. A.; Jayaram, B.; Beveridge, D. L. *J. Am. Chem. Soc.* **1997**, *119*, 59-69.

(9) MacKerell, A. D., Jr. *J. Phys. Chem. B* **1997**, *101*, 646-650.

(10) Auffinger, P.; Westhof, E. *Encyc. Comp. Chem.* **1997**, *In press*.

(11) MacKerell, A. D., Jr.; Wiórkiewicz-Kuczera, J.; Karplus, M. *J. Am. Chem. Soc.* **1995**, *117*, 11946-11975.

(12) Cornell, W. D.; Cieplak, P.; Bayly, C. I.; Gould, I. R.; Merz, J., K.M.; Ferguson, D. M.; Spellmeyer, D. C.; Fox, T.; Caldwell, J. W.; Kollman, P. A. *J. Amer. Chem. Soc.* **1995**, *117*, 5179-5197.

(13) Lee, H.; Darden, T.; Pedersen, L. G. *J. Chem. Phys.* **1995**, *102*, 3830-3834.

(14) York, D. M.; Yang, W.; Lee, H.; Darden, T.; Pedersen, L. G. *J. Am. Chem. Soc.* **1995**, *117*, 5001-5002.

(15) Cheatham, T. E., III; Miller, J. L.; Fox, T.; Darden, T. A.; Kollman, P. A. *J. Am. Chem. Soc.* **1995**, *117*, 4193-4194.

(16) Weerasinghe, S.; Smith, P. E.; Mohan, V.; Cheng, Y.-K.; Pettitt, B. M. *J. Am. Chem. Soc.* **1995**, *117*, 2147-2158.

(17) Ewald, P. P. *Ann. Phys.* **1921**, *64*, 253-287.

(18) Darden, T. A.; York, D.; Pedersen, L. G. *J. Chem. Phys.* **1993**, *98*, 10089-10092.

(19) Norberg, J.; Nilsson, L. *J. Biomol. NMR* **1996**, *7*, 305-314.

(20) Cheatham, I., T.E.; Kollman, P. A. *J. Amer. Chem. Soc.* **1997**, *119*, 4805-4825.

(21) Berman, H. M.; Olson, W. K.; Beveridge, D. L.; Westbrook, J.; Gelbin, A.; Demeny, T.; Hsieh, S.-H.; Srinivasan, A. R.; Schneider, B. *Biophys. J.* **1992**, *63*, 751-759.

(22) Brooks, B. R.; Bruccoleri, R. E.; Olafson, B. D.; States, D. J.; Swaminathan, S.; Karplus, M. *J. Comput. Chem.* **1983**, *4*, 187-217.

(23) Arnott, S.; Hukins, D. W. L. *J. Mol. Biol.* **1973**, *81*, 93-105.

(24) Jorgensen, W. L. *J. Am. Chem. Soc.* **1981**, *103*, 335.

(25) Reiher, W. E., III. Theoretical Studies of Hydrogen Bonding. Ph.D., Harvard University, 1985.

(26) Ryckaert, J. P.; Ciccotti, G.; Berendsen, H. J. C. *J. Comp. Phys.* **1977**, *23*, 327-341.

(27) Ravishanker, G.; Swaminathan, S.; Beveridge, D. L.; Lavery, R.; Sklenar, H. *J. Biomol. Str. Dyn.* **1989**, *6*, 669-699.

(28) Saenger, W. *Principles of Nucleic Acid Structure*; Springer-Verlag: New York, 1984.

(29) Cheatham, I., T.E.; Kollman, P. A. *Structure* **1997**, *In Press*.

Chapter 19

Modeling Duplex DNA Oligonucleotides with Modified Pyrimidine Bases

John Miller[1], Michael Cooney[1], Karol Miaskiewicz[2], and Roman Osman[3]

[1]**Molecular Biosciences Department, Pacific Northwest Laboratory, 902 Battelle Boulevard, MS 7–56, Richland, WA 99352**
[2]**The DASGroup, Inc., 1732 Lyter Drive, Second Floor, Johnstown, PA 15905**
[3]**Department of Physiology and Biophysics, Mount Sinai School of Medicine, One Gustave L. Levy Place, New York, NY 10029–6574**

An approach to modeling the structure of modified nucleic acids that combines quantum-chemical calculations with molecular dynamics (MD) simulation is illustrated by studies of DNA oligonucleotides containing pyrimidine base damage. For all of the lesions studied thus far, we have found the most stable stereoisomer to be energetically well separated from other possible lesion configurations. This fact coupled with the large number of conformational degrees of freedom in duplex DNA means that the equilibrium conformations of lesions in an oligonucleotide do not differ greatly from their conformations in isolation. MD simulations of damaged oligonucleotides with explicit inclusion of water and counterions as well as accurate calculation of the electrostatic potential energy generally show that major deviations from the B-DNA conformation are localized to the immediate neighborhood of the damaged site and mainly reflect adaptation of the native DNA conformation to the most energetically favored configuration of the lesion in isolation.

Pyrimidine bases are major targets for DNA modifications that are strongly implicated in aging and age-dependent diseases including cancer (1-4). Absorption of UV light leads to several DNA lesions, including *cis, syn* cyclobutane pyrimidine dimers and 6-4 photoproducts (5). Free radicals from exdogenous and endogenous sources add easily to the C(5)C(6) bond and lead to stable products with a saturated or fragmented pyrimidine ring. The former includes 5,6-dihydropyrimidines and pyrimidine glycols that have been detected upon radiolysis of polynucleotides and DNA both *in vitro*

(6,7) and *in vivo* (8,9) as well as 5-hydroxy-6-dihydropyrimidines and pyrimidine hydrates (6-hydroxy-5,6-dihydropyrimidines) that are also common products of UV damage to DNA (10,11). Similar products to those seen in DNA radiolysis are also observed in the damage caused by oxidative chemicals including some antibiotics (12).

Pyrimidine DNA lesions can have significant biological consequences including mutagenesis and inhibition of replication. Since UV-induced mutations occur predominantly at dipyrimidine sequences, pyrimidine dimers and 6-4 photoproducts are believed to be responsible for many of the adverse biological effects of UV radiation. Surprisingly, the biological consequences of ring-saturated pyrimidine lesions can be very different. Strong inhibition of DNA polymerase I activity has been observed for thymine glycol (13-16), which is in sharp contrast to the chemically similar lesion 5,6-dihydrothymine that is at most a very weak block for polymerase I action (17). Similarly, only slight inhibition of DNA replication by UV-induced thymine hydrate was detected (18). The 5-triphosphate of dihydrothymine can serve as an efficient substrate for DNA polymerase I but the 5-triphosphate of thymidine glycol is not incorporated into DNA by this polymerase (19).

Additional evidence that DNA structural perturbations by chemically related pyrimidine lesions are very different comes from the activity of the UvrABC nuclease complex toward these lesions. This repair pathway, which is generally considered to be responsible for removing bulky adducts that cause extensive distortions to DNA structure (20,21), recognizes and incises DNA that contains thymine glycol (21,22) but it is not active toward 5,6-dihydrothymine (22). Cyclobutane pyrimidine dimers are also removed by the nucleotide excision repair pathway, which contributed to the idea that this lesion induced large global distortions in duplex DNA. However, recent structural studies, both experimental (23-27) and theoretical (28,29), have concluded that major distortions of DNA structure are limited to the local environment of the thymine dimer and that global changes, such as bending of the helical axis, are relatively small.

Theoretical studies of the effects of damaged bases on the conformation and dynamics of DNA, tested whenever possible by experimental data, can make a significant contribution to our understanding of the biological consequences of DNA lesions. Electronic-structure calculations show the energetic basis for the conformational preferences of modified pyrimidines in isolation. We incorporated the most stable stereoisomers of ring-saturated thymine lesions were into the dodecamer $d(CGCGAATTCGCG)_2$ at T7 (30) and the *cis, syn* thymine dimer was studied at position T7-T8 (29). Effects of a flexible TpA step on the 3' side of the dimer were investigated using the oligonucleotide d(GCACGAATTAAG):d(CTTAATTCGTGC) with and without a dimer at T8-T9 (31).

Conformational Preferences of Ring-Saturated Thymine Derivatives

Ring-saturated thymine lesions can exist in several stereoisomeric forms which may be energetically distinct depending on their environment. The stereochemistry of thymine glycol (Tg) and thymine hydrate are particularly complex because they have two asymmetric carbon centers. Each asymmetic carbon gives rise to two enantiomers and, since the ring is nonplanar, each enantiomer can exist in two conformations with different substituents in the equatorial and axial orientations. In the absence of an external perturbation, such as deoxyribose or other components of DNA, the eight stereoisomers of Tg can be grouped into four pairs of degenerate enantiomers distinguished by the orientation of the C(5) and C(6) hydroxyl groups relative to the mean plane of the distorted thymine ring. *Trans* isomers have both OH groups either equatorial (5-OH_{eq} 6-OH_{eq}) or axial (5-OH_{ax} 6-OH_{ax}) while *cis* isomers have one equatorial and one axial OH (5-OH_{eq} 6-OH_{ax} and 5-OH_{ax} 6-OH_{eq}). The stereochemistry of thymine hydrates is analogous to that of Tg but with 5-OH_{eq} and 5-OH_{ax} replaced by 5-Met_{eq} and 5-Met_{ax} , respectively.

For ring-saturated pyrimidine derivatives like 5,6-dihydrothymine (dhT) with one asymmetric carbon center, one need only consider two pairs of degenerate enantiomers if the lesions are isolated. These can be denoted by the equatorial or axial orientation of a substituent at the asymmetric carbon center. We have chosen 5-Met_{eq} and 5-Met_{ax} to denote the energetically distinct conformers of dhT and 5-OH_{eq} and 5-OH_{ax} to label the energies that we calculate for stereoisomers of 5-hydroxy-6-dihydrothymine. All of our quantum-chemical calculations were performed using the GAUSSIAN90 (32) package with geometries optimized without constraints at the HF/6-31G level and single-point calculations at MP2/6-31G* level performed on the optimized structures. Solvation energies were calculated using a representation of the solvent as a dielectric continuum and the solute as a cavity in the solvent (33). The charge distribution of the solute is described as a collection of point charges that induce a reaction field in the continuum. The point charges were obtained from a Mulliken population analysis of a 6-31G wave function. Our results for ring-saturated thymine derivatives are summarized in Table I.

Table I. Relative stability (kcal/mol) of thymine-derivative stereoisomers

Derivative	*Conformation*	*MP2 (6-31G*)*	*MP2+hydration*
dihydrothymine	5-Met_{eq}	0.00	0.00
	5-Met_{ax}	0.85	0.45
5-hydroxy-6-dihydrothymine	5-OH_{eq}	0.00	0.00
	5-OH_{ax}	2.99	1.57
thymine hydrate	5-Met_{eq} 6-OH_{ax}	0.00	0.00
	5-Met_{ax} 6-OH_{ax}	0.64	0.41
	5-Met_{eq} 6-OH_{eq}	3.43	3.68
	5-Met_{ax} 6-OH_{eq}	4.49	4.61
thymine glycol	5-OH_{eq} 6-OH_{ax}	0.00	0.00
	5-OH_{eq} 6-OH_{eq}	4.08	4.52
	5-OH_{ax} 6-OH_{ax}	6.06	5.40
	5-OH_{ax} 6-OH_{eq}	8.93	4.90

Like native thymine, dhT prefers a conformation with the methyl group at C(5) in an equatorial orientation. The energy difference for this structural preference is about an order of magnitude smaller than that found for the other ring-saturated pyrimidine lesions which involve addition of hydroxyl groups rather than hydrogen atoms. These OH groups shows different stereochemical preference at C(5) and C(6) with a pseudoaxial orientation of OH preferred at C(6) and pseudoequatorial OH energetically favored at C(5). Interactions between polar groups in the equatorial plane are primarily responsible for these preferences. The interaction between the equatorial OH group at C(5) and the adjacent C(4)O(4) carbonyl bond in thymine can be clearly seen from the structure of 5-OH_{eq} 5,6-dihydrouracil shown in Figure 1a. The dipoles of the OH and C(4)O(4) bonds have an ideal antiparallel orientation for a favorable electrostatic interaction. The same interaction would occur in 5-OH_{eq} 6-OH_{ax} Tg but with more drastic implications for nucleic acid structure because the methyl group shifted to the pseudoaxial orientation in this case is too large to be accommodated in the native 3.4 Å separation between base pairs. The OH group in pyrimidine derivatives with an equatorial orientation at C(6) experiences an unfavorable electrostatic interaction with the N(1)H(1) dipole since the direction of the N(1)H(1) bond dipole is opposite to that of the C(4)O(4) bond. To minimize this unfavorable interaction the OH bond dipole is rotated away from the N(1)H(1) bond as is illustrated by the structure of 6-OH_{eq} 5,6-dihydrouracil in Figure 1b.

The calculated conformational preferences described above agree very well with x-ray crystallographic data on *cis*-thymidine glycol (34) which show both the 5-CH_3 and 6-OH groups in the pseudoaxial orientation. In addition, an axial orientation of OH at C(6) is suggested by NMR spectra on a variety of 5,6-saturated derivatives of uracil (35) and thymine (36). The structural preferences calculated for dhT are also in good agreement with x-ray (37) and NMR data (38) which indicate that the methyl group at C(5) prefers the pseudoequatorial orientation.

Configurations of modified bases that are enantiomers in isolation need not have the same effect on DNA conformation, a fact that is well illustrated by the *cis* isomers of Tg. Two stereoisomers of thymidine glycol are released from oxidized polynucleotides, (-)*cis*-5R,6S and (+)*cis*-5S,6R. The crystal structure of the former (34) indicates that the modified nucleotide in DNA has C(5) of the nonplanar thymine ring on the 5' side of the damaged strand, as shown schematically in Figure 2a. The crystal structure of (+)*cis*-5S,6R-thymidine glycol has not been determined; however, since the isomers have almost identical NMR parameters and HPLC elution times, it is reasonable to speculate that this oxidation product results from the DNA lesion shown schematically in Figure 2b with C(5) on the 3' side of the damaged strand.

Both of the DNA lesions shown in Figure 2 have the energetically-favored 5-OH_{eq} 6-OH_{ax} configuration (see Table 1) but their effect on the local DNA structure should be different because there is more room to accommodate the bulky axial methyl group when it is pointing toward the 3' end of the damaged strand. The smaller OH group of C(6) is more easily tolerated under the adjacent base on the 5' side of the modified thymine. Oxidation of thymine in single-stranded DNA produces the 5R,6S isomer with a 5-fold greater yield than the 5S,6R isomer (39). A similar preference is expected for production of ring-saturated thymine lesions by free radical processes acting on duplex DNA (40); hence, most of our MD simulations of oligonucleotides

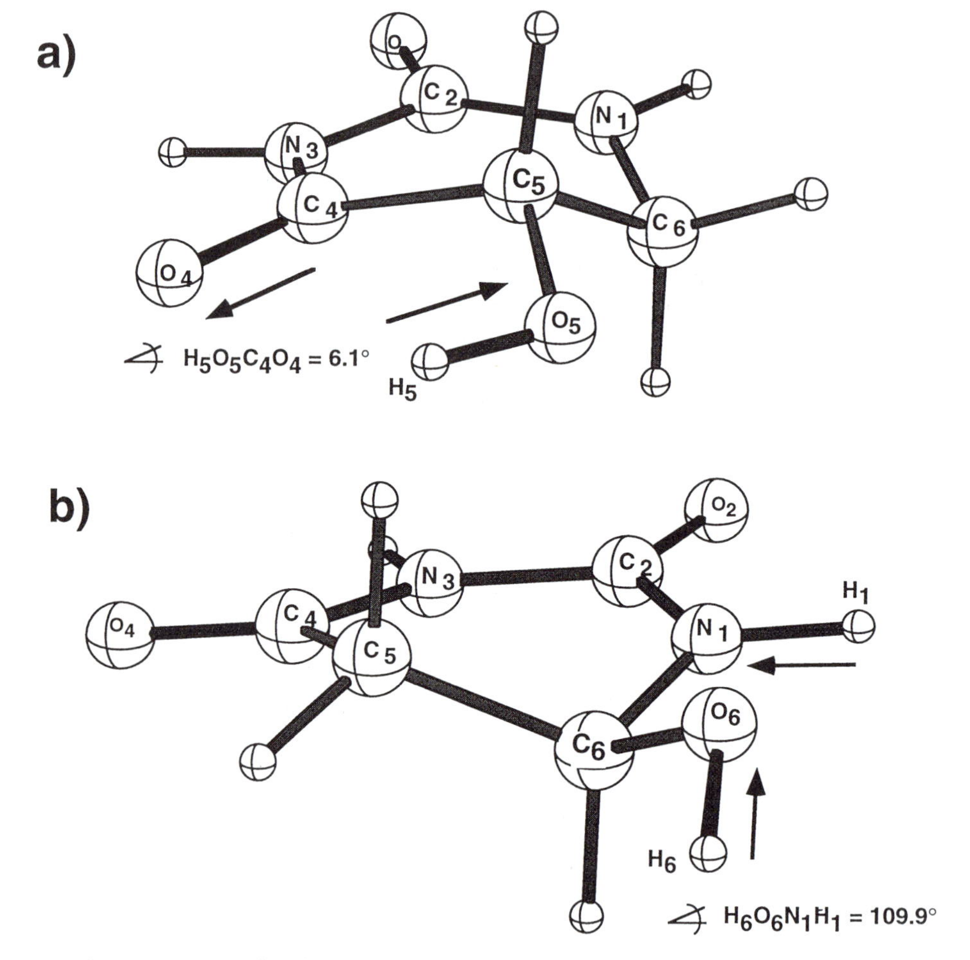

Figure 1. Interaction between pseudoequatorial substituents in ring-saturated uracil derivatives. (Reproduced with permission from ref. 30. Copyright 1994 New York Academy of Sciences)

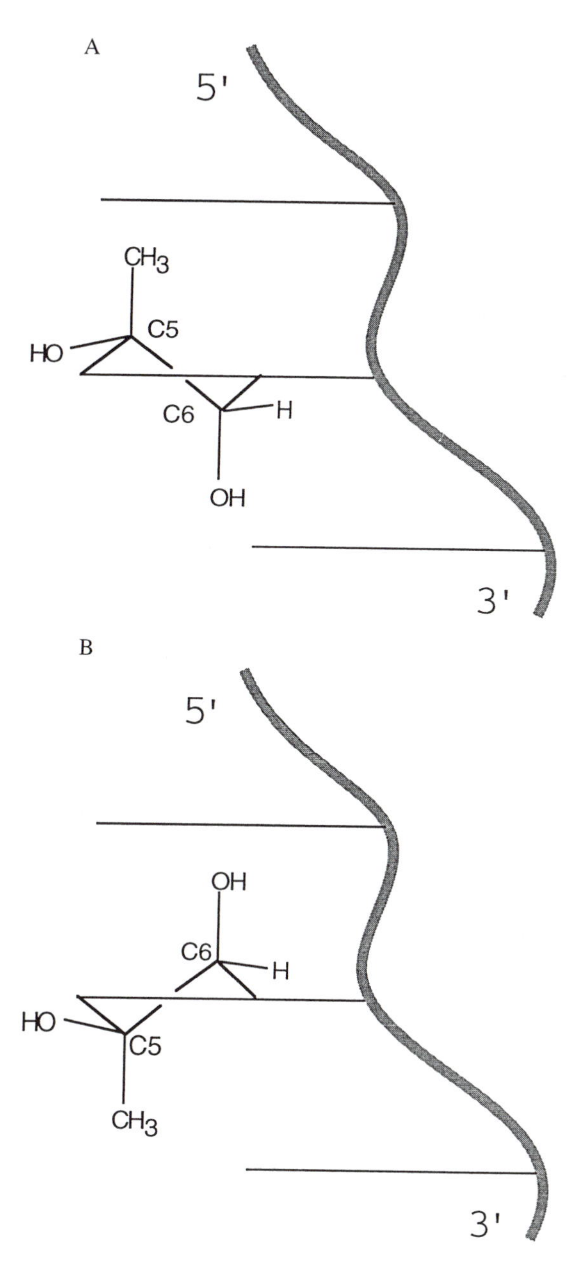

Figure 2. Schematic of 5R-6S (**A**) and 5S-6R (**B**) *cis* isomers of thymine glycol in DNA. (Reproduced with permission from ref. 30. Copyright 1994 New York Academy of Sciences)

containing these lesions have started from structures analogous to that shown in Figure 2a.

Molecular Dynamics Simulation of DNA Containing Ring-Saturated Thymine Derivatives

The dodecamer $d(CGCGAATTCGCG)_2$ was chosen for most of our work on thymine base damage because it has been the subject of numerous experimental and theoretical studies (reviewed in 41). Distortions of this oligonucleotide by the ring-saturated thymine derivatives listed in Table 1 were modeled with AMBER4.0 (42) using the all-atom force-field parameters developed for nucleic acids by Weiner et al. (43). Atomic charges on the modified thymine base were obtained by a fit to the electrostatic potential evaluated from the STO-3G wavefunction, which is consistent with the standard residual atomic charges in this force field. The AMBER *CM* atom type of C(5) and C(6) in native thymine was changed to *CT* in ring-saturated derivatives. Atom types *OH* and *HO* were used for extra hydroxyl groups and atom type *HC* was applied to extra hydrogen substituents at C(5) and C(6). Existing experimental data are not sufficient to test the validity of force-field parameters implied by this assignment of atom types to ring-saturated thymine derivatives; nevertheless, comparison with *ab initio* quantum chemical calculations suggests that the AMBER force field gives a reasonable description of this type of molecule. Ring puckering that results from this parameterization is in good agreement with that predicted by *ab initio* calculations and the four lowest vibrational modes, which are mainly ring deformation modes, have the same energy ordering in AMBER and *ab initio* calculations (40).

A starting structure for MD simulation was developed by inserting the most stable isomer of a lesion into the standard right-handed B-DNA geometry of the oligomer. Negative charges of phosphate groups were neutralized by placing 22 Na^+ ions around the DNA dodecamer containing the modified thymine base. The DNA and counterions were surrounded by a rectangular box of TIP3P water molecules with periodic boundary conditions. Even though the constant volume approximation was used, the water density in these simulations remained close to unity. A cutoff distance of 10 Å was imposed on all non-bonded interactions. After minimization, the system was heated to 300K in ten 0.2 ps runs followed by a 1.0 ps run at 300K. Equilibration at 300K was followed by 120 ps of production dynamics using the constant temperature algorithm. Results from the final 60 ps of the simulations were used for structural analysis using the CURVES algorithm (44) and its implementation in DIALS and WINDOWS (45). Convergence was tested by plotting root mean square (RMS) deviations of backbone dihedral angles and helicoidal parameters from their average value in the 60 to 120 ps interval.

DNA distortions observed in our MD simulations of the four ring-saturated thymine derivatives listed in Table 1 were qualitatively similar; however their magnitudes are greater for Tg and 5-hydroxy-6-dhT due to the axial methyl group at C(5). Variations in rise for several dinucleotide steps near the modified thymine base T*7 are compared in Figure 3. For Tg and 5-hydroxy-6-dhT, the average rise between A6 and T*7 is about twice the value expected for B-DNA. A less dramatic increase in rise at this step is observed for 6-hydroxy-5,6-dhT, while for dhT the increase is negligible. In spite of the increases in rise at the A6-T*7 step, Figure 4 shows that base pairing between T*7 and A18 is not compromised for any of the ring-saturated derivatives. The only significant impact on base pairing, indicated by the length and angle of Watson-Crick hydrogen bonds, occurs at A6-T19 and is largest for 5-hydroxy-6-dhT

Simulations with the axial methyl group of Tg oriented in the 3' direction of the damaged strand predict that perturbations in rise occur primarily between Tg7 and T8 and are about half as large as the distortion in rise between A6 and Tg7 when the methyl group is oriented in the 5' direction. Hydrogen bonds in the T8-A17 base pair are not affected by these smaller changes in rise so that the major effect on base pairing occurs at the lesion site, Tg7-A18, and is comparable to that observed for the A6-T19 base pair when the methyl group of Tg7 is pointing in the 5' direction. These results clearly show the importance of the asymmetric environment for perturbations of DNA by stereoisomers that are degenerate enantiomers in isolation.

Molecular Dynamics Simulation of DNA Distortions by a *Cis, Syn* Thymine Dimer

Simulations of CGCGAATTCGCG with a *cis, syn* thymine dimer at T7-T8 were carried out by the methods outlined above for ring-saturated thymine derivatives but subsequently repeated using the updated AMBER force-field (46) and the particle mesh Ewald (PME) method (47) to calculate the electrostatic potential energy without introduction of a cutoff distance on Coulomb interactions. The two methods gave qualitatively similar results with major distortions limited to the immediate neighborhood of the dimer and the global structure retaining its B-DNA character (29). Figure 5 shows changes in backbone torsion angles, the glycosyl bond dihedral χ and sugar puckering ϕ. The dimer induced a *syn* conformation for χ at T7 and a slight reduction in backbone torsion β. Sugar pucker at T7 remained in the B-DNA-like ^{2}E state while at T8 it was almost as rigidly fixed in an A-DNA-like ^{3}E conformation. Brief transitions to the B_{II} conformation (48) were detected in the time course of ε and ζ backbone torsions at C3, A5 and C9 in both the modified and the native dodecamer.

Intrabase-pair parameters (Figure 6) and base-pairing properties (Figure 7) point out the predominance of distortions at the 5' thymine relative to the 3' thymine of the dimer. Increased buckle and large negative propeller twist are particularly evident in the T7-A18 base pair, where the increased length and decreased angle of one of the Watson-Crick hydrogen bonds are consistent with imino-proton NMR spectra (23) indicating a more severe weakening of base pairing at the 5' thymine. This asymmetry may be related to the observation (49) that adenine complementary to the 5' thymine of a *cis, syn* thymine dimer is rotated out of the helix in the crystal structure of endonuclease V bound to dimer-containing DNA.

If localized perturbations of DNA structure reflect adaptation to a more rigid dinucleotide step at the dimer, as indicated by the fluctuations in torsional angles shown in Figure 5, then a flexible TpA step on the 3' side of the dimer might allow a more symmetric distribution of distortions than that exhibited by simulations on

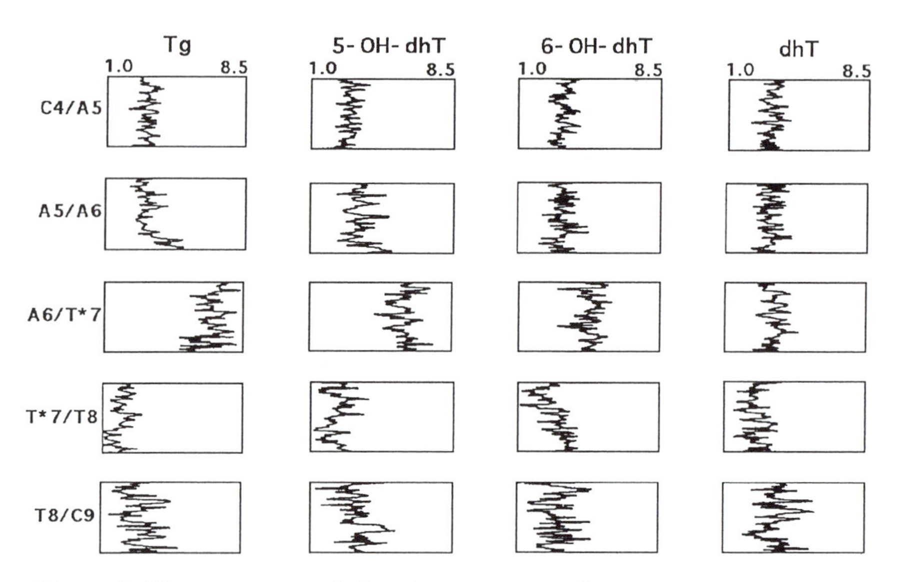

Figure 3. Time sequence of the rise parameter (Å) in MD simulations of CGCGAATTCGCG with T7 replaced by ring-saturated thymine derivatives. (Reproduced with permission from ref. 30. Copyright 1994 New York Academy of Sciences)

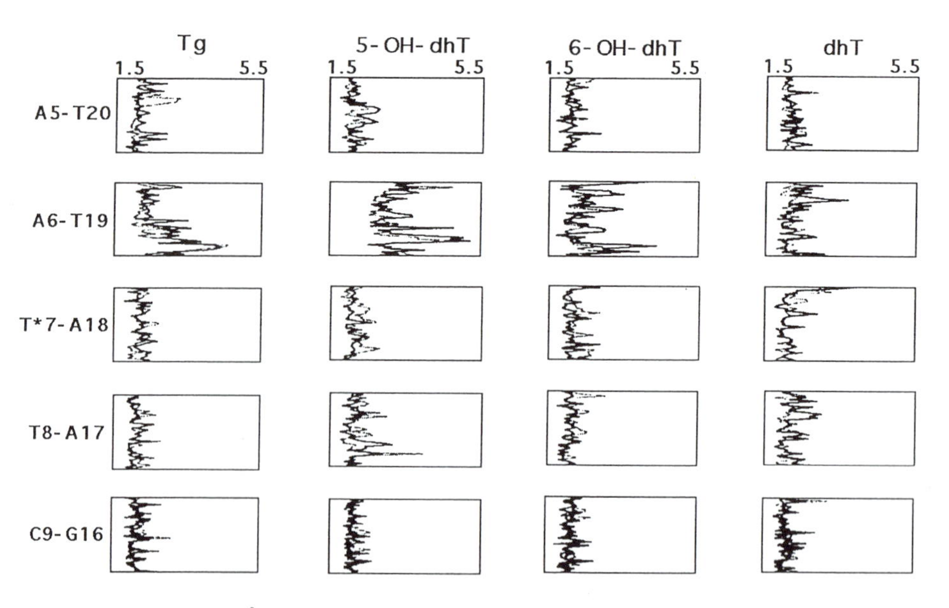

Figure 4. Length (Å) of Watson-Crick hydrogen bonds in CGCGAATTCGCG with T7 replaced by ring-saturated thymine derivatives. (Reproduced with permission from ref. 30. Copyright 1994 New York Academy of Sciences)

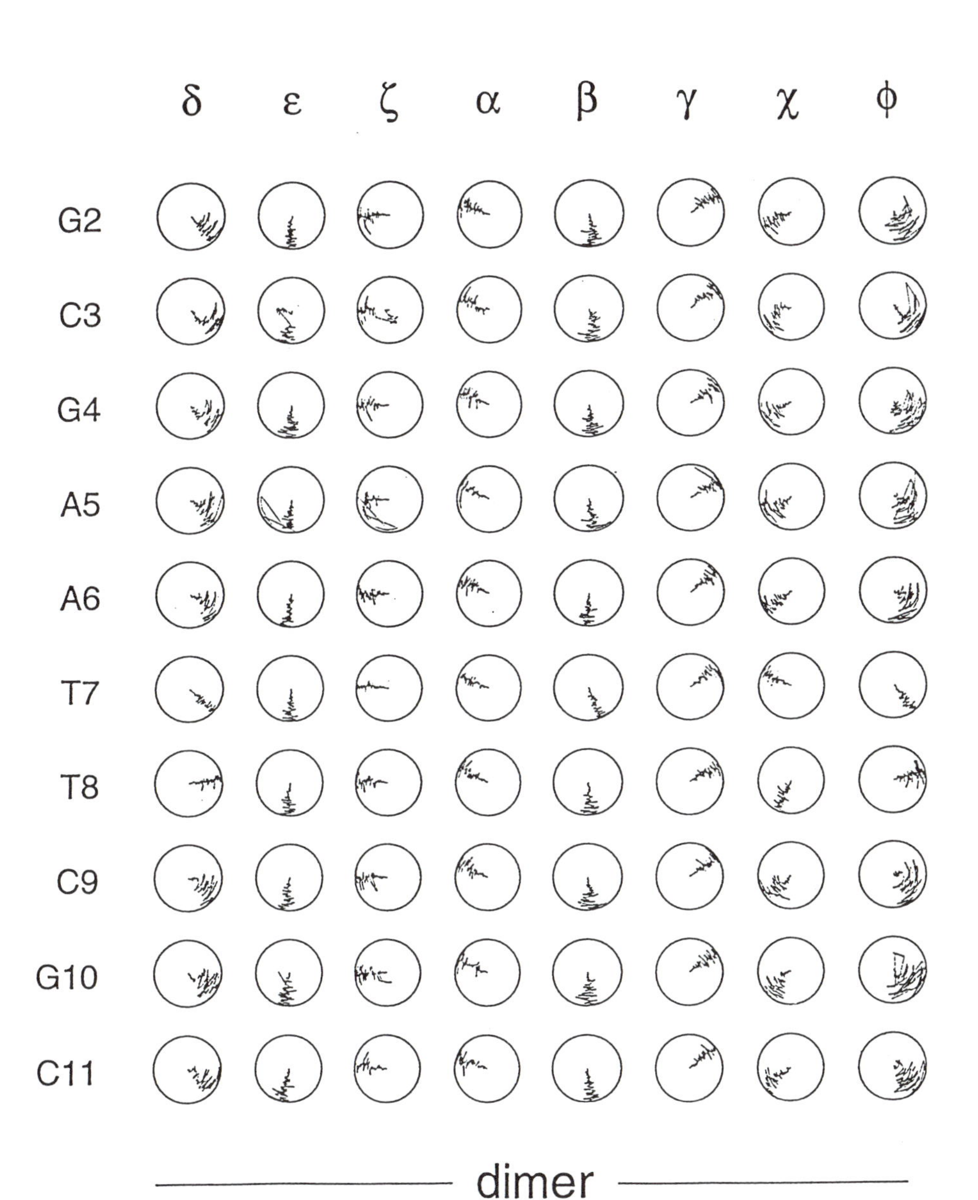

Figure 5. Backbone dihedral angles in the damaged strand of CGCGAATTCGCG from MD simulations with a *cis, syn* dimer at T7-T8.

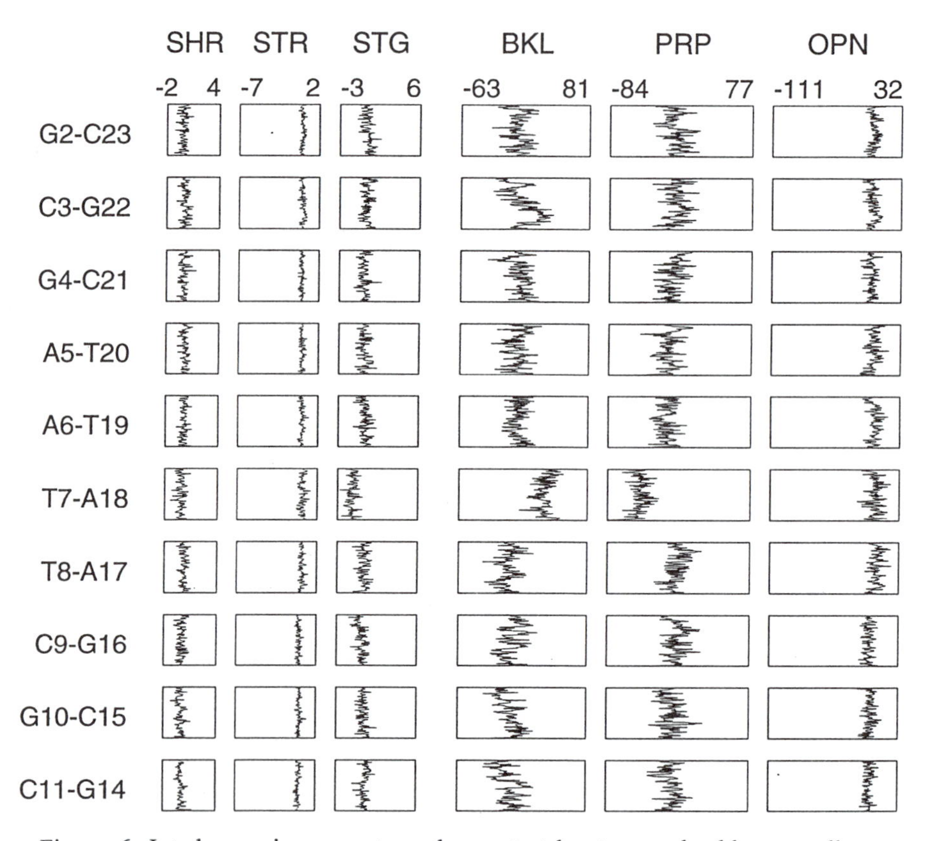

Figure 6. Intrabase-pair parameters shear, stretch, stagger, buckle, propeller twist and opening of CGCGAATTCGCG simulated with a *cis, syn* dimer at T7-T8. (Reproduced with permission from ref. 29. Copyright 1996 American Chemical Society)

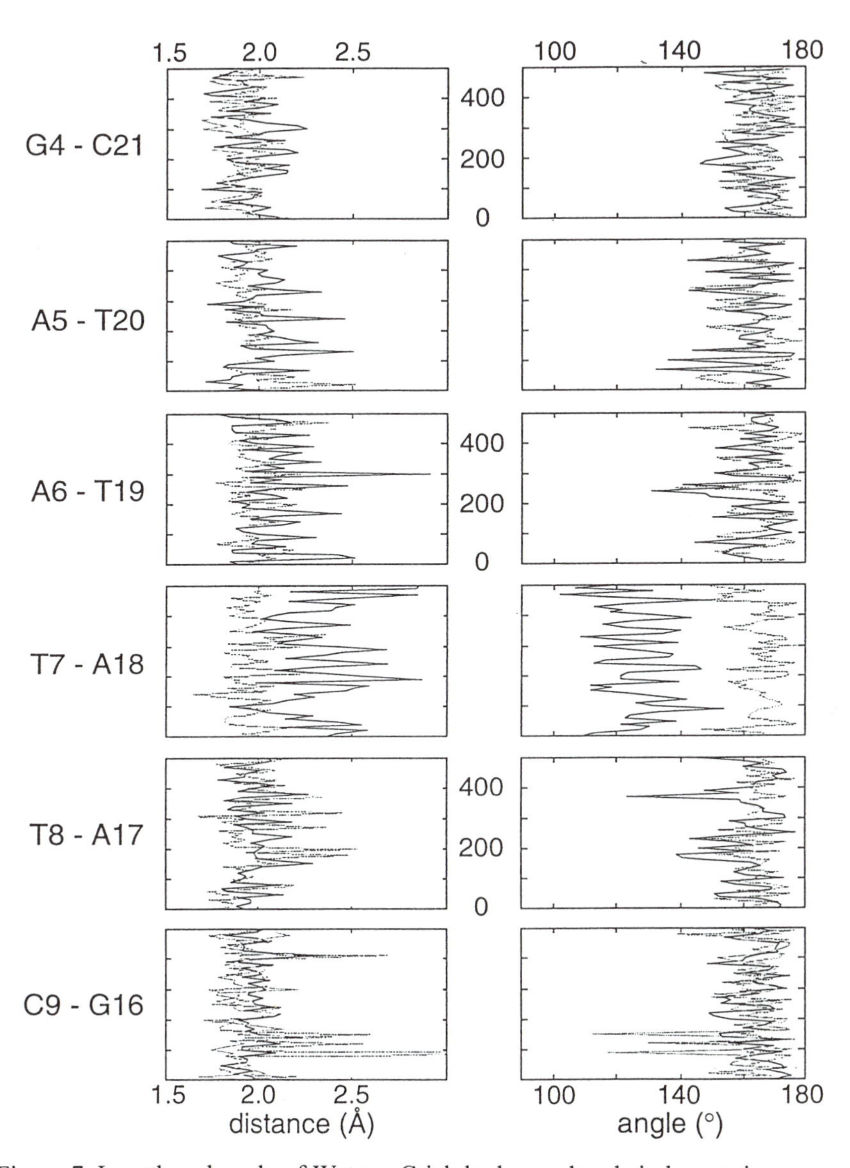

Figure 7. Length and angle of Watson-Crick hydrogen bonds in base pairs near the *cis, syn* thymine dimer in MD simulations of CGCGAATTCGCG. (Reproduced with permission from ref. 29. Copyright 1996 American Chemical Society)

$d(CGCGAATTCGCG)_2$. The flexibility of TpA steps in response to changes in twist and roll (50,51) is particularly relevant since our simulations on $d(CGCGAATTCGCG)_2$ indicate that significant reduction in twist and bending toward the major groove accompany dimer formation at the TpT step. To test this hypothesis, we carried out simulations on the oligonucleotide GCACGAATTAAG with a *cis, syn* dimer at T8-T9, which shares a 6 bp run ...CGAATT... with CGCGAATTCGCG (31).

Some of the differences in our results with these two sequences appear to be associated with a flexible TpA step on the 3' side of the dimer in GCACGAATTAAG. The average twist is about 20° at the thymine dimer in both sequences. Average twist returns to a value of 33° at the TpC step on the 3' side of the dimer in CGCGAATTCGCG, which is somewhat closer to the B-DNA value than the 29° found for the TpA step on the 3' side of the dimer in GCACGAATTAAG. This 4° difference in twist is in the direction expected for a flexible TpA step even though it may not be statistically significant. A larger difference was found for average roll, which is about 22° at the dimer in both cases, returns to a value near zero (B-DNA result) in the TpC step of CGCGAATTCGCG but remains at 11° at the TpA step in GCACGAATTAAG. While these differences suggest some dependence on sequence context, the 5' thymine of the dimer was still the main focus of distortions calculated for both oligonucleotides.

Parameters of the CURVES algorithm (44) that we use to describe irregular DNA structures are global in the sense that they are defined relative to an overall helical axis. This distinguishes them from local structural parameters that specify the position of a base or base pair relative to the preceding base or base pair in a locally defined coordinate system. The helical axis of an irregular DNA fragment is determined by minimizing a function of the changes in orientation of successive nucleotides relative to a global helical coordinate system as well as kinks and dislocations of helical axis segments. We have found that the 3-dimensional space curve resulting from this optimization procedure is well approximated by a circular arc, which allows the deviation from a straight helical axis to be characterized by a magnitude and direction of bending (29).

Figure 8 illustrates the application of this technique for analysis of global bending induced by a *cis, syn* thymine dimer in $d(CGCGAATTCGCG)_2$. The magnitude of curvature, measured as the angle between radii that encompass the DNA, has a similar distribution in the presence and absence of the dimer but the mean is 11° greater in the former. This amount of excess bending due to the dimer is in good agreement with the results of gel electrophoresis (52) and NMR (53) experiments. The direction of global curvature is defined by the angle between vectors projected into the plane of the central A6-T19 base pair, one of which is the radius of the fitted circle through the central base pair. The other vector is the pseudo dyad axis of the central base pair which connects the midpoint of a line between C(1) atoms to the midpoint of a line between pyrimidine C(6) and purine C(8). The direction of these vectors are chosen so that bending toward the major groove has a direction angle near zero. The distribution of curvature direction in Figure 8 shows that dimer formation shifts the mean direction of curvature by about 40° and decreases the width of the distribution by about half. These changes probably reflected increased rigidity of the T7-T8 dinucleotide step due to dimerization.

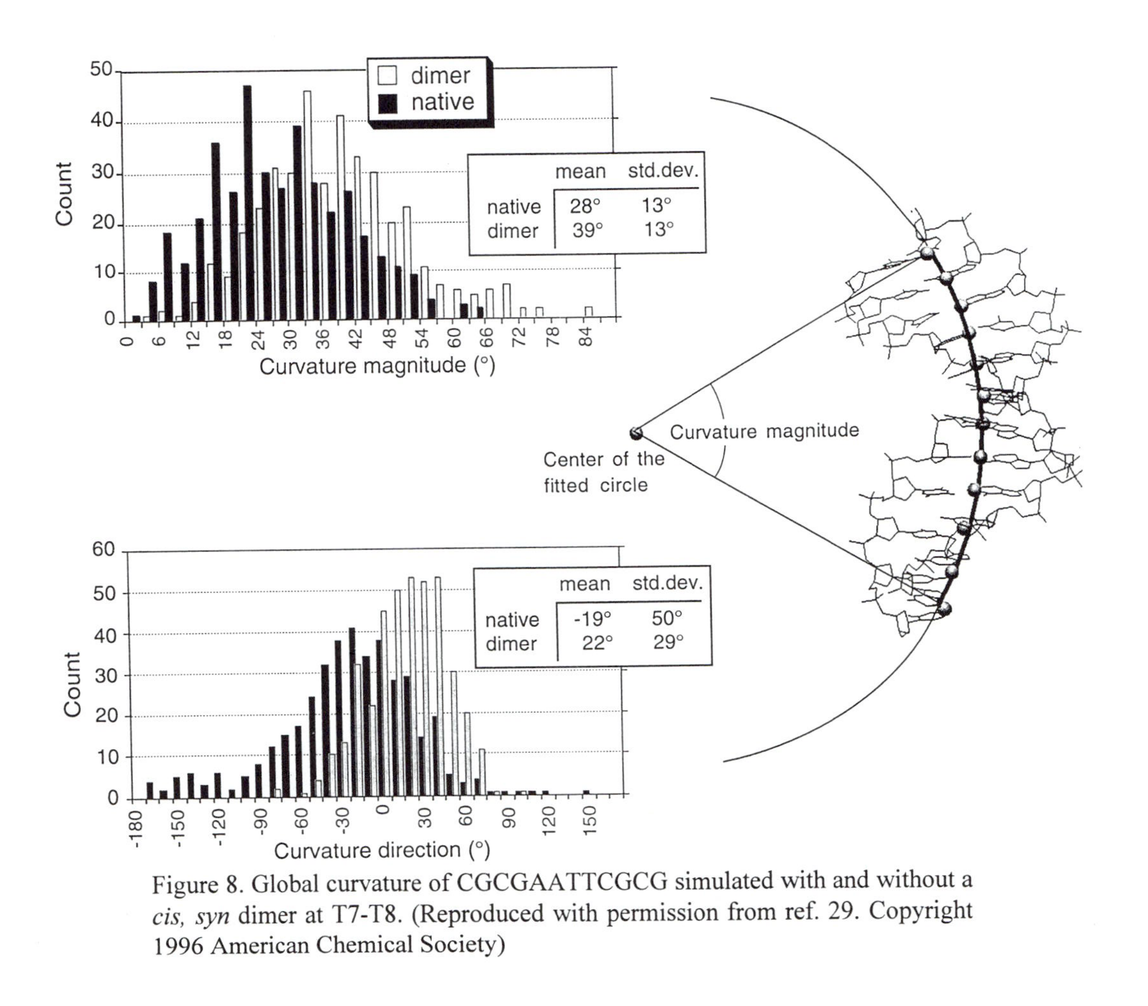

Figure 8. Global curvature of CGCGAATTCGCG simulated with and without a *cis, syn* dimer at T7-T8. (Reproduced with permission from ref. 29. Copyright 1996 American Chemical Society)

Comparison of Ring-Saturated and Thymine-Dimer Lesions

Formation of the *cis, syn* cyclobutane thymine dimer induces distortions of the ring structure similar to those caused by ring-saturation from hydrogen or hydroxyl addition; however, the effects on base pairing are less dramatic than those found for thymine glycol and 5-hydroxy-6-dihydrothymine where the methyl group is displaced from its native pseudoequatorial orientation by OH addition at C(5). All three of these lesions (dimer, Tg, and 5-OH-6-dhT) induce major changes in the helicoidal parameters of the dinucleotide step at the damage site but for the dimer, these changes are more effectively transferred to distortion of backbone dihedral angles; particularly sugar puckering. The distribution of simulated DNA distortions induced by a thymine dimer over more conformational degrees of freedom makes then appear more subtle; nevertheless, like thymine glycol, *cis, syn* thymine dimers are substrates for the nucleotide excision repair pathway that recognizes bulky lesions. Compromise of base-stacking interactions is an effect common to both of these lesions, which could lead to rotation of the damage site out of the double helix as a consequence of normal DNA metabolism. This type of biological processing may be a crucial step in damage recognition.

Future Directions

The NMR group at the Pacific Northwest National Laboratory is engaged in a collaboration with Professor John-Stephen Taylor at Washington University in St. Louis to determine solution structures of GCACGAATTAAG with and without a dimer at T8-T9. Preliminary results kindly provided by these researchers suggest two areas where improvements in our computation methods are needed. Force-field parameters for modified nucleic acids are a potentially important source of error in calculations of the type described in this paper. For ring-saturated thymine derivatives, we addressed this issue by comparing force-field calculations with *ab initio* quantum calculations (40). NMR data on GCACGAA$\underline{TT}$AAG, where underlining denotes the presence of a dimer, detected an unusually short H(1) to H(8) distance in nucleotide A10 and a short H(1) to H(6) distance in its complementary T15 nucleotide. These distances relaxed to typical values of about 3.8 Å early in the equilibration phase of our simulation procedure. This suggests that parameters which affect propagation of distortions on the 3' side of the dimer may not be adequately represented in the standard AMBER force field (46).

Subtle differences in starting structure also appear to effect the results of our simulations on GCACGAA$\underline{TT}$AAG. Heteronuclear ^{31}P-^{1}H NMR data from Taylors laboratory and a solution structure based on proton NMR from our own laboratory indicate a B_{II} configuration of the backbone at the T9-A10 dinucleotide step. Starting structures for GCACGAA$\underline{TT}$AAG obtained by inserting a *cis, syn* thymine dimer into standard B-DNA have a B_{I} configuration at T9-A10 that was maintained throughout a 1 ns MD simulation (31); however, when the NMR results from our laboratory are used as a starting structure, a B_{II} state at T9-A10 predominates.

Even though large structural changes are not associated with the B_I-B_{II} transition, changes in the CURVES parameters X-displacement, rise, twist and roll (54) are as great as the sequence-dependent effects discussed above in relation to a flexible TpA step on the 3' side of the dimer. Hence, simulations started by inserting a modified base into standard B-DNA may not be adequate to predict fine-structure changes of this magnitude as a consequence of base modification. Greater rigidity due to dimer formation and the slower evolution of structures when solvent and electrostatic interactions are treated accurately (55) increases the challenge of adequately exploring the conformation space of fine-structure changes associated with *cis, syn* thymine-dimer induction. Application of random starting structures and/or simulated annealing may prove useful in this area.

Acknowledgments. Work supported by the Office of Biological and Environmental Research (OBER) of the US Department of Energy under contract DE-AC06-76RLO 1830 and DOE grant DE-FG02-88ER60675. JHM expresses his gratitude to Michael Kennedy for sharing his NMR results prior to publication and to Marvin Lien for assistance in manuscript preparation.

References

1. Lutz, W. K. *Mutation Res.* **1990**, *238*, 287-295.
2. Ames, B. N. *Mutation Res.* **1989**, *148*, 41-46.
3. Ames, B. N. *Environ. Mol. Mutagen.* **1989**, *14*, 66-77.
4. Ames, B. N.; Gold, L. S. *Mutation Res.* **1991**, *250*, 3-16.
5. Cadet, J.; Vigny, P. In *Bioorganic Photochemistry*; Morison, H., Ed.; John Wiley and Sons: New York, 1990, Vol. 1; pp 1-272.
6. Teoule, R. *Int. J. Radiat. Biol.* **1987**, *51*, 573-589.
7. Dizdaroglu, M. *Free Radical Biol. Med.* **1991**, *10*, 225-242.
8. Leadon, S. A.; Hanawalt, P. C. *Mutation Res.* **1983**, *112*, 191.
9. Breimer, L.H.; Lindahl, T. *Biochemistry* **1985**, *24*, 4018.
10. Boorstein, R. J.; Hilbert, T.P.; Cunningham, R.P.; Teebor, G.W. *Biochemistry* **1990**, *29*, 10455-10460.
11. Ganguly, T.; Duker, N. J. *Nucleic acids Res.* **1991**, *14*, 3319-3323.
12. Quinlan, G. J.; Gutteridge, J. M. C. *Biochem. Pharmacol.* **1991**, *42*, 1595-1599.
13. Ide, H.; Kow, Y. W.; Wallace, S. S. *Nucleic Acids Res.* **1985**, *13*, 8035-8052.
14. Clark, J. M.; Beardsley, G. P. *Nucleic Acids Res.* **1986**, *14*, 737-749.
15. Clark, J. M.; Beardsley, G. P. *Biochemistry* **1987**, *26*, 5398-5403.
16. Hayes, R. C.; Leclerc, J. E. *Nucleic Acids Res.* **1986**, *14*, 1045-1061.
17. Ide, H.; Petrullo, L.. A.; Hatahet, Z.; Wallace, S. S.; *J. Biol. Chem.* **1991**, *266*, 1469-1477.
18. Ganguly, T.; Duker, N. J. *Mutation Res.* **1992**, *293*, 71-77.
19. Ide, H.; Melamede, R. J..; Wallace, S. S .*Biochemistry* **1987**, *26*, 964-969.
20. Sancar, A.: Sancar, G. B. *Annu. Rev. Biochem.* **1988**, *57*, 29-67.
21. Lin, J.; Sancar, A. *Biochemistry* **1989**, *26*, 7979-7984.
22. Kow, Y. W.; Wallace, S. S.; Van Houten, B. *Mutation Res.* **1990**, *235*, 147-156.
23. Kemmink, J.; Boelens, R.; Koning, T.;van der Marel, G. A.; van Boom, J. H.; Kaptein, R. *Nucleic Acids Res.* **1987**, *15*, 4645-4653.
24. Kemmink, J.; Boelens, R.; Kaptein, R. *Eur. Biophys. J.* **1987**, *14*, 293-299.
25. Taylor, J.-S.; Garrett, D. S.; Brockie, I. R.; Svoboda, D. L.; Telser, J. *Biochemistry* **1990**, *29*, 8858-8866.

26. Lee, B. J.; Sakashita, H.; Ohkudo, T.; Ikehara,M.; Doi,T.; Morikawa, K.; Kyogoku, Y.; Osafune, T.; Iwai, S.; Ohtsuka, E. *Biochemistry* **1994**, *33*, 57-64.
27. Kim, J.-K.; Patel, D.; Choi, B.-S. *Photochem. Photobiol.* **1995**, *62*, 44-50.
28. Rao, S. N.; Kollman, P. A. *Bull. Chem. Soc. Jpn.* **1993**, *66*, 3133-3134.
29. Miaskiewicz, K.; Miller, J.; Cooney, M.; Osman, R. *J. Am. Chem. Soc.* **1996**, *118*, 9156-9163.
30. Miller, J.; Miaskiewicz, K.; Osman, R. In *Dna Damage: Effects on DNA Structure and Protein Reccognition*; Wallace, S.S.; Van Houten, B; Kow, Y. W., Ed; Annals of the New York Academy of Sciences Vol. 726; pp 71-91.
31. Cooney, M. G.; Miller, J. H. *Nucleic Acids Res.* **1997**, *25*, 1432-1436.
32. Frisch, M. J.; Head-Gordon, M.; Trucks, G. W.; Foresman, J. B.; Schlegel, H. B.; Raghavachari, K.; Robb, M.; Binkley, J. S.; Gonzalez, C.; Defrees, D. J.; Fox, D. J.; Whiteside, R. A.; Seeger, R. A. R.; Melius, C. F.; Baker, J.; Martin, R. L.; Kahn, L. R.; Stewart, J. J. P.; Topiol, S.; Pople, J. A. *Gaussian 90* **1990**, Gaussian, Inc. Pittsburgh, PA.
33. Rashin, A; Honig, B. *J. Phys. Chem.* **1985**, *89*, 5588-5593.
34. Hruska, F. E.; Sebastian, R.; Grand, A.; Voituriez, L.; Cadet, J. *Can. J. Chem.* **1987**, *65*, 2618-2623.
35. Ducolomb, R.; Cadet, J.; Taieb, C.; Teoule, R. *Biochim. Biophys. Acta* **1976**, *432*, 18-27.
36. Cadet, J.; Ducolomb, R.; Hruska, F. E. *Biochim. Biophys. Acta* **1979**, *563*, 206-215.
37. Konnert, J.; Karle, I. L.; Karle, J. *Acta Crystallogr.* **1970**, *B26*, 770-778.
38. Cadet, J; Voituriez, L; Hruska, F. E.; Kan, S.-L.; De Leeuw, F. A. A. M.; Altona, C. *Can. J. Chem.* **1985**, *63*, 2861-2868.
39. Koa, J. Y.; Goljer, I.; Phan, T. A.; Bolton, P. H. *J. Biol. Chem.* **1993**, *268*, 17787-17793.
40. Miaskiewicz, K.; Miller, J.; Ornstein, R.; Osman, R. *Biopolymers* **1995**, *35*, 113-124.
41. Young, M. A.; Nirmala, R.; Srinivasan, J.; McConnell, K. J.; Ravishanker, G.; Beveridge, D. L.; Berman, H. M. In *Structural Biology: The State of the Art*; Sarma, R.H.; Sarma, M. H. Ed.; Adenine Press, Schenectady, NY, 1994; pp. 197-214.
42. Pearlman, D. A.; Case, D. A.; Cadwell, J. C.; Siebel, G. L.; Chandra Singh, U.; Weiner, P.; Kollman, P. A. *AMBER4.0* **1991**, University of California, San Francisco, CA.
43. Weiner, S. J.; Kollman, P.A.; Nguygen, D. T.; Case, D. A. *J. Comput. Chem.* **1986**, *7*, 230-252.
44. Lavery, R. and Sklenar, H. *J. Biomol. Struct. Dyn.* **1988**, 6, 655-667.
45. Ravishanker, G.; Swaminatham, S.; Beveridge, D. L.; Lavery, R.; Sklenar, H. *J. Biomol. Struct. Dyn.* **1989**, 6, 669-699.
46. Cornell, W.D.; Cieplak, P.; Bayly, C. I.; Gould, I. R.; Merz, K.; Ferguson, D. M.; Spellmeyer, D. C.; Fox, T.; Caldwell, J. W.; Kollman, P. A. *J. Am. Chem. Soc.* **1995**, *117*, 5179-5197.
47. Darden, T.: York, D.; Pedersen, L. *J. Chem. Phys.* **1993**, *98*, 10089-10092.
48. Gupta, G.; Bansal, M.; Sasiekharan, V. *Proc. Natl.Acad. Sci.* **1980**, *77*, 6486-6490.
49. Vassylyev, D.G.; Kashiwagi, T.; Mikami, Y.; Ariyoshi, M.; Iwai, S.; Ohtsuka, E.; Morikawa, K. *Cell* **1995**, *83*, 773-782.
50. Poncin, M.; Piazzola, D.; Lavery, R. *Biopolymers* **1992**, *32*, 1077-1103.
51. Suzuki, M.; Yagi, N.; Finch, J.T. *FEBS Letters* **1996**, *379*, 148-152.
52. Wang, C.-I.; Taylor, J.-S. *Proc. Natl. Acad. U.S.A.* **1991**, *88*, 9072-9076.
53. Kim, J.-K.; Patel, D.; Choi, B.-S. *Photochem. Photobiol.* **1995**, *62*, 44-50.
54. Hartman, B.; Piazzola, D.; Lavery, R. *Nucleic Acids Res.* **1993**, *21*, 561-568.
55. Cheatham, III, T. E.; Miller, J. L.; Fox, T.; Darden, T. A.; Kollman, P. A. *J. Am. Chem. Soc.* **1995**, *117*, 4193-4194.

Chapter 20

How the TATA Box Selects Its Protein Partner

Nina Pastor[1], Leonardo Pardo[1,2], and Harel Weinstein[1]

[1]**Department of Physiology and Biophysics, Mount Sinai School of Medicine, One Gustave L. Levy Place, New York, NY 10029–6574**
[2]**Laboratorio de Medicina Computacional, Unidad de Bioestadistica, Facultad de Medicina, Universidad Autónoma de Barcelona, 08193 Bellaterra, Barcelona, España**

The TATA box-binding protein (TBP) binds in the minor groove of the TATA promoter sequence, drastically bending and unwinding the DNA helix. To explore the role of the TATA sequence in determining the specificity of TBP binding to DNA, we studied the solution structure and dynamics of 7 DNA dodecamer sequences, 6 of which bind to TBP. Molecular dynamics simulations were carried out with the CHARMM23 potential, explicit waters, one Na+/phosphate, and periodic boundary conditions. A comparison of the structural parameters of these dodecamers with DNA in the crystal structures of TBP/DNA complexes shows that the probability of various sequences for adopting transient conformations mimicking the distortions required for TBP binding, correlates with TBP-binding ability. The results indicate how the propensity of various TATA sequences to prepare structurally for TBP binding, constitutes a key selectivity determinant.

The TATA box-binding protein (TBP) is a basic transcription factor absolutely required for transcription by the three nuclear RNA polymerases (*1*). RNA polymerase II transcription of many protein coding genes requires direct contact of the DNA by TBP, at a sequence called the TATA box, which is located approximately 30 basepairs upstream of the transcription initiation site (*2*). This complex directs the assembly of the remaining basic transcription factors into a preinitiation complex (*3*).

TBP has been cloned from many organisms, ranging from archeabacteria (*4, 5*) to humans (see (*6*) for a review). Crystallographic determinations of TBP structures from archeabacteria (*7*), yeast (*8*), and plants (*6*) reveal the same architecture: a molecular saddle with a near two-fold symmetry axis. TBP has also been crystallized in complex with DNA (*9-12*), and in ternary complexes with two other basic transcription factors, TFIIA (*13, 14*) and TFIIB (*15*). In all the crystallized complexes, TBP binds to 8 basepairs in the minor groove of the DNA encoding a characteristic TA repeat. The underside of the TBP saddle is the binding interface with the DNA. This interface is mostly hydrophobic, but 6 hydrogen bonds have been identified in the crystal structures.

They involve the interactions between Thr and Asn residues and the two central basepairs of the TATA box. While TBP does not change its conformation drastically upon binding to DNA, the geometry of the DNA, which is practically identical in all the crystallized complexes, has two kinks that are caused by the partial insertion of two pairs of Phe residues at the first and last basepair steps of the TATA box. These insertions result in a bend of ~90° between the DNA preceding the TATA box and the stretch following it. The DNA is unwound in these complexes, and the minor groove is widened.

TBP binds the minor groove of AT rich DNA with a clear sequence preference (the consensus TATA box has the sequence T A T A t/a A t/a X (*16*)). Understanding the basis for this selectivity is complicated by the fact that AT and TA basepairs are very similar in the minor groove, which precludes direct readout as a means for specific sequence recognition (*17*). Because TBP distorts the DNA, it is likely that sequence dependent DNA bendability plays an important role in determining the observed sequence specificity (*18*).

Two manifestations of DNA bendability can be imagined that are not mutually exclusive: One is that certain basepair steps may have an average geometry that is biased towards the conformation induced by TBP. This is the same idea as Calladine's "static wedges" (*19*). The other option is that even for basepair steps with a structure that is straight on average, some might make more frequent transient excursions (*20*) than others into the distorted conformation imposed by TBP, and hence offer a transient "prepared state" for TBP binding.

To distinguish which of these two mechanisms might be used by TBP to select its binding sites, we have carried out molecular dynamics (MD) simulations to explore the conformations of double stranded DNA oligomers that contain TATA boxes. The simulation cell contains one DNA dodecamer, 22 sodium ions to achieve electroneutrality and a 14Å water layer. The sequences (identified by the coding strand only) were chosen to include 3 known binding sites for wild type TBP (*21*) (**mlp** [d(CTATAAAAGGGC)], **at** [d(ATATATATATAT)] and **6t** [d(CTATATAAGGGC)]), 2 binding sites for mutant TBPs (*22, 23*) (**2c** [d(CCATAAAAGGGC)] and **7g** [d(CTATAAGAGGGC)]), one inverted TATA box (**r28** [d(CTTTTATAGGGC)]), and a sequence that is not bound by TBP and serves as a negative control (**gc** [d(GCGCGCGCGCGC)]).

Materials

Statistical analyses of structural parameters were carried out for the following coordinates of structures derived from X-ray crystallography or NMR spectroscopy:

a) TBP/DNA complexes: NDB (*24*) entries PDT009, PDT032, PDT034, PDT012, PDT036, and PDT024

b) NMR derived duplex DNA structures: PDB (*25*) entries 1BUF, 1UQA, 1UQB, 1UQC, 1UQD, 1UQE, 1UQF, 1UQG, 1D42, 1D70, 1D18, 1D19, 1D20, and 142D

c) A-DNA structures: NDB (*24*) entries ADH008, ADH010, ADH026, ADH027,ADH033, ADH038, ADH039, ADH047, ADH070, ADJ049, ADJ050, ADJ051, ADJ067, ADJ069, ADJ075, and ADL025

d) B-DNA structures: NDB (*24*) entries BDF068, BDJ017, BDJ019, BDJ025, BDJ031, BDJ036, BDJB44, BDJ051, BDJ052, BDJ060, BDJ081, and BDL020

Computational Methods

The initial DNA structures were built in the standard B-DNA conformation within QUANTA (Molecular Simulations, Inc 1992). The sodium ions were placed along the O-P-O bisector, 5Å away from the P atom. The systems thus assembled were solvated with the overlay procedure within Insight II (Biosym Technologies 1993), and the final simulation cells were trimmed to hexagonal prisms of 24Å side and 72Å length with Simulaid (*26*).

The simulations were done with the CHARMM23 potential (*27*), as described in (*28*), in the NVE ensemble, using periodic boundary conditions for a hexagonal prism. SHAKE was applied to all hydrogen-containing bonds. We used a spherical cutoff of 13Å, with a switching function for the van der Waals term and a shifting function for the electrostatic term. The nonbonded interaction lists were updated every 10 steps (15 fs). The water was equilibrated first (6 ps heating from 0K to 300K, and 30 ps dynamics), keeping both DNA and sodium ions fixed. After energy minimization of the whole system (250 steps of steepest descent minimization followed by 250 steps using the Adopted Basis Set Newton-Raphson method), it was slowly heated to 300K (10 ps), and equilibrated for 30 ps. The production phase lasted 510 ps for all the simulations, except for that of **mlp**, which was extended to 2080 ps.

The conformational analysis for DNA was carried out with the CURVES (*29, 30*) algorithm included in the Dials and Windows package (*31*), and with CHARMM23 (*32*).

Validation

The Importance of Electrostatic Interactions. A molecule with the high charge density of DNA requires special attention to the approximations used in the calculations. In particular, we were concerned about the effect that a finite cutoff would have on the ionic environment and on the DNA structure itself. To test the effect of the spherical cutoff, we repeated the simulation of **mlp** using the AMBER program (*33*) and AMBER 4.1 (*34*) all atom potential and the treatment of electrostatic environment with Ewald sums representing an infinite cutoff. This simulation was carried out in the NVT ensemble, and SHAKE was applied to all hydrogen-containing bonds. A cutoff value of 9Å was used for the van der Waals interactions, and the algorithm used for the electrostatic interactions was the particle mesh Ewald algorithm implemented in AMBER 4.1 (*35*). In this run, water was heated during 15 ps and equilibrated for 85 ps; then, the system was energy minimized, heated from 0K to 300K in 15 ps, and equilibrated for 50 ps. The production run extended to a total of 2 ns.

Figure 1 presents a comparison between the CHARMM run and the AMBER run for the resulting radial distribution functions of the three pairs of charged species in the **mlp** system: sodium-sodium, sodium-phosphorus (as the center of the phosphate

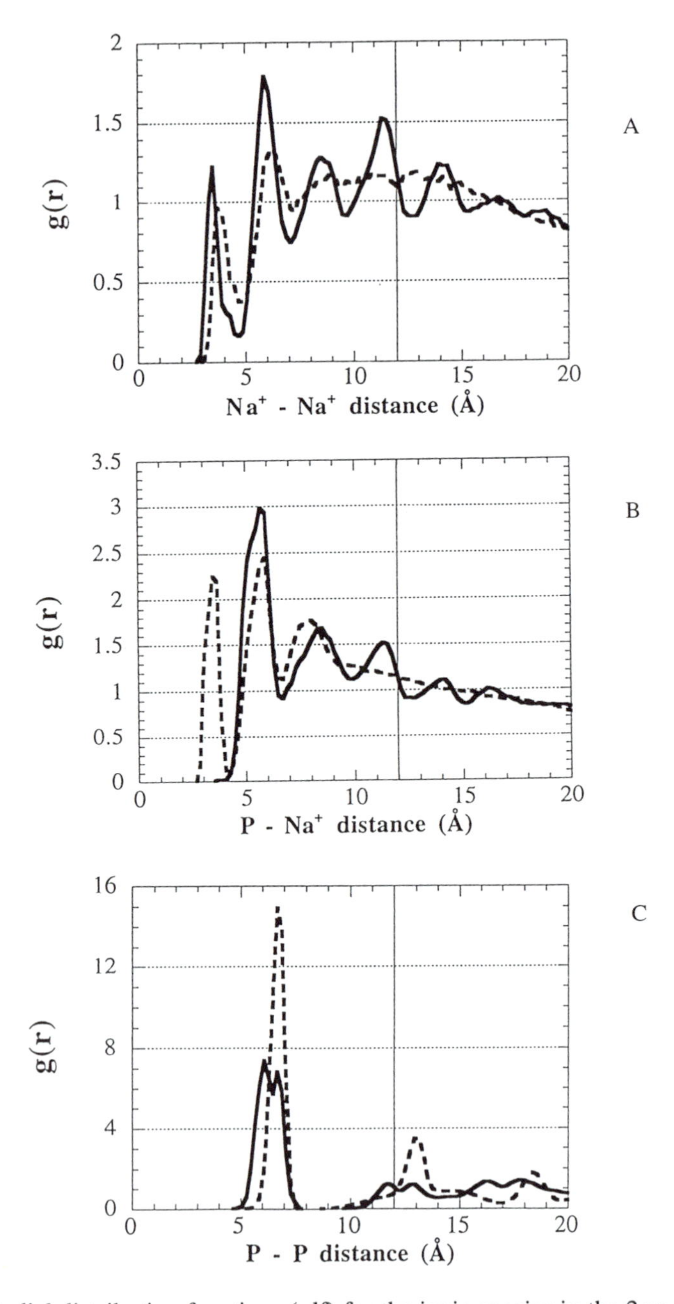

Figure 1. Radial distribution functions (rdf) for the ionic species in the 2 ns **mlp** simulations carried out with CHARMM23 (solid line) and AMBER 4.1 (broken line). **A**: sodium - sodium rdf; **B**: phosphorus - sodium rdf; **C**: phosphorus - phosphorus rdf.

group), and phosphorus-phosphorus. The evidence for problems in handling electrostatics has been shown to be the artificial accumulation of ion pairs exactly at the cutoff distance (*36*). The shift function makes the interaction zero exactly at the distance of 12Å marked by the vertical line in the plots shown in Figure 1. It is clearly evident that there is no accumulation of pairs in any case, suggesting that this cutoff in the CHARMM simulations is sufficiently long so as not to distort the DNA or impose an artificial structure on the ion distribution. There are differences between the results from CHARMM and AMBER: CHARMM produces more structure on the sodium-sodium rdf, while the structure appears to end for AMBER at ~9Å, which is the van der Waals cutoff. For the P-sodium rdf, AMBER pulls the sodiums to within contact distance with the phosphates, while CHARMM prefers the solvent separated arrangement of the counterion; this difference correlates with a larger negative charge on the phosphate in AMBER than in CHARMM, and could also be due to differences in the Lennard-Jones parameters used for the sodium and phosphate in the two force fields. From the P-P rdf, it appears that AMBER is better than CHARMM at keeping the B-DNA distances. All these results lead to the conclusion that the cutoff does not distort the DNA (from the P-P rdf), and it does not impose any artifactual structure on the ion distribution around DNA (from the Na-P rdf).

Comparison to Structural Data from Experiments. To validate the DNA conformations, we compared the structures generated by CHARMM to those found in high resolution X-ray structures and NMR structures refined using a forcefield. The compared properties included consecutive P - P distances, sugar puckers and glycosidic bond torsional angles, and basepair step geometric parameters. Figure 2 shows the distribution of P - P distances: the CHARMM runs produce a major peak at 6.8Å, very close to the center of the distributions for NMR structures and B-DNA. The secondary peak is intermediate between A-DNA and B-DNA characteristic values. The 2D histograms for sugar pucker and glycosidic bond torsional angles, in Figure 3, indicate both A-DNA and B-DNA like structures, shown by the black isopopulation contours. The gray area represents the wide range of conformations generated by CHARMM. The overlap with experimental data, indicated by the asterisks, is very satisfactory. The basepair step geometrical parameters are represented in the 2D histograms in Figure 4. They reveal a satisfactory overlap between the experimental data (asterisks) and the calculated profiles represented by the isopopulation contours in black on the gray background showing the ranges of values from the conformations generated by CHARMM. All these comparisons show that the CHARMM simulations reproduce both the average properties of DNA and the observed correlations among these structural parameters.

Results and Discussion

Global Conformational Analysis. The extreme widening of the minor groove and sugar puckers displayed in DNA bound by TBP prompted the Burley (*37*) and Shakked

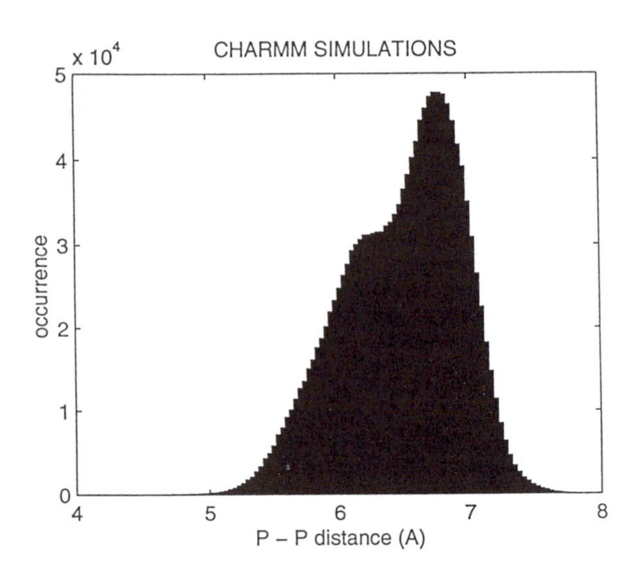

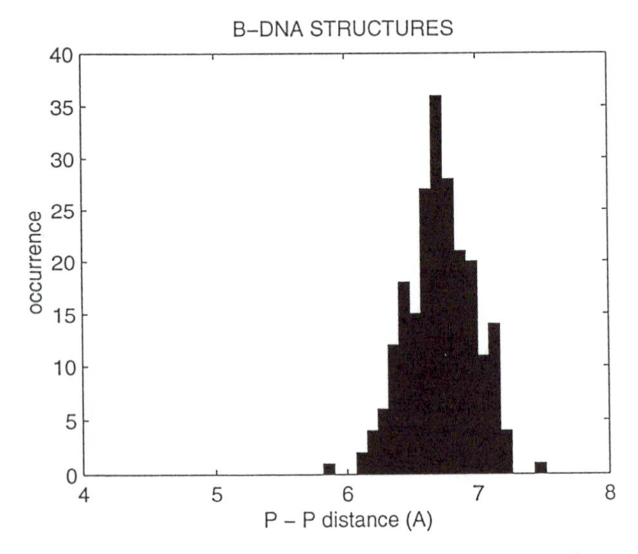

Figure 2. Histograms of consecutive phosphorus - phosphorus distances for the structures generated during the simulations with the CHARMM23 potential, structures determined by NMR spectroscopy, and high resolution B-DNA and A-DNA crystal structures, as listed in the Materials section.

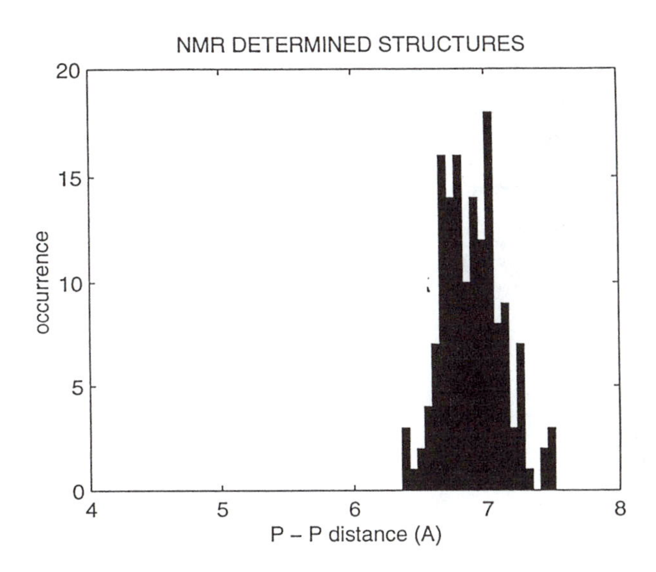

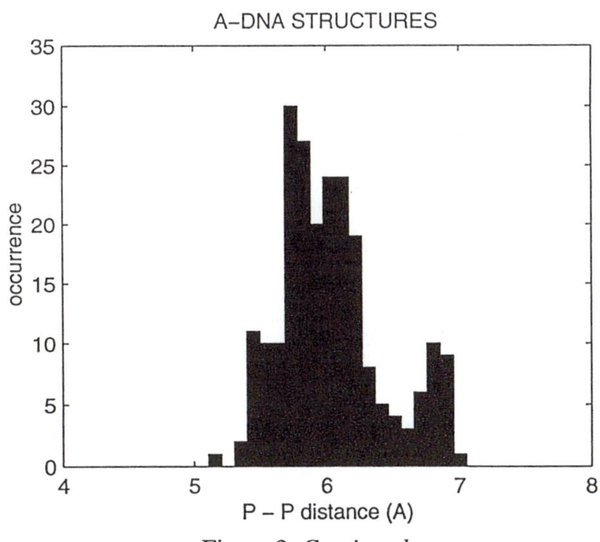

Figure 2. *Continued.*

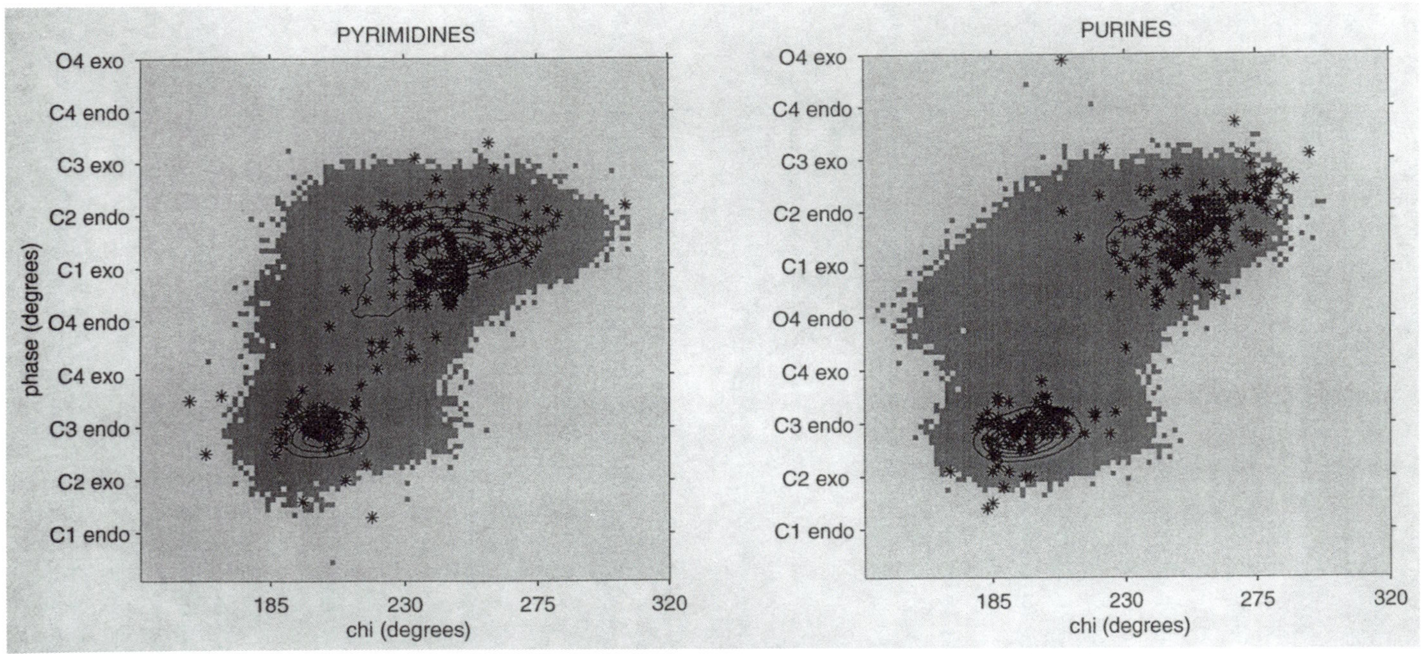

Figure 3. Two dimensional histograms of the sugar pucker phase angle and the glycosidic bond torsional angle for pyrimidine and purine nucleotides. Gray represents the space sampled by the simulations, and the black lines are isopopulation contours. The black asterisks are data from structures determined by NMR spectroscopy, and high resolution B-DNA and A-DNA crystal structures (listed in the Materials section).

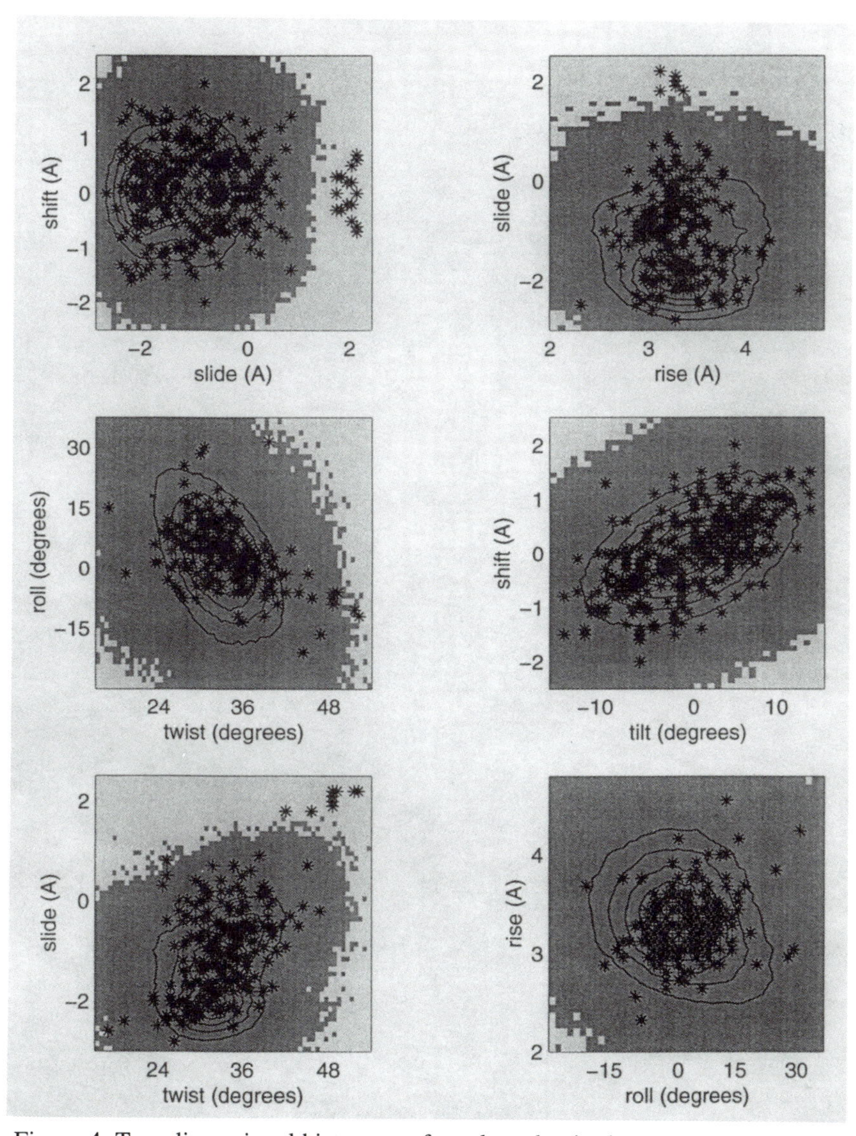

Figure 4. Two dimensional histograms for selected pairwise combinations of the six geometrical parameters describing the conformation of basepair steps. Gray represents the space sampled by the simulations, and the black lines are isopopulation contours. The black asterisks are data from structures determined by NMR spectroscopy, and high resolution B-DNA and A-DNA crystal structures (listed in the Materials section).

Continued on next page

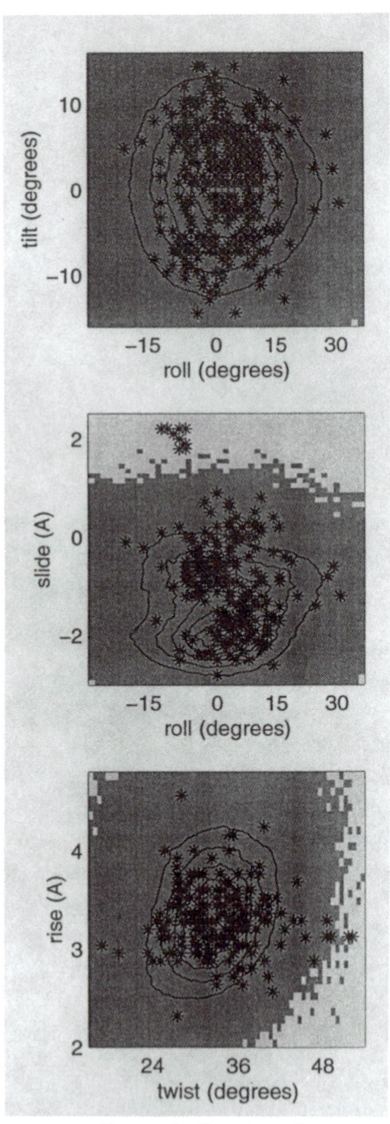

Figure 4. *Continued.*

(*38*) labs to propose an intermediate in the binding reaction that would look like A-DNA. As there are no experimentally derived structures for any of the sequences we studied, the simulation results were examined for evidence of such a conformational intermediate for the DNA alone, in the absence of TBP. The 2D root mean square (rms) distance plot for the 2 ns run of **mlp** indicated that at least three distinct structures were visited during the simulation trajectory (data not shown). As a first characterization of these structures, we calculated the rms difference from A-DNA and B-DNA for the 7 simulated dodecamers as a function of time. The plots for all of them (Figure 5) show a progressive departure from B-DNA and a concomitant approximation to A-DNA, converging after 550 ps towards an rms value around 4Å from either canonical structure. There are nevertheless some differences in the distance from A-DNA and B-DNA at which different sequences stabilize. For example, **at** stabilizes at 5Å from A-DNA, while **2c** keeps drifting towards A-DNA even after 550 ps. To test whether this is a feature of the sequences or of the potential, 2 ns runs for **mlp**, done with AMBER and CHARMM, were compared (Figure 5). After 2 ns, both simulations are far away from B-DNA (~ 5Å rms), but the CHARMM simulation has brought the structure closer to A-DNA (final rms of 2.5Å). While these rms values are too large to classify the structures as either A-DNA or B-DNA, the results clearly differ from both these canonical structures.

Another prominent characteristic of DNA bound by TBP is the ~90° bend. To explore the extent of bending, the DNA helix axes were calculated for the average structures obtained from the intervals in the simulation in which the structures showed little or no drift to another structure. These intervals were determined for each simulation from the 2D rms plots as the regions in which differences < 1.8Å. The superimposition of the helical axes for the average structures of the three known binding sites (**mlp**, **6t**, **at**) and the negative control (**gc**) revealed them to be indistinguishable from each other and to lack any significant bending (data not shown).

The absence of a distinguishing feature among the calculated average structures of the various dodecamers moved the search for the selectivity determinants for TBP binding to the analysis of more local properties of the DNA oligomers, both bound to TBP and free.

Local Conformational Properties and Analysis. Overall, TBP-bound DNA is found to exhibit low *twist* and positive *roll*; the first and last basepair steps have high *rise*, as a consequence of the insertion of Phe residues from TBP. To look for the appearance of such specific local properties in the simulations, all the basepair step conformations generated in the simulations were scanned to identify those steps that have average values biased towards these characteristics. RY steps (R stands for purine and Y for pyrimidine) were thus found to share the properties of low *twist* and positive *roll*, while YR steps were found to have high *rise*.

These findings lead to a proposal for a canonical binding site that is based on the

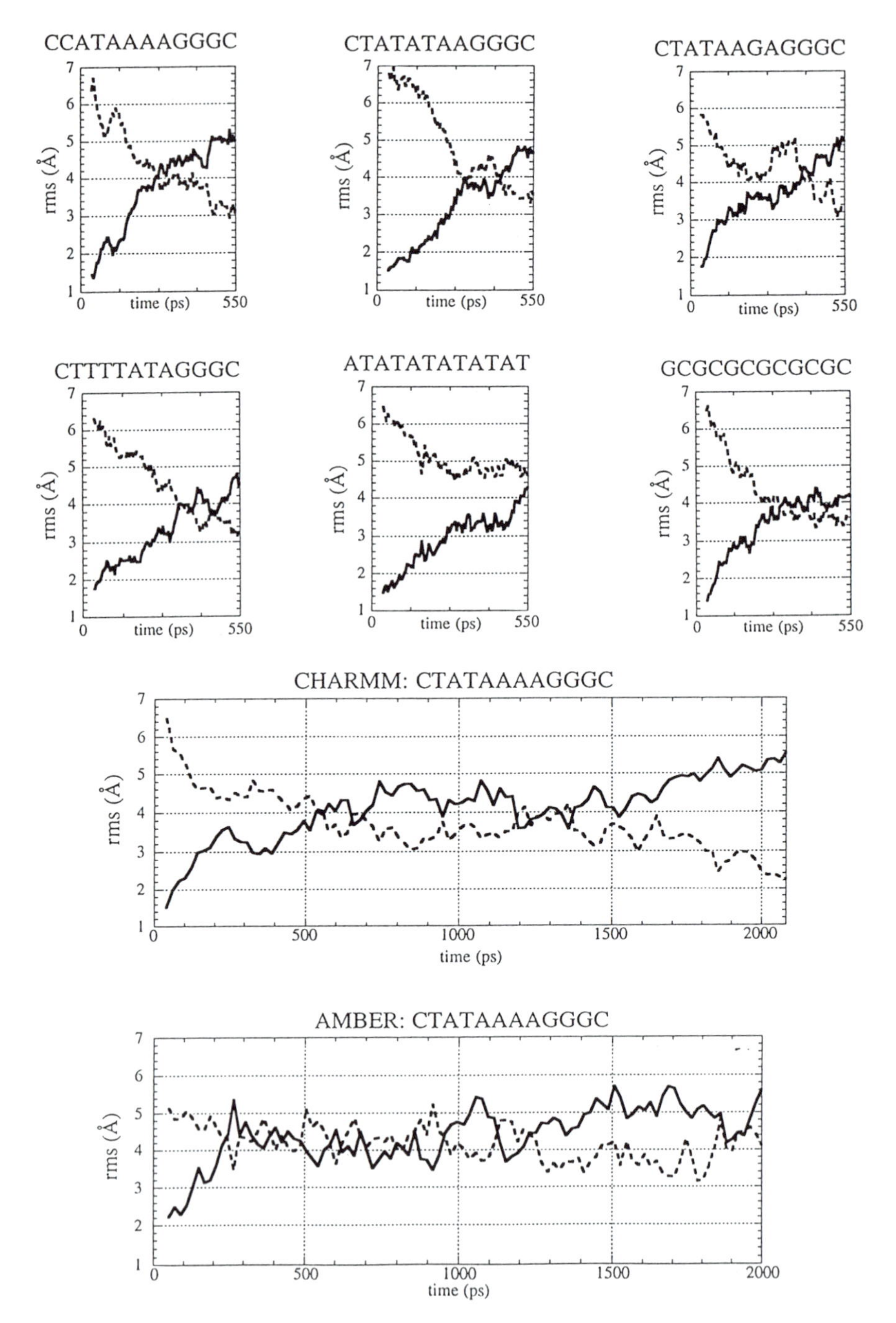

Figure 5. Time evolution of the root mean square (rms) distance from B-DNA (solid line) and A-DNA (broken line) for the heavy atoms of the simulated DNA dodecamers.

average properties of basepair steps and on the local requirements for TBP binding identified from the complexes with TATA box DNA. To satisfy these requirements, YR steps should occur at the ends of the binding site, while RY steps would be needed throughout. Therefore, the combination that best accommodates the requirements is an alternating YR sequence. Notably, this combination of conformational properties of the basepair steps, does not permit a distinction between alternating sequences that contain GC basepairs from those that contain only AT basepairs, as shown by the similarity in the geometrical parameters of these steps, compiled in Table I. The fact that TBP binds poorly, if at all, to GC containing sequences indicates that the poor binding is very probably due to the steric clash between the TBP sidechains and the exocyclic amino group of the G base, and not to conformational preferences of the DNA.

Table I. Basepair Step Parameters for AT, TA, GC and CG:

step	*shift*	*slide*	*rise*	*tilt*	*roll*	*twist*
AT	.1 ± .7	-1.1 ± .5	3.0 ± .4	2.0 ± 6	10.6 ± 11	30.4 ± 5
GC	.0 ± .5	-.8 ± .4	2.7 ± .3	.2 ± 5	21.7 ± 10	27.2 ± 5
TA	.2 ± 1.0	-1.0 ± .7	3.6 ± .5	1.2 ± 8	3.2 ± 13	33.2 ± 6
CG	.3 ± .9	-.7 ± .7	4.0 ± .4	1.8 ± 6	-3.2 ± 12	33.9 ± 5

Transient Properties. The general consensus of the canonical TBP binding site defined above can be refined further to yield more specific sequence characteristics. This refinement is based on the premise that TBP binding would be facilitated if the DNA adopted even transiently a set of local conformations corresponding to the geometries observed in the complexes with TBP. Such sparsely populated conformations can be identified from the results of the simulations, which offer 673,200 conformations for the basepair steps that constitute the simulated dodecamers.

To identify specifically the targets for the search of conformational space, the basepair step parameters for the 8 DNA oligomers found in crystal complexes with TBP were calculated, and 99% confidence intervals were constructed from the distribution of the geometrical parameters. These intervals, shown in Table II, define the range of conformational variability compatible with the formation of a complex with TBP. To compare the special geometrical properties determined from the experimental data for DNA/TBP complexes to those of "general sequence" DNA, we utilized the results from the simulations. The range contained between the mean ± one standard deviation of the values observed for all the basepair steps from these simulations was taken to define the properties of "general sequence" DNA. The comparison between the properties of DNA in the complexes with those of the general sequence DNA identifies parameters that are outside the thermally accessible interval for general sequence DNA (shown in the last line of Table II). The parameters that achieve values outside the interval accessible for general sequence DNA are expected to be useful selectivity indicators because good TBP substrates will be more likely than poor binding sequences to visit these conformational ranges. The parameters identified in this manner are underlined in Table II.

We used these selected parameters and their conformational ranges as filters for the conformations generated in the simulations, and counted the number of times that each different basepair step visited each of these discriminant conformational ranges. Those properties that were found to have as the most frequent visitor a step that binds to TBP were considered as "selectivity determinants" which are probably used by TBP to distinguish productive binding sites from other binding sites. From this selection scheme, only two selectivity determinant sites emerged: positive *slide* is apparently required at basepair steps 3 and 5, which flank the central basepair step. This agrees with the finding by Suzuki and collaborators (*40*), who point out that positive *slide* at these two positions is required to align the floor of the minor groove for recognition by the sidechains of TBP.

Table II. Discriminant Geometry Requirements of Basepair Step Parameters for DNA in Complexes with TBP:

step	*shift*	*slide*	*rise*	*tilt*	*roll*	*twist*
1	.2 ± .5	-1.9 ± .4	**4.9 ± .4**	-0.6 ± 3	**40.9 ± 5**	**19.8 ± 3**
2	**-1.3 ± .5**	-1.5 ± .1	3.4 ± .2	1.7 ± 2	16.5 ± 3	**16.4 ± 1**
3	.2 ± .3	**1.3 ± .1**	3.3 ± .2	3.4 ± 2	7.6 ± 3	**25.5 ±2**
4	.3 ± .4	**1.2 ± .3**	3.4 ± .3	-1.8 ± 1	**26.2 ± 3**	**8.1 ± 6**
5	-.4 ± .3	**2.0 ± .3**	3.5 ± .4	1.0 ± 3	**25.9 ± 4**	**22.4 ± 3**
6	.4 ± .3	**1.2 ± .5**	3.3 ± .3	0.8 ± 5	**24.2 ± 4**	**22.7 ± 4**
7	-.1 ± .7	**.8 ± 1.0**	**5.4 ± .6**	1.7 ± 4	**44.6 ± 6**	22.5 ± 6

DNA (General Sequence):

	shift	*slide*	*rise*	*tilt*	*roll*	*twist*
	.1 ± .8	-1.5 ± .8	3.3 ± .5	0.8 ± 7	2.5 ± 12	32.1 ± 5

Note: steps are listed 5' to 3' along the TATA box; the **values** indicated in this manner are outside the thermally accessible conformational range for general sequence DNA, defined as in (*39*).

The sequence that can access most efficiently the particular range of positive *slide* required at these steps is TA. Because the populations are extremely small (0.23% and 0.01% for the steps 5' and 3' to the central basepair step, respectively), it is possible that the slow binding of TBP (*41-47*) may be explained by the low effective concentration of DNA competent for binding that is suggested by such a small population.

Relations of Selectivity Determinants to Measured Affinities. Combining the findings from the analysis of the average properties of the basepair steps with those from the transient conformational properties, we propose a canonical TBP binding site of the form: YRTATAYR. The considerations underlying this definition of the canonical binding site further lead to the prediction that the closer a DNA sequence is to this consensus, the better its TBP binding ability should be.

The prediction is probed by the relation between the adherence of the calculated sequences to the canonical TBP binding site, and the available data for binding affinity determined by Wong and Bateman (*21*). As shown in Table III, **at** and **6t** match the consensus almost completely, and are found to have the highest affinity for TBP. Because **mlp** is missing the second TA of the canonical sequence, it should bind less well; its affinity is found to be 3 to 4 times lower than that of **at** or **6t**. While **2c** and **7g** are not bound by wild type TBP, we predict that the binding constants measured with the appropriate TBP mutants should be in the same range as that for **mlp**.

Table III. Comparison to Equilibrium Binding Constants

consensus sequence	**YRTATAYR**	$K_{eq}(10^{-9})$(from (*21*))
sequences:		
mlp	TATAAAAG	3.7
2c	CATAAAAG	n.a.
7g	TATAAGAG	n.a.
6t	TATATAAG	1.1
at	TATATATA	1.4

the sequences indicate both **average** and transient specificity determinants; n.a. : not available

Conclusions

The selectivity determinants for TBP binding rely on specific aspects of DNA bendability. Expressed as a bias in the average basepair step geometry (positive *roll* and low *twist*, and high *rise*), and in more frequent transient excursions into a specific basepair step geometry (positive *slide*), these DNA bendability properties appear to be used by promoter sequence DNA to prepare for TBP binding. The results show that the sequence dependence of these calculated bendability properties translates into sequence selectivity for TBP that is correlated with the measurable binding affinity. The mechanism by which these properties of the DNA sequences determine the kinetic and thermodynamic properties of TBP binding constitutes the subject of ongoing investigations.

Acknowledgements

Simulations were carried out at the Cornell National Supercomputer Facility (sponsored by the National Science Foundation and IBM). This work as supported by a Fulbright/CONACyT (México) scholarship to NP, grants from DGICYT: PB95-0624 and PR95-275 (LP), and the Association for International Cancer Research (HW).

Literature Cited

1. Cormack, B.P. and Struhl, K., *Cell* **1992**, *69*, 685-96.
2. Burley, S.K. and Roeder, R.G., *Annu. Rev. Biochem.* **1996**, *65*, 769-99.
3. Orphanides,G., Lagrange,T.,and Reinberg,D.,*Genes Dev.* **1996**, *10*, 2657-83.
4. Marsh, T.L., Reich, C.I., Whitelock, R.B., and Olsen, G.J., *Proc. Natl .Acad. Sci .USA* **1994**, *91*, 4180-4.
5. Rowlands, T., Baumann, P., and Jackson, S.P., *Science* **1994**, *264*, 1326-9.
6. Nikolov, D.B. and Burley, S.K., *Nature Struct. Biol.* **1994**, *1*, 621-37.
7. DeDecker, B.S., O'Brien, R., Fleming, P.J., Geiger, J.H., Jackson, S.P., and Sigler, P.B., *J. Mol. Biol.* **1996**, *264*, 1072-84.
8. Chasman, D.I., Flaherty, K.M., Sharp, P.A., and Kornberg, R.D., *Proc. Natl. Acad. Sci. USA* **1993**, *90*, 8174-8.
9. Juo, Z.S., Chiu, T.K., Leiberman, P.M., Baikalov, I., Berk, A.J., and Dickerson, R.E., *J. Mol. Biol.* **1996**, *261*, 239-54.
10. Kim,Y., Geiger, J.H., Hahn, S., and Sigler, P.B., *Nature* **1993**, *365*, 512-20.
11. Kim, J.L. and Burley, S.K., *Nature Struct. Biol.* **1994**, *1*, 638-53.
12. Nikolov, D.B., Chen, H., Halay, E.D., Hoffman, A., Roeder, R.G., and Burley, S.K., *Proc. Natl. Acad. Sci. USA* **1996**, *93*, 4862-7.
13. Geiger, J.H., Hahn, S., Lee, S., and Sigler, P.B., *Science* **1996**, *272*, 830-6.
14. Tan, S., Hunziker, Y., Sargent, D.F., and Richmond,T.J., *Nature* **1996**, *381*, 127-51.
15. Nikolov, D.B., Chen, H., Halay, E.D., Usheva, A.A., Hisatake, K., Lee, D.K., Roeder, R.G., and Burley, S.K., *Nature* **1995**, *377*, 119-28.
16. Breathnach, R. and Chambon, P., *Annu. Rev. Biochem.* **1981**, *50*, 349-83.
17. White, S., Baird, E.E., and Dervan, P.B., *Biochemistry* **1996**, *1996*, 12532-7.
18. Harrington, R.E. and Winicov, I., *Prog.Nucleic Acid Res.Mol. Biol.* **1994**, *47*, 195-270.
19. Calladine, C.R. and Drew, H.R., *J. Mol. Biol.* **1986**, *192*, 907-18.
20. Hagerman, P.J., *Biochim. Biophys. Acta* **1992**, *1131*, 125-32.
21. Wong, J.M. and Bateman, E., *Nucleic Acids Res.* **1994**, *22*, 1890-6.
22. Arndt, K.M., Ricupero, S.L., Eisenmann, D.M., and Winston, F., *Mol. Cell. Biol.* **1992**, *12*, 2372-82.
23. Arndt, K.M., Wobbe, C.R., Ricupero, H.S., Struhl, K., and Winston, F., *Mol. Cell. Biol.* **1994**, *14*, 3719-28.
24. Berman, H.M., Olson, W.K., Beveridge, D.L., Westbrook, J., Gelbin, A., Demeny, T., Hsieh, S.-H., Srinivasan, A.R., and Schneider, B., *Biophys. J.* **1992**, *63*, 751-759.
25. Bernstein, F.C., Koetzle, T.F., Williams, G.J., Meyer, E.E., Jr., Brice, M.D., Rodgers, J.R., Kennard, O., Shimanouchi, T., and Tasumi, M., *J. Mol. Biol.* **1977**, *112*, 535-42.
26. Mezei, M., *J. Comp. Chem.* **1997**, *18*, 812-5.

27. MacKerell Jr, A.D., Wiorkiewicz-Kuczera, J., and Karplus, M., *J. Am. Chem. Soc.* **1995**, *117*, 11946-75.
28. Pastor, N., Pardo, L., and Weinstein, H., *Biophys. J.* **1997**, *73*, in press.
29. Lavery, R. and Sklenar, H., *J. Biomol. Struct. Dyn.* **1988**, *6*, 63-91.
30. Lavery, R. and Sklenar, H., *J. Biomol. Struct. Dyn.* **1989**, *6*, 655-67.
31. Ravishanker, G., Swaminathan, S., Beveridge, D.L., Lavery, R., and Sklenar, H., *J. Biomol. Struct. Dyn.* **1989**, *6*, 669-99.
32. Brooks, B.R., Bruccoleri, R.E., Olafson, B.D., States, D.J., Swaminathan, S., and Karplus, M., *J. Comp. Chem.* **1983**, *4*, 187-217.
33. Pearlman, D.A., Case, D.A., Caldwell, J.W., Ross, W.S., Cheatham III, T.E., Ferguson, D.M., Seibel, G.L., Singh, U.C., Weiner, P.K., and Kollman, P.A., *University of California, San Francisco* **1995**.
34. Cornell, W.D., Cieplak, P., Bayly, C.I., Gould, I.R., Merz Jr., K.M., Ferguson, D.M., Spellmeyer, D.C., Fox, T., Caldwell, J.W., and Kollman, P.A., *J. Am. Chem. Soc.* **1995**, *117*, 5179-97.
35. Darden,T.A., York,D., and Pedersen,L., *J. Chem. Phys.* **1993**, *98*, 10089-92.
36. Neumann, N. and Steinhauser, O., *Mol. Phys.* **1980**, *39*, 437-54.
37. Kim, J.L., Nikolov, D.B., and Burley, S.K., *Nature* **1993**, *365*, 520-7.
38. Guzikevich-Guerstein,G. and Shakked, Z., *Nature Struct. Biol.* **1996**, *3*, 32-7.
39. Olson, W.K., Babcock, M.S., Gorin, A., Liu, G., Marky, N.L., Martino, J.A., Pedersen, S.C., Srinivasan, A.R., Tobias, I., Westcott, T.P., and Zhang, P., *Biophys. Chem.* **1995**, *55*, 7-29.
40. Suzuki, M., Allen, M.D., Yagi, N., and Finch, J.T.,*Nucleic Acids Res.* **1996**, *24*, 2767-73.
41. Coleman, R.A. and Pugh, B.F., *J. Biol. Chem.* **1995**, *270*, 13850-9.
42. Coleman, R.A., Taggart, A.K., Benjamin, L.R., and Pugh, B.F., *J. Biol. Chem.* **1995**, *270*, 13842-9.
43. Hoopes, B.C., LeBlanc, J.F., and Hawley, D.K., *J. Biol. Chem.* **1992**, *267*, 11539-47.
44. Parkhurst, K.M., Brenowitz, M., and Parkhurst, L.J., *Biochemistry* **1996**, *35*, 7459-65.
45. Perez-Howard,G.M., Weil, P.A., and Beechem, J.M., *Biochemistry* **1995**, *34*, 8005-17.
46. Petri, V., Hsieh, M., and Brenowitz, M., *Biochemistry* **1995**, *34*, 9977-84.
47. Starr,D.B., Hoopes,B.C.,and Hawley,D.K., *J. Mol. Biol.* **1995**, *250*, 434-46.

Chapter 21

RNA Tectonics and Modular Modeling of RNA

Eric Westhof[1], Benoît Masquida, and Luc Jaeger

Institut de Biologie Moléculaire et Cellulaire du Centre National de la Recherche Scientifique, UPR 9002, 15 rue René Descartes, F–67084 Strasbourg Cedex, France

Our understanding of the structural, folding and catalytic properties of RNA molecules has increased enormously in recent years. Here, we emphasize the mosaic structure of RNA and describe the application of modular assembly in the modeling of RNA molecules. The hierarchy between levels of organization is more apparent in the RNA world than in the protein universe. Information about secondary structure and overall architecture can be obtained by methods rooted in evolutionary biology. The more than one hundred known sequences of group I introns reflect a sampling of positions which can change without affecting 3D folding and catalytic function. Their alignments therefore filter mutational noise and strengthen relevant signals. Compensatory base changes maintaining Watson-Crick pairs yield the 2D structure, while other covariations reveal 3D contacts. Covariations due to historical contingencies, and not to 3D contacts, can be distinguished by classifying the sequence-objects in groups and partitioning the covariations among them. Structural constraints appear in all groups and historical covariations in some only. With a library of preformed modules, 3D models of catalytic RNAs have been assembled and insight into their protein or ligand binding properties gained.

According to the Concise Oxford Dictionary, from the Greek word meaning "carpenter", "tectonics" is the "art of producing useful and beautiful buildings". The word is used in several fields of science, especially in geology where plate tectonics refers to the movements of elements of the earth crust. The knowledge of the building blocks and of the rules governing the assembly of biological macromolecules into complex objects forms the basis for modeling and designing new objects. Although the building blocks of RNA appear, at first sight, well defined, the rules of interaction and assembly between them are only beginning to be delineated. Here, we discuss RNA architecture and folding with the aim of presenting an RNA logic useful for modeling complex RNA molecules.

[1]Corresponding author

Mosaic Structure of RNA.

The naturally occurring complex objects of biology, be they the hemoglobin protein, the ribosome, or the human eye, did not evolve as integral entities. Instead biological evolution rather proceeds by assembling modular units, modifiable independently, sometimes redundant and often previously selected for other functions. The presence of RNA molecules at every single crucial and decisive step in the life of a cell, attests to RNA ancestral glory and roles in the origin and evolution of life. While some RNA molecules have transient lifetimes, others are structurally central in intricate life machineries like the ribosomes and spliceosomes. But, as in other biological objects, the complexity of the structures formed by RNA molecules can be resolved into separable pieces or modules as in a three-dimensional mosaic. Large RNA molecules are well-known for being able to work in *trans*, i.e. for functioning upon re-association of cut-out parts. The dissociation of complex RNA systems, such as autocatalytic introns, into parts, or modules, is facilitated by a property of nucleic acids : Watson-Crick pairing between complementary bases. The complementary Watson-Crick base pairs, with *cis* glycosyl bonds, form the only set of pairs which are isosteric in antiparallel helices. Thus, they allow formation of helices with regular, or quasi-regular, sugar-phosphate backbones. In single-stranded molecules, stacking and base pairing drive the folding of the chain on itself through the formation of helical regions linked by non-helical elements, hairpin loops, internal bulges, and multiple junctions (see Figure 1). The folded structure is usually discussed first in terms of a secondary structure which schematically represents the base-paired segments in a planar drawing. We use the terms "architecture" or "three-dimensional structure" (or "assembly" and "self-assembly") when meaning or describing a native "folded" state and the term "folding" when discussing how the RNA molecule actually folds kinetically.

In contrast to proteins and as noticed several years ago by Paul Sigler (1), secondary structure is defined by H-bonds between side chains in RNA and not by H-bonds between backbone atoms, as for example in α−helices. Thus, in RNA molecules, secondary structure determination is best achieved through phylogenetic sequence comparisons when more than one sequence is available (for reviews, see (2,3)) or by chemical probing of the sites involved in Watson-Crick pairing with the help of computer prediction (4,5), Both approaches are complementary and both contain information potentially extending beyond secondary structures.

Small assemblies of secondary structure elements are known to form autonomous and functional entities. For example the hammerhead ribozyme (Figure 2), for which two crystal structures are available, is made of a triad of helices (6-8). In the next level of organization, the tertiary structure, the secondary structure elements are associated through numerous van der Waals contacts, specific H-bonds via the formation of a small number of additional Watson-Crick pairs and/or unusual pairs involving hairpin loops or internal bulges. The partitioning of energy levels between secondary and tertiary structures is a reasonable assumption in large RNAs considering the relative energies and the clear identification of the secondary structure elements.

In summary, an ensemble of observations (9,10) bears out a view of RNA folding whereby 3D architecture results from the cooperative compaction of separate, pre-formed, and stable sub-structures, which might undergo only minor and local rearrangement during the process. Several recurrent 3D motifs, which control and direct RNA-RNA recognition and RNA assembly, have now been identified. We first discuss the building blocks and some assembly rules recently uncovered without due consideration to the selecting and stabilizing roles of proteins.

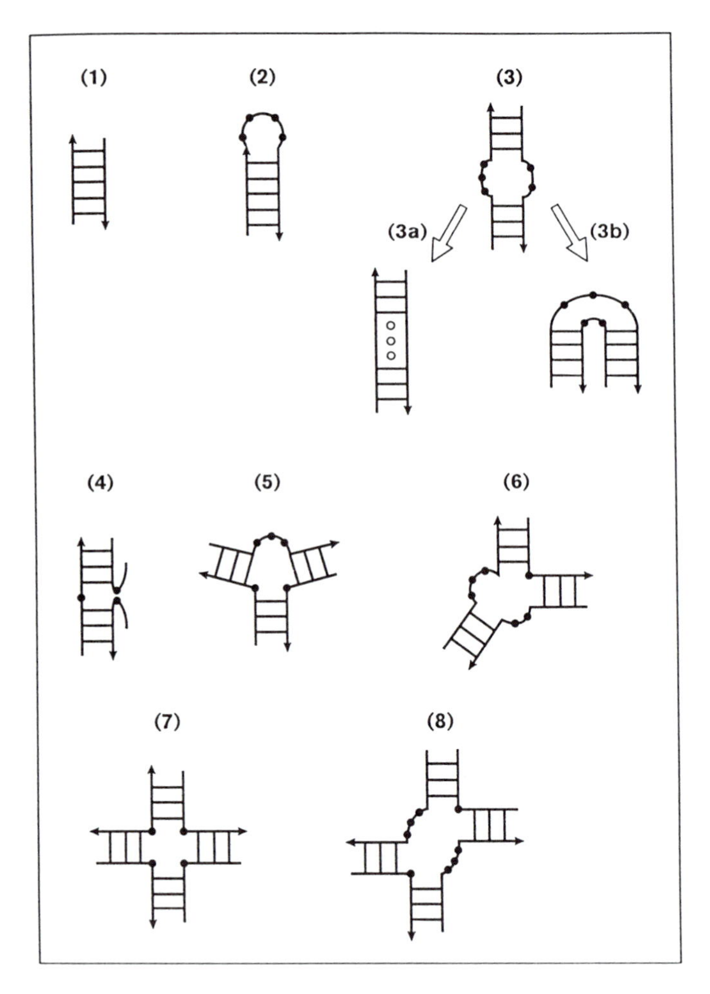

Figure 1. Schematic drawings of typical RNA secondary structure elements. The arrows indicate the 5' to 3' direction. Bars denote Watson-Crick base pairs and dots phosphodiesters linking helices. (1) An RNA helix; (2) An RNA hairpin with four unpaired residues; (3) An asymmetric internal bulge between two helices; (4) Two stacked helices; (5) A three-way junction with one single-stranded region; (6) A three-way junction with two single-stranded regions; (7) A four-way junction; (8) A four-way junction with two single-stranded regions.

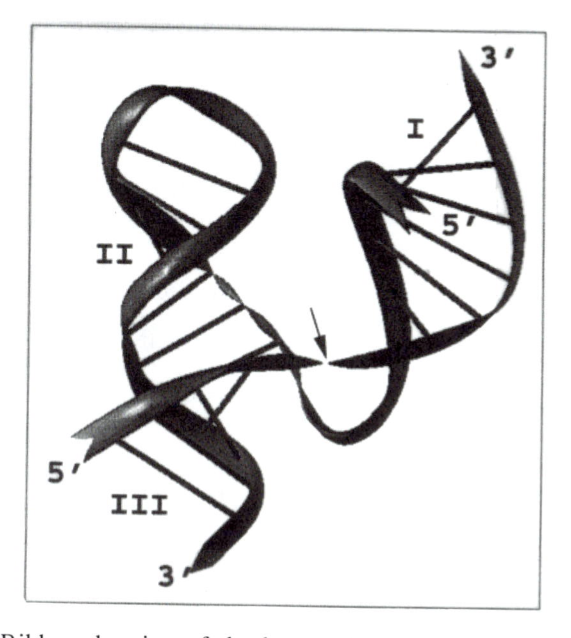

Figure 2. Ribbon drawing of the hammerhead ribozyme after the crystal structure of (7). The cleavable bond is indicated by an arrow. Helices I, II, and III are indicated. In that structure, the substrate (DNA for cleavage inhibition) pairs to the ribozyme core, forming helices I and III.

Mosaic Modeling of RNA

Levinthal's paradox applies to RNA as well as to proteins (11,12). The number of possible conformers is even more astronomical in RNA than in proteins. Indeed, while there are three variable torsion angles in the polypeptide chain (and between zero and five in the side chains), there are six variable torsion angles in polynucleotides (with one for the side chain). Pauling reduced the number to two in polypeptide chains (13) and, for nucleic acids, Sundaralingam reduced severely two of them and contained the others to preferred domains (14). Thus, the torsion angles about the C-O bonds stick around 180°, while the sugar rings adopt either the C3'-*endo* or the C2'-*endo* puckers and the base is either in an *anti* or a *syn* orientation with respect to the sugar. The two polymers adopt opposing strategies for inter-residue flexibility. In proteins, the torsion angles on either side of the peptide bond constitute the main flexible links while, in nucleic acids, the phosphodiester bonds themselves direct chain re-orientations. Despite those conformational restrictions, the number of possible conformers for an RNA of a biologically relevant size is still untractable.

The introduction of modular units, hierarchically organized and folded, circumvents most, if not all, of the numerical nightmares inherent in mathematical modeling of macromolecules. Three-dimensional modeling based on a modular and hierarchical approach to RNA assembly has now been applied to several structured RNAs (9,15-17). The hierarchical framework implies that secondary structure elements form first, with the folding process moving progressively from small-scale elements to the larger scales of sub-domains and domains. As such, this scenario does not imply that there is an interdependence between the sub-domains, i.e. that some secondary elements should form before others. In mosaic modeling of RNA, 3D motifs are first built and then assembled synchronically. Although this assembly process is often performed without due consideration to its actual time dependence, it is more than often pertinent to follow and exploit folding hypotheses as an insightful guide during modeling (for a review on folding kinetics, see (18). Fundamentally, the hierarchical folding of RNA assumes that tertiary structure forms once all the secondary structure elements are present. This distinction is feasible because the secondary structure is easily identifiable and energetically the main component of an RNA architecture while tertiary structure contributes only minimally to the stability of the native state in terms of free energy. One consequence is that RNA molecules can be observed with stable secondary structures in the absence of tertiary interactions. Such intermediate states, akin to the molten globule state of proteins, have been observed, under some experimental conditions, in a group I intron (9) and in the RNase P ribozyme of *E. coli* (19).

Assembly Rules of RNA Mosaic Units

The very advantage of RNA, its conspicuous 2D structure, can be also the Achille's heel of the structuralist. Indeed, the drawing of RNA secondary structures can be misleading. A long time ago, Struther Arnott (20) remarked that the cloverleaf structure of tRNA impeded early attempts at assembling a 3D model. Besides, in some instances, a case especially blatant for pseudoknots, the representation is arbitrary. Figure 3 shows three different drawings of a pseudoknot in the central core the tRNA-like domain present at the 3'-end of Tobacco Mosaic Virus (TMV) (21). One view, bottom left, reveals the hidden similarity in pairing schemes between the central core of TMV and the catalytic domain of Hepatitis Delta Virus (HDV) (22). The biological relevance of such similarities is difficult to evaluate, but the structural insight they bring along makes them valuable. The 3'-end of the RNA genome of some plant viruses, which folds as a tRNA-like domain due to the presence of an additional pseudoknot preceding the -CCA terminal part, has itself still an unknown function and much debated origin (23).

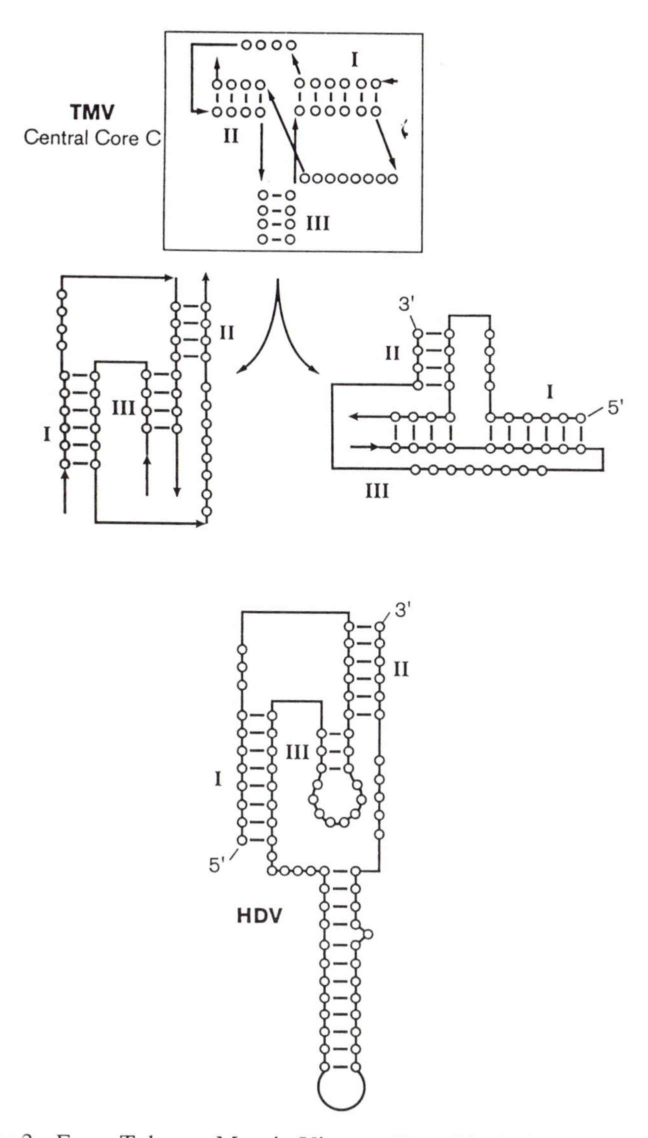

Figure 3. From Tobacco Mosaic Virus to Hepatitis Delta Virus: the 3'-end of the TMV RNA genome contains a tRNA-like domain in which a pseudoknotted three-way junction is present (top); two other drawings of the corresponding secondary structure are shown below; the drawing at the left reveals the hidden similarity with the proposed secondary structure of the catalytic core of the Hepatitis Delta Virus shown at the bottom. In HDV, the cleaved phosphodiester is at the 5'-end of helix I, while the loop of hairpin III and the junction between helices I and II are crucial for catalysis. It is presently unknown whether this similarity is purely formal or reflects common biological function and/or origins. Adapted from (21).

A recurring theme in the highly structured hairpins and internal bulges is that stacking between bases is maximized thereby promoting hydrogen bonding between bases or between bases and the sugar-phosphate backbone. The same forces are at play at junctions between helices where, at the same time, they face the opposing electrostatic forces of bringing close together negatively charged backbones and the steric repulsions due to the bulkiness of RNA helices. As in tRNA (see Figure 4), two helices with at least one contiguous strand will tend to stack co-axially (24,25), a conformation which maximizes stacking and H-bonding while minimizing electrostatic and steric repulsion. Such interplay between opposing forces rationalizes the fact that junctions with three helices do not occur in structured RNAs without at least one single-stranded stretch, the archetype being the three-way junction of 5S rRNAs. A three-way junction with two single-stranded stretches occurs in the hammerhead ribozyme. On the other hand, in junctions where four helices meet, it is frequent to observe continuity in base pairing (26,27). In tRNAs, there are two single-stranded stretches linking the two sets of co-axial helices which stack on top of each other. In one stack, the two helices form a quasi-continuous helix with minor distortions (the amino acid and thymine helices). In the second stack, the loop-loop interaction between the thymine and the dihydrouridine loops imposes, at the junction between the two helices, a much larger twist angle than within a standard RNA helix (around 45° instead of 33°) (28,29).

Any complex secondary structure of a structured RNA, like the 16S rRNA molecule, contains several examples of junctions formed by converging helices. Could one use the mosaic view of RNA structure to dissect a complex RNA molecule? Idealized junctions are shown in Figure 1: the 5S rRNA 3-way junction (16,30), drawing (5) in Figure 1; the hammerhead 3-way junction (7,8), drawing (6); the Holliday-like 4-way junction (26,27), drawing (7); the tRNA 4-way junction (31), drawing (8). Although it is often easy to recognize the type of junction, it is less apparent to orient the helices with respect to each other according to a chosen module when it is inserted within a complex structure. It is, however, useful to be able to make structural hypotheses on the relative arrangements of helices in a large RNA when discussing data or suggesting new experiments. The following discussion illustrates this point.

Right-handed RNA helices are highly asymmetric objects made of two strands of opposite polarities, each constituted of chiral units. Single-stranded ends introduce a new element of asymmetry. Because of the right-handed rotation between nucleotides in helices, single-stranded nucleotides tend to leave or enter helices in a right-handed fashion. Thus, the 5'-strand of a helix would face the shallow groove and the 3'-strand the deep groove of a adjacent and stacked helix. Consider the 5S rRNA molecule for which there is evidence that helices II and III (Figure 5) are roughly co-axial (indicated by the double arrowed arc in Figure 5) with the single-strand facing the shallow groove of helix III. Depending on the relative positions of the hairpin loops with respect to the single-stranded stretch, co-axial stacking of helices I and III (at the left) or of helices I and II (at the right) can be derived.

The high thermodynamic stability of RNA helices is coupled with a low information content of RNA secondary structure helices, which leads in RNA molecules to a tendency to fall into kinetic traps during folding and simultaneously to a difficulty in specifying a thermodynamically favored tertiary structure. For a complex RNA molecule, the dependence of the 3D architecture on the presence of the extended and correct secondary structure might therefore be a necessity to guarantee the absence of kinetically trapped misfoldings. Some three-dimensional RNA motifs are especially important both for guaranteeing the absence of kinetic misfoldings and for specifying a precise three-dimensional structure. One of them, the pseudoknot (32-34), is governed by Watson-Crick base pairing between a hairpin loop and a single-stranded stretch or between two single-stranded stretches. Consequently, a pseudoknot can be considered either secondary structure (it is constituted of an RNA helix with standard Watson-Crick pairs) or tertiary structure (a prerequisite for its formation being the presence of at least one hairpin). We prefer to consider a

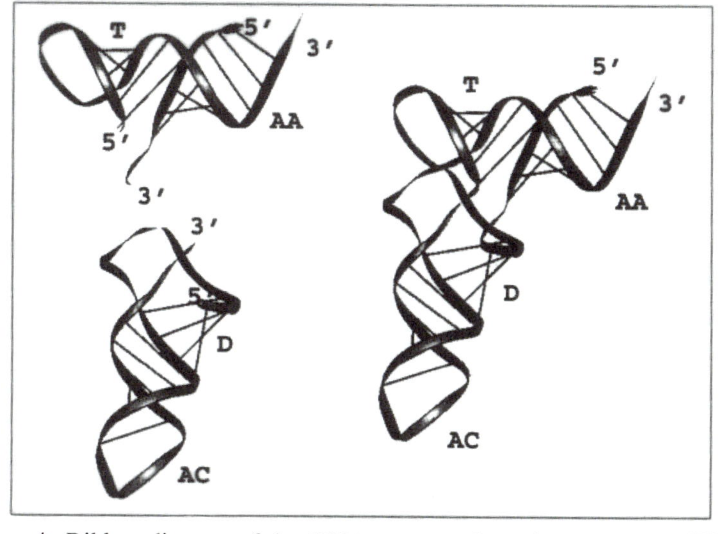

Figure 4. Ribbon diagram of the tRNA structure, here the yeast $tRNA^{asp}$ (29), on the right and its decomposition into two domains made of co-axially stacked helices on the left. Notice how the 3'-end of the anticodon helix (AC) runs over the deep groove of the dihydrouridine helix (D). Note the rotation angle between the AC and D helices is large, while the one between the acceptor helix (AA) and the thymine (T) helix is usual for RNA helices.

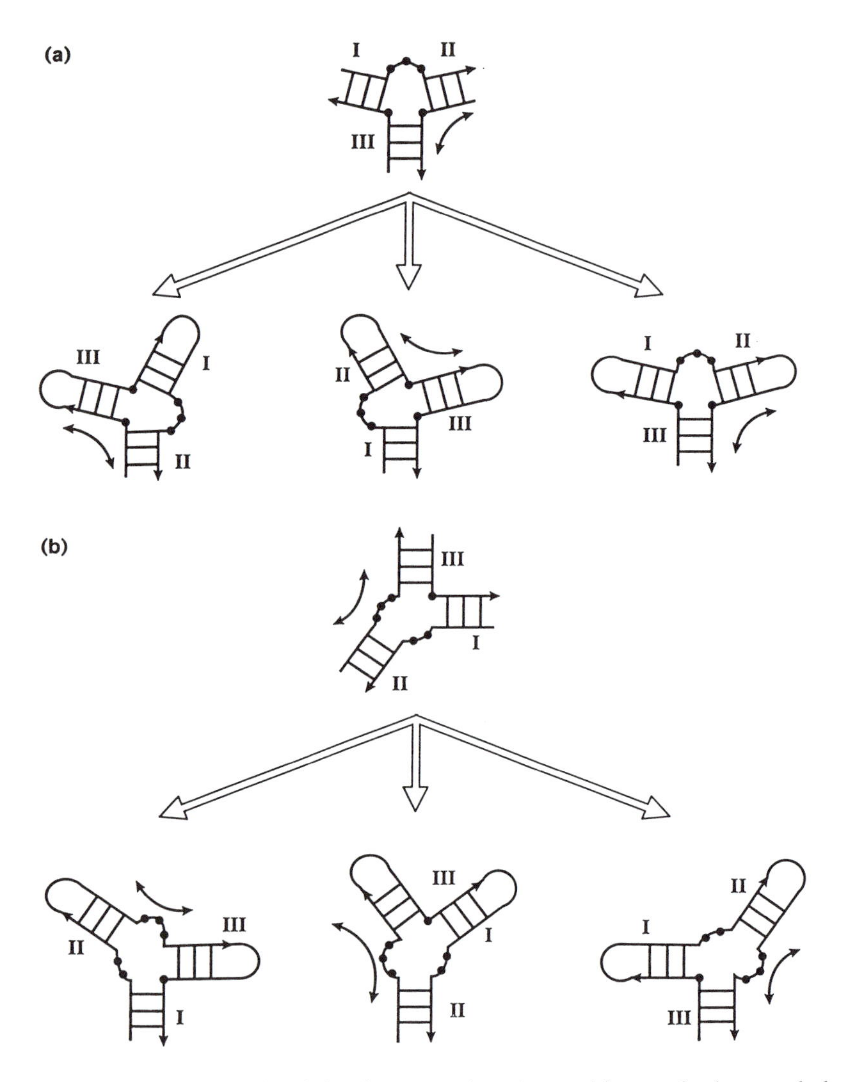

Figure 5. (Top) Analysis of the three-way junctions with one single-stranded region. The starting archetype is the 5S rRNA. One assumes the validity of co-axial stacking between helices II and III as present in the proposed models of 5S rRNAs (indicated by the arc with double arrows). In a large RNA, three different situations could occur (from left to right). Symmetry and rotations perpendicular to the plane yield another six drawings. (Bottom) Analysis of the three-way junctions with two single-stranded regions. The starting archetype is the hammerhead ribozyme where helices II and III stack on top of each other with non-canonical base pairings between residues of the joining single-stranded regions. One assumes that the co-axial stacking (indicated by the arc with double arrows) is maintained in a large RNA. Three different situations could occur (from left to right). Again, symmetry and rotations perpendicular to the plane yield another six drawings.

pseudoknot as a special three-dimensional motif involving Watson-Crick pairs. Recent evidence shows that in group I introns, the cores of which are centered around a pseudoknot (helices P3 and P7), one helix forms after the other one, i.e. helix P7 follows helix P3 (35), In this example, helix P7 constitutes a criterion for the validity of the correctness of the previously formed secondary structure elements. Thus, the origin of frequent occurrences of pseudoknots in large structured RNAs, like the autocatalytic group I intron (36) or RNase P ribozymes (37), might follow from the stringent constraints they impose and the control they exercise on the RNA folding pathway.

The interaction between two hairpin loops, called a loop-loop motif, constitutes a pseudoknot when the interaction occurs intramolecularly and a "kissing" complex (38,39) when the interaction occurs intermolecularly between two RNA molecules. In a large RNA structure, loop-loop interactions can only occur once the two interacting hairpins are formed and properly positioned with respect to each other so that Watson-Crick pairing can take place. Such interactions finalize the 3D structure, like -GNRA-loops interacting in the shallow groove of helices, and are thus better considered as RNA-RNA anchors.

RNA-RNA Anchors and RNA Recognition Motifs

The secondary structure indicates the positions of base paired helices. Those are linked by single-stranded regions which can form hairpins, internal bulges within helices, or link helices (see Figure 1). The complexity and design variability of such structures are stunning and rival those present in proteins.

Particularly stable hairpins (e.g. -UNCG- loops) (40) could form nucleation points for folding. Hairpins are often sites of protein binding (e.g. loop B of U1 snRNA) (41) but, in the architecture of structured RNAs, hairpins play a crucial role as RNA-RNA anchors (see Figure 6). Two hairpins can interact with each other, either solely through Watson-Crick base pairing (loop-loop motif) (42) or with a mixture of base pairing and intercalation as in the intra-molecular association between the thymine and dihydrouridine loops of tRNAs (see Figure 4). A hairpin can base pair with a single-stranded stretch forming an additional helix, leading to a pseudoknot. And finally, a family of tetraloop hairpins, the -GNRA- family, binds specifically to the shallow groove of another helix (43,44), while -GAAA- tetraloops recognizes specifically an internal bulge (45).

Internal bulges, frequently protein binding sites (e.g. Rev protein (46), S8 ribosomal protein (47)), present variable flexibility (48,49), some of them being almost helical because of unusual base pairing (e.g. loop E of 5S rRNA (16,50), see Figure 1, (3a)) while others lead to 180° turn (28) (e.g. J5/5a in *Tetrahymena*, see Figure 1, (3b)). Internal bulges can also serve as RNA anchor points. For example, in the *Tetrahymena* ribozyme the J4/5 internal loop recognizes specifically the invariant G-U base pair at the cleavage site (17,51). How the base sequence governs the conformation of a given internal bulge, and thus its potential structural and biological role, is unknown.

Group I introns have served as a test system for studying the folding and structure of complex RNAs with catalytic functions. The universally conserved catalytic core of group I introns (~120 nt) is generally not sufficient for maintaining the active structure of the ribozyme. Most self-splicing group I introns require non-conserved peripheral elements which stabilize the overall structure by RNA-RNA interactions. Phylogenetic and experimental evidence has shown that RNA-RNA intramolecular contacts are dominated by loop-loop Watson-Crick complementary pairs and by GNRA/helix contacts with the shallow groove of helices. At this time, three-dimensional models of group I introns belonging to four different subgroups are available (52). They all emphasize the modular and hierarchical organization of the architecture of group I introns and the widespread use of loop-loop base-pairings and -GNRA- tetraloop contacts for stabilizing structured RNA molecules.

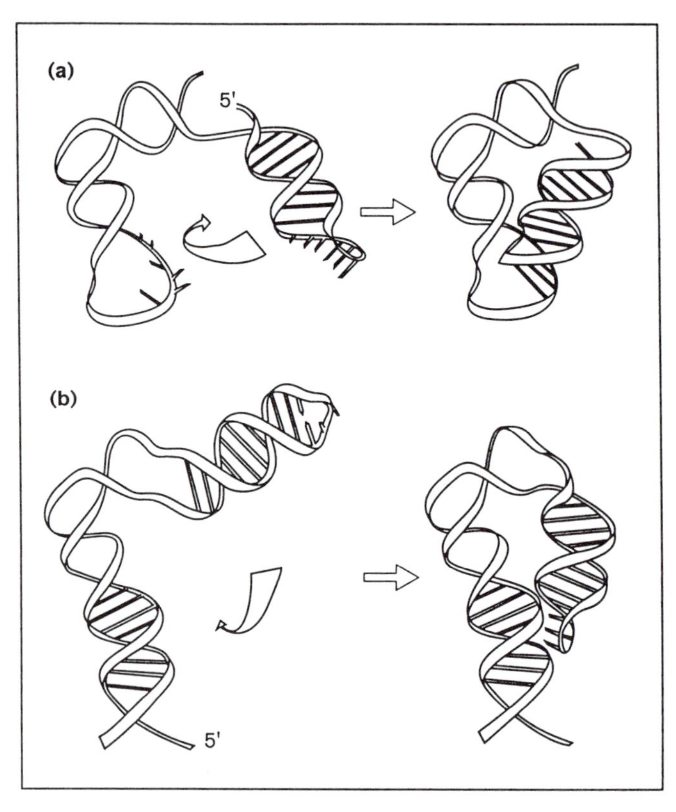

Figure 6. Schematic drawings illustrating two RNA-RNA anchors. Both motifs introduce new chiralities. (Top) The single-stranded residues of a hairpin loop form Watson-Crick base pairs with an unpaired complementary sequence in another structural domain, thereby molding a pseudoknot. As in standard RNA helices, the base pairing occurs with antiparallel polarities of the strands. (Bottom) A -GNRA-tetraloop in a hairpin binds to the shallow groove of a helix belonging to another domain so that an A of the tetraloop interacts with either a G-C or an A-U pair and a G with only an A-U pair. The purines in the loop and in the helix are oriented with a parallel polarity.

Motif Swap between RNA Motifs.

One of the most striking evidence for the existence of modularity in RNA structure is the possibility of swapping one structural motif for another. Loop-loop interaction and -GNRA-/helix contact have been successfully exchanged experimentally in group I introns (12,28,43,53). Very recently, phylogenetic evidence has shown that a similar swap occurs in bacterial RNase P RNAs (12,28,43,54).

Acknowledgments

This paper is an abridged and actualized version of a review published in **Folding & Design** (**1**, R78-R88, 1996) which was adapted to focus on the relations between RNA tectonics and RNA modeling. We are very grateful to the publishers and editors of Current Biology Ltd. for giving us permission to adapt our review.

Literature Cited

1. Sigler, P. B. *Ann. Rev. Biophys. Bioeng.* **1975**, *4*, 477-527.
2. Westhof, E.; Michel, F. *Prediction and experimental investigation of RNA secondary and tertiary foldings*; Westhof, E.; Michel, F., Ed.; IRL Press at Oxford University Press, 1994, pp 25-51.
3. Woese, C. R.; Pace, N. R. *Probing RNA structure, function, and history by comparative analysis*; Woese, C. R.; Pace, N. R., Ed.; Cold Spring Harbor Laboratory Press: Cold Spring Harbor, NY, 1993, pp 91-117.
4. Ehresmann, C.; Baudin, F.; Mougel, M.; Romby, P.; Ebel, J. P.; Ehresmann, B. *Nucleic Acids Res.* **1987**, *15*, 53-71.
5. Krol, A.; Carbon, P. *Methods Enzymol.* **1989**, *180*, 212-227.
6. Symons, R. H. *Anun. Rev. Biochem.* **1992**, *61*, 641-71.
7. Pley, H. W.; Flaherty, K. M.; McKay, D. B. *Nature.* **1994**, *372*, 68-74.
8. Scott, W. G.; Finch, J. T.; Klug, A. *Cell.* **1995**, *81*, 991-1002.
9. Jaeger, L.; Westhof, E.; Michel, F. *J. Mol. Biol.* **1993**, *234*, 331-346.
10. Tanner, M. A.; Cech, T. R. *RNA* **1996**, *2*, 74-83.
11. Levinthal, C. *J. Chim. Phys.* **1968**, *65*, 44-47.
12. Zwanzig, R.; Szabo, A.; Bagchi, B. *Proc. Natl. Acad. Sci. U.S.A.* **1992**, *89*, 20-22.
13. Pauling, L.; Corey, R. B.; Branson, H. R. *Proc; Natl. Acad. Sci. U.S.A.* **1951**, *37*, 205-211.
14. Sundaralingam, M. *The concept of a conformationally 'rigid' nucleotide and its significance in polynucleotide conformational analysis*; Sundaralingam, M., Ed.: Jerusalem, 1973, pp 417-456.
15. Tanner, N. K.; Schaff, S.; Thill, G.; Petit-Koskas, E.; Crain-Denoyelle, A.; Westhof, E. *Current Biology* **1994**, *4*, 488-98.
16. Brunel, C.; Romby, P.; Westhof, E.; Ehresmann, C.; Ehresmann, B. *J. Mol. Biol.* **1991**, *221*, 293-308.
17. Michel, F.; Westhof, E. *J. Mol. Biol.* **1990**, *216*, 585-610.
18. Draper, D. E. *Nature Struct. Biol.* **1996**, *3*, 397-400.
19. Westhof, E.; Wesolowski, D.; Altman, S. *J. Mol. Biol.* **1996**, *258*, 600-613.
20. Arnott, S. *The structure of transfer RNA*; Arnott, S., Ed.; Pergamon Press: Oxford, 1971; Vol. 22, pp 181-213.
21. Felden, B.; Florentz, C.; Giegé, R.; Westhof, E. *RNA* **1996**, *2*, 201-212.
22. Perrotta, A. T.; Been, M. D. *Nature* **1991**, *350*, 434-436.
23. Felden, B.; Florentz, C.; Giegé, R.; Westhof, E. *J. Mol. Biol.* **1994**, *235*, 508-531.
24. Kim, S.-H.; Cech, T. R. *Proc. Natl. Acad. Sci. U.S.A.* **1987**, *84*, 8788-8792.
25. Walter, A. E.; Turner, D. H.; Kim, J.; Lyttle, M. H.; Müller, P.; Mathews, D. H.; Zuker, M. *Proc. Natl. Acad. Sci. U.S.A.* **1994**, *91*, 9218-9222.

26. Krol, A.; Westhof, E.; Bach, M.; Lührmann, R.; Ebel, J.-P.; Carbon, P. *Nucleic Acids Res.* **1990**, *18*, 3803-3811.
27. Duckett, D. R.; Murchie, A. I. H.; Lilley, D. M. J. *Cell* **1995**, *83*, 1027-36.
28. Wang, Y.; Murphy, F. L.; Cech, T. R.; Griffith, J. D. *J. Mol. Biol.* **1994**, *236*, 64-71.
29. Westhof, E.; Dumas, P.; Moras, D. *J. Mol. Biol.* **1985**, *184*, 119-145.
30. Westhof, E.; Romby, P.; Romaniuk, P. J.; Ebel, J. P.; Ehresmann, C.; Ehresmann, B. *J. Mol. Biol.* **1989**, *207*, 417-31.
31. Saenger, W. *tRNA - A treasury of stereochemical information*; Saenger, W., Ed.; Springer Verlag: New York, 1984, pp 331-349.
32. Pleij, C. W. A. *TIBS* **1990**, 143-7.
33. Westhof, E.; Jaeger, L. *Curr. Op. Struct. Biol.* **1992**, *2*, 327-333.
34. Shen, L. X.; Tinoco, I. *J. Mol. Biol.* **1995**, *247*, 963-978.
35. Zarrinkar, P. P.; Williamson, J. R. *Nature Struct. Biol.* **1996**, *3*, 432-438.
36. Michel, F.; Jaquier, A.; Dujon, B. *Biochimie* **1982**, *64*, 867-881.
37. James, B. D.; Olsen, G. J.; Liu, N. R.; Pace, N. R. *Cell* **1988**, *52*, 19-26.
38. Persson, C.; Wagner, E. G. H.; Nordström, K. *EMBO* **1990**, *9*, 3767-3775.
39. Tomizawa, J. *J. Mol. Biol.* **1990**, *212*, 683-694.
40. Tuerk, C.; Gauss, P.; Thermes, C.; Groebe, D. R.; Gayle, M.; Guild, N.; Stormo, G.; d'Aubenton-Carafa, Y.; Uhlenbeck, O. C.; Tinoco, I.; Brody, E.; Gold, L. *Proc. Natl. Acad. Sci. U.S.A.* **1988**, *85*, 1364-1367.
41. Oubridge, C.; Ito, N.; Evans, P. R.; Teo, C.-H.; Nagai, K. *Nature* **1994**, *372*, 432-8.
42. Gregorian, R. S. J.; Crothers, M. D. *J. Mol. Biol.* **1995**, *248*, 968-84.
43. Jaeger, L.; Michel, F.; Westhof, E. *J. Mol. Biol.* **1994**, *236*, 1271-1276.
44. Pley, H. W.; Flaherty, K. M.; McKay, D. B. *Nature.* **1994**, *372*, 111-113.
45. Costa, M.; Michel, F. *EMBO J.* **1995**, *14*, 1276-85.
46. Iwai, S.; Pritchard, C.; Mann, D. A.; Karn, J.; Gait, M. J. *Nucleic Acids Res.* **1992**, *24*, 6465-6472.
47. Mougel, M.; Allmang, C.; Eyermann, F.; Cachia, C.; Ehresmann, B.; Ehresmann, C. *Eur. J. Biochem.* **1993**, *215*, 787-92.
48. Bhattacharyya, A.; Murchie, A. I. H.; Lilley, D. M. *Nature* **1990**, *343*, 484-487.
49. Zacharias, M.; Hagerman, P. J. *J. Mol. Biol.* **1996**, *257*, 276-89.
50. Wimberly, B.; Varani, G.; Tinoco, I., Jr *Biochemistry* **1993**, *32*, 1078-1087.
51. Wang, J. F.; Downs, W. D.; Cech, T. R. *Science* **1993**, *260*, 504-508.
52. Lehnert, V.; Jaeger, L.; Michel, F.; Westhof, E. *Chem. & Biol.* **1996**, *3*, 993-1009.
53. Jaeger, L.; Westhof, E.; Michel, F. *Biochimie* **1996**, *78*, 466-473.
54. Massire, C.; Jaeger, L.; Westhof, E. *RNA* **1997**, *3*, 553-556.

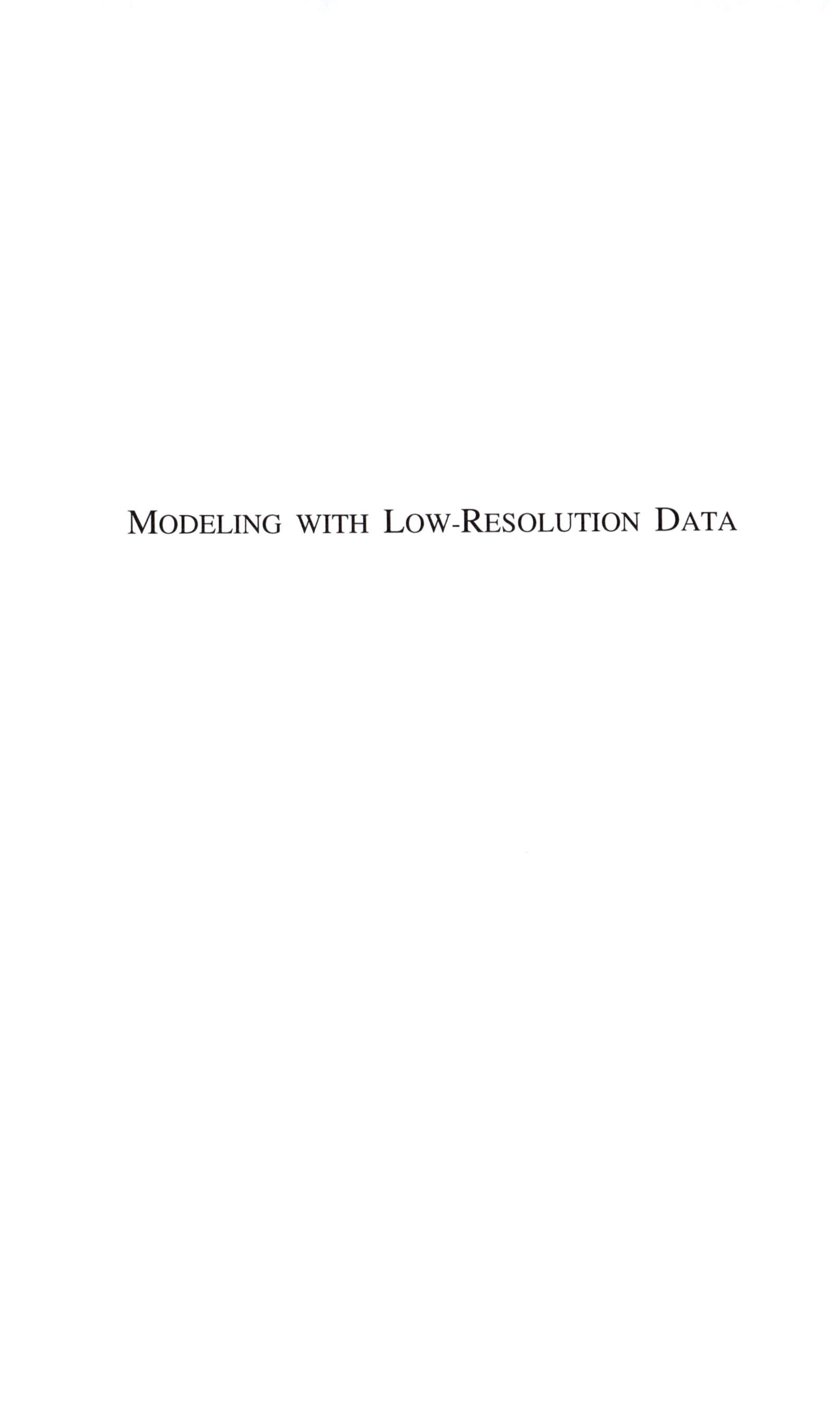

Modeling with Low-Resolution Data

Chapter 22

Hairpin Ribozyme Structure and Dynamics

A. R. Banerjee[1], A. Berzal-Herranz[2], J. Bond[1], S. Butcher[3], J. A. Esteban[4], J. E. Heckman[1], B. Sargueil[5], N. Walter[1], and J. M. Burke[1]

[1]Department of Microbiology and Molecular Genetics, The Markey Center for Molecular Genetics, University of Vermont, Burling, VT 05405

A wide range of experimental methods have been used to explore the structure and function of the hairpin ribozyme. In vitro selection methods has provided comprehensive information on RNA secondary structure and the identity of essential bases within the two internal loops of the ribozyme-substrate complex. Crosslinking analysis has identified a tertiary structure motif within ribozyme loop B that is also found in rRNA. The ribozyme contains two domains that fold independently. A sharp bend at the junction between the two domains is necessary for the two domains to interact in the active complex. Biphasic kinetic behavior in pre-steady state cleavage kinetics results from partitioning of the ribozyme between the active, bent conformer, and an inactive conformer in which the two domains are coaxially stacked, thus preventing their interaction. Novel strategies have been used to reconstruct the ribozyme in such a way that folding into the active conformer is strongly favored.

The hairpin ribozyme is a small (50 nt) catalytic RNA molecule that catalyzes a reversible RNA cleavage reaction, using a magnesium-dependent transesterification reaction which generates 5'-OH and 2',3'-cyclic phosphate termini (*1-3*, Gait, M.J. et al. *Antisense and Nucleic Acids Drug Development*, in press). In nature, the ribozyme is responsible for RNA processing reactions that are essential for replication of RNA molecules associated with plant RNA viruses. Trans-acting ribozymes have been developed through deletion of the natural sequences that are not germane to RNA processing. Such ribozymes provide excellent model systems for studying the biochemistry of RNA processing, and biological

[2]Current address: Instituto de Parasitologia y Biomedicinia, CSIC, Ventanilla 11, 18001 Granada, Spain

[3]Current address: Molecular Biology Institute, University of California at Los Angeles, East Los Angeles, CA 90095–1570

[5]Current address: Centre National de la Recherche Scientifique, CGM, Avenue de Laterasse, 91190 Gif-Sur-Yvette, France

catalysis by RNA. Because catalytic activity is a direct consequence of the folded structure of RNA, ribozymes are excellent model systems for studying RNA structure and dynamics.

Results

In vitro selection. To elucidate the secondary structure of the ribozyme-substrate complex, and to identify essential bases, we developed a rapid and powerful in vitro selection method (*4*). This method essentially permits investigators to do both genetics and molecular evolution in the test tube, by selectively replicating molecules with user-defined catalytic properties from highly complex populations (> 10^{12}) of variants generated through chemical synthesis. Results define a secondary structure consisting of four short helical elements consisting of canonical base pairs (Figure 1) (Gait, M.J. et al. *Antisense and Nucleic Acids Drug Development*, in press). Two internal loops are present, one of which contains the substrate cleavage site. Interestingly, the identity of nearly all bases within the loops is important for activity; single base substitutions at each of these sites strongly inhibit catalysis (*5-7*).

Photoreactive motif within internal loop B is shared with viroids and rRNA. A covalent crosslink between G21 and U42 is induced with high efficiency upon modest UV irradiation of the unmodified RNA (*8*). It is likely that the photoreactive species reflects the active conformation of the molecule, because structural modifications that increase photocrosslinking efficiency to greater than 90% are accompanied by a corresponding increase in catalytic activity. Similar high-efficiency crosslinks are found in RNA loops within viroids, and within 5S and 28S rRNA. In all cases, the loops have strikingly similar sequences and apparently utilize identical crosslinking sites. Structural analysis by NMR spectroscopy shows that the structures of the two rRNA loops are essentially identical; both structures include noncanonical base pairing and a local reversal of chain direction lead to base stacking across the loop; this cross-strand stacking leads to the UV-induced crosslink (*9-10*). Chemical modification analysis of the hairpin ribozyme is entirely consistent with that expected on the basis of the rRNA NMR structures. Therefore, we have modeled the three-dimensional structure of the relevant sequences within loop B by homology to the rRNA models. The structure of the segment of loop B between the photoreactive motif and helix 4 is unknown.

Escape from conformational hell. RNA molecules in general, and ribozymes in particular, are notorious for the formation of multiple conformations, usually one active species and numerous inactive conformers. This makes both the prediction and experimental determination of biologically-relevant three dimensional structures very difficult. We devoted considerable effort to the identification of specific misfolded structures and to redesign of the ribozyme so as to minimize their formation.

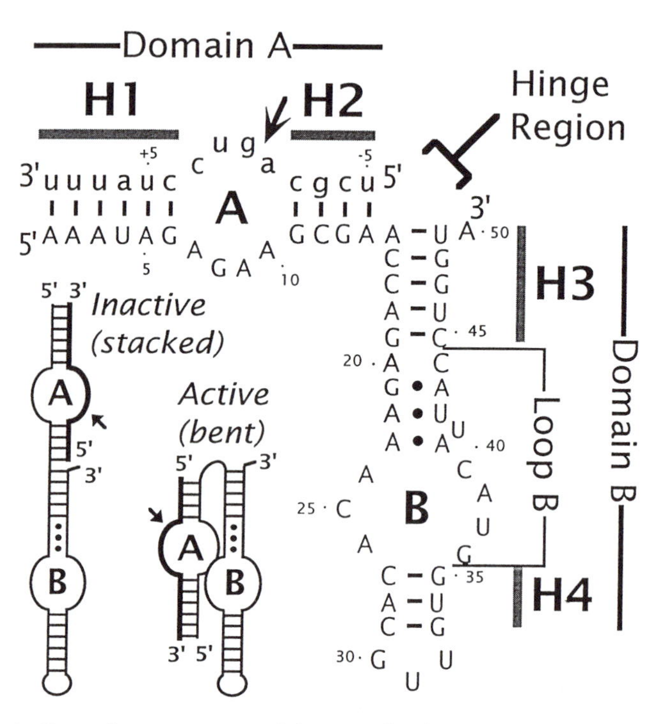

Figure 1. Secondary structure of the complex between the hairpin ribozyme and its cognate substrate. Ribozyme and substrate sequences are in upper-case and lower-case, respectively. Ribozyme nucleotides are numbered from 1 to 50. Substrate nucleotides are numbered with negative numbers 5' to the cleavage site, and vice versa. A short arrow indicates the cleavage-ligation site. The four helical domains are designated as H1-H4, and the two internal loops as A and B. Proposed non-canonical base pairs at the top of loop B (13) are indicated. Domain A is defined as the helical element comprising H1, loop A and H2. Similarly, domain B stands for the H3-loop B-H4 element. Three base pairs (one in H1 and two in H2) of the naturally occurring sequences have been changed to minimize self-complementarity of the substrate (6, 9, 11). A rate-enhancing U39C mutation was also introduced. The putative hinge region is indicated. **Inset.** Schematic cartoon showing the two alternative conformations of the hairpin ribozyme. See text for details.

Specific problems with the wild type ribozyme were identified and corrected. First, the wild type substrate was found to assume a hairpin structure at low RNA concentrations, and to form dimers at higher concentrations. We solved this problem by redesigning the substrate and the ribozyme's substrate-binding domain so as to prevent this misfolding (Heckman, J.E. et al., University of Vermont, unpublished data). Second, helix 4 of the ribozyme is metastable. Consequently, the ribozyme can dimerize even at relatively low concentrations, through the formation of intermolecular versions of domain B (*8*). Although these dimers are catalytically active, we chose to engineer modified ribozymes that have increased activity and do not dimerize, through stabilization of helix 4 (*11*).

In vitro selection experiments showed that the mutation U39C was recovered as a second-site revertant of several partially inactivating mutations within the ribozyme. This mutation is a general activating mutation that may assist in formation of the active structure within loop B. C39 is now routinely incorporated into all ribozymes in our lab.

Two independently-folding domains. A variety of methods were applied to examine the structure of intact hairpin ribozymes, ribozyme-substrate complexes, and ribozyme derivatives in which some sequences were deleted. Crosslinking studies, gel mobility analysis, and chemical modification experiments suggested that the ribozyme consists of two independently-folding domains. Domain A is a duplex between the substrate and substrate-binding strand (helix 1, loop A and helix 2), while domain B consists of helix 3, loop B and helix 4.

Studies in which the effect of inserting variable-length linkers between the 5' end of the substrate and the 3' end of the ribozyme showed that coaxial stacking of helices 2 and 3 prevented intramolecular cleavage activity, and strongly suggested that a sharp bend must occur between these helices to form the active configuration (*12-13*).

We were able to experimentally demonstrate a functional interaction between the two domains, through reconstitution of cleavage activity (*14*). Success of this experiment depended on having previously solved the misfolding problems described above. The separated domains are able to associate in solution to assemble a catalytically-proficient complex. The observed rate of the reaction approaches k_{cat} of the unmodified ribozyme, but only at very high RNA concentrations. This demonstrates the presence of tertiary interactions between the two domains, but suggests that they are quite weak. The identity of these interactions has not yet been determined.

Pre-steady state kinetic analysis. To increase the depth of our understanding of the steps involved in the catalytic cycle, we have recently completed a comprehensive analysis of the rates of individual steps of the cleavage and ligation reaction pathway (15). In addition to utilizing the well-established kinetic methods for ribozymes that involve quantitative analysis of reactions using radiolabeled substrates, we developed real-time assays using fluorescent substrates and substrate analogs (16). These assays monitor quenching and dequenching of 3'-fluorescein labeled substrates and analogs upon binding and dissociation, as well as loss of a fluorescence energy transfer signal upon cleavage of substrate containing both 3'-fluorescein and 5'-hexachlorofluorescein labels.

Our pre-steady state kinetic analysis was conducted with at 25°C in a simple buffer containing 12 mM magnesium ions, and yielded very interesting results (*15*), several of which have been independently determined (*17*) (Figure 2). Binding of substrate is very fast (4 x 10^8 M^{-1} min^{-1}), essentially the same rate as that observed for the diffusion-limited formation of an analogous A-RNA duplex. In contrast, substrate dissociation is very slow (<0.01 min^{-1}), so that the substrate essentially binds and then waits for a reaction to happen. Cleavage is the rate-limiting step of the reaction, and occurs at a rate similar to that of other small ribozymes under similar conditions (0.2 min^{-1}). The rate of the ligation reaction (3 min^{-1}) greatly exceeds that for cleavage; this is decidedly different from other small ribozymes like the hammerhead. Because dissociation of cleavage products is very fast (>>3.0 min^{-1}), cleavage reactions proceed nearly to completion under typical laboratory conditions.

Cleavage and ligation kinetics follow biphasic behavior. The time courses of single-turnover cleavage and ligation reactions are described by double-exponential equations; the fit of the observed data to single-exponential equations is much less precise (Esteban, J. et al., *J. Biol. Chem.*, in press). Curve-fitting provides rate constants for both phases of the reaction, as well as coefficients describing partitioning between "fast" and "slow" forms. These observations strongly suggest that two populations of ribozymes coexist, each with distinctive kinetic characteristics. All four combinations of ribozymes and substrates show biphasic kinetics, although the distribution between "fast" and "slow" forms varies from construct to construct. Increasing ribozyme activity by stabilizing domain B is accompanied by an increase in the fraction of substrate that reacts in the fast phase of the reaction. When the kinetics of ligation reactions were characterized, biphasic kinetic behavior was also observed.

A chase experiment was conducted in order to explore the basis for biphasic kinetics (*15*). In this experiment, cleavage kinetics were monitored in reactions initiated with a trace quantity of radiolabeled substrate bound to the ribozyme, with and without a chase step consisting of a large molar excess of unlabeled substrate. We found that the chase step specifically ablated the slow phase of the reaction, but did not alter the rate or extent of the fast reaction phase. From this result, we inferred that the slow phase of the reaction results from substrate that is bound to an inactive form of the ribozyme. For this substrate to react, it must dissociate and then bind to an active form. Measurements of the interchange between active and inactive forms of the ribozyme indicate that exchange between the two forms is insignificant over the time course of these reaction.

Structural basis for heterogeneous kinetics. The kinetic model described above strongly implicates the presence of two conformational isomers of the

$$Rz + S$$

$k_{off} \approx 0.008\ min^{-1}$ VERY SLOW ⇅ $k_{on} = 3 \times 10^{8} M^{-1}\ min^{-1}$ ($K_D \approx 30$ pM)

$$Rz \bullet S$$

$k_{lig} = 3.0\ min^{-1}$ FAST ⇅ $k_{cleav} = 0.2\ min^{-1}$ SLOW

$$Rz \bullet 5'P \bullet 3'P$$

k_{on} ($K_M = 1\ \mu M$) ⇅ $k_{off} >> 3.0\ min^{-1}$ VERY FAST

$$Rz + 5'P + 3'P$$

Figure 2. Kinetic mechanism of the hairpin ribozyme. Rate constants were derived from pre-steady-state and multiple turnover reactions, using radiolabeled and fluorescent substrates as described (Esteban et al., in press)

hairpin ribozyme. Both conformers can clearly bind substrate, but one has catalytic activity and the other does not. What is the structural basis for this difference in activity? Linker-insertion studies, in which the 5' end of the substrate was covalently tethered to the 3' end of the ribozyme, had shown that when helix 2 was forced to coaxially stack upon helix 3, self cleavage activity was eliminated (*12-13*). These results, together with the domain separation experiment described above, suggested that the active form of the ribozyme must contain a sharp bend at the interface between helices 2 and 3. We reasoned that the ribozyme-substrate complex might partition between two conformers, an inactive, stacked form stabilized by stacking of helices 2 and 3, and an active, bent form stabilized by as-yet unidentified tertiary interactions between domains A and B.

We tested this hypothesis in two ways. First, we noted that the naturally-occurring base pairs at the helical interface (A-U A-U) are expected to contribute relatively weak stacking energies. Substitution of these base pairs with ones that more strongly favor stacking results in a significant decrease in the rate of the reaction, and decreases the amount of substrate that reacts in the fast phase of the reaction. Second, we developed a enzymatic method to physically separate the putative stacked and bent conformers of the ribozyme-substrate complex (Esteban, J. et al., University of Vermont, unpublished data). In this assay, T4 RNA ligase was expected to covalently link substrate to the 3' end of the ribozyme in the putative stacked form, and was not expected to react with the bent configuration. When T4 RNA ligase was added to a single-turnover cleavage reaction, there was no affect on the fast phase of the reaction. However, the slow phase of the reaction was completely quenched, so that the time course could be accurately described as a single exponential with an associated rate and extent identical to that of the fast phase of the reaction in the absence of RNA ligase. The fraction of substrate that normally reacts in the slow phase of the reaction was instead trapped in a covalent complex.

Reengineering the hairpin ribozyme for homogeneous folding. The results described above clearly demonstrate that the slow phase of the cleavage reaction catalyzed by the hairpin ribozyme results from the formation of an inactive conformer in which the two domains are stacked upon one another, and so cannot interact to form the functional complex. We used two distinct strategies in an attempt to develop modified versions of the ribozyme that would fold uniformly into the active structure (Esteban, J. et al., Cold Spring Harbor, unpublished data) (Heckman, J. et al., University of Vermont, unpublished data). First, we designed a hairpin ribozyme in which a three-way junction replaces the H2-H3 interface in a manner that is expected to inhibit their coaxial stacking. Three cytidine residues were kept unpaired in each strand at the connections with H2 and H3 to increase the flexibility of the junction. A second approach was explored by introducing a novel non-nucleotidic linker into the "hinge" region, replacing A14 with an ortho-hydroxymethyl benzene moiety, expected to impose a sharp dihedral angle between helices 2 and 3. Molecular modeling studies indicate that both modified ribozymes are capable of forming structures similar to that of the active, bent form of the unmodified hairpin ribozyme.

To evaluate partitioning of the modified ribozymes between active and inactive conformations, we carried out cleavage reactions under single-turnover conditions in the presence and absence of RNA ligase. Results showed that the reactions catalyzed by the new ribozymes were essentially monophasic. The very small amount of residual slow phase was completely eliminated by ligase. We believe that the development of conformationally homogeneous ribozymes will be useful for simplifying in vitro kinetic studies, for structural work aimed at defining interdomain contacts, and for optimizing the activity of ribozymes in cellular RNA-inactivation work.

Conclusion

Considerable progress has been made towards understanding the structure and activity of the hairpin ribozyme. Recent results have been achieved using a wide range of laboratory methods accessible to most biochemistry and molecular biology laboratories. While the hairpin and hammerhead ribozymes catalyze analogous reactions, it is now quite apparent that the two ribozymes accomplish this task using very different structures and catalytic strategies. We have defined alternative conformers using kinetic methods, and then developed and tested structural models for the kinetic heterogeneity. The use of this information to rationally modify the ribozyme to eliminate both the structural and kinetic heterogeneity may be very beneficial for structural determination and modeling, and may serve as a guide to others studying the biological activity of structurally heterogeneous RNA molecules.

Literature Cited

1. Burke, J. M. *Biochem. Soc. Trans.* **1996,** *24,* 608-615.
2. Burke, J. M. *Nucleic Acids & Molecular Biology* **1994,** *8,* 105-118.
3. Burke, J. M.; Butcher, S. E.; Sargueil, B. Nucleic Acids & Molecular Biology **1996,** *10,* 129-143.
4. Berzal-Herranz A, Joseph S & Burke JM. 1992. In vitro selection of active hairpin ribozymes by sequential RNA-catalyzed cleavage and ligation reactions. Genes Dev. 6, 129-134.
5. Berzal-Herranz, A.; Joseph, S.; Chowrira, B. M.; Butcher, S. E.; Burke, J. M. *EMBO J.* **1993,** *12,* 2567-2574.
6. Grasby, J. A.; Mersmann, K.; Singh, M.; Gait, M. J. *Biochemistry* **1995,** *34 (12),* 4068-76.
7. Joseph, S.; Berzal-Herranz, A.; Chowrira, B. M.; Butcher, S. E.; Burke, J. M. *Genes Dev.* **1993,** *7,* 130-138.
8. Butcher, S. E.; Burke, J. M. *Biochemistry* **1994,** *33,* 992-999.
9. Wimberly, B.; Varani, G.; Tinoco, I. *Biochemistry* **1993,** 32, 1078-1087.

10. Szewczak, A. A.; Moore, P. B.; Chan, Y. L.; Wool, I. G. *Proc. Natl. Acad. Sci. USA* **1993,** *90,* 9581-9585.
11. Sargueil, B.; Pecchia, D. B.; Burke, J. M. *Biochemistry* **1995,** *34,* 7739-7748.
12. Feldstein, P. A.; Bruening, G. *Nucleic Acids Res.* **1993,** *21,* 1991-1998.
13. Komatsu, Y.; Kanzaki ,I.; Koizumi, M.; Ohtsuka, E. *J. Mol. Biol.* **1995,** *252,* 296-304.
14. Butcher, S. E.; Heckman, J. E.; Burke, J. M. *J. Biol. Chem.* **1995,** *270,* 29648-29651.
15. Esteban J.E., Banerjee A.R., Burke, J.M. 1997. Kinetic mechanism of the hairpin ribozyme: Identification and characterization of two nonexchangeable conformations. J. Biol. Chem. 272, 13629-13639.
16. Hegg, L. A.; Fedor, M. J. *Biochemistry* **1995,** *34,* 15813-15828.
17. Walter, N. G.; Burke, J. M. *RNA* **1997,** *3,* 392-404.

Chapter 23

Molecular Modeling Studies on the Ribosome

Stephen C. Harvey, Margaret S. VanLoock, Thomas R. Rasterwood, and Robert K.-Z. Tan

Department of Biochemistry and Molecular Genetics, University of Alabama at Birmingham, Birmingham, AL 35294–0005

Macromolecular models should have a level of resolution commensurate with the quality and quantity of available data. We have developed a package of programs for building and refining models of nucleic acids and protein-nucleic acid complexes, with different levels of detail in different parts of the molecule. The method incorporates data from a variety of experiments. Our RNA models have atomic detail in some regions and lower resolution (typically one pseudoatom per nucleotide) in others. We are using this approach to refine models for the 30S subunit of the *E. coli* ribosome, with emphasis on the decoding site, where the 16S rRNA holds the mRNA and two tRNAs.

Molecular dynamics (MD), Monte Carlo, energy minimization, and related algorithms are used for refining macromolecular structures, using data from x-ray crystallography and NMR. MD can also be employed to simulate macromolecular motions over timescales ranging from femtoseconds to nanoseconds (*1*).

The principal advantage of these methods lies in the great detail that they provide: the position of every atom is specified, and, in the case of MD, these positions are monitored at intervals of time that are short by comparison to the period of the highest frequency motions, the vibrations of covalent bondlengths. Unfortunately, this also represents the major weakness of these methods, because the level of detail means that these algorithms are computationally very demanding.

In recent years, various "reduced representations" have been developed for modeling large nucleic acids, particularly for large double-helical DNA molecules (*2-6*). All of these methods replace the atomic description by a lower resolution description of the structure.

Pseudoatom Representation of Large Nucleic Acids

Our pseudoatom approach follows the "united atom" tradition of molecular mechanics models (*1*). In this method, most of the hydrogen atoms in all-atom models are eliminated from the model by combining them with the heavy atoms to which they are covalently attached. For example, the four atoms of a methyl group can be represented by a single united atom whose van der Waals radius and total charge are chosen so that the united atom reproduces with reasonable accuracy the average behavior of the methyl group in its interactions with other chemical groups in the molecule. Since hydrogen atoms comprise about half the atoms in a macromolecule, and since the number of nonbonded interactions that must be calculated scales as the square of the number of atoms in the model, the united atom representation provides a substantial reduction in the computational burden for modeling macromolecular systems.

To model large closed circular DNA molecules, we use three point masses to represent each basepair (*6*). We chose this representation because three points define a plane, so we can monitor the variations in the angles between successive basepair planes, i.e., variations in roll, tilt, and twist. The three point masses do not correspond exactly to sets of atoms in the real molecule, so they are not properly called united atoms. Consequently, we call the point masses "pseudoatoms". (A united atom representation might use one pseudoatom for each phosphate group, one for each deoxyribose, and one for each base, but this would require six united atoms per basepair, and it would be difficult to monitor the roll, tilt, and twist angles that are of interest in our studies.)

We extended this approach to low resolution models for RNAs, because of our interest in developing approaches for refining the structure of the ribosome (*7-8*). The *E. coli* ribosome contains 55 proteins and three large RNAs, and its total molecular mass is about 2.5×10^6. As a consequence, its structure cannot be determined by NMR, and it remains an unsolved challenge to the x-ray crystallographers. There is, however, a large and growing body of low resolution data on ribosomal structure, from such experimental techniques as footprinting by chemical and enzymatic probes, crosslinking, cryo-electron microscopy, immunoelectron microscopy, and neutron diffraction. Our goal was to develop molecular mechanics refinement protocols to build three-dimensional models compatible with the low resolution data. The methods are similar to those used in the refinement of structures by x-ray crystallography and NMR. The data are of lower resolution, and there are not enough data to define the structure uniquely, so a low resolution representation, rather than an all-atom description, is appropriate. We want to specify the position of each nucleotide, using the protein positions from neutron diffraction as a scaffold. Thus, we chose a pseudoatom representation with one pseudoatom per nucleotide and one large spherical pseudoatom per protein.

Our low resolution RNA modeling protocol (*7*) generates a series of models for the structure of interest. The method permits us to identify conflicts in the experimental data, by a covariation analysis of the extent to which each experimental constraint is or is not satisfied by each model. By examining the variation in position of a given nucleotide in the different models, the protocol provides a statistical statement about the uncertainty in position of each nucleotide, in analogy with the temperature factors obtained from x-ray crystallography.

Our current low resolution model of the ribosome (*8*) has an average uncertainty of about ±15Å in the position of each nucleotide, providing additional *post hoc* justification for the pseudoatomic representation. Some regions of the model are reasonably tightly constrained, with uncertainties of less than ±10Å, while others have uncertainties of more than ±40Å. The latter are poorly defined, because of the absence of data to identify which ribosomal proteins or which other parts of the rRNA must lie nearby. The principal results of this study are the definition of the overall organization of the small ribosomal subunit, along with quantitative estimates of the positional uncertainties associated with the model.

The RNA pseudoatom method has also been applied to the building and refinement of models for the catalytic RNA of RNase P (*9*).

Motivation for Mixed Resolution Models

While the methods described above are useful for describing the global organization of the ribosome, there are many important questions that can only be answered by inspecting structures at atomic resolution. Among these are: What is the recognition mechanism at the decoding site where the messenger RNA is read, i.e., how do the two tRNAs, the mRNA, and the ribosomal RNAs and proteins interact? What is the catalytic mechanism by which the two aminoacylated ends of the tRNAs are juxtaposed so that the growing peptide chain adds the next amino acid? How is the energy of GTP hydrolysis converted into the mechanical energy required to move the mRNA forward three nucleotides? Similarly, what drives the tRNA into the A-site, then successively to the P-site, the E-site, and off the ribosome? What are the specific interactions by which various antibiotics interrupt bacterial ribosome function?

To answer these questions, one has two options. One can wait until data at atomic resolution are available for the entire ribosome. Alternatively, the limited data that are currently available can be used to examine parts of the ribosomal structure at atomic resolution, using the low resolution models described in the previous section as a framework for the overall organization of the structure. The result will be a picture of the structure with low resolution description in most regions, and with atomic detail in areas of particular importance. We are developing this strategy, beginning with atomic detail in the decoding site of the small ribosomal subunit.

Atomic Resolution Models for the Ribosomal Decoding Site

The Complex Between mRNA and the Two tRNAs. With no data except the crystal structure of transfer RNA, one can define the range of possible structures for the complex between messenger RNA and the tRNAs at the A- and P-sites, because of two simple facts. First, we know that the two tRNA anticodons are close together, because they interact simultaneously with successive codons on the mRNA. Second, the 3' termini of the two tRNAs are also close together, because the P-site tRNA bears the peptide chain that is to be covalently linked to the amino acid on the A-site tRNA.

Broadly speaking, there are two possible configurations for the mRNA/tRNA complex, called R and S for Rich and Sundaralingam, who first proposed them (*10,11*). The reader can visualize these two models by representing the A-site tRNA with the thumb and index finger of the left hand -- keeping these at a right angle to one another -- and by representing the P-site tRNA with the thumb and index finger of the right hand, in a similar conformation. The tips of the thumbs must be kept in contact, representing the juxtaposition of the two anticodons when they are bound to the mRNA. At the same time, the tips of the index fingers should be kept together, to represent the contact between amino acids covalently bound at the 3' terminus of each tRNA. Several experiments have shown that the two tRNAs lie in fairly close contact across their entire L-shaped bodies. You can achieve this in two ways. One is to bring your palms together and interlace the middle, ring, and little fingers of your two hands, so that your thumbs and index fingers point toward you; this gives an S model. Or you can bring the backs of your hands together, so that the thumbs and index fingers point away from you; this represents the R model.

An examination of simple physical models for the R and S configurations convinced us that S models were to be favored. They lay the mRNA in a fairly smooth right-handed wrapping around the anticodon regions of the two tRNAs, while R models leave the 5' end of the mRNA buried between the two tRNAs, requiring a very sharp reversal of the backbone to prevent tangling during the translocation step. Therefore, our first investigation of the complex between mRNA and two tRNAs at atomic resolution examined S models (*12*). A similar solution was proposed by McDonald and Rein (*13*). Nevertheless, Valery

Lim and coworkers concluded that the initial crosslinking data between the tRNAs and the ribosomal subunits favored the R model (*14*).

We re-examined this question, reporting a series of R and S models (*15*) that we built using the MC-SYM program(*16-17*). We found that only the S model is compatible with distances within the tRNA/mRNA complex measured by fluorescence energy transfer (*18-19*) and chemical crosslinking (*20-21*). It is further favored by a photocrosslink between nucleotide 34 of the P-site tRNA and residue C1400 of the 16S RNA (*22-24*). It also explains the effects of substitutions in the anticodon loop of the P-site tRNA on the binding of tRNA at the A-site (*25*). Further crosslinking studies between the tRNAs and the ribosome also show that S models are correct (*26*).

Models for the Decoding Region of the 16S rRNA. The task of reading the mRNA codons at the A and P-sites by tRNA ("decoding") falls primarily to the small ribosomal subunit. In the *Escherichia coli* 30S subunit, the 16S ribosomal RNA contains a highly conserved sequence in the 1400-1500 region (*27*). This region has been examined extensively by chemical and enzymatic probing, crosslinking, and mutational experiments. Together these have demonstrated that nucleotides C1399-C1409 and G1492-1504 are the heart of the decoding center.

Purohit and Stern (*28*) demonstrated that an RNA oligonucleotide containing about 30 nucleotides and including the 1400-1500 decoding segment of the 16S rRNA can bind antibiotics. They found that the protection patterns induced by these drugs are similar to those induced in the intact 16S rRNA. They also found that the oligonucleotide could bind tRNA anticodon stem-loop analogues and poly(U) mRNA analogues, and that these, too, give similar protection patterns as in the intact 16S rRNA. The implications of the Purohit and Stern experiments are (1) that the heart of the decoding site has a definite three-dimensional structure, and (2) that this structure does not require proteins or other auxiliary factors to recognize and bind the mRNA.

The Purohit and Stern oligonucleotide is relatively small, it contains a fairly extensive secondary structure, and the chemical protection data provide a substantial body of data on the exposure of several of the bases. As a consequence, this molecule falls into the class where reasonable predictions can be made about the three dimensional structure. Given the importance of the decoding site, such predictions can suggest sequence-structure relationships that might be tested experimentally.

We decided to model the decoding site using a combination of manual methods and MC-SYM (*16-17*). The models produced by MC-SYM incorporate base pairings, both canonical and non-canonical, along with information from probing experiments. Our model of the mRNA/tRNA complex (*15*) can be used as a starting point, if we can identify the general organization of the structure and specify a few key contacts between the decoding site and the mRNA/tRNA complex.

The locations of the A- and P-sites in the rRNA are known (*22,27-29*). This provides the general orientation of the decoding site with respect to the mRNA/tRNA complex. Thus, the only remaining question about overall organization is whether the mRNA/tRNA complex is more likely to lie in the major or minor groove of the rRNA decoding site. The results of probing experiments (*29*) suggest that the major groove is the more likely binding site. This conclusion is supported by the large number of internal bulge loops in this region and the observation that such bulges tend to open the major groove in RNA duplexes (*30*).

There are two sets of key interactions between the decoding site and the mRNA/tRNA complex that allow us to enter into detailed modeling at atomic resolution.

First, the ribose-phosphate backbone of the mRNA interacts with the base pairing face of the A1492 and A1493 (*28*). We propose that a specific set of hydrogen bonds from N1 and N6 of A1492 and A1493 to the O2' hydroxyl of the mRNA backbone and O2 (pyrimidines) or N3 (purines) holds the mRNA in the A-site. We also propose that these bonds are stabilized when the base pairing between codon and anticodon give a standard A-RNA type helix; this would lead to enhanced fidelity of messenger decoding. These interactions would explain the observation that the observed chemical protections in the A-site are primarily mRNA-dependent (*31-32*).

Second, ultraviolet radiation causes a spontaneous photocrosslink between a pyrimidine at the wobble base of the P-site tRNA and C1400 in the rRNA decoding site (*22*). This is another reason to believe that the tRNA is bound in the major groove in the P-site, because the 5-6 double bond of C1400 involved in the UV dimerization faces the major groove. The myriad of possible interactions in the major groove would enhance the binding of the tRNA in the P-site, holding it in position for initiation and/or elongation steps. This would explain the observation that the P-site protections and thus binding of tRNA is not mRNA dependent (*31-32*).

Taken together, these data and assumptions have led us to a specific three-dimensional model of the decoding site when mRNA and two tRNAs are bound (Figure 1).

Comparison of the Model with Experimental Results

Our model might appear to be in conflict with the experimentally determined NMR structure of the A-site with a bound aminoglycoside antibiotic (*33*). In that structure, the drug binds in the major groove, blocking the site where we have positioned the interaction between A1492, A1493 and the message. Further, A1492 and A1493 are swung

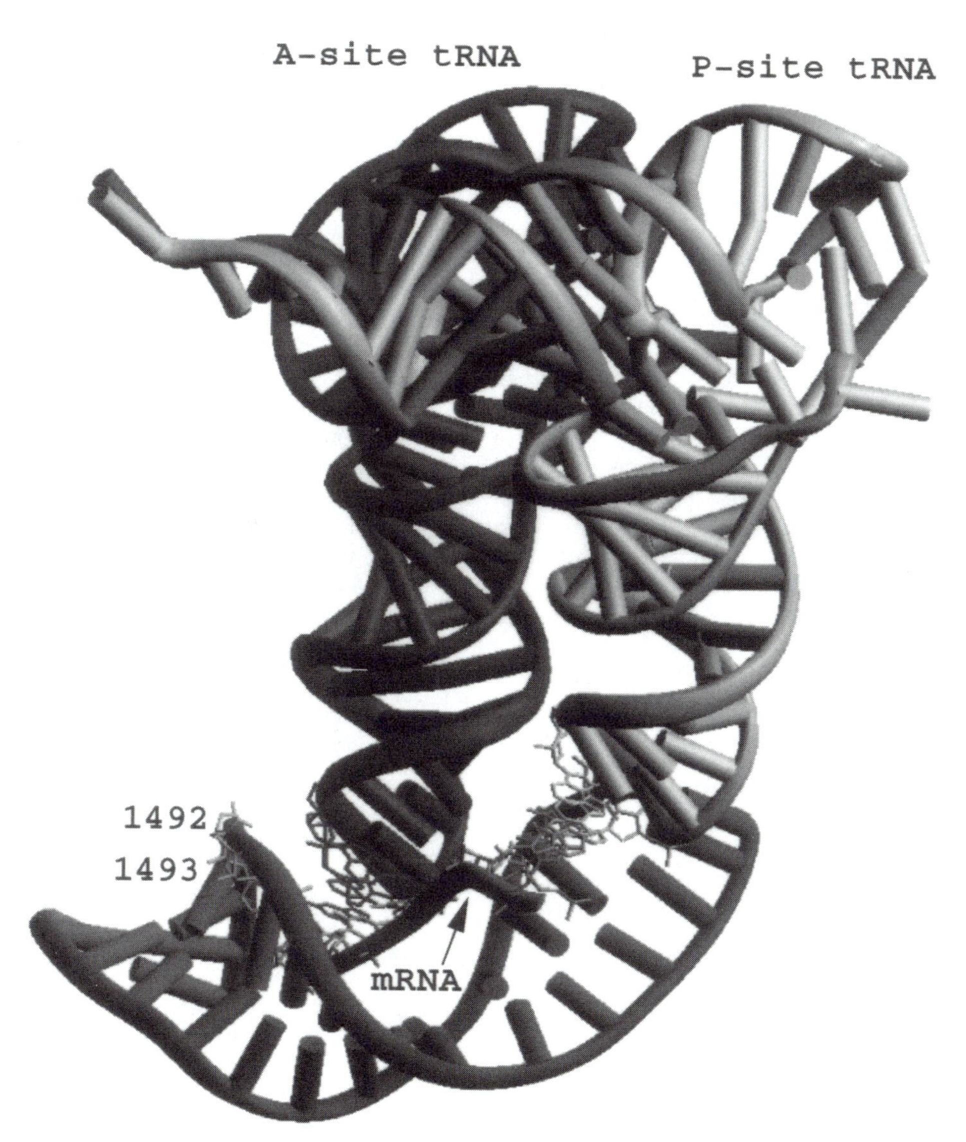

Figure 1. The A-site tRNA and P-site tRNA bind to adjacent codons on the mRNA, which is held in the major groove of the decoding region of the 16S rRNA. A specific set of hydrogen bonds from A1492 and A1493 to two nucleotides in the mRNA hold the latter into the decoding site, while the wobble base of the P-site tRNA sits directly on top of C1400. This view shows A1492, A1493, the six nucleotides of the A- and P-sites in the mRNA, and the six anticodonic nucleotides in atomic detail. Although the rest of the structure is shown at lower resolution, the actual model specifies the positions of every atom.

into the minor groove in the NMR structure. The authors of the NMR study propose that the mRNA/tRNA complex binds to the minor groove of the 16S decoding site.

It is known that translation can take place when the antibiotic is bound to the decoding site, although with reduced fidelity. Thus, a question arises about how the mRNA is held to the decoding site in the presence of the antibiotic. We propose that drug binding displaces A1492 and A1493, and that the mRNA/tRNA complex remains in the major groove and forms a series of new hydrogen bonds. Some of these occur directly between the drug and the mRNA/tRNA complex, while others are between the 16S rRNA backbone and the complex. In this model, more hydrogen bonds are formed to the mRNA/tRNA complex when drug is bound than in the drug-free case. This would offer an explanation of the effects of the drug on translation fidelity, and, we believe that the hypothesis removes the apparent contradiction between our model and the NMR results. We have developed this hypothesis into a second model that includes mRNA, two tRNAs, the 16S decoding site, and a bound paromomycin molecule.

It should also be mentioned that we have examined alternative conformational possibilities, particularly those in which the mRNA and tRNAs are bound in the minor groove of the decoding site, as suggested by the NMR data. Detailed studies using MC-SYM and manual methods have convinced us that, if the mRNA is bound to A1492 and A1493 in the minor groove as proposed by Fourmy et al. (*33*), there will be an unacceptably large distance between C1400 in the P-site and the wobble base of the P-site tRNA, unless the decoding site rRNA adopts an extraordinarily distorted structure. In the absence of distortion, this distance is over 25Å. By contrast, it is easy to find models using the major groove docking mode where the wobble base is adjacent to C1400, as required for photocrosslinking.

Future Directions

We are now incorporating the atomic model for the complex between the 16S decoding site, the mRNA and the two tRNAs into our low resolution model for the entire 30S subunit (*8*). From this beginning mixed resolution model, we will work outward from the decoding site, including additional atomic detail as the availability of data and the reasonableness of assumptions permit. The evolving model will incorporate the structures of proteins that have been studied by x-ray crystallography, fragments of RNA whose structures have been determined by crystallography or NMR, and some speculation. Until the structure of the ribosome is determined by crystallography, we believe this offers the best opportunity to understand structure-function relationships in this critical macromolecular assembly.

Acknowledgments

Supported by a grant to SCH from the National Institutes of Health (GM-53827). We are grateful to François Major for providing MC-SYM, and to Joseph D. Puglisi for providing data prior to publication. We also thank them and Joachim Frank and Richard Brimacombe for ongoing stimulating discussions.

Literature Cited

1. McCammon, J.A.; Harvey, S.C. *Dynamics of Proteins and Nucleic Acids.* Cambridge University Press: Cambridge, UK., 1987.
2. Vologodskii, A.V. *Topology and Physics of Circular DNA.* CRC Press: Boca Raton, FL., 1992.
3. Malhotra, A.; Gabb, H.A.; Harvey, S.C. *Curr. Opin. Struct. Biol.* **1993,** *3,* 241-246.
4. Schlick, T. *Curr. Opin. Struct. Biol.* **1995,** *5,* 245-262.
5. Olson, W.K. *Curr. Opin. Struct. Biol.* **1996,** *6,* 242-256.
6. Tan, R.K.Z.; Harvey, S.C. *J. Mol. Biol.* **1989,** *205,* 573-591.
7. Malhotra, A.; Tan, R.K.Z.; Harvey, S.C. *Biophys. J.* **1994,** *66,* 1777-1795.
8. Malhotra, A.; Harvey, S.C. *J. Mol. Biol.* **1994,** *240,* 308-340.
9. Harris, M.E.; Nolan, J.M.; Malhotra, A.; Brown, J.W.; Harvey, S.C.; Pace, N.R. *EMBO J.* **1994,** *13,* 3953-3963.
10. Rich, A. In: *Ribosomes.* (Nomura, M., Ed., Cold Spring Harbor Laboratory, New York) pp 871-884, 1974.
11. Sundaralingam, M.; Brennan, T.; Yathindra, N.; Ichikawa, T. In: *Structure and Conformation of Nucleic Acids and Protein-Nucleic Acid Interactions.* (Sundaralingam, M. & Rao, S. T., Eds., University Park Press, Baltimore) pp 101-115, 1975.
12. Prabhakaran, M.; Harvey, S. C. *J. Biomol. Struct. Dyns.* **1989,** *7,* 167-179.
13. McDonald, J.J.; Rein, R. *J. Biomolec. Struct. Dyns.* **1987,** *4,* 729-744.
15. Easterwood, T.R.; Major, F.; Malhotra, A.; Harvey, S.C. *Nucleic Acids Res.* **1994,** *22,* 3779-3786.
16. Major, F.; Turcotte, M.; Gautheret, D.; Lapalme, G.; Fillion, E.; Cedergren, R. *Science* **1991,** *253,* 1255-1260.
17. Gautheret, D.; Major, F.; Cedergren, R. *J. Mol. Biol.* **1993,** *229,* 1049-1064.
18. Johnson, A.E.; Adkins, H.J.; Matthews, E.A.; Cantor, C.R. *J. Mol. Biol.* **1982,** *156,* 113-140.
19. Paulsen, H.; Robertson, J.M. ; Wintermeyer, W. *J. Mol. Biol.* **1983,** *167,* 411-426.
20. Odom, O.W.; Craig, B.B.; Hardesty, B.A. *Biopolymers* **1978,** *17,* 2909-2931.
21. Fairclough, R.H.; Cantor, C.R.; Wintermeyer, W.; Zachau, H.G. *J. Mol. Biol.* **1979,** *132,* 557-573.

22. Prince, J.B.; Taylor B.H.; Thurlow, D.L.; Ofengand, J.; Zimmerman, R.A. *Proc. Natl. Acad. Sci. USA* **1982,** *79*, 5450-5454.
23. Gornicki, P.; Ciesiolka, J.; Ofengand, J. *Biochemistry* **1985,** *24*, 4924-4930.
24. Ciesiolka, J.; Gornicki, P.; Ofengand, J. *Biochemistry* **1985,** *24*, 4931-4938.
25. Smith, D.; Yarus, M. *Proc. Natl. Acad. Sci. USA* **1989,** *86*, 4397-4401.
26. Rinke-Appel, J.; Junke, N.; Brimacombe, R.; Lavrik, I.; Dokudovskaya, S.; Dontsova, O.; Bogdanov, A. *Nucleic Acids Res.* **1994,** *22*, 3018-3025.
27. Gutell, R.R.; Larsen, N.; Woese, C.R. *Microbiol. Revs.* **1994,** *58*, 10-26.
28. Purohit, P.; Stern, S. *Nature* **1994,** *370*, 659-662.
29. Douthwaite, S.; Christensen, A.; Garrett, R.A. *J. Mol. Biol.* **1983,** *169*, 249-279.
30. Weeks, K.; Crothers, D.M. *Science* **1993,** *261*, 1574-1577.
31. Moazed, D.; Noller, H.F. *Cell* **1986,** *47*, 985-994.
32. Moazed, D.; Noller, H.F. *J. Mol. Biol.* **1990,** *211*, 135-145.
33. Fourmy, D.; Recht, M.; Blanchard, S. D.; Puglisi, J. D. *Science* **1996,** *274*, 1367-1371.

Chapter 24

Modeling Unusual Nucleic Acid Structures

Thomas J. Macke and David A. Case

Department of Molecular Biology, Research Institute of Scripps Clinic, La Jolla, CA 92307

Abstract. We describe nab, a computer language that facilitates the construction of three-dimensional molecular models. The language was designed to construct models of helical and non-helical nucleic acids from a few dozen to a few hundred nucleotides in size, and uses a combination of rigid body transformations and distance geometry to create candidate structures that match input criteria. This provides a flexible way to describe nucleic acid structures at an atomic level of resolution. In addition to providing high-level control over fragment transformations and bounds manipulation for distance geometry, nab incorporates molecular mechanics minimization and molecular dynamics code (in both three and four spatial dimensions) for refinement of structures, along with close links to continuum solvation models and the AVS visualization system. Two examples are given to illustrate how molecular models may be constructed with this language: *(a)* an RNA psuedo-knot, which is created by distance geometry using data from the Nucleic Acid Database; *(b)* a simple model for the nucleosome core fragment, which illustrates laying out helical DNA along a three-dimensional space curve.

1. Introduction.

This paper describes the development and initial applications of nab, a computer language for modeling biological macromolecules. It was developed to create atomic-level models of nucleic acid structures such as stem-loops, pseudoknots, multi-armed junctions and catalytic RNAs, and to investigate biological processes that involve nucleic acids, such as hybridization, branch migration at junctions, and DNA replication.

Using a computer language to model polynucleotides follows logically from the fundamental nature of nucleic acids, which can be described as "conflicted" or "contradictory" molecules. Each repeating unit contains seven rotatable bonds (creating a very flexible backbone), but also contains a rigid, planar base which can participate in a limited number of regular interactions, such as base pairing and stacking. The result of these opposing tendencies is a family of molecules that have the potential to adopt a virtually unlimited number of conformations, yet have very strong preferences for regular helical structures and for certain types of loops.

The controlled flexibility of nucleic acids makes them difficult to model. On one hand, the limited range of regular interactions for the bases permits the use of simplified and more abstract geometric representations. The most common of these is the replacement of each base by a plane, reducing the representation of a molecule to the set of transformations that relate the planes to each other. On the other hand, the flexible backbone makes it likely that there are entire families of nucleic acid structures that satisfy the constraints of any particular modeling problem. Families of structures must be created and compared to the model's constraints. From this we can see that modeling nucleic acids involves not just chemical knowledge but also three processes—abstraction, iteration and testing—that are the basis of programming.

Molecular computation languages are not a new idea. Many interactive molecular modeling programs include a scripting language that can be used to direct conformational searches or control structure refinement. In addition, non-interactive molecular modeling codes like X-PLOR *(1)*, yammp *(2)* and MC-SYM *(3)* are actually specialized languages in which a molecular modeling problem is expressed as the appropriate program. Execution of that program creates the desired structures. Here we briefly describe some past approaches to nucleic acid modeling, to provide a context for `nab`.

1.1. Conformation build-up procedures

MC-SYM *(3-5)* is a high level molecular description language used to describe single stranded RNA molecules in terms of functional constraints. It then uses those constraints to generate structures that are consistent with that description. MC-SYM structures are created from a small library of conformers for each of the four nucleotides, along with transformation matrices for each base. Building up conformers from these starting blocks can quickly generate a very large tree of structures. The key to MC-SYM's success is its ability to prune this tree, and the user has considerable flexibility in designing this pruning process.

In a related approach, Erie *et al.* *(6)* used a Monte-Carlo build-up procedure based on sets of low energy dinucleotide conformers to construct longer low energy single stranded sequences that would be suitable for incorporation into larger structures. Sets of low energy dinucleotide conformers were created by selecting one value from each of the sterically allowed ranges for the six backbone torsion angles and χ. Instead of an exhaustive build- up search over a small set of conformers, this method samples a much larger region of conformational space by randomly combining members of a larger set of initial conformers. Unlike strict build-up procedures, any member of the initial set is allowed to follow any other member, even if their corresponding torsion angles do not exactly match, a concession to the extreme flexibility of the nucleic acid backbone. A key feature determined the probabilities of the initial conformers so that the probability of each created structure accurately reflected its energy.

1.2. Base-first strategies

An alternative approach that works well for some problems is the "base-first" strategy, which lays out the bases in desired locations, and attempts to find conformations of the sugar-phosphate backbone to connect them. Rigid-body transformations often provide a good way to place the bases. One solution to the backbone problem would be to determine the relationship between the helicoidal parameters of the bases and the associated backbone/sugar torsions. Work along these lines suggests that the relationship is complicated and non-linear *(7)*. However, considerable simplification can be achieved if instead of using the complete relationship between all the helicoidal parameters and the entire backbone, the problem is limited to describing the relationship between the helicoidal parameters and the backbone/sugar torsion angles of single nucleotides and then using this information to drive a constraint minimizer that tries to connect adjacent nucleotides. This is the approach used in JUMNA *(8)*, which decomposes the problem of building a model nucleic acid structure into the constraint satisfaction problem of connecting adjacent flexible nucleotides. The sequence is decomposed into 3'-nucleotide monophosphates. Each nucleotide has as independent variables its six helicoidal parameters, its glycosidic torsion angle, three sugar angles, two sugar torsions and two backbone torsions. JUMNA seeks to adjust these independent variables to satisfy the constraints involving sugar ring and backbone closure.

Even constructing the base locations can be a non-trivial modeling task, especially for non-standard structures. Recognizing that coordinate frames should be chosen to provide a simple description of the transformations to be used, Gabarro-Arpa *et al. (9)* devised "Object Command Language" (OCL), a small computer language that is used to associate parts of molecules called objects, with arbitrary coordinate frames defined by sets of their atoms or numerical points. OCL can "link" objects, allowing other objects' positions and orientations to be described in the frame of some reference object. Information describing these frames and links is written out and used by the program MORCAD *(10)* which does the actual object transformations.

OCL contains several elements of a molecular modeling language. Users can create and operate on sets of atoms called objects. Objects are built by naming their component atoms and to simplify creation of larger objects, expressions, `IF` statements, an interated `FOR` loop and limited I/O are provided. Another nice feature is the equivalence between a literal 3-D point and the position represented by an atom's name. OCL includes numerous built-in functions on 3-vectors like the dot and cross products as well as specialized molecular modeling functions like creating a vector that is normal to an object. However, OCL is limited because these language elements can only be assembled into functions that define coordinate frames for molecules that will be operated on by MORCAD. Functions producing values of other data types and stand-alone OCL programs are not possible.

2. Design of nab

Our primary goals for `nab` were: *(1)* The new language had to be strongly geometric. *(2)* It needed to minimize the use of residue libraries by allowing chemical information to be added at run time. *(3)* It had to offer easy connectivity to other computational chemistry packages. *(4)* It needed to provide reasonable user extensibility, because if the language was successful, other investigators would want to use it in unexpected ways. *(5)* It had to be simple and easy to learn and operate, or at least as simple as is compatible with the creation of complicated nucleic acid models.

`nab` is implemented as a compiler, which provides considerable freedom to experiment with the language's constructs. Unlike most compilers, which produce either machine code or some form of symbolic intermediate code, the `nab` compiler converts the `nab` source into C. C is a sufficiently low level language that this conversion is relatively efficient yet platform-independent. Since `nab` is really C, it provides easy access to other codes via C's separate compilation facilities.

`nab` has a C- or awk-like syntax with several new types that provide molecular and geometric operations. The molecular types are `atom`, `residue` and `molecule`, which provide the usual three level molecular hierarchy, and `bounds`, for use by `nab`'s distance geometry package. Two purely geometric types, `point` (holding 3 floating point numbers) and `matrix` (holding a 4×4 transformation matrix), also are provided. `nab` has awk-like string operations as well as several in-line point and 3-vector operations like dot and cross product. The language provides some special syntax for logically looping over molecules. These "for-in" loops along with a short explanation of how they work are shown in the Table 1.

Table 1. Molecular Loops

```
molecule m;
residue  r;
atom     a;
```

`for( a in m ) ... ;`	Set the atom variable `a` to each atom in the molecule `m`.
`for( r in m ) for( a in r ) ... ;`	Two nested for-in loops. The first sets the residue variable `r` to each residue in the molecule `m`. Then the second or inner loops sets the atom variable `a` to each atom of the current residue `r`.

Table 2. Code statistics.

Component	*Number*	*Unit*
Lexical Analysis	64	lex rules
Grammar	67	yacc productions
Compiler	4593*	lines of C
Low level run time support	19998	lines of C
High level run time support	1734	lines of nab

* Does not include C source generated by lex or yacc.

nab was developed using the standard Unix compiler tools *lex* and *yacc*. The compiler itself is written in C. A run-time library provides operations such as initialization of a bounds matrix or transformation of part of a molecule. Support routines provide more "chemical" support, such as converting a string into a Watson/Crick duplex or a peptide in an extended conformation. nab is a small language—less than half the size of a C compiler (see Table 2).

In addition to insuring that every nab expression has the proper syntax or form, the compiler also checks that the semantics or meaning of every operation in every expression is correct. To do this nab associates three attributes with every value (literal, variable, or function return value). These are the *class*: literal, variable, function call or expression; the *kind*: scalar, array, dynamic array or hashed array; and the *type*: int, float, etc. Each operator has restrictions on the attributes of its inputs or operands and the attributes of its result (if valid) are a function of the attributes of the operands. The nab compiler performs this checking by a series of lookups in a set of tables that are indexed by the operator. There are three tables (one for each attribute) for each operator. These tables are maintained in an easy to read and to update symbolic form which is automatically converted into the C code required to validate each operator's inputs and determine the attributes of its output. This makes nab easy to extend.

nab also implements a form of regular expressions that we call atom regular expressions, which provide a uniform and convenient method for working on parts of molecules. Many of the general programming features of the awk language have been incorporated in nab. These include regular expression pattern matching, hashed arrays (i.e. arrays with strings as indices), the splitting of strings into fields, and string manipulations. The power of such a language derives both from its syntax, which can be carefully constructed to allow molecular manipulation tasks to be described in an intuitive fashion, and from its default library, which can provide a consistent interface to a variety of molecular modelling tasks. For example, the atom regular expressions in nab allow a compact and consistent method for choosing subsets of atoms. Similar features are present in other modeling codes, but here they are a part of the language, and so can be constructed and manipulated with awk-like string handling capabilities as part of some other manipulations.

Support is also present for compiling nab code into an AVS (Application Visualization System) module rather than to a stand-alone program. In combination with the AVS Geometry Viewer (or other AVS modules) this allows one to fairly easily build interactive programs that manipulate and display fairly complex molecular transformations *(11)*.

3. Methods for structure creation

As a structure-generating tool, nab provides three methods for building models. They are rigid-body transformations, metric matrix distance geometry, and molecular mechanics. The first two methods are good initial methods, but almost always create structures with some distortion that must be removed. On the other hand, molecular mechanics is a poor initial method but very good at refinement. Thus the three methods work well together.

3.1. Rigid body transformations

Rigid-body transformations create model structures by applying coordinate transformations to members of a set of standard residues to move them to new positions and orientations where they are incorporated into the growing model structure. The method is especially suited to helical nucleic acid molecules with their highly regular structures. It is less satisfactory for more irregular structures where internal rearrangement is required to remove bad covalent or non-bonded geometry, or where it may not be obvious how to place the bases.

nab uses the matrix type to hold a 4×4 transformation matrix. Transformations are applied to residues and molecules to move them into new orientations or positions. nab does *not* require that transformations applied to parts of residues or molecules be chemically valid. It simply transforms the coordinates of the selected atoms leaving it to the user to correct (or ignore) any chemically incorrect geometry caused by the transformation.

Every nab molecule includes a frame, or "handle" that can be used to position two molecules in a generalization of superimposition. Traditionally,when a molecule is superimposed on a reference molecule, the user first forms a correspondance between a set of atoms in the first molecule and another set of atoms in the reference molecule. The superimposition algorithm then determines the transformation that will minimize the rmsd between corresponding atoms. Because superimposition is based on actual atom positions, it requires that the two molecules have a common substructure, and it can only place one molecule on top of another and not at an arbitrary point in space.

The nab frame is a way around these limitations. A frame is composed of three orthonormal vectors originally aligned along the axes of a right handed coordinate frame centered on the origin. nab provides two builtin functions setframe() and setframep() that are used to reposition this frame based on vectors defined by atom expressions or arbitrary 3-D points, respectively. To position two molecules via their frames, the user moves the frames so that when they are superimposed via the nab builtin alignframe(), the two molecules have the desired orientation. This is a generalization of the methods described above for OCL.

3.2. Distance geometry

nab's second initial structure-creation method is *metric matrix distance geometry (12)*, which can be a very powerful method of creating initial structures. It has two main strengths. First, since it uses internal coordinates, the initial position of atoms about which nothing is known may be left unspecified. This has the effect that distance geometry models use only the information the modeler considers valid. No assumptions are required concerning the positions of unspecified atoms. The second advantage is that much structural information is in the form of distances. These include constraints from NMR or fluorescence energy transfer experiments, implied propinquities from chemical probing and footprinting, and tertiary interactions inferred from sequence analysis. Distance geometry provides a way to formally incorporate this information, or other assumptions, into the model-building process.

Distance geometry converts a molecule represented as a set of interatomic distances into a 3-D structure. nab has several builtin functions that are used together to provide metric matrix distance geometry. A bounds object contains the molecule's interatomic distance bounds matrix and a list of its chiral centers and their volumes. The function newbounds() creates a bounds object containing a distance bounds matrix containing initial upper and lower bounds for every pair of atoms, and a list of the molecule's chiral centers and their volumes. Distance bounds for pairs of atoms involving only a single residue are derived from that residue's coordinates. The 1,2 and 1,3 distance bounds are set to the actual distance between the atoms. The 1,4 distance lower bound is set to the larger of the sum of the two atoms Van der Waals radii or their *syn* (torsion angle = 0°) distance, and the upper bound is set to their *anti* (torsion angle = 180°) distance. newbounds() also initializes the list of the molecule's chiral centers. Each chiral center is an ordered list of four atoms and the volume of the tetrahedron those four atoms enclose. Each entry in a nab residue library contains a list of the chiral centers composed entirely of atoms in that residue.

Once a bounds object has been initialized, the modeler can use functions to tighten, loosen or set other distance bounds and chiralities that correspond to experimental measurements or parts of the model's hypothesis. The functions andbounds() and orbounds() allow logical manipulation of bounds. setbounds_from_db allows distance information from a model structure or a database to be incorporated into a part of the current molecule's bounds object, facilitating transfer of information between partially-built structures.

These primitive functions can be incorporated into higher-level routines. For example the functions stack() and watsoncrick() set the bounds between the two specifed bases to what they would be if they were stacked in a strand or base-paired in a standard Watson/Crick duplex, with ranges of allowed distances derived from an analysis of structures in the Nucleic Acid Database.

After all experimental and model constraints have been entered into the `bounds` object, the function `tsmooth()` applies "triangle smoothing" to pull in the large upper bounds, since the maximum distance between two atoms can not exceed the sum of the upper bounds of the shortest path between them. Random pairwise metrization *(13)* can also be used to help ensure consistency of the bounds and to improve the sampling of conformational space. The function `embed()` finally takes the smoothed bounds and converts them into a 3-D object. The newly embedded coordinates are subject to conjugate gradient refinement against the distance and chirality information contained in `bounds`. The call to `embed()` is usually placed in a loop to explore the diversity of the structures the bounds represent.

3.3. Molecular mechanics

The final structure creation method that `nab` offers is *molecular mechanics*. This includes both energy minimization and molecular dynamics – simulated annealing. Since this method requires a good estimate of the initial position of every atom in structure, it is not suitable for creating initial structures. However, given a reasonable initial structure, it can be used to remove bad initial geometry and to explore the conformational space around the initial structure. This makes is a good method for refining structures created either by rigid body transformations or distance geometry. `nab` has its own 3-D/4-D molecular mechanics package that implements several AMBER force fields and reads AMBER parameter and topology files. Solvation effects can also be modelled with generalized Born continuum models.

4. Examples of use

4.1. RNA pseudoknot

A pseudoknot is a single stranded nucleic acid molecule that contains two improperly nested hairpin loops as shown Figure 1 *(14)*. Shen and Tinoco *(15)* used the molecular mechanics program X-PLOR to determine the three dimensional structure of a 34 nucleotide RNA sequence that folds into a pseudoknot and promotes −1 frame shifting in mouse mammary tumor virus. NMR distance and angle constraints were converted into a three dimensional structure using a restrained molecular dynamics protocol. A simplified version of this is constructed in program 1. The model assumptions simply include stacking and base-pairing relations in the stem regions, including stacking between stems (lines 13 and 18).

Figure 1. Single stranded RNA *(top)* folded into a pseudoknot *(bottom). The black and dark gray base pairs can be stacked.*

Program 1 uses distance geometry followed by minimization to create a model of a pseudoknot. Distance geometry code begins in line 9 with the call to `newbounds()` and ends on line 29 with the call to `embed()`. The structure created with distance geometry is further refined with molecular mechanics in lines 31-36. This program produces a useful, topologically-correct model of a short pseudoknot, and illustrates how `nab` can be used to manipulate bounds and to create complex structures. Running the program multiple times, with different random numbers for the embed step, would produce a family of similar structures that satisfy the distance constraints in different ways. Such a family could be used to study the extent to which a given set of input distance bounds provides limits on the corresponding three-dimensional structures. Studies of similar models for longer sequences, including comparisons to NMR data, will be presented elsewhere.

The refinement technique in Program 1 just uses conjugate gradient minimization in three-dimensional space. NAB also supports refinements that allow the molecule to temporarily move into four spatial dimensions. The use of a fourth dimension allows bad initial contacts or intertwinings to be more easily removed by expanding the available configurational space *(16)*. At the end, the value of the fourth coordinate is forced back to zero, yielding again a three-dimensional object. This capability, in conjunction with molecular dynamics-based simulated annealing, greatly extends the refinement capabilities over the simple protocol outlined in Program 1.

4.2. Nucleosome Model

While the DNA duplex is locally rather stiff, many DNA molecules are sufficiently long that they can be bent into a wide variety of both open and closed curves. Some examples would be simple closed circles, supercoiled closed circles that have relaxed into circles with twists, and the nucleosome core fragment, where the duplex itself is wound into a short helix.

The overall strategy for wrapping DNA around a curve is to create the curve, find the points on the curve that contain the base pair origins, place the base pairs at

```
1   // Program 1 - Create a "pseudoknot"
2
3   bounds b;
4   molecule m;
5   point xyz[ 1000 ];
6   float energy;
7
8   m = link_na( "", "gcggaaacgccgcguaagcg", "arna.std.rlb", "" );
9   b = newbounds( m, "" );
10
11  stack( b, m, ":1:", ":2:" );
12  stack( b, m, ":2:", ":3:" );
13  stack( b, m, ":3:", ":18:" );
14  stack( b, m, ":18:", ":19:" );
15  stack( b, m, ":19:", ":20:" );
16  stack( b, m, ":8:", ":9:" );
17  stack( b, m, ":9:", ":10:" );
18  stack( b, m, ":10:", ":11:" );
19  stack( b, m, ":11:", ":12:" );
20  stack( b, m, ":12:", ":13:" );
21
22  watsoncrick( b, m, ":1:", ":13:" );
23  watsoncrick( b, m, ":2:", ":12:" );
24  watsoncrick( b, m, ":3:", ":11:" );
25  watsoncrick( b, m, ":18:", ":10:" );
26  watsoncrick( b, m, ":19:", ":9:" );
27  watsoncrick( b, m, ":20:", ":8:" );
28
29  tsmooth( b ); embed( b, m, "" );
30
31  leap( m ); readparm( m, "prmtop" );
32  mme_init( m, NULL, "::z", NULL, NULL );
33  setxyz_from_mol( m, NULL, xyz );
34  conjgrad( xyz, 3*m.natoms, energy, mme, 0.1, 10.0, 100 );
35  setmol_from_xyz( m, NULL, xyz );
36  putpdb( "pseudoknot.pdb", m );
```

these points, oriented so that their helical axes are tangent to the curve, and finally rotate the base pairs so that they have the correct helical twist. In the example below, the simplifying assumption is made that the rise is constant at 3.38Å.

The nucleosome core fragment *(17)* is composed of duplex DNA wound in a left handed helix around a cental protein core. A typical core fragment has about 145 base pairs of duplex DNA forming about 1.75 superhelical turns. Measurements of the overall dimensions of the core fragment indicate that there is very little space between adjacent wraps of the duplex. A side view of a schematic of core particle is shown in Fig. 2.

Computing the points at which to place the base pairs on a helix requires us to spiral an inelastic wire (representing the helical axis of the bent duplex) around a cylinder (representing the protein core). The system is described by four numbers of which only three are independent. They are the number of base pairs n, the

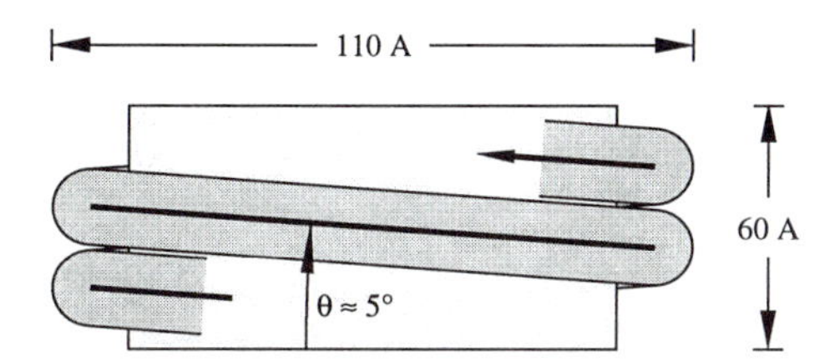

Figure 2. Schematic of core particle (side view).

number of turns the DNA makes around the protein core t, the "winding" angle θ (which controls how quickly the the helix advances along the axis of the core) and the helix radius r. Both the number of base pairs and the number of turns around the core can be measured. This leaves two choices for the third parameter. Since the relationship of the winding angle to the overall particle geometry seems more clear than that of the radius, this code lets the user specify the number of turns, the number of base pairs and the winding angle, then computes the helical radius and the displacement along the helix axis for each base pair:

$$d = 3.38\sin(\theta); \quad \phi = 360t/(n-1) \tag{1}$$

$$r = \frac{3.38(n-1)\cos(\theta)}{2\pi t} \tag{2}$$

where d and ϕ are the displacement along and rotation about the protein core axis for each base pair.

These relationships are easily derived. Let the nucleosome core particle be oriented so that its helical axis is along the global Y-axis and the lower cap of the protein core is in the XZ plane. Consider the circle that is the projection of the helical axis of the DNA duplex onto the XZ plane. As the duplex spirals along the core particle it will go around the circle t times, for a total rotation of $360t°$. The duplex contains $n-1$ steps, resulting in $360t/(n-1)°$ of rotation between successive base pairs.

Finding the radius of the superhelix is a little tricky. In general a single turn of the helix will not contain an integral number of base pairs. For example, using typical numbers of 1.75 turns and 145 base pairs requires ≈ 82.9 base pairs to make one turn. An approximate solution can be found by considering the ideal superhelix that the DNA duplex is wrapped around. Let L be the arc length of this helix. Then $L\cos(\theta)$ is the arc length of its projection into the XZ plane. Since this projection is an overwound circle, L is also equal to $2\pi rt$, where t is the number of turns and r is the unknown radius. Now L is not known but is approximately $3.38(n-1)$. Substituting and solving for r gives Eq. (2).

The resulting nab code is shown in Program 2. This code requires three arguments—the number of turns, the number of base pairs and the winding angle. In

```
1  // Program 2. Create simple nucleosome model.
2  #define PI  3.141593
3  #define RISE    3.38
4  #define TWIST   36.0
5  int          b, nbp; int getbase();
6  float        nt, theta, phi, rad, dy, ttw, len, plen, side;
7  molecule     m, m1;
8  matrix       matdx, matrx, maty, matry, mattw;
9  string       sbase, abase;
10
11 nt = atof( argv[ 2 ] );      // number of turns
12 nbp = atoi( argv[ 3 ] );     // number of base pairs
13 theta = atof( argv[ 4 ] ); // winding angle
14
15 dy = RISE * sin( theta );
16 phi = 360.0 * nt / ( nbp-1 );
17 rad = (( nbp-1 )*RISE*cos( theta ))/( 2*PI*nt );
18
19 matdx = newtransform( rad, 0.0, 0.0, 0.0, 0.0, 0.0 );
20 matrx = newtransform( 0.0, 0.0, 0.0, -theta, 0.0, 0.0 );
21
22 m = newmolecule();
23 addstrand( m, "A" ); addstrand( m, "B" );
24 ttw = 0.0;
25 for( b = 1; b <= nbp; b = b + 1 ){
26     getbase( b, sbase, abase );
27     m1 = wc_helix( sbase, "", abase, "",
28         2.25, -4.96, 0.0, 0.0 );
29     mattw = newtransform( 0., 0., 0., 0., 0., ttw );
30     transformmol( mattw, m1, NULL );
31     transformmol( matrx, m1, NULL );
32     transformmol( matdx, m1, NULL );
33     maty = newtransform( 0.,dy*(b-1),0., 0.,-phi*(b-1),0.);
34     transformmol( maty, m1, NULL );
35
36     mergestr( m, "A", "last", m1, "sense", "first" );
37     mergestr( m, "B", "first", m1, "anti", "last" );
38     if( b > 1 ){
39         connectres( m, "A", b - 1, "O3'", b, "P" );
40         connectres( m, "B", 1, "O3'", 2, "P" );
41     }
42     ttw += TWIST; if( ttw >= 360.0 ) ttw -= 360.0;
43 }
44 putpdb( "nuc.pdb", m );
```

lines 15-17, the helical rise (dy), twist (phi) and radius (rad) are computed according to the formulas developed above.

Two constant transformation matrices, matdx and matrx are created in lines 19-20. matdx is used to move the newly created base pair along the X-axis to the circle that is the helix's projection onto the XZ plane. matrx is used to rotate the new base pair about the X-axis so it will be tangent to the local helix of spirally wound duplex. The model of the nucleosome will be built in the molecule `m` which is created and given two strands `"A"` and `"B"` in line 23. The variable `ttw` will hold the total local helical twist for each base pair.

The molecule is created in the loop in lines 25-43. The user specified function `getbase()` takes the number of the current base pair (b) and returns two strings that specify the actual nucleotides to use at this position. These two strings are converted into a single base pair using the `nab` builtin `wc_helix()`. The new base pair is in the XY plane with its origin at the global origin and its helical axis along Z oriented so that the 5'-3' direction is positive.

Each base pair must be rotated about its Z-axis so that when it is added to the global helix it has the correct amount of helical twist with respect to the previous base. This rotation is performed in lines 29-30. Once the base pair has the correct helical twist it must be rotated about the X-axis so that its local origin will be tangent to the global helical axes (line 31).

The properly-oriented base is next moved into place on the global helix in two stages in lines 32-34. It is first moved along the X-axis (line 32) so it intersects the circle in the XZ plane that is the projection of the duplex's helical axis onto that plane. Then it is simultaneously rotated about and displaced along the global Y-axis to move it to its final place in the nucleosome. Since both these movements are with respect to the same axis, they can be combined into a single transformation.

The newly positioned base pair in `m1` is added to the growing molecule in `m` using two calls to the `nab` builtin `mergestr()`. Note that since the two strands of a DNA duplex are antiparallel, the base of the `"sense"` strand of molecule `m1` is added *after* the last base of the `"A"` strand of molecule `m` and the base of the `"anti"` strand of molecule `m1` is *before* the first base of the `"B"` strand of molecule `m`. For all base pairs except the first one, the new base pair must be bonded to its predecessor. Finally, the total twist (`ttw`) is updated and adjusted to remain in the interval [0,360) in line 42. After all base pairs have been created, the loop exits, and the molecule is written out. The coordinates are saved in PDB format using the `nab` builtin `putpdb()`.

5. Conclusions

Our hope is that nab will serve to formalize the step-by-step process that is used to build complex model structures. It will facilitate the management and use of higher level symbolic constraints. Writing a program to create a structure forces one to make explicit more of the model's assumptions in the program itself. And an nab description can serve as a way to exhibit a model's salient features, much like helical parameters are used to characterize duplexes. So far, nab has been used to construct models for synthetic Holliday junctions *(18)*, calcyclin dimers *(19)*, HMG-protein/DNA complexes *(20)*, active sites of Rieske iron-sulfur proteins *(21)*, and supercoiled DNA *(22)*. An extended description of the development and implementation of the language is available *(22)*, and the Users' Manual provides examples for constructing triple helices, tetraplexes, RNA bulges and four-arm junctions.

The code itself can be downloaded by anonymous ftp (see `http://www.scripps.edu/case` for pointers to the most recent version.)

6. Acknowledgments

This work was partially supported by NIH grant GM45811. We thank Steve Harvey, Gerry Joyce and Jayashree Srinivasan for many useful comments, and Paul Beroza, Doree Sitkoff, Jarrod Smith and Neill White for serving as code testers, and for contributing to the code and documentation.

7. References.

1. Brünger, A.T., *X-PLOR Version 3.1. A System for Crystallography and NMR,* Yale University (1992).
2. Tan, R.K.-Z.; Harvey, S.C. *J. Computat. Chem.* **1993,** *14,* 455-470.
3. Major, F.; Turcotte, M.; Gautheret, D.; Lapalme, G.; Fillion, E.; Cedergren, R. *Science* **1991,** *253,* 1255-1260.
4. Gautheret, D.; Major, F.; Cedergren, R. *J. Mol. Biol.* **1993,** *229,* 1049-1064.
5. Turcotte, M.; Lapalme, G.; Major, F. *J. Funct. Program.* **1995,** *5,* 443-460.
6. Erie, D.A.; Breslauer, K.J.; Olson, W.K. *Biopolymers* **1993,** *33,* 75-105.
7. Zhurkin, V. B.; Lysov, Yu. P.; Ivanov, V. I. *Biopolymers* **1978,** *17,* 277-312.
8. Lavery, R.; Zakrzewska, K.; Skelnar, H. *Comp. Phys. Commun.* **1995,** *91,* 135-158.
9. Gabarro-Arpa, J.; Cognet, J.A.H.; Le Bret, M. *J. Mol. Graphics.* **1992,** *10,* 166-173.
10. Le Bret, M.; Gabarro-Arpa, J.; Gilbert, J. C.; Lemarechal, C. *J. Chim. Phys.* **1991,** *88,* 2489-2496.
11. Duncan, B.S.; Macke, T.J.; Olson, A.J. *J. Mol. Graphics* **1995,** *13,* 271-282.
12. Crippen, G.M.; Havel, T.F., *Distance Geometry and Molecular Conformation,* Research Studies Press, Taunton, England (1988).
13. Hodsdon, M.E.; Ponder, J.W.; Cistola, D.P. *J. Mol. Biol.* **1996,** *264,* 585-602.
14. Wyatt, J.R.; Puglisi, J.D.; Jr., I. Tinoco *J. Mol. Biol.* **1990,** *214,* 455-470.
15. Shen, L.X.; Tinoco, I., Jr *J. Mol. Biol.* **1995,** *247,* 963-978.
16. van Schaik, R.C.; Berendsen, H.J.C.; Torda, A.E.; van Gunsteren, W.F. *J. Mol. Biol.* **1993,** *234,* 751-762.
17. Lewin, B., in *Genes IV*, Cell Press, Cambridge, Mass. (1990). pp. 409-425.
18. Macke, T.; Chen, S.-M.; Chazin, W.J., in *Structure and Function, Volume 1: Nucleic Acids*, ed. R.H. Sarma; M.H. Sarma,Adenine Press, Albany (1992). pp. 213-227.

19. Potts, B.C.M.; Smith, J.; Akke, M.; Macke, T.J.; Okazaki, K.; Hidaka, H.; Case, D.A.; Chazin, W.J. *Nature Struct. Biol.* **1995,** *2,* 790-796.

20. Love, J.J.; Li, X.; Case, D.A.; Giese, K.; Grosschedl, R.; Wright, P.E. *Nature* **1995,** *376,* 791-795.

21. Gurbiel, R.J.; Doan, P.E.; Gassner, G.T.; Macke, T.J.; Case, D.A.; Ohnishi, T.; Fee, J.A.; Ballou, D.P.; Hoffman, B.M. *Biochemistry* **1996,** *35,* 7834-7845.

22. Macke, T.J., *NAB, a Language for Molecular Manipulation*, Ph.D. thesis, The Scripps Research Institute 1996.

Chapter 25

Computer RNA Three-Dimensional Modeling from Low-Resolution Data and Multiple-Sequence Information

François Major, Sébastien Lemieux, and Abdelmjid Ftouhi

Département d'Informatique et de Recherche Opérationnelle, Université de Montréal, Montréal, Québec H3C 3J7, Canada

The problem of modeling three-dimensional structures of ribonucleic acids is expressed in terms of the constraint satisfaction problem. Three-dimensional structures are represented by constraint graphs, where vertices represent nucleotides and edges represent structural constraints. A formalism to help rationalize a series of modeling experiments in the context of low resolution and multiple-sequence data was developed. From secondary structure and low resolution data, several structural hypotheses corresponding to different constraint graphs can be derived. In presence of several structurally related sequences, the application of three-dimensional modeling to each sequence and hypotheses produces a sequence-structure relation that can be analyzed using fuzzy set theory, given the imprecision and uncertainty involved in the modeling process.

The popularity of computer modeling of RNA three-dimensional structure can be explained by the desire to rapidly understand the function of newly discovered RNAs and by the difficulties of applying high resolution structure determination techniques, such as X-ray crystallography and nuclear magnetic resonance spectroscopy. Computer modeling implies the interpretation of experimental data, the formation of structural hypotheses, and the building of three-dimensional models. Such models offer a simultaneous view of many aspects of the molecule and allow one to design more precise and incisive experiments which, in turn, generate new structural data and hypotheses leading to new modeling experiments. Thus, models are dynamic objects that represent the quantity and quality of structural knowledge on a molecule at a given time. The iterative use of modeling and low resolution experimental methods should converge on a highly defined and accurate model.

FM is a fellow of the Canadian Genome Analysis and Technology program and the MRC of Canada. This work has been supported by the MRC of Canada.

Most three-dimensional modeling projects begin with primary and secondary structure, low resolution data and multiple-sequence data *(1)* from which many different structural hypotheses can be derived. One way to support a structural hypothesis consists in building a consistent three-dimensional model compatible with each available sequence *(2, 3, 4)*. A systematic verification consists in building all possible models for each active sequence. This creates a relation, $R \subseteq S \times H$, where R, the set of relations, is a subset of binary relations between S, the set of sequences, and H, the set of structural hypotheses; $(S_i, H_j) \in R$ if and only if the sequence S_i generates at least one three-dimensional model that satisfies the structural hypothesis H_j. Each structural hypothesis, H_j, is associated with a set, E_{H_j}, that contains all three-dimensional models consistent with H_j. Computer programs such as MC-SYM *(5)* which transform constraint graphs into three-dimensional models can be used, although the formalism presented here is independent of any particular modeling method.

Uncertainty in modeling lies in the fact that a three-dimensional model can either support a structural hypothesis or can be the result of modeling artifacts. Computer modeling is subject to imprecision in the low resolution data, subjectivity in the generation of three-dimensional models, and uncertainty in the formation of structural hypotheses. The theory of possibility, based on fuzzy logic, is used to classify structural hypotheses according to their likelihood to contain multiple-sequence data consistent conformations based upon the sequence-structure relation, R.

In this article we present the constraint graph representation used by MC-SYM to transform structural data into three-dimensional models. Then, we discuss the sequence-structure relation and the theory of possibility to assign plausibility coefficients to each structural hypothesis. Finally, we discuss the application of this technique to the lead-activated ribozyme and indicate how modeling was used iteratively with experimentation to derive its active structure.

RNA Conformational Space

Here, we consider a *RNA three-dimensional structure* as the assembly of its constituent nucleotides in three-dimensional space. We introduce a RNA *conformational search space* defined by molecular contacts (or constraints). The molecular contacts are used in operators that position and orient the nucleotides in three-dimensional space.

A *molecular contact* is formed between two nucleotides, A and B, if they are connected through a phosphodiester bond or if they share a hydrogen bond between their nitrogen bases. The combination of all molecular contacts constitutes the *contact graph* of the RNA. It is self evident from the definition of a molecular contact that in all known RNA three-dimensional structures, every nucleotides make at least one molecular contact with another one. Thus, all RNAs contain at least one *path* of molecular contacts that connects all its constituent nucleotides, which does not contain any cycle, a *spanning tree* of the *nucleotide contact graph*.

In the following, we first present how contact graphs define the conformational search space of RNAs, to position and orient all nucleotides in three-dimensions. Then, a database of spatial relations based on molecular contacts, as observed among pairs of nucleotides in known structures, is introduced.

RNA Conformational Search Space Defined by Molecular Contacts. The premise to use molecular contacts in defining the conformational space of RNAs relies on the fact that molecular contacts contain all the information critical to the global fold of the molecule. Consider the best characterized case of an RNA double-helix. The spatial relation between two bases involved in a Watson-Crick pairing can be used, in conjunction with a canonical base stacking geometry, as a good approximation to position and orient double-helical strands in three-dimensions.

The spatial information is encoded by homogeneous transformation matrices *(6)*. The *local referential* of a nucleotide, A, can be represented by an homogeneous transformation matrix, R_A. R_A is determined by the coordinates of three atoms in A from which three right handed unary orthogonal vectors can be derived. The Cartesian coordinates of the first selected atom, for instance, can be chosen as the origin of the residue (see Figure 1). The *spatial relation* between two nucleotides, A and B, is an homogeneous coordinate transformation matrix, $T_{A \to B} = R_A^{-1} R_B$. In this way, the spatial relations between any pair of nucleotides forming molecular contacts in the known three-dimensional structures can be extracted and used as building blocks of RNA three-dimensional structures.

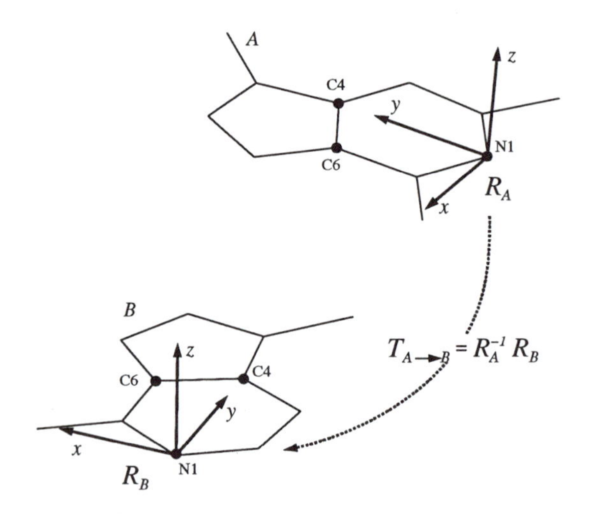

Figure 1. Spatial relation between two nitrogen bases. The axis systems define the local referential of each base and the dotted arrow represents the homogeneous transformation matrix encoding the relation.

An observed contact between nucleotides A and B can be reproduced between any pair of nucleotides, let's say A' and B', by applying the homogeneous transformation matrix $R_{B'}^{-1} T_{A \to B} R_{A'}$ to the atomic coordinates of B' to position and orient B' with respect to A' as observed between nucleotides A and B; or symmetrically, by applying the homogeneous transformation matrix $R_{A'}^{-1} T_{A \to B}^{-1} R_{B'}$ to the atomic coordinates of A'. The final result of this manipulation is, in either case, that the new observed spatial relation between A' and B' is exactly the same as that observed between A and B, thus reproducing the same molecular contact in the newly built model.

A *transformational set* is a set of homogeneous transformation matrices associated with a molecular contact type defined by the nature of the nucleotides in contact. For instance, there are four types of RNA bases determining ten different types of pairs by considering that symmetric pairs are likely to share the same types of molecular contacts, and two main types of molecular contacts: paired and connected. Connected nucleotides can be either stacked or not. In practice, we consider only two types of bases (purines and pyrimidines) for contact defined by a phosphodiester bond. This partition of the different molecular contacts gives a possibility of (10 + 6 = 16) different transformational sets.

The number of spanning trees, pairs of nucleotide contacts and homogeneous transformation matrices associated with a contact graph determine the conformational search space *size* of a RNA. The number of homogeneous transformation matrices associated with a molecular contact type is given by the number of occurrences observed in all available RNA three-dimensional structures in the Protein DataBank (PDB) *(7)*, Nucleic acids DataBase (NDB) *(8)* and other personally communicated structures.

The transformations were extracted and classified among the 16 different types of contacts. Those sets were then sorted in such a way that any subset composed of the first n elements represents the most efficient sampling of the addressed space. This property is achieved by selecting, as the first element of the set, the one that minimizes the sum of its distances with all other elements. This element is then considered the most "common" example. The next elements are those that maximizes their distances with all previously included elements. This sorting method supposes the existence of a distance metric to evaluate the difference between two homogeneous transformation matrices. The simplest metric is to sum the squares of the differences between the corresponding matrix elements, the Euclidean distance metric.

Starting from the contact graph, an efficient way to build three-dimensional models is to first determine a reference nucleotide that will be placed arbitrarily in three-dimensional space. From that, a spanning tree of the contact graph is expanded, determining which contacts will be used in the building procedure. For each molecular contact appearing in the spanning tree, the corresponding transformational set is used to systematically search the conformational space for valid three-dimensional models. Molecular contacts represented by edges that are not considered in the selected spanning tree are replaced in the simulation by geometrical constraints to guarantee their satisfaction in the final three-dimensional models.

The computer program MC-SYM is currently used to perform this search. Since the number of spanning trees of a fully connected graph composed of N vertices is N^{N-2}, the problem of selecting the one that is the most likely to generate complete structures is still open. The fuzzy logic approach presented here was developed in part to deal with the inaccuracy introduced by the approximation made while selecting a specific spanning tree.

The Sequence-Structure Relation. Structural hypotheses are derived from available structural data and are distinguished by their patterns of base pairing and stacking. For each structural hypothesis and sequence, a three-dimensional modeling simulation is performed, for instance using the MC-SYM program. In fact, any three-dimensional scheme determining if a three-dimensional model can be built for a given sequence and constraint graph is acceptable. The sequence-structure relation is established by associating the sequences to their consistent structural hypotheses; a link is created if and only if a three-dimensional model can be built. An MC-SYM input script describes the constraint graph and a sequence. By changing the latter, one can easily verify if a different sequence is compatible with the constraint graph.

Terminology and Notation of the Uncertainty Principle

In the history of mathematics, *uncertainty* was approached in the XVIIth century by Pascal and Fermat who introduced the notion of probability. However, probabilities do not allow one to process subjective beliefs nor imprecise or vague knowledge, such as in computer modeling of three-dimensional structure. Subjectivity and imprecision were only considered from 1965, when Zadeh, known for his work in systems theory, introduced the notion of *fuzzy set*. The concept of fuzziness introduces partial membership to classes, admitting intermediary situations between no and full membership. Zadeh's *theory of possibility*, introduced in 1977, constitutes a framework allowing for the representation of such *uncertain* concepts of non-probabilistic nature *(9)*. The concept of fuzzy set allows one to consider imprecision and uncertainty in a single formalism and to quantitatively measure the preference of one hypothesis versus another. Note, however, that Bayesian probabilities could have been used instead.

Consider a finite reference set, X. Events can be defined by subsets of X to which can be assigned coefficients between 0 and 1 evaluating their possibility to occur. In order to define these coefficients, a measure of possibility is introduced, Π, which is a function defined over the power set of X (the set of all subsets composed of the elements of X), $\mathcal{P}(X)$, the parts of X which take their values in $[0, 1]$, such that:

$$\Pi(\emptyset) = 0, \quad \Pi(X) = 1, \tag{1}$$

$$\begin{aligned} &\forall A_1 \in \mathcal{P}(X), \quad A_2 \in \mathcal{P}(X), \ldots \\ &\Pi(\textstyle\bigcup_{i=1,2,\ldots} A_i) = \max_{i=1,2,\ldots} \Pi(A_i), \end{aligned} \tag{2}$$

where $\emptyset$ is the empty set and max indicates the maximum value of all values. The possibility associated to the empty set is zero. The possibility of X is one. The possiblity of a series of events (union) is the maximum possibility among these events.

The functions of *belief* concern a quantification of credibility attached to the events. Shafer's *theory of evidence* considers a finite universe of reference, X, upon which are determined belief coefficients obtained by distributing a global mass of belief equal to 1 among all possible events *(10)*. A mass, m, can be defined as follows:

$$m : \mathcal{P}(X) \longrightarrow [0, 1]$$

such that

$$m(\emptyset) = 0 \quad \text{and} \quad \sum_{A \in \mathcal{P}(X)} m(A) = 1.$$

For each set $A \in \mathcal{P}(X)$, the value $m(A)$ represents the degree with which a group of observers believe in the realization of an event from the elements of A. This value,

$m(A)$, involves only a single set, the set A, and does not involve any other information for the subsets of A. If there exists additional evidence which confirms the realization of the same event in a subset of A, $B \subset A$, it must be expressed by another value, $m(B)$. Every non empty part A of X, for which $m(A) \neq 0$, is called a *focal element* corresponding to an event believed by the observers. The *belief measure* of such a part A of X is defined by considering all the focal elements implying A:

$$Bel(A) = \sum_{B|B \subseteq A} m(B),$$

that is, the belief of a part A of X is defined by the sum of all parts, B such that A contains B. The *plausibility measure* of A is defined by taking all focal elements related to A:

$$Pl(A) = \sum_{B|B \cap A \neq \emptyset} m(B),$$

that is, the plausibility of a part A of X is defined by the sum of all parts, B such that B intersects with, or contains any element of, A. The above measures verifies the following relations:

$$Pl(A) = 1 - Bel(\bar{A}) \quad \text{and} \quad Bel(A) \leq Pl(A).$$

where $\bar{A}$ indicates the complement of A in X.

The range $[Bel(A), Pl(A)]$ embeds the imprecise probability, $P(A)$, for any part A of X. A particular case of the mass m is remarkable: consider that all focal elements are singletons of X, that is, beliefs only concern elementary events. Then, every part A of X is such that $Bel(A) = Pl(A)$ and this common value is equal to the probability, $P(A)$.

Calculating Possibilities. Consider the sequence-structure relation in Figure 2. A uniform probability of $\frac{1}{5}$ is assigned to each sequence. Note that the uniform distribution is not a requirement of the mathematical model. It was assumed that all sequences could adopt the same conformation. The possibility for each structural hypothesis to contain the conformation was computed using Zadeh's theory of possibility.

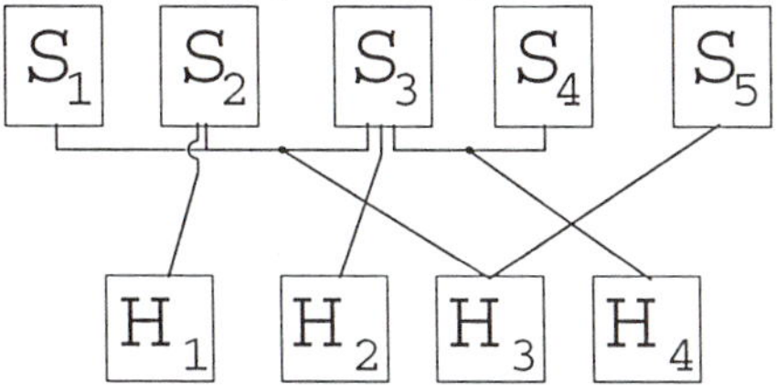

Figure 2. Mapping active variants and structural hypotheses. The relation, $R \subseteq S \times H$, for an hypothetical example. The sequence variants are identified by S_i for i = 1,2,3,4,5 and the structural hypotheses by H_j for j = 1,2,3,4. The lines indicate which sequences are compatible with which structural hypotheses as determined by the generation of consistent models by the MC-SYM program. Here $S_{1,2,3,5}$ were found consistent with the three-dimensional models of H_3.

Consider X as the set containing all conformations generated by MC-SYM for all sequence variants. From the sequence-structure relation, R, the focal elements are

$$S_{1,5}, S_2, S_3 \text{ and } S_4,$$

where $S_{1,5} = S_1 \cap S_5$ and S_1, S_2, S_3, S_4, S_5 are subsets of X which contain the three-dimensional conformations associated with the active structure.

From R we have:

$$\begin{aligned} S_{1,5} &= E_{H_3} \\ S_2 &= E_{H_1} \cup E_{H_3} \\ S_3 &= E_{H_2} \cup E_{H_3} \cup E_{H_4} \\ S_4 &= E_{H_4} \end{aligned}$$

where E_{H_i}, $i = 1, 2, 3, 4$, represents the set of conformations that satisfy hypothesis H_i.

A belief coefficient of possibility to contain the conformation is assigned to each focal element:

$$\begin{aligned} m(S_{1,5}) &= \frac{2}{5} \\ m(S_2) &= m(S_3) = m(S_4) = \frac{1}{5}. \end{aligned}$$

The basic probabilities (masses) were assigned by considering that any of the available sequences could adopt the active conformation. The possibility distribution, π, is then

$$\forall x \in X \;\; \pi(x) = 1,$$

which is equivalent in the case of the probability distribution, p, to

$$\forall x \in X \;\; p(x) = \frac{1}{|X|}.$$

However, given the biological supposition that the active conformation should be found among the structures common to all sequences, the belief coefficients were assigned according to how many sequences are compatible with the hypothesis, that is, for $S_{1,5}$, the belief coefficient of sequences S_1, and S_5,

$$m(S_{1,5}) = \frac{2}{5},$$

based on the fact that two sequences in the set of five sequences were found compatible with a particular subset of the structural hypothesis. Figure 3 shows the focal elements $S_{1,5}, S_2, S_3 \text{and} S_4$. It is now possible to define the intervals of probabilities (possibilities) for each part of the conformational space.

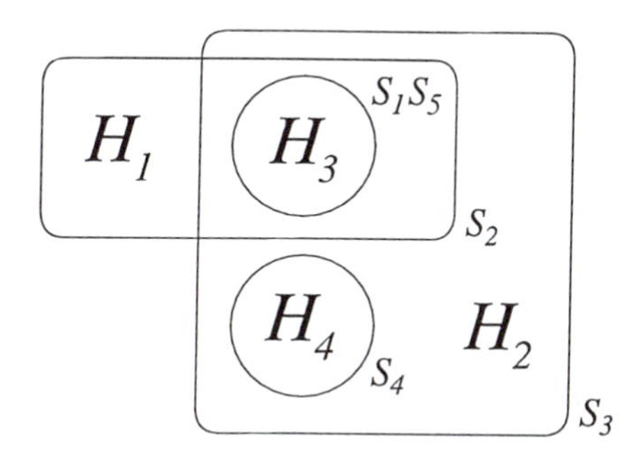

Figure 3. The focal elements. The focal elements $S_{1,5}$, S_2, S_3 and S_4, and the consistent structural hypotheses H_1, H_2, H_3 and H_4.

This situation allows to deduce the intervals of probabilities of each structural hypothesis, that is, the belief measures:

$$
\begin{aligned}
Bel(E_{H_1}) &= Bel(E_{H_2}) = 0 \\
Bel(E_{H_3}) &= \sum_{B|B \subseteq E_{H_3}} m(B) = m(S_{1,5}) = \frac{2}{5} \\
Bel(E_{H_4}) &= \sum_{B|B \subseteq E_{H_4}} m(B) = m(S_4) = \frac{1}{5}
\end{aligned}
$$

and the plausibility measures:

$$
\begin{aligned}
Pl(E_{H_1}) &= \sum_{B|B \cap E_{H_1} \neq \emptyset} m(B) \\
&= m(S_2) = \frac{1}{5} \\
Pl(E_{H_2}) &= \sum_{B|B \cap E_{H_2} \neq \emptyset} m(B) \\
&= m(S_3) = \frac{1}{5} \\
Pl(E_{H_3}) &= \sum_{B|B \cap E_{H_3} \neq \emptyset} m(B) \\
&= m(S_{1,5} + m(S_2) + m(S_3) = \frac{4}{5} \\
Pl(E_{H_4}) &= \sum_{B|B \cap E_{H_4} \neq \emptyset} m(B) \\
&= m(S_3) + m(S_4) = \frac{2}{5}
\end{aligned}
$$

Table I. Belief and Plausibility Measures for H_1, H_2, H_3 and H_5. $Bel(A)$ is the belief value. $Pl(A)$ is the plausibility value.

A	$Bel(A)$	$Pl(A)$
E_{H1}	0	$\frac{1}{5}$
E_{H2}	0	$\frac{1}{5}$
E_{H3}	$\frac{2}{5}$	$\frac{4}{5}$
E_{H5}	$\frac{1}{5}$	$\frac{2}{5}$

which are summarized in Table I. According to the belief and plausibility coefficients calculated for all the hypotheses, H_3, with an imprecise probability over the range $[Bel(E_{H_3}), Pl(E_{H_3})]$, is the one that seems the most likely to be shared by all variant sequences. This does not necessarily indicate, without any doubt, that the actual three-dimensional structure will be found in those of H_3. It simply indicates that among the current evaluated hypotheses H_3 is the one that best reflects the combined results of the modeling experiments. With that information in hand, the models generated under H_3 should be carefully examined and used in the design of future laboratory experiments, either to confirm the hypothesis or produce new structural data and hypotheses.

Application to the Leadzyme

The Pb^{2+} cleavage of a specific ribophosphodiester bond in yeast $tRNA^{Phe}$ is the classical model of metal-assisted RNA catalysis. *In vitro* selection experiments have identified $tRNA^{Phe}$ variants a derivative of which, named the leadzyme, is very active in cleavage by Pb^{2+} *(11)*. The leadzyme consists of an RNA duplex with an asymmetric internal loop of six nucleotides *(12)*. Cleavage of the leadzyme domain produces two fragments: one with a terminal 5'-hydroxyl group, and the other with a 3'-phosphomonoester presumably generated via a 2',3'-cyclic phosphodiester intermediate. The two-step reaction mechanism of the leadzyme is reminiscent of protein ribonucleases and distinguishes it from other ribozymes, such as the hammerhead, the hairpin and the hepatitis δ domains, which produce 2'-3'-cyclic phosphates *(13, 14, 15, 16)*. The detailed three-dimensional structure was a requisite to the understanding of the particularities of this reaction.

Modeling of the leadzyme was initiated with a series of structural hypotheses derived from the primary and secondary structures. A list of active analogous sequences was previously isolated by *in vitro* selection experiments *(11)*. The program MC-SYM was used to establish the sequence-structure relation, $R \subseteq S \times H$ by generating conformational libraries for the wild-type sequence and all sequence analogs. The fuzzy logic model was applied to these libraries to identify the most plausible hypothesis that was then experimentally evaluated. Activity data of leadzyme variants that incorporated modified nucleotides into the catalytic core *(17)* led to a new structural hypothesis and a second round of computer modeling. The final model is consistent with all available structural data and provided insight into the catalytic reaction of this ribozyme. The details about the active conformation and the three-dimensional modeling of the leadzyme are reported in a manuscript in preparation, available from the authors.

Conclusion

A mathematical model based on fuzzy logic was developed for the selection of structural hypotheses that are more likely to contain active conformations consistent with a series of analogous sequences. The application of this model is especially useful when multiple-sequence data are available, that however do not reveal sufficient structural aspects to initiate three-dimensional modeling. The MC-SYM program or any other RNA modeling approach can be used to produce the sequence-structure relation. The fuzzy logic model was incorporated in the iteration of computer modeling, hypothesis formation and experimental work. This protocol was successfully applied to the three-dimensional modeling of the leadzyme. The fuzzy logic model made possible the identification of a structural hypothesis that was used in the design of laboratory experiments which, in turn, generated structural data that produced a final consistent model.

Literature Cited

1. Major, F.; Gautheret, D. In *Encyclopedia of Molecular Biology and Molecular Medicine*; Myers, R.A., Ed.; VCH Publishers Inc.: NY, 1996, Vol. 5; pp 371–388.

2. Brown, J.; Nolan, J.; Haas, E.; Rubio, M.; Major, F.; Pace, N. *Proc. Natl. Acad. Sci.* **1996**, 93, pp. 3001–3006.

3. Gautheret, D.; Koonings, D.; Gutell, R. *J. Mol. Biol.* **1994**, 242, pp. 1–8.

4. Michel, F.; Westhof, E. *J. Mol. Biol.* **1990**, 216, pp. 585–610.

5. Major, F.; Turcotte, M.; Gautheret, D.; Lapalme, G.; Fillion, E.; Cedergren, R. *Science* **1991**, 253, pp. 1255–1260.

6. Paul, R. P. Robot Manipulators: Mathematics, Programming, and Control; MIT Press: Cambridge, MA, 1981.

7. Bernstein, F. C.; Koetzle, T. F.; Williams, G.J. B.; Meyer, E.F. J.; Brice, M. D.; Rodgers, J. R.; Kennard, O.; Shimanouchi, T.; Tasumi, M. *Eur. J. Biochem.***1977**, 80, pp. 319–324.

8. Berman, H.; Olson, W.; Beveridge, D.; Westbrook, J.; Gelbin, A.; Demeny, T.; Hsieh, S.-H.; Srinivasan, A.; Schneider, B. *Biophys. J.* **1992**, 63, pp. 751–759.

9. Zadeh, L. In Fuzzy sets and systems I; North-Holland Publishing Company: Amsterdam, Holland, 1977, pp. 3–28.

10. Shafer, G. A mathematical theory of evidence; Princeton Univ. Press: Princeton, NJ, 1976.

11. Pan, T.; Uhlenbeck, O. *Biochemistry* **1992**, 31, pp. 3887–3895.

12. Pan, T.; Uhlenbeck, O. *Nature* **1992**, 358, pp. 560–563.

13. Buzayan, J.; Gerlach, W.; Bruening, G. *Proc. Natl. Acad. Sci.* **1986**, 83, pp. 8859–8862.

14. Hutchins, C.; Rathjen, P.; Forster, A.; Symons, R. *Nucl. Acids Res.* **1986**, 14, pp. 3627–3640.

15. Forster, A. C.; Symons, R. H. *Cell* **1987**, 49, pp. 211–220.

16. Epstein, L.; Gall, J. *Cell* **1987**, 48, pp. 535–543.

17. Chartrand, P.; Usman, N.; Cedergren, R. *Biochemistry* **1997**, 36, pp. 3145–3150.

Chapter 26

Comparative Modeling of the Three-Dimensional Structure of Signal Recognition Particle RNA

Christian Zwieb[1], Krishne Gowda[1], Niels Larsen[2], and Florian Müller[3]

[1]**Department of Molecular Biology, University of Texas Health Science Center, P.O. Box 2003, Tyler, TX 75710**
[2]**Department of Microbiology, Giltner Hall, Room 180, Michigan State University, East Lansing, MI 48824**
[3]**Max Planck Institute for Molecular Genetics, Ihnestrasse 73, D–14195 Berlin, Germany**

Comparative sequence analysis and ERNA-3D software were used to model the three-dimensional structures of several signal recognition particle RNAs. RNA secondary structures were established by allowing only phylogenetically-supported base pairs. The folding of the RNA molecules was constrained further to include a pseudoknot and a tertiary interaction. Founded by the concept that all SRP RNAs must be shaped similarly in three dimensions, helical sections were oriented coaxially where a continuous helical stack was formed in the RNA of another species. Finally, RNA helices were placed at distances that preserved the connectivity of the molecule with the smallest number of single-stranded nucleotide residues as identified from the aligned sequences. Representative models of the three-dimensional structures of an eukaryote, an archaeon, and three bacterial SRP RNAs are presented.

In the investigation of the structural and functional properties of RNA molecules, comparisons of RNA sequences are an extremely powerful method. This approach requires only a critical number of carefully aligned sequences, but hardly any experimental effort. A prime example for the success of comparative sequence analysis is the structure of the ribosomal RNAs, where the comparison of sequences from more than 3000 species (*1*) identified Watson-Crick interactions at the level of individual nucleotides.

A similarly refined level of detail is now emerging from the analysis of several non-ribosomal RNAs. Of these, the RNA from signal recognition particle (SRP) (*2*) is a particularly interesting molecule. Because of its relatively small size, the SRP is an excellent model system for the study of RNA protein interactions and of the assembly of other ribonucleoprotein particles, such as ribosomes and spliceosomes. Furthermore, SRP serves as a convenient testing ground where computational tools for the modeling of the three-dimensional structures of larger RNA molecules can be developed.

The SRP exists either free in the cytosol, bound to ribosomes, or associated with the membrane. Its function is to facilitate the co-translational translocation of proteins across lipid bilayers (reviewed by (*3*)). SRP has a dumbbell-like shape, with a small and a large domain that are separated by a slender adapter (*4*). Preliminary model building (*5*) indicated that the SRP RNA must be folded to fit within the electron microscopic dimensions (220-240 Å by 50-60 Å).

Of the six SRP proteins present in the mammalian SRP, the heterodimer of proteins SRP9 and SRP14 associates with the small SRP domain, whereas the large domain includes polypeptides SRP19, SRP54, and SRP68/72 (*6*). SRP RNA is present throughout the particle (*7*); therefore, a detailed knowledge of its structure is essential to understand structure and function of the SRP.

To explore the folding of SRP RNA in three dimensions, we have used the computer program ERNA-3D (*8*), on a SGI Indigo2 workstation (IRIX version 5.3). Among the advantages of the ERNA-3D software is its ability to automatically generate representations of A-form-RNA directly from the specified base paired regions, whereas the inter-helical sections single strands are derived from reiterated rotations along the sugar-phosphate backbone. A convenient feature of ERNA-3D is to assist the modeling process from the perspective of the user, as is possible with a physical model, but to allow manipulations in space and real time with substantially greater speed and accuracy.

The first three-dimensional model of SRP RNA generated with ERNA-3D was the human SRP RNA. This model incorporated data from enzymatic and chemical modification, electron microscopy, and site-directed mutagenesis (*9*). A pseudoknot in the small SRP domain (a pairing of 12-UGGC-15 with 33-GCUA-36), and a tertiary interaction in the large SRP domain (198-GA-199 with 232-GU-233) were identified. Nevertheless, the spatial distribution of many helical sections, for example in the adapter region (helix 5), remained ambiguous.

SRP RNA secondary structures derived by comparative sequence analysis

The availability of phylogenetically supported secondary structures is a prerequisite in any serious RNA modeling attempt. The current SRP RNA secondary structures are derived from the alignment of nearly 100 sequences (*2,10*) and are minimal in that they contain only those base pairs that are supported by compensatory base changes (CBCs). A CBC is observed when a base pair in one organism differs by *both* bases when compared to the equivalent base pair in another organism. During evolution, mutations that introduced an unstable base pair would not have been compensated for, unless a base pair was required; thus, CBCs provide evidence for its existence. Invariant regions of the RNA provide neither positive nor negative evidence for base pairs; therefore, CBCs can only be identified in the variable regions. As another caveat, certain CBC-supported base pairs may form only in one of several functional RNA conformations.

Thus derived secondary structure models are presented in Figure 1 for the SRP RNAs of five species. They include the smallest known SRP RNA (77 nucleotides) from *Mycoplasma mycoides* (Myc. myc.), a typical bacterial SRP RNA (also referred to as 4.5S RNA) from *Escherichia coli* (Esc. col.), an example of the substantially larger SRP RNA from the *Bacilli* (*Bacillus subtilis*, Bac. sub.) and *Clostridium* (*11*), an archaeal SRP RNA from *Methanococcus jannaschii* (Met. jan.), and a typical eukaryotic SRP RNA (*Homo sapiens*, Hom. sap.). Base pairs supported by CBCs are shown in juxtaposition and are connected with a line; G-U "pairings" are marked by small circles. Bases of unsupported pairings at helical ends are placed adjacently with no line between them. Finally, when there was more negative than positive evidence for a base pair, the base symbols were spaced apart.

A common SRP RNA nomenclature

Eight helices were identified and were numbered from one to eight, beginning at the 5'-end. No single SRP RNA molecule possesses all eight helices, as eukaryotic SRP RNAs lack a helix 1, and the archaea miss helix 7. Bacterial SRP RNAs are confined to helix 8 and a conserved portion of helix 5 (Myc. myc. and Esc. col., Figure 1), with the exception of the SRP RNA of the *Bacilli* and *Clostridium* (Figure 1). Interestingly, helix 8 contains the binding site for protein SRP54, or its homologue (*ffh* or P48), at a bent or "knuckle" that is common to all SRP RNAs (*9,12*) (see solid black arrowheads in Figure 1). In agreement with this finding, protein SRP54 is found in all three phylogenetic domains (*2*), and even in chloroplasts (*13*).

For historical reasons, the mammalian SRP has been studied to the greatest extent. (SRP was first isolated from the pancreas of dogs.) Consequently, helical sections of the SRP RNA helices were named with suffixed letters first in the human SRP RNA. According to the general nomenclature, the small SRP domain includes helices 1 to 4 and sections 5a and 5b; the adapter consists of several helix 5-sections (5c to 5f), and the large SRP domain is composed of sections 5g to 5k, 6a to 6c, and 8a to 8c (see Hom. sap., Figure 1). To preserve the helix nomenclature for all SRP RNAs, composite names were given to helical section that are continuously stacked in other species. As a result, the SRP RNA of *M. jannaschii* includes the compounded helices 5bcd, 5gh, 5ij, and 6bc (Met. jan., Figure 1). Likewise, there is a helix 5fgh in the bacterial SRP RNA. Because of the location of this helix at the termini of the RNA molecule, this assignment could be somewhat arbitrary.

Tertiary interactions

For additional constraint of the model, we have expanded the comparative sequence analysis approach to identify paired loops that are apart in the primary and secondary structure, but close in three dimensions. Of the two proposed interactions, CBCs strongly supports the formation of a pseudoknot between the loops of helices 3 and 4 in the SRP RNAs of the archaea, in *Clostridium,* and the *Bacilli* (Figure 1). A similar interaction is allowed in the eukaryotic SRP RNAs, however with less stringent base pairings.

The other tertiary interaction we have considered is located in the large SRP domain between the tip of helix 8 (positions 198 and 199 in the human SRP RNA) and an internal loop of helix 5 (positions 232 and 233, Figure 1). This pairing was identified first by the computer program Consensus Matrix (*14*) and is present in all known SRP RNAs. Because of the relatively high degree of conservation of the participating bases, the tertiary interaction is more difficult to support by CBCs. We noticed, however, that the terminal loop of helix 8 is a tetranucleotide loop (tetraloop) with the consensus sequence GNRA in all but the plant SRP RNAs. This GNRA-motif is structurally similar to a U-turn (or UNR-motif) (*15*), ideally suited for participation in tertiary interactions (*16*).

Coaxially oriented SRP RNA helices

In simplified terms, folded RNA molecules consist of rigid "limbs" (for the helical sections) that are joined by potentially flexible single-strands (*17*). As the CBC approach uses aligned sequences to determine RNA secondary structure, the "limb/joint" approach takes into account several RNA secondary structures to extract common features of the three-dimensional structure. A considerable restriction of a 3D model can thus be achieved by comparing the size of the "limbs" and by identifying the coaxial helical sections. The approach is particularly useful when

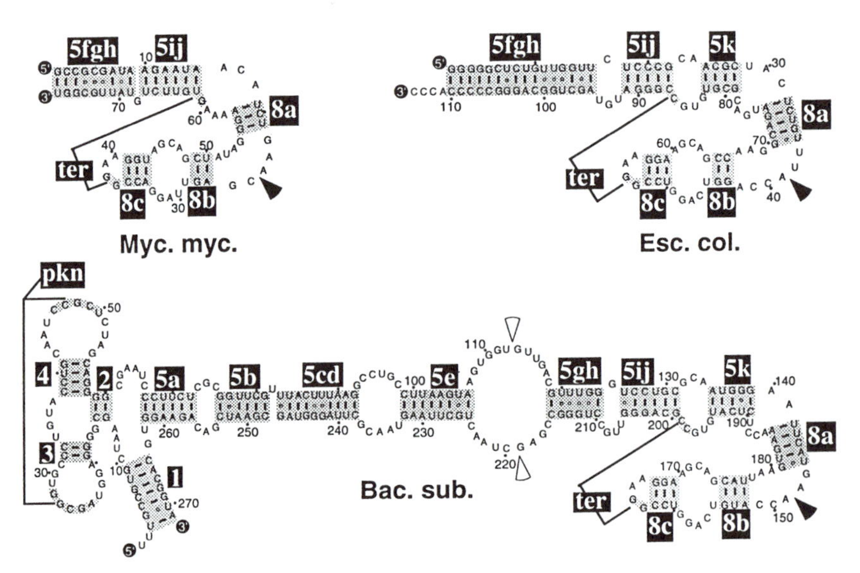

Figure 1. **Secondary structures of SRP RNAs.** Structures are from *Mycoplasma mycoides* (Myc. Myc., GenBank accession X53678), *Escherichia coli* (Esc. col., X01074), *Bacillus subtilis* (Bac. sub., X06802) *Methanococcus jannaschii* (Met. jan., (*21*)), and *Homo sapiens* (Hom. sap., V00588). Base pairings are supported by comparative sequence analysis of SRP RNA sequences in the SRP database (*2*). The 5'- and 3'-ends of the RNA molecule are labeled as such; helices are marked 1 to 8 according to the nomenclature of Larsen and Zwieb (*10*); residues are labeled in ten-nucleotide increments with base paired sections highlighted in gray. Helices 5, 6, and 8 are given suffices **a** through **k** in helix 5, and **a** through **c** in helices 6 and 8. Pairing between the loops of helices 3 and 4 indicate a pseudoknot (**pkn**) in the small SRP domain; a postulated tertiary interaction (**ter**) between the tip of helix 8 and an internal loop between sections 5j and 5k, is shown in each structure. The solid-black arrowheads mark a "knuckle" in helix 8 that acts as a binding site for protein SRP54; open arrowheads between helical sections 5f and 5g indicate sites that are hypersensitive towards Micrococcal nuclease (*22*).

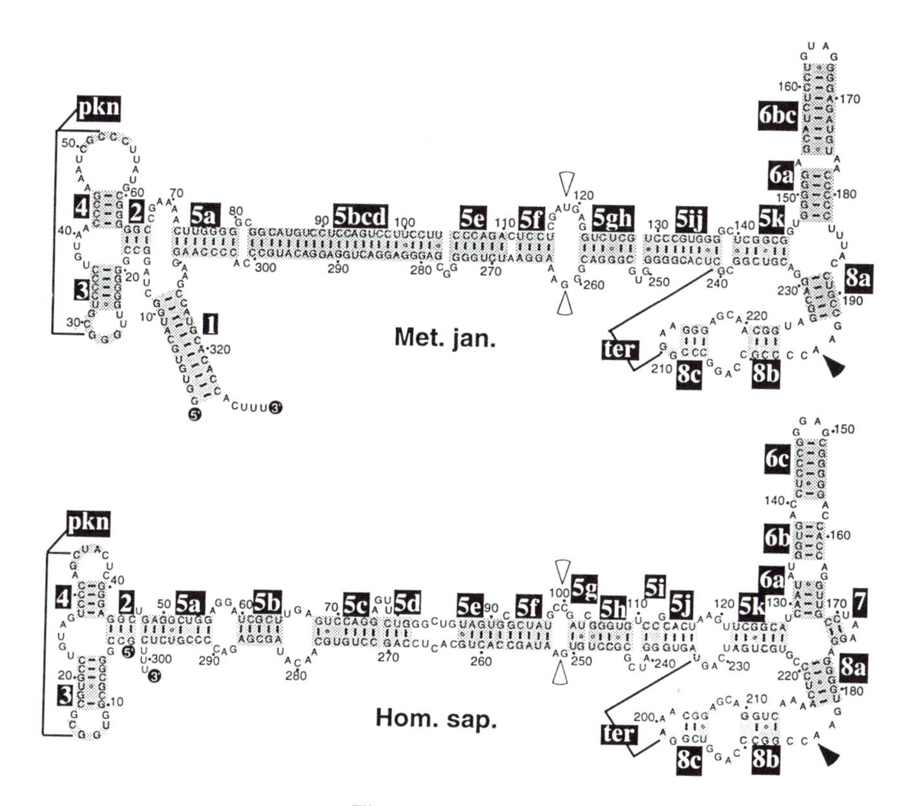

Figure 1. *Continued.*

RNAs are included that display a high degree of base pairing, such as the structures from certain thermophilic organisms (e.g. *M. jannaschii*, Figure 1).

As a result of additional folding constraints, helical stacking may occur when there is only one immediate transition between helices. For example, stacking between helix 4 and helix 2 at A43 and G44 of the human SRP RNA (Figure 1) is favored because of a nearby phylogenetically well-supported pseudoknot.

Distances between helical sections

We adhered to the general idea that one should construct *minimal* 3D models that include only supported features. Consequently, we considered the distances between helical regions as variable, with the maximal distance allowed by the number of nucleotide residues between helical sections. (At which point the sugar-phosphate backbone would break.) However, when there was substantial comparative support for a helix at the equivalent RNA region of another species, we tended to preserve the A-form helical character of the single-stranded regions.

In the integrative part of the RNA modeling process, we gave emphasis to the overall three-dimensional molecular design. This bias is supported by the general finding that biological activity is determined be the size and shape of folded molecules, and that an identical function often correlates with an identical, or at least similar, three-dimensional structure. Since all SRP RNA molecules, presumably, perform a similar function in co-translational protein targeting, we postulate a high degree of three-dimensional similarity which is achieved only by superimposing corresponding helical sections, thus generating a "three-dimensional alignment".

Modeling of the three-dimensional structure of SRP RNA with ERNA-3D

As a foundation for comparing the SRP RNAs in three dimensions, we used the human SRP RNA model, generated earlier with ERNA-3D (*9*). Next, we focused on the SRP RNA of *M. jannaschii* to take advantage of the large number base pairs compounded into helical sections 5bcd, 5gh, 5ij, and 6bc. A simple textual input file was generated to contain the *M. jannaschii* SRP RNA sequence, information about the paired residues, and about positions of the helical sections. The positions of the helical sections were copied from the PDB-formatted (*18*) molecular structure file of the human SRP RNA model (*9*).

ERNA-3D automatically generated A-form RNA for the helical regions, adjusted some of the single strands, and preserved a large part of the three-dimensional helical arrangement. Using wireless LCD stereo glasses (Stereographics), the model was inspected visually with the help of the 3D viewing capability of ERNA-3D, and by using several simple representations, such as cylinders for the helical sections and tubes for the sugar-phosphate backbone (see Figure 2). By giving commands directly to the program or by using the on-screen cursor-box, helices were adjusted using the criteria discussed above.

The cylinder representations were particularly useful for comparing 3D models from the various SRP RNAs. First we superimposed cylinders from the *M. jannaschii* structure and the atom coordinates of the human SRP RNA to improve the human model. We then used the cylinder representations of the human SRP RNA to adjust the atom coordinates of *M. jannaschii* SRP RNA. These two sequences illustrate well how comparative modeling is mutually beneficial. For example, a coaxial orientation of helices 5b, 5c, and 5d in the human SRP RNA model is supported by the extended helix (5bcd) of the *M. jannaschii* SRP RNA. On the other hand, the *M. jannaschii* sequence inserts "extra" nucleotides in the loop of helix 4 and between helix 2 and helix 5a; however, the high degree of potential flexibility that would normally result from these insertions, is contained by comparison with the human model.

In similar fashion, we generated the *B. subtilis* SRP RNA model by using the

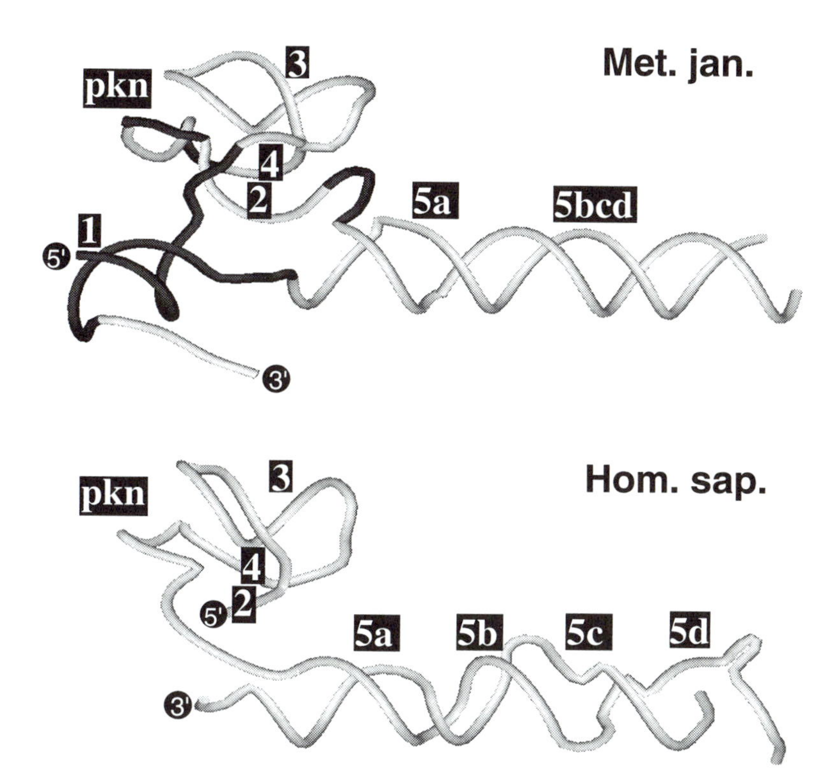

Figure 2. **3D comparison of the small SRP RNA domain.** Models were constructed on a Silicon Graphics Indigo 2 Extreme workstation with the program ERNA-3D (*8*) with considerations as described in the text. Views of the small domains of the SRP RNAs of *Methanococcus jannaschii* (Met. jan.) and *Homo sapiens* (Hom. sap.) are shown with the sugar-phosphate backbones as narrow tubes and the base paired helical sections as cylinders. Labeling of the RNA ends, the helices, and the pseudoknot are precisely as in Figure 1. RNA sections of the *M. jannaschii* model that correspond to nucleotides absent in the human model are shown dark-gray. (The PDB coordinates of the three-dimensional models are available from the SRP database at the internet address http://www.uthct.edu/SRPDB/SRPDB.html (*2*).)

helix coordinates of the *M. jannaschii* SRP RNA and by reiterating the process described above, thus effectively comparing three SRP RNA models. The positions of helix 1 and of helix 8a were adjusted to accommodate that there are only two nucleotide residues between helix 5a and helix 1, and between helix 5k and helix 8a (Figure 1). As expected from their small size, the bacterial SRP RNAs from *M. mycoides* and *E. coli* (Figure 1) gave no new three-dimensional insights.

It is unclear why a "streamlined" molecule, such as the SRP RNA from *M. jannaschii,* would possess additional nucleotides in the small domain. Perhaps, the extra nucleotides in helix 4 may provide added stability by A-U pairing or by using non-Watson-Crick interactions similar to what is observed in a group I ribozyme domain (*19*). We noted that proteins SRP9/14 are sandwiched between the knot and helix 5 (9,*20*), and suggest that nucleotides 67-CGAAAA-72 of the *M. jannaschii* SRP RNA (Figures 1) may be mimicking SRP9/14. This RNA region could provide a similar stabilization of the small domain as SRP9/14, however with the use of nucleotide instead of amino acid residues. This suggestion is consistent with the spatial location of the nucleotides 67-CGAAAA-72 between helix 2 and helix 5a (Figure 2). In conclusion, the archaeal and bacterial SRP9/14 proteins would be considerably different from their eukaryotic homologues, or these proteins may be entirely absent. Indeed, in a search of the complete genomes of *M. jannaschii* (*21*) and *B. subtilis*, we were unable to identify significant matches with mammalian SRP9 or SRP14 sequences (not shown).

Structure of the SRP

The three-dimensional structure of the SRP RNA determines to a large part the overall size and shape of the SRP. The revised human SRP RNA model has a maximum length of 240 Å (the distance between U38 and C170) and a maximum width of 83 Å (the distance between G182 and G148). This approximates what has been determined in the electron microscope for the canine SRP (*4,7*) with the model being slightly larger. The differences in the order of 20 to 30 Å may reflect ambiguities or dehydration in the electron microscopic determinations and/or our model building.

We considered a model in which "backfolding" occurs in helix 5 at Micrococcal nuclease hypersensitive sites (see open arrowheads in Figure 1). Although this arrangement is more consistent with the overall shape of the SRP, it would shorten excessively the long axis to 155 Å (the distance between U38 and G98). To resolve this discrepancy, it will be necessary to carry out additional experiments such as low-dose EM.

Outlook

With the huge amount of data generated by the large-scale sequencing projects, we anticipate an increased need to extract reliably RNA structure models from aligned sequences. Using the SRP RNA as an example, we have shown that it may be possible to develop a set of simple rules, that would propel the approach of comparative sequence analysis into the third dimension. Ideas about the three-dimensional structure of macromolecules may provide clues for functions that may be illusive at the level of the primary structure. The development of powerful computational tools will be crucial for automating the comparative 3D modeling process. Hopefully, this will be achieved to a degree that, in the future, accurate models will emerge at a single command from the user.

Acknowledgments

We thank Richard Brimacombe for his continuing encouragement and the critical reading of the manuscript. This work was supported by NIH grant GM-49034 to C.Z.

References

1. Maidak, B.; Olsen, G.; Larsen, N.; Overbeek, R.; McCaughey, M.; Woese, C. *Nucleic Acids Res.* **1997**, *25*, pp. 109-110.
2. Zwieb, C.; Larsen, N. *Nucleic Acids Res.* **1997**, *25*, pp. 107-108.
3. Lütcke, H. *Eur. J. Biochem.* **1995**, *228*, pp. 531-550.
4. Andrews, D.; Walter, P.; Ottensmeyer, F. *Proc. Natl. Acad. Sci. U.S.A.* **1985**, *82*, pp. 785-789.
5. Zwieb, C.; Schüler, D. *Biochem. and Cell Biol.* **1989**, *67*, pp. 434-442.
6. Walter, P.; Blobel, G. *Cell* **1983**, *34*, pp. 525-533.
7. Andrews, D.; Walter, P.; Ottensmeyer, F. *EMBO J.* **1987**, *6*, pp. 3471-3477.
8. Müller, F.; Döring, T.; Erdemir, T.; Greuer, B.; Jünke, N.; Osswald, M.; Rinke-Appel, L.; Stade, K.; Thamm, S.; Brimacombe, R. *Biochem. Cell Biol.* **1995**, *73*, pp. 767-773.
9. Zwieb, C.; Müller, F.; Larsen, N. *Folding & Design* **1996**, *1*, pp. 315-324.
10. Larsen, N.; Zwieb, C. *Nucleic Acids Res.* **1991**, *19*, pp. 209-215.
11. Nakamura, K.; Hashizume, E.; Shibata, T.; Nakamura, Y.; Mala, S.; Yamane, K. *Microbiology* **1995**, *141*, pp. 2965-2975.
12. Lentzen, G.; Moine, H.; Ehresmann, C.; Ehresmann, B.; Wintermeyer, W. *RNA* **1996**, *2*, pp. 244-253.
13. Franklin, A.; Hoffman, N. *J. Biol. Chem.* **1993**, *268*, pp. 22175-22180.
14. Davis, J. P.; Janjic, N.; Pribnow, D.; Zichi, D. A. *Nucleic Acids Res.* **1995**, *23*, pp. 4471-4479.
15. Jucker, F.; Pardi, A. *RNA* **1995**, *1*, pp. 219-222.
16. Quigley, G.; Rich, A. *Science* **1976**, *194*, pp. 796-806.
17. Zwieb, C. *Prog. Nucl. Acid Res. Mol. Biol.* **1989**, *37*, pp. 207-234.
18. Bernstein, F.; Koetzle, T.; Williams, G.; Meyer, E. J.; Brice, M.; Rodgers, J.; Kennard, O.; Shimanouchi, T.; Tasumi, M. *J. Mol. Biol.* **1977**, *112*, pp. 535-542.
19. Cate, J.; Gooding, A.; Podell, E.; Zhou, K.; Golden, B.; Kundrot, C.; Cech, T.; Doudna, J. *Science* **1996**, *273*, pp. 1678-1685.
20. Strub, K.; Moss, J.; Walter, P. *Mol Cell Biol* **1991**, *11*, pp. 3949-3959.
21. Bult, C.; White, O.; Olsen, G.; Zhou, L.; Fleischmann, R.; Sutton, G.; Blake, J.; FitzGerald, L.; Clayton, R.; Gocayne, J.; Kerlavage, A.; Dougherty, B.; Tomb, J.; Adams, M.; Reich, C.; Overbeek, R.; Kirkness, E.; Weinstock, K.; Merrick, J.; Glodek, A.; Scott, J.; Geoghagen, N.; Weidman, J.; Fuhrmann, J.; Venter, J.; et al. *Science* **1996**, *273*, pp. 1058-1073.
22. Gundelfinger, E. D.; Krause, E.; Melli, M.; Dobberstein, B. *Nucleic Acids Res.* **1983**, *11*, pp. 7363-7374.

Author Index

Affiliation Index

Subject Index